797,885 Books

are available to read at

www.ForgottenBooks.com

Forgotten Books' App

Available for mobile, tablet & eReader

ISBN 978-1-330-69396-4
PIBN 10093274

This book is a reproduction of an important historical work. Forgotten Books uses state-of-the-art technology to digitally reconstruct the work, preserving the original format whilst repairing imperfections present in the aged copy. In rare cases, an imperfection in the original, such as a blemish or missing page, may be replicated in our edition. We do, however, repair the vast majority of imperfections successfully; any imperfections that remain are intentionally left to preserve the state of such historical works.

Forgotten Books is a registered trademark of FB &c Ltd.
Copyright © 2017 FB &c Ltd.
FB &c Ltd, Dalton House, 60 Windsor Avenue, London, SW19 2RR.
Company number 08720141. Registered in England and Wales.

For support please visit www.forgottenbooks.com

1 MONTH OF
FREE
READING

at

www.ForgottenBooks.com

By purchasing this book you are eligible for one month membership to ForgottenBooks.com, giving you unlimited access to our entire collection of over 700,000 titles via our web site and mobile apps.

To claim your free month visit:

www.forgottenbooks.com/free93274

* Offer is valid for 45 days from date of purchase. Terms and conditions apply.

English
Français
Deutsche
Italiano
Español
Português

www.forgottenbooks.com

Mythology Photography **Fiction** Fishing Christianity **Art** Cooking Essays Buddhism Freemasonry Medicine **Biology** Music **Ancient Egypt** Evolution Carpentry Physics Dance Geology **Mathematics** Fitness Shakespeare **Folklore** Yoga Marketing **Confidence** Immortality Biographies Poetry **Psychology** Witchcraft Electronics Chemistry History **Law** Accounting **Philosophy** Anthropology Alchemy Drama Quantum Mechanics Atheism Sexual Health **Ancient History** **Entrepreneurship** Languages Sport Paleontology Needlework Islam **Metaphysics** Investment Archaeology Parenting Statistics Criminology **Motivational**

LOEB CLASSICAL LIBRARY

THEOPHRASTUS

ENQUIRY INTO PLANTS

BOOKS 1–5

Translated by

ARTHUR HORT

THEOPHRASTUS of Eresus in Lesbos, born about 370 BC, is the author of the most important botanical works that have survived from classical antiquity. He was in turn the student, collaborator, and successor of Aristotle. Like his predecessor he was interested in all aspects of human knowledge and experience, especially natural science. His writings on plants form a counterpart to Aristotle's zoological works.

In the *Enquiry into Plants* Theophrastus classifies and describes varieties—covering trees, plants of particular regions, shrubs, herbaceous plants, and cereals; in the last of the nine books he focuses on plant juices and medicinal properties of herbs. The Loeb edition is in two volumes; the second contains two additional treatises: *On Odors* and *Weather Signs*.

In *De Causis Plantarum* Theophrastus turns to plant physiology. Books One and Two are concerned with generation, sprouting, flowering and fruiting, and the effects of climate. In Books Three and Four Theophrastus studies cultivation and agricultural methods. In Books Five and Six he discusses plant breeding; diseases and other causes of death; and distinctive flavors and odors.

808375
vol. 1

The New York
Public Library
Astor, Lenox and Tilden Foundations

The Branch Libraries
MID-MANHATTAN LIBRARY
Literature & Language Dept.
455 Fifth Avenue
New York, N.Y. 10016

MM
LL

Books and non-print media may be returned to any branch of The New York Public Library. Music scores, orchestral sets and certain materials must be returned to branch from which borrowed.

All materials must be returned by the last date stamped on the card. Fines are charged for overdue items.

Form #0692

THE LOEB CLASSICAL LIBRARY

FOUNDED BY JAMES LOEB

EDITED BY

G. P. GOOLD

PREVIOUS EDITORS

T. E. PAGE E. CAPPS

W. H. D. ROUSE L. A. POST

E. H. WARMINGTON

THEOPHRASTUS

ENQUIRY INTO PLANTS

I

LCL 70

THEOPHRASTUS

ENQUIRY INTO PLANTS

BOOKS I–V

WITH AN ENGLISH TRANSLATION BY

ARTHUR HORT

HARVARD UNIVERSITY PRESS
CAMBRIDGE, MASSACHUSETTS
LONDON, ENGLAND

First published 1916
Reprinted 1948, 1961, 1968, 1990, 1999

LOEB CLASSICAL LIBRARY® is a registered trademark
of the President and Fellows of Harvard College

ISBN 0-674-99077-3

Printed in Great Britain by St Edmundsbury Press Ltd,
Bury St Edmunds, Suffolk, on acid-free paper.
Bound by Hunter & Foulis Ltd, Edinburgh, Scotland.

CONTENTS

BOOK II

OF PROPAGATION, ESPECIALLY OF TREES

BOOK III

OF WILD TREES

CONTENTS

BOOK IV

OF THE TREES AND PLANTS SPECIAL TO PARTICULAR DISTRICTS AND POSITIONS

CONTENTS

BOOK V

OF THE TIMBER OF VARIOUS TREES AND ITS USES

PREFACE

This is, I believe, the first attempt at an English translation of the 'Enquiry into Plants.' That it should be found entirely satisfactory is not to be expected, since the translator is not, as he should be, a botanist; moreover, in the present state at least of the text, the Greek of Theophrastus is sometimes singularly elusive. I should never have undertaken such a responsibility without the encouragement of that veteran student of plant-lore the Rev. Canon Ellacombe, who first suggested that I should make the attempt and introduced me to the book. It is a great grief that he did not live to see the completion of the work which he set me. If I had thought it essential that a translator of Theophrastus should himself grapple with the difficulties of identifying the plants which he mentions, I must have declined a task which has otherwise proved quite onerous enough. However the kindness and the expert knowledge of Sir William Thiselton-Dyer came to my rescue; to him I not only owe gratitude for constant help throughout; the identifications in the Index of Plants are entirely his work, compared with which the compilation of the Index itself was

but mechanical labour. And he has greatly increased my debt and the reader's by reading the proofs of my translation and of the Index. This is perhaps the place to add a note on the translation of the plant-names in the text:—where possible, I have given an English equivalent, though I am conscious that such names as 'Christ's thorn,' 'Michaelmas daisy' must read oddly in a translation of a work written 300 years before Christ; to print Linnean binary names would have been at least equally incongruous. Where an English name was not obvious, although the plant is British or known in British gardens, I have usually consulted Britten and Holland's Dictionary of Plant-names. Where no English equivalent could be found, *i.e.* chiefly where the plant is not either British or familiar in this country, I have either transliterated the Greek name (as *arakhidna*) or given a literal rendering of it in inverted commas (as 'foxbrush' for ἀλωπέκουρος); but the derivation of Greek plant-names being often obscure, I have not used this device unless the meaning seemed to be beyond question. In some cases it has been necessary to preserve the Greek name and to give the English name after it in brackets. This seemed desirable wherever the author has apparently used more than one name for the same plant, the explanation doubtless being that he was drawing on different local authorities; thus κέρασος and λακάρη both probably represent 'bird-cherry,' the latter being the Macedonian name for the tree.

Apart from this reason, in a few places (as 3.8.2; 3.10.3.) it seemed necessary to give both the Greek and the English name in order to bring out some particular point. On the other hand one Greek name often covers several plants, *e.g.* λωτός; in such cases I hope that a reference to the Index will make all clear. Inverted commas indicate that the rendering is a literal translation of the Greek word; the identification of the plant will be found in the Index. Thus φελλόδρυς is rendered 'cork-oak,' though 'holm-oak' would be the correct rendering,—cork-oak (*quercus Suber*) being what Theophrastus calls φελλός, which is accordingly rendered cork-oak without commas. As to the spelling of proper names, consistency without pedantry seems unattainable. One cannot write names such as Arcadia or Alexander otherwise than as they are commonly written; but I cannot bring myself to Latinise a Greek name if it can be helped, wherefore I have simply transliterated the less familiar names; the line drawn must of course be arbitrary.

The text printed is in the main that of Wimmer's second edition (see Introd. p. xiv). The textual notes are not intended as a complete apparatus criticus; to provide a satisfactory apparatus it would probably be necessary to collate the manuscripts afresh. I have had to be content with giving Wimmer's statements as to MS. authority; this I have done wherever any question of interpretation depended on the reading; but I have not thought it necessary to record mere

variations of spelling. Where the textual notes go beyond bare citation of the readings of the MSS., Ald., Gaza, and Pliny, it is usually because I have there departed from Wimmer's text. The references to Pliny will, I hope, be found fairly complete. I am indebted for most of them to Schneider, but I have verified these and all other references.

I venture to hope that this translation, with its references and Index of Plants, may assist some competent scholar-botanist to produce an edition worthy of the author.

Besides those already mentioned I have to thank also my friends Professor D'Arcy Thompson, C.B., Litt.D. of Dundee, Mr. A. W. Hill of Kew, Mr. E. A. Bowles for help of various kinds, and the Rev. F. W. Galpin for his learned exposition of a passage which otherwise would have been dark indeed to me—the description of the manufacture of the reed mouth-pieces of wood-wind instruments in Book IV. Sir John Sandys, Public Orator of Cambridge University, was good enough to give me valuable help in matters of bibliography.

INTRODUCTION

I.—Bibliography and Abbreviations used

A. *Textual Authorities*

Wimmer divides the authorities on which the text of the περὶ φυτῶν ἱστορία is based into three classes:—

First Class:

U. Codex Urbinas: in the Vatican. Collated by Bekker and Amati; far the best extant MS., but evidently founded on a much corrupted copy. See note on 9. 8. 1.

P_2. Codex Parisiensis: at Paris. Contains considerable excerpts; evidently founded on a good MS.; considered by Wimmer second only in authority to U.

(Of other collections of excerpts may be mentioned one at Munich, called after Pletho.)

Second Class:

M (M_1, M_2). Codices Medicei: at Florence. Agree so closely that they may be regarded as a single MS.; considered by Wimmer much inferior to U, but of higher authority than Ald.

P. Codex Parisiensis: at Paris. Considered by Wimmer somewhat inferior to M and V, and more on a level with Ald.

mP. Margin of the above. A note in the MS. states that the marginal notes are not *scholia*, but *variae lectiones aut emendationes.*

V. Codex Vindobonensis: at Vienna. Contains the first five books and two chapters of the sixth; closely resembles M in style and readings.

Third Class:

Ald. Editio Aldina: the *editio princeps,* printed at Venice 1495–8. Believed by Wimmer to be founded on a single MS., and that an inferior one to those enumerated above, and also to that used by Gaza. Its readings seem often to show signs of a deliberate attempt to produce a smooth text: hence the value of this edition as witness to an independent MS. authority is much impaired.

(Bas. Editio Basiliensis: printed at Bâle, 1541. A careful copy of Ald., in which a number of printer's errors are corrected and a few new ones introduced (Wimmer).

Cam. Editio Camotiana (or Aldina minor, altera): printed at Venice, 1552. Also copied from Ald., but less carefully corrected than Bas.; the editor Camotius, in a few passages,

altered the text to accord with Gaza's version.)

G. The Latin version of Theodore Gaza,[1] the Greek refugee: first printed at Treviso (Tarvisium) in 1483. A wonderful work for the time at which it appeared. Its present value is due to the fact that the translation was made from a different MS. to any now known. Unfortunately however this does not seem to have been a better text than that on which the Aldine edition was based. Moreover Gaza did not stick to his authority, but adopted freely Pliny's versions of Theophrastus, emending where he could not follow Pliny. There are several editions of Gaza's work: thus

G.Par.G.Bas. indicate respectively editions published at Paris in 1529 and at Bâle in 1534 and 1550. Wimmer has no doubt that the Tarvisian is the earliest edition, and he gives its readings, whereas Schneider often took those of G.Bas.

Vin.Vo.Cod.Cas. indicate readings which Schneider believed to have MS. authority, but which are really anonymous emendations from the margins of MSS. used by his predecessors, and all, in Wimmer's opinion

[1] See Sandys, *History of Classical Scholarship*, ii. p. 62, etc.

traceable to Gaza's version. Schneider's so-called Codex Casauboni he knew, according to Wimmer, only from Hofmann's edition.

B. *Editions*

H. Editio Heinsii, printed at Leyden, 1613: founded on Cam. and very carelessly printed, repeating the misprints of that edition and adding many others. In the preface Daniel Heins[1] pretends to have had access to a critical edition and to a Heidelberg MS.; this claim appears to be entirely fictitious. The book indeed contains what Wimmer calls a *farrago emendationum*; he remarks that 'all the good things in it Heinsius owed to the wit of others, while all its faults and follies we owe to Heinsius.' Schneider calls it *editio omnium pessima.*

Bod. Editio Bodaei (viz. of Joannes Bodaeus à Stapel), printed at Amsterdam, 1644. The text of Heinsius is closely followed; the margin contains a number of emendations taken from the margin of Bas. and from Scaliger, Robertus Constantinus, and Salmasius, with a few due to the editor himself. The commentary, according to Sir William Thiselton-Dyer, is 'botanically monumental and fundamental.'

[1] See Sandys, *op. cit.* p. 313 etc.

St. Stackhouse, Oxford, 1813: a prettily printed edition with some illustrations: text founded on Ald. The editor seems to have been a fair botanist, but an indifferent scholar, though occasionally he hits on a certain emendation. The notes are short and generally of slight value. The book is however of interest, as being apparently the only work on the 'Enquiry' hitherto published in England.

Sch. J. G. Schneider (and Linck), Leipzig: vols. i.–iv. published in 1818, vol. v. in 1821; contains also the *περὶ αἰτιῶν* and the fragments, and a reprint of Gaza's version (corrected). The fifth, or supplementary, volume, written during the author's last illness, takes account of the Codex Urbinas, which, unfortunately for Schneider, did not become known till his edition was finished. It is remarkable in how many places he anticipated by acute emendation the readings of U. The fifth volume also gives an account of criticisms of the earlier volumes by the eminent Greek Adamantios Koraës[1] and Kurt Sprengel. This is a monumental edition, despite the verbosity of the notes, somewhat careless references and reproduction of the MSS. readings, and an imperfect comprehension of the compressed style of Theophrastus, which leads to a good deal of wild emendation or rewriting of the text. For the first time we find an attempt at

[1] See Sandys, *op. cit.* iii. pp. 361 foll.

providing a critical text, founded not on the Aldine edition, but on comparison of the manuscripts then known; the Medicean and Viennese had been collated a few years before by J. Th. Schneider. We find also full use made of the ancient authors, Athenaeus, Plutarch, Pliny, Dioscorides, Nicander, Galen, etc., who quoted or adapted passages of Theophrastus, and copious references, often illuminating, to those who illustrate him, as Varro, Columella, Palladius, Aelian, the *Geoponica*.

Spr. Kurt Sprengel, Halle, 1822. This is not an edition of the text, but a copious commentary with German translation. Sprengel was a better botanist than scholar; Wimmer speaks disparagingly of his knowledge of Greek and of the translation. (See note prefixed to the Index of Plants.)

W Fr. Wimmer: (1) An edition with introduction, analysis, critical notes, and Sprengel's identifications of the plant-names; Breslau, 1842.

(2) A further revised text with new Latin translation, apparatus criticus, and full indices; the Index Plantarum gives the identifications of Sprengel and Fraas; Didot Library, Paris, n.d.

(3) A reprint of this text in Teubner's series, 1854.

These three books are an indispensable supplement to Schneider's great work. The notes in the edition of

1842 are in the main critical, but the editor's remarks on the interpretation of thorny passages are often extremely acute, and always worth attention. The mass of material collected by Schneider is put into an accessible form. Wimmer is far more conservative in textual criticism than Schneider, and has a better appreciation of Theophrastus' elliptical and somewhat peculiar idiom, though some of his emendations appear to rest on little basis. A collation of the Paris MSS. (P and P_2) was made for Wimmer; for the readings of U and M he relied on Schneider, who, in his fifth volume, had compared U with Bodaeus' edition. A fresh collation of the rather exiguous manuscript authorities is perhaps required before anything like a definitive text can be provided. Wimmer's Latin translation is not very helpful, since it slurs the difficulties: the Didot edition, in which it appears, is disfigured with numerous misprints.

(Sandys' *History of Classical Scholarship* (ii. p. 380) mentions translations into Latin and Italian by Bandini; of this work I know nothing.)

C. *Other Commentators*

Seal. J. C. Scaliger: *Commentarii et animadversiones* on the περὶ φυτῶν ἱστορία posthumously published by his son Sylvius at Leyden, 1584. (He also wrote a commentary on the περὶ αἰτιῶν, which was edited by Robertus Constantinus and pub-

lished at Geneva in 1566.) The most accurate and brilliant scholar who has contributed to the elucidation of Theophrastus.

R.Const. Robertus Constantinus (see above). Added notes of his own, many of them valuable, which are given with Scaliger's in Bodaeus' edition.

Salm. Salmasius (Claude de Saumaise). Made many happy corrections of Theophrastus' text in his *Exercitationes Plinianae.*

Palm. Jacobus Palmerins (Jacques de Paulmier). His *Exercitationes in optimos auctores Graecos* (Leyden, 1668) contain a certain number of acute emendations; Wimmer considers that he had a good understanding of Theophrastus' style.

Meurs. Johannes Meursius (Jan de Meurs). Author of some critical notes on Theophrastus published at Leyden in 1640; also of a book on Crete.

Dalec. Jean Jacques D'Aléchamps: the botanist. Author of *Historia plantarum universalis,* Lyons, 1587, and editor of Pliny's *Natural History.*

Mold. J. J. P. Moldenhauer. Author of *Tentamen in Historiam plantarum Theophrasti,* Hamburg, 1791. This book, which I have not been able to see and know only from Wimmer's citations, contains, according to him, very valuable notes on the extremely difficult Introduction to the 'Historia' (Book I. chaps. i.–ii.).

II.—Theophrastus' Life and Works

Such information as we possess concerning the life of Theophrastus comes mainly from Diogenes Laertius' *Lives of the Philosophers,* compiled at least four hundred years after Theophrastus' death; it is given therefore here for what it may be worth; there is no intrinsic improbability in most of what Diogenes records.

He was born in 370 B.C. at Eresos in Lesbos; at an early age he went to Athens and there became a pupil of Plato. It may be surmised that it was from him that he first learnt the importance of that principle of classification which runs through all his extant works, including even the brochure known as the 'Characters' (if it is rightly ascribed to him), and which is ordinarily considered as characteristic of the teaching of his second master Aristotle. But in Plato's own later speculations classification had a very important place, since it was by grouping things in their 'natural kinds' that, according to his later metaphysic, men were to arrive at an adumbration of the 'ideal forms' of which these kinds are the phenomenal counterpart, and which constitute the world of reality. Whether Theophrastus gathered the principle of classification from Plato or from his fellow-pupil Aristotle, it appears in his hands to have been for the first time systematically applied to the vegetable world. Throughout his botanical

works the constant implied question is 'What is its *difference*?', 'What is its essential nature?', viz. 'What are the characteristic features in virtue of which a plant may be distinguished from other plants, and which make up its own 'nature' or essential character?

Theophrastus appears to have been only Aristotle's junior by fifteen years. On Plato's death he became Aristotle's pupil, but, the difference in age not being very great, he and his second master appear to have been on practically equal terms. We are assured that Aristotle was deeply attached to his friend; while as earnest of an equally deep attachment on the other side Theophrastus took Aristotle's son under his particular care after his father's death. Aristotle died at the age of sixty-three, leaving to his favourite pupil his books, including the autographs of his own works, and his garden in the grounds of the Lyceum. The first of these bequests, if the information is correct, is of great historical importance; it may well be that we owe to Theophrastus the publication of some at least of his master's voluminous works. And as to the garden it is evident that it was here that the first systematic botanist made many of the observations which are recorded in his botanical works. Diogenes has preserved his will, and there is nothing in the terms of this interesting document to suggest that it is not authentic. Of special interest is the provision made for the maintenance of the garden;

it is bequeathed to certain specified friends and to those who will spend their time with them in learning and philosophy; the testator is to be buried in it without extravagant expense, a custodian is appointed, and provision is made for the emancipation of various gardeners, so soon as they have earned their freedom by long enough service.

According to Diogenes Theophrastus died at the age of eighty-five. He is made indeed to say in the probably spurious Preface to the 'Characters' that he is writing in his ninety-ninth year; while St. Jerome's Chronicle asserts that he lived to the age of 107. Accepting Diogenes' date, we may take it that he died about 285 B.C.; it is said that he complained that "we die just when we are beginning to live." His life must indeed have been a remarkably full and interesting one, when we consider that he enjoyed the personal friendship of two such men as Plato and Aristotle, and that he had witnessed the whole of the careers of Philip and Alexander of Macedon. To Alexander indeed he was directly indebted; the great conqueror had not been for nothing the pupil of the encyclopaedic Aristotle. He took with him to the East scientifically trained observers, the results of whose observations were at Theophrastus' disposal. Hence it is that his descriptions of plants are not limited to the flora of Greece and the Levant; to the reports of Alexander's followers he owed his accounts of such plants as the cotton-plant, banyan, pepper, cinnamon, myrrh and

frankincense. It has been a subject of some controversy whence he derived his accounts of plants whose habitat was nearer home. Kirchner, in an able tract, combats the contention of Sprengel that his observations even of the Greek flora were not made at first hand. Now at this period the Peripatetic School must have been a very important educational institution; Diogenes says that under Theophrastus it numbered two thousand pupils. Moreover we may fairly assume that Alexander, from his connexion with Aristotle, was interested in it, while we are told that at a later time Demetrius Phalereus assisted it financially. May we not hazard a guess that a number of the students were appropriately employed in the collection of facts and observations? The assumption that a number of 'travelling students' were so employed would at all events explain certain references in Theophrastus' botanical works. He says constantly 'The Macedonians say,' 'The men of Mount Ida say' and so forth. Now it seems hardly probable that he is quoting from written treatises by Macedonian or Idaean writers. It is at least a plausible suggestion that in such references he is referring to reports of the districts in question contributed by students of the school. In that case 'The Macedonians say' would mean 'This is what our representative was told in Macedonia.' It is further noticeable that the tense used is sometimes past, *e.g.* 'The men of Mount Ida *said*'; an obvious explanation of this is

supplied by the above conjecture. It is even possible that in one place (3. 12. 4.) the name of one of these students has been preserved.

Theophrastus, like his master, was a very voluminous writer; Diogenes gives a list of 227 treatises from his pen, covering most topics of human interest, as Religion, Politics, Ethics, Education, Rhetoric, Mathematics, Astronomy, Logic, Meteorology and other natural sciences. His oratorical works enjoyed a high reputation in antiquity. Diogenes attributes to him ten works on Rhetoric, of which one On Style was known to Cicero, who adopted from it the classification of styles into the 'grand,' the 'plain,' and the 'intermediate.'[1] Of one or two other lost works we have some knowledge. Thus the substance of an essay on Piety is preserved in Porphyry *de Abstinentia.*[2] The principal works still extant are the nine books of the Enquiry into Plants, and the six books on the Causes of Plants; these seem to be complete. We have also considerable fragments of treatises entitled :—of Sense-perception and objects of Sense, of Stones, of Fire, of Odours, of Winds, of Weather-Signs, of Weariness, of Dizziness, of Sweat, Metaphysics, besides a number of unassigned excerpts. The style of these works, as of the botanical books, suggests that, as in the case of Aristotle, what we possess consists of notes for lectures or notes taken of lectures. There is no literary charm; the sen-

[1] Sandys, i. p. 99.
[2] Bernays, Theophrastus, 1866.

tences are mostly compressed and highly elliptical, to the point sometimes of obscurity. It follows that translation, as with Aristotle, must be to some extent paraphrase. The thirty sketches of 'Characters' ascribed to Theophrastus, which have found many imitators, and which are well known in this country through Sir R. Jebb's brilliant translation, stand on a quite different footing; the object of this curious and amusing work is discussed in Sir R. Jebb's Introduction and in the more recent edition of Edmonds and Austen. Well may Aristotle, as we are assured, have commended his pupil's diligence. It is said that, when he retired from the headship of the school, he handed it over to Theophrastus. We are further told that the latter was once prosecuted for impiety, but the attack failed; also that he was once banished from Athens for a year, it does not appear under what circumstances. He was considered an attractive and lively lecturer. Diogenes' sketch ends with the quotation of some sayings attributed to him, of which the most noteworthy are 'Nothing costs us so dear as the waste of time,' 'One had better trust an unbridled horse than an undigested harangue.' He was followed to his grave, which we may hope was, in accordance with his own wish, in some peaceful corner of the Lyceum garden, by a great assemblage of his fellow townsmen.

INTRODUCTION

The principal references in the notes are to the following ancient authors :—

Apollon.	Apollonius, *Historia Miraculorum.*
Arist.	Aristotle. Bekker, Berlin, 1831.
Arr.	Arrian. Hercher (Teubner).
Athen.	Athenaeus. Dindorf, Leipzig, 1827.
Col.	Columella, *de re rustica.* Schneider, Leipzig, 1794.
Diod.	Diodorus.
Diosc.	Pedanius Dioscurides, *de materia medica.* Wellmann, Berlin, 1907.
Geop.	*Geoponica.* Beckh (Teubner), 1895.
Nic.	Nicander, *Theriaca.* Schneider, Leipzig, 1816.
Pall.	Palladius, *de re rustica.* Schneider, Leipzig, 1795.
Paus.	Pausanias. Schubart (Teubner), Leipzig, 1881.
Plin.	Plinius, *Naturalis Historia.* Mayhoff (Teubner), 1887. (Reference by book and section.)
Plut.	Plutarch. Hercher (Teubner), Leipzig, 1872.
Scyl.	Scylax, *Periplus.* Vossius, Amsterdam 1639.

THEOPHRASTUS

ENQUIRY INTO PLANTS

BOOK I

ΘΕΟΦΡΑΣΤΟΥ

ΠΕΡΙ ΦΥΤΩΝ ΙΣΤΟΡΙΑΣ

Α

I. Τῶν φυτῶν τὰς διαφορὰς καὶ τὴν ἄλλην φύσιν ληπτέον κατά τε τὰ μέρη καὶ τὰ πάθη καὶ τὰς γενέσεις καὶ τοὺς βίους· ἤθη γὰρ καὶ πράξεις οὐκ ἔχουσιν ὥσπερ τὰ ζῶα. εἰσὶ δ' αἱ μὲν κατὰ τὴν γένεσιν καὶ τὰ πάθη καὶ τοὺς βίους εὐθεωρητότεραι καὶ ῥᾴους, αἱ δὲ κατὰ τὰ μέρη πλείους ἔχουσι ποικιλίας. αὐτὸ γὰρ τοῦτο πρῶτον οὐχ ἱκανῶς ἀφώρισται τὰ ποῖα δεῖ μέρη καὶ μὴ μέρη καλεῖν, ἀλλ' ἔχει τινὰ ἀπορίαν.

2 Τὸ μὲν οὖν μέρος ἅτε ἐκ τῆς ἰδίας φύσεως ὂν ἀεὶ δοκεῖ διαμένειν ἢ ἁπλῶς ἢ ὅταν γένηται, καθάπερ ἐν τοῖς ζώοις τὰ ὕστερον γενησόμενα, πλὴν εἴ τι

---

[1] τὰ ins. Sch., om. Ald.H.

[2] πάθη, a more general word than δυνάμεις, 'virtues': *cf.* 1. 5. 4; 8. 4. 2; it seems to mean here something like 'behaviour,' in relation to environment. Instances of πάθη are given 4. 2. 11; 4. 14. 6.

[3] ἔχουσι conj. H.; ἔχουσαι W. with Ald.

THEOPHRASTUS

ENQUIRY INTO PLANTS

BOOK I

Of the Parts of Plants and their Composition. Of Classification.

Introductory: How plants are to be classified; difficulty of defining what are the essential 'parts' of a plant, especially if plants are assumed to correspond to animals.

I. In considering the distinctive characters of plants and their nature generally one must take into account their[1] parts, their qualities,[2] the ways in which their life originates, and the course which it follows in each case: (conduct and activities we do not find in them, as we do in animals). Now the differences in the way in which their life originates, in their qualities and in their life-history are comparatively easy to observe and are simpler, while those shewn[3] in their 'parts' present more complexity. Indeed it has not even been satisfactorily determined what ought and what ought not to be called 'parts,' and some difficulty is involved in making the distinction.

Now it appears that by a 'part,' seeing that it is something which belongs to the plant's characteristic nature, we mean something which is permanent either absolutely or when once it has appeared (like those parts of animals which remain for a time undeveloped)

διὰ νόσον ἢ γῆρας ἢ πήρωσιν ἀποβάλλεται. τῶν δ' ἐν τοῖς φυτοῖς ἔνια τοιαῦτ' ἐστὶν ὥστ' ἐπέτειον ἔχειν τὴν οὐσίαν, οἷον ἄνθος βρύον φύλλον καρπός, ἁπλῶς ὅσα πρὸ τῶν καρπῶν ἢ ἅμα γίνεται τοῖς καρποῖς· ἔτι δὲ αὐτὸς ὁ βλαστός· αἰεὶ γὰρ ἐπίφυσιν λαμβάνει τὰ δένδρα κατ' ἐνιαυτὸν ὁμοίως ἔν τε τοῖς ἄνω καὶ ἐν τοῖς περὶ τὰς ῥίζας· ὥστε, εἰ μέν τις ταῦτα θήσει μέρη, τό τε πλῆθος ἀόριστον ἔσται καὶ οὐδέποτε τὸ αὐτὸ τῶν μορίων· εἰ δ' αὖ μὴ μέρη, συμβήσεται, δι' ὧν τέλεια γίνεται καὶ φαίνεται, ταῦτα μὴ εἶναι μέρη· βλαστάνοντα γὰρ καὶ θάλλοντα καὶ καρπὸν ἔχοντα πάντα καλλίω καὶ τελειότερα καὶ δοκεῖ καὶ ἔστιν. αἱ μὲν οὖν ἀπορίαι σχεδόν εἰσιν αὗται.

3 Τάχα δὲ οὐχ ὁμοίως ἅπαντα ζητητέον οὔτε ἐν τοῖς ἄλλοις οὔθ' ὅσα πρὸς τὴν γένεσιν, αὐτά τε τὰ γεννώμενα μέρη θετέον οἷον τοὺς καρπούς. οὐδὲ γὰρ τὰ ἔμβρυα τῶν ζώων. εἰ δὲ ἐν τῇ ὥρᾳ ὄψει τοῦτό γε κάλλιστον,

---

[1] *i.e.* the male inflorescence of some trees; the term is of course wider than 'catkin.'

[2] *i.e.* flower, catkin, leaf, fruit, shoot.

—permanent, that is, unless it be lost by disease, age or mutilation. However some of the parts of plants are such that their existence is limited to a year, for instance, flower, 'catkin,'[1] leaf, fruit, in fact all those parts which are antecedent to the fruit or else appear along with it. Also the new shoot itself must be included with these; for trees always make fresh growth every year alike in the parts above ground and in those which pertain to the roots. So that if one sets these[2] down as 'parts,' the number of parts will be indeterminate and constantly changing; if on the other hand these are not to be called 'parts,' the result will be that things which are essential if the plant is to reach its perfection, and which are its conspicuous features, are nevertheless not 'parts'; for any plant always appears to be, as indeed it is, more comely and more perfect when it makes new growth, blooms, and bears fruit. Such, we may say, are the difficulties involved in defining a 'part.'

But perhaps we should not expect to find in plants a complete correspondence with animals in regard to those things which concern reproduction any more than in other respects; and so we should reckon as 'parts' even those things to which the plant gives birth, for instance their fruits, although[3] we do not so reckon the unborn young of animals. (However, if such[4] a product seems fairest to the eye, because the plant is then in its prime, we can draw no inference from this in

[3] οὐδὲ γὰρ: οὐδὲ seems to mean no more than οὐ (*cf. neque enim = non enim*); γὰρ refers back to the beginning of the §.

[4] ἐν τῇ ὥρᾳ ὄψει τοῦτό γε I conj.; τῇ ὥρᾳ ὄψει τό γε vulg. W.; τοῦτο, *i.e.* flower or fruit.

οὐδὲν σημεῖον, ἐπεὶ καὶ τῶν ζώων εὐθενεῖ τὰ κύοντα.

Πολλὰ δὲ καὶ τὰ μέρη κατ' ἐνιαυτὸν ἀποβάλλει, καθάπερ οἵ τε ἔλαφοι τὰ κέρατα καὶ τὰ φωλεύοντα τὰ πτερὰ καὶ τρίχας τετράποδα· ὥστ' οὐδὲν ἄτοπον ἄλλως τε καὶ ὅμοιον ὂν τῷ φυλλοβολεῖν τὸ πάθος.

Ὡσαύτως δ' οὐδὲ τὰ πρὸς τὴν γένεσιν· ἐπεὶ καὶ ἐν τοῖς ζώοις τὰ μὲν συνεκτίκτεται τὰ δ' ἀποκαθαίρεται καθάπερ ἀλλότρια τῆς φύσεως. ἔοικε δὲ παραπλησίως καὶ τὰ περὶ τὴν βλάστησιν ἔχειν. ἡ γάρ τοι βλάστησις γενέσεως χάριν ἐστὶ τῆς τελείας.

4 *Ὅλως δὲ καθάπερ εἴπομεν οὐδὲ πάντα ὁμοίως καὶ ἐπὶ τῶν ζώων ληπτέον. δι' ὃ καὶ ὁ ἀριθμὸς ἀόριστος· πανταχῇ γὰρ βλαστητικὸν ἅτε καὶ πανταχῇ ζῶν. ὥστε ταῦτα μὲν οὕτως ὑποληπτέον οὐ μόνον εἰς τὰ νῦν ἀλλὰ καὶ τῶν μελλόντων χάριν· ὅσα γὰρ μὴ οἷόν τε ἀφομοιοῦν περίεργον τὸ γλίχεσθαι πάντως, ἵνα μὴ καὶ τὴν οἰκείαν ἀποβάλλωμεν θεωρίαν. ἡ δὲ ἱστορία τῶν φυτῶν ἐστιν ὡς ἁπλῶς εἰπεῖν ἢ κατὰ*

[1] εὐθενεῖ conj. Sch., εὐθετεῖ UMVAld. *i.e.* we do not argue from the fact that animals are at their handsomest in the breeding season that the young is therefore 'part' of the animal.

[2] Lit. 'which are in holes,' in allusion to the well-known belief that animals (especially birds) which are out of sight in the winter are hiding in holes; the text is supported by [Arist.] *de plantis* 1. 3, the author of which had evidently read this passage; but possibly some such words as τάς τε φολίδας καὶ have dropped out after φωλεύοντα.

support of our argument, since even among animals those that are with young are at their best.[1])

Again many plants shed their parts every year, even as stags shed their horns, birds which hibernate[2] their feathers, four-footed beasts their hair: so that it is not strange that the parts of plants should not be permanent, especially as what thus occurs in animals and the shedding of leaves in plants are analogous processes.

In like manner the parts concerned with reproduction are not permanent in plants; for even in animals there are things which are separated from the parent when the young is born, and there are other things[3] which are cleansed away, as though neither of these belonged to the animal's essential nature. And so too it appears to be with the growth of plants; for of course growth leads up to reproduction as the completion of the process.[4]

And in general, as we have said, we must not assume that in all respects there is complete correspondence between plants and animals. And that is why the number also of parts is indeterminate; for a plant has the power of growth in all its parts, inasmuch as it has life in all its parts. Wherefore we should assume the truth to be as I have said, not only in regard to the matters now before us, but in view also of those which will come before us presently; for it is waste of time to take great pains to make comparisons where that is impossible, and in so doing we may lose sight also of our proper subject of enquiry. The enquiry into plants, to put it generally, may

[3] *i.e.* the embryo is not the only thing derived from the parent animal which is not a 'part' of it; there is also the food-supply produced with the young, and the after-birth.

[4] *cf.* *C.P.* 1. 11. 8.

τὰ ἔξω μόρια καὶ τὴν ὅλην μορφὴν ἢ κατὰ τὰ ἐντός, ὥσπερ ἐπὶ τῶν ζώων τὰ ἐκ τῶν ἀνατομῶν.

5 Ληπτέον δ' ἐν αὐτοῖς ποῖά τε πᾶσιν ὑπάρχει ταὐτὰ καὶ ποῖα ἴδια καθ' ἕκαστον γένος, ἔτι δὲ τῶν αὐτῶν ποῖα ὅμοια· λέγω δ' οἷον φύλλον ῥίζα φλοιός. οὐ δεῖ δὲ οὐδὲ τοῦτο λανθάνειν εἴ τι κατ' ἀναλογίαν θεωρητέον, ὥσπερ ἐπὶ τῶν ζώων, τὴν ἀναφορὰν ποιουμένους δῆλον ὅτι πρὸς τὰ ἐμφερέστατα καὶ τελειότατα. καὶ ἁπλῶς δὲ ὅσα τῶν ἐν φυτοῖς ἀφομοιωτέον τῷ ἐν τοῖς ζώοις, ὡς ἄν τίς τῷ γ' ἀνάλογον ἀφομοιοῖ. ταῦτα μὲν οὖν διωρίσθω τὸν τρόπον τοῦτον.

6 Αἱ δὲ τῶν μερῶν διαφοραὶ σχεδὸν ὡς τύπῳ λαβεῖν εἰσιν ἐν τρισίν, ἢ τῷ τὰ μὲν ἔχειν τὰ δὲ μή, καθάπερ φύλλα καὶ καρπόν, ἢ τῷ μὴ ὅμοια μηδὲ ἴσα, ἢ τρίτον τῷ μὴ ὁμοίως. τούτων δὲ ἡ μὲν ἀνομοιότης ὁρίζεται σχήματι χρώματι πυκνότητι μανότητι τραχύτητι λειότητι καὶ τοῖς ἄλλοις πάθεσιν, ἔτι δὲ ὅσαι διαφοραὶ τῶν χυλῶν. ἡ δὲ ἀνισότης ὑπεροχῇ καὶ ἐλλείψει κατὰ πλῆθος ἢ μέγεθος. ὡς δ' εἰπεῖν τύπῳ

1 A very obscure sentence; so W. renders the MSS. text.
2 *i.e.* 'inequality' might include 'unlikeness.'

either take account of the external parts and the form of the plant generally, or else of their internal parts: the latter method corresponds to the study of animals by dissection.

Further we must consider which parts belong to all plants alike, which are peculiar to some one kind, and which of those which belong to all alike are themselves alike in all cases; for instance, leaves roots bark. And again, if in some cases analogy ought to be considered (for instance, an analogy presented by animals), we must keep this also in view; and in that case we must of course make the closest resemblances and the most perfectly developed examples our standard; [1] and, finally, the ways in which the parts of plants are affected must be compared to the corresponding effects in the case of animals, so far as one can in any given case find an analogy for comparison. So let these definitions stand.

The essential parts of plants, and the materials of which they are made.

Now the differences in regard to parts, to take a general view, are of three kinds: either one plant may possess them and another not (for instance, leaves and fruit), or in one plant they may be unlike in appearance or size to those of another, or, thirdly, they may be differently arranged. Now the unlikeness between them is seen in form, colour, closeness of arrangement or its opposite, roughness or its opposite, and the other qualities; and again there are the various differences of flavour. The inequality is seen in excess or defect as to number or size, or, to speak generally, [2] all the above-mentioned differences too

κἀκεῖνα πάντα καθ' ὑπεροχὴν καὶ ἔλλειψιν· τὸ
7 γὰρ μᾶλλον καὶ ἧττον ὑπεροχὴ καὶ ἔλλειψις· τὸ
δὲ μὴ ὁμοίως τῇ θέσει διαφέρει· λέγω δ' οἷον τὸ
τοὺς καρποὺς τὰ μὲν ἐπάνω τὰ δ' ὑποκάτω τῶν
φύλλων ἔχειν καὶ αὐτοῦ τοῦ δένδρου τὰ μὲν ἐξ
ἄκρου τὰ δὲ ἐκ τῶν πλαγίων, ἔνια δὲ καὶ ἐκ τοῦ
στελέχους, οἷον ἡ Αἰγυπτία συκάμινος, καὶ ὅσα δὴ
καὶ ὑπὸ γῆς φέρει καρπόν, οἷον ἥ τε ἀραχίδνα καὶ
τὸ ἐν Αἰγύπτῳ καλούμενον οὔϊγγον, καὶ εἰ τὰ μὲν
ἔχει μίσχον τὰ δὲ μή. καὶ ἐπὶ τῶν ἀνθέων ὁμοίως·
τὰ μὲν γὰρ περὶ αὐτὸν τὸν καρπὸν τὰ δὲ ἄλλως.
ὅλως δὲ τὸ τῆς θέσεως ἐν τούτοις καὶ τοῖς φύλλοις
καὶ ἐν τοῖς βλαστοῖς ληπτέον.

8 Διαφέρει δὲ ἔνια καὶ τῇ τάξει· τὰ μὲν ὡς
ἔτυχε, τῆς δ' ἐλάτης οἱ κλῶνες κατ' ἀλλήλους
ἑκατέρωθεν· τῶν δὲ καὶ οἱ ὄζοι δι' ἴσου τε καὶ
κατ' ἀριθμὸν ἴσοι, καθάπερ τῶν τριόζων.

Ὥστε τὰς μὲν διαφορὰς ἐκ τούτων ληπτέον ἐξ
ὧν καὶ ἡ ὅλη μορφὴ συνδηλοῦται καθ' ἕκαστον.

9 Αὐτὰ δὲ τὰ μέρη διαριθμησαμένους πειρατέον
περὶ ἑκάστου λέγειν. ἔστι δὲ πρῶτα μὲν καὶ
μέγιστα καὶ κοινὰ τῶν πλείστων τάδε, ῥίζα
καυλὸς ἀκρεμὼν κλάδος, εἰς ἃ διέλοιτ' ἄν τις

---

[1] *cf. C.P.* 5. 1. 9.

[2] *cf.* 1. 6. 11. T. extends the term *καρπός* so as to include any succulent edible part of a plant.

[3] T. does not consider that *καρπός* was necessarily anteceded by a flower.

are included under excess and defect: for the 'more' and the 'less' are the same thing as excess and defect, whereas 'differently arranged' implies a difference of position; for instance, the fruit may be above or below the leaves,[1] and, as to position on the tree itself, the fruit may grow on the apex of it or on the side branches, and in some cases even on the trunk, as in the sycamore; while some plants again even bear their fruit underground, for instance *arakhidna*[2] and the plant called in Egypt *uingon*; again in some plants the fruit has a stalk, in some it has none. There is a like difference in the floral organs: in some cases they actually surround the fruit, in others they are differently placed[3]: in fact it is in regard to the fruit, the leaves, and the shoots that the question of position has to be considered.

Or again there are differences as to symmetry[4]: in some cases the arrangement is irregular, while the branches of the silver-fir are arranged opposite one another; and in some cases the branches are at equal distances apart, and correspond in number, as where they are in three rows.[5]

Wherefore the differences between plants must be observed in these particulars, since taken together they shew forth the general character of each plant.

But, before we attempt to speak about each, we must make a list of the parts themselves. Now the primary and most important parts, which are also common to most, are these—root, stem, branch, twig; these are the parts into which we might divide the plant, regarding them as members,[6] corresponding to

[4] Plin. 16. 122. [5] *i.e.* ternate.

[6] *i.e.* if we wished to make an anatomical division. μέλη conj. Sch. *cf.* 1. 2. 7; μέρη Ald.

ὥσπερ εἰς μέλη, καθάπερ ἐπὶ τῶν ζώων. ἕκαστόι τε γὰρ ἀνόμοιον καὶ ἐξ ἁπάντων τούτων τὰ ὅλα.

Ἔστι δὲ ῥίζα μὲν δι᾽ οὗ τὴν τροφὴν ἐπάγεται, καυλὸς δὲ εἰς ὃ φέρεται. καυλὸν δὲ λέγω τὸ ὑπὲρ γῆς πεφυκὸς ἐφ᾽ ἕν· τοῦτο γὰρ κοινότατον ὁμοίως ἐπετείοις καὶ χρονίοις, ὃ ἐπὶ τῶν δένδρων καλεῖται στέλεχος· ἀκρεμόνας δὲ τοὺς ἀπὸ τούτου σχιζομένους, οὓς ἔνιοι καλοῦσιν ὄζους. κλάδον δὲ τὸ βλάστημα τὸ ἐκ τούτων ἐφ᾽ ἕν, οἷον μάλιστα τὸ ἐπέτειον.

Καὶ ταῦτα μὲν οἰκειότερα τῶν δένδρων.
10 ὁ δὲ καυλός, ὥσπερ εἴρηται, κοινότερος· ἔχει δὲ οὐ πάντα οὐδὲ τοῦτον, οἷον ἔνια τῶν ποιωδῶν. τὰ δ᾽ ἔχει μὲν οὐκ ἀεὶ δὲ ἀλλ᾽ ἐπέτειον, καὶ ὅσα χρονιώτερα ταῖς ῥίζαις. ὅλως δὲ πολύχουν τὸ φυτὸν καὶ ποικίλον καὶ χαλεπὸν εἰπεῖν καθόλου· σημεῖον δὲ τὸ μηδὲν εἶναι κοινὸν λαβεῖν ὃ πᾶσιν ὑπάρχει, καθάπερ τοῖς ζώοις
11 στόμα καὶ κοιλία. τὰ δὲ ἀναλογίᾳ ταὐτὰ τὰ δ᾽ ἄλλον τρόπον. οὔτε γὰρ ῥίζαν πάντ᾽ ἔχει οὔτε καυλὸν οὔτε ἀκρεμόνα οὔτε κλάδον οὔτε φύλλον οὔτε ἄνθος οὔτε καρπὸν οὔτ᾽ αὖ φλοιὸν ἢ μήτραν ἢ ἶνας ἢ φλέβας, οἷον μύκης ὕδνον· ἐν τούτοις δὲ ἡ οὐσία καὶ ἐν τοῖς τοιούτοις· ἀλλὰ μάλιστα ταῦτα

---

[1] *i.e.* before it begins to divide. [2] Or 'knots.'
[3] ἐφ᾽ conj. W.; ὑφ᾽ P_2P_3Ald.
[4] χρονιώτερα conj. Sch.; χρονιώτερον Ald.H.
[5] ἀναλογίᾳ conj. Sch.; ἀναλογία UAld.H.

the members of animals: for each of these is distinct in character from the rest, and together they make up the whole.

The root is that by which the plant draws its nourishment, the stem that to which it is conducted. And by the 'stem' I mean that part which grows above ground and is single[1]; for that is the part which occurs most generally both in annuals and in long-lived plants; and in the case of trees it is called the 'trunk.' By 'branches' I mean the parts which split off from the stem and are called by some 'boughs.'[2] By 'twig' I mean the growth which springs from the branch regarded as a single whole,[3] and especially such an annual growth.

Now these parts belong more particularly to trees. The stem however, as has been said, is more general, though not all plants possess even this, for instance, some herbaceous plants are stemless; others again have it, not permanently, but as an annual growth, including some whose roots live beyond the year.[4] In fact your plant is a thing various and manifold, and so it is difficult to describe in general terms: in proof whereof we have the fact that we cannot here seize on any universal character which is common to all, as a mouth and a stomach are common to all animals; whereas in plants some characters are the same in all, merely in the sense that all have analogous[5] characters, while others correspond otherwise. For not all plants have root, stem, branch, twig, leaf, flower or fruit, or again bark, core, fibres or veins; for instance, fungi and truffles; and yet these and such like characters belong to a plant's essential nature. However, as has been said, these

ὑπάρχει, καθάπερ εἴρηται, τοῖς δένδροις κἀκείνων οἰκειότερος ὁ μερισμός· πρὸς ἃ καὶ τὴν ἀναφορὰν τῶν ἄλλων ποιεῖσθαι δίκαιον.

12 Σχεδὸν δὲ καὶ τὰς ἄλλας μορφὰς ἑκάστων ταῦτα διασημαίνει. διαφέρουσι γὰρ πλήθει τῷ τούτων καὶ ὀλιγότητι καὶ πυκνότητι καὶ μανότητι καὶ τῷ ἐφ' ἓν ἢ εἰς πλείω σχίζεσθαι καὶ τοῖς ἄλλοις τοῖς ὁμοίοις. ἔστι δὲ ἕκαστον τῶν εἰρημένων οὐχ ὁμοιομερές· λέγω δὲ οὐχ ὁμοιομερὲς ὅτι ἐκ τῶν αὐτῶν μὲν ὁτιοῦν μέρος σύγκειται τῆς ῥίζης καὶ τοῦ στελέχους, ἀλλ' οὐ λέγεται στέλεχος τὸ ληφθὲν ἀλλὰ μόριον, ὡς ἐν τοῖς τῶν ζώων μέλεσίν ἐστιν. ἐκ τῶν αὐτῶν μὲν γὰρ ὁτιοῦν τῆς κνήμης ἢ τοῦ ἀγκῶνος, οὐχ ὁμώνυμον δὲ καθάπερ σὰρξ καὶ ὀστοῦν, ἀλλ' ἀνώνυμον· οὐδὲ δὴ τῶν ἄλλων οὐδενὸς ὅσα μονοειδῆ τῶν ὀργανικῶν· ἁπάντων γὰρ τῶν τοιούτων ἀνώνυμα τὰ μέρη. τῶν δὲ πολυειδῶν ὠνομασμένα καθάπερ ποδὸς χειρὸς κεφαλῆς, οἷον δάκτυλος ῥὶς ὀφθαλμός. καὶ τὰ μὲν μέγιστα μέρη σχεδὸν ταῦτά ἐστιν.

II. Ἄλλα δὲ ἐξ ὧν ταῦτα φλοιὸς ξύλον μήτρα, ὅσα ἔχει μήτραν. πάντα δ' ὁμοιομερῆ. καὶ τὰ τούτων δὲ ἔτι πρότερα καὶ ἐξ ὧν ταῦτα, ὑγρὸν ἶς

---

[1] There is no exact English equivalent for ὁμοιομερές, which denotes a whole composed of parts, each of which is, as it were, a miniature of the whole. *cf.* Arist. *H.A.* 1. 1.

[2] *i.e.* any part taken of flesh or bone may be called 'flesh' or 'bone.'

[3] *e.g.* bark; *cf.* 1. 2. 1.

[4] *e.g.* fruit.

characters belong especially to trees, and our classification of characters belongs more particularly to these; and it is right to make these the standard in treating of the others.

Trees moreover shew forth fairly well the other features also which distinguish plants; for they exhibit differences in the number or fewness of these which they possess, as to the closeness or openness of their growth, as to their being single or divided, and in other like respects. Moreover each of the characters mentioned is not 'composed of like parts'[1]; by which I mean that though any given part of the root or trunk is composed of the same elements as the whole, yet the part so taken is not itself called 'trunk,' but 'a portion of a trunk.' The case is the same with the members of an animal's body; to wit, any part of the leg or arm is composed of the same elements as the whole, yet it does not bear the same name (as it does in the case of flesh or bone[2]); it has no special name. Nor again have subdivisions of any of those other organic parts[3] which are uniform special names, subdivisions of all such being nameless. But the subdivisions of those parts[4] which are compound have names, as have those of the foot, hand, and head, for instance, toe, finger, nose or eye. Such then are the largest[5] parts of the plant.

II. Again there are the things of which such parts are composed, namely bark, wood, and core (in the case of those plants which have it[6]), and these are all 'composed of like parts.' Further there are the things which are even prior to these, from which

[5] *i.e.* the 'compound' parts.
[6] ξύλον μήτρα conj. W. from G. μήτρα ξύλον MSS. ξύλον, ὅσα conj. W.; ξύλα, ἢ ὅσα Ald.H.

φλὲψ σάρξ· ἀρχαὶ γὰρ αὗται· πλὴν εἴ τις λέγοι τὰς τῶν στοιχείων δυνάμεις, αὗται δὲ κοιναὶ πάντων. ἡ μὲν οὖν οὐσία καὶ ἡ ὅλη φύσις ἐν τούτοις.

Ἄλλα δ' ἐστὶν ὥσπερ ἐπέτεια μέρη τὰ πρὸς τὴν καρποτοκίαν, οἷον φύλλον ἄνθος μίσχος· τοῦτο δ' ἐστὶν ᾧ συνήρτηται πρὸς τὸ φυτὸν τὸ φύλλον καὶ ὁ καρπός· ἔτι δὲ [ἕλιξ] βρύον, οἷς ὑπάρχει, καὶ ἐπὶ πᾶσι σπέρμα τὸ τοῦ καρποῦ· καρπὸς δ' ἐστὶ τὸ συγκείμενον σπέρμα μετὰ τοῦ περικαρπίου. παρὰ δὲ ταῦτα ἐνίων ἴδια ἄττα, καθάπερ ἡ κηκὶς δρυὸς καὶ ἡ ἕλιξ ἀμπέλου.

2 Καὶ τοῖς μὲν δένδρεσιν ἔστιν οὕτως διαλαβεῖν. τοῖς δ' ἐπετείοις δῆλον ὡς ἅπαντα ἐπέτεια· μέχρι γὰρ τῶν καρπῶν ἡ φύσις. ὅσα δὴ ἐπετειόκαρπα καὶ ὅσα διετίζει, καθάπερ σέλινον καὶ ἄλλ' ἄττα, καὶ ὅσα δὲ πλείω χρόνον ἔχει, τούτοις ἅπασι καὶ ὁ καυλὸς ἀκολουθήσει κατὰ λόγον· ὅταν γὰρ σπερμοφορεῖν μέλλωσι, τότε ἐκκαυλοῦσιν, ὡς ἕνεκα τοῦ σπέρματος ὄντων τῶν καυλῶν.

Ταῦτα μὲν οὖν ταύτῃ διῃρήσθω. τῶν δὲ ἄρτι εἰρημένων μερῶν πειρατέον ἕκαστον εἰπεῖν τί ἐστιν ὡς ἐν τύπῳ λέγοντας.

3 Τὸ μὲν οὖν ὑγρὸν φανερόν· ὃ δὴ καλοῦσί τινες ἁπλῶς ἐν ἅπασιν ὀπόν, ὥσπερ καὶ Μενέστωρ, οἱ

---

[1] οὐσία conj. Sch. (but he retracted it); συνουσία MSS. (?) Ald.

[2] This definition is quoted by Hesych. *s.v.* μίσχος.

[3] ? om. ἕλιξ, which is mentioned below.

[4] τὸ συγκείμενον σπέρμα, lit. 'the compound seed,' *i.e.* as many seeds as are contained in one περικάρπιον.

they are derived—sap, fibre, veins, flesh: for these are elementary substances—unless one should prefer to call them the active principles of the elements; and they are common to all the parts of the plant. Thus the essence[1] and entire material of plants consist in these.

Again there are other as it were annual parts, which help towards the production of the fruit, as leaf, flower, stalk (that is, the part by which the leaf and the fruit are attached to the plant),[2] and again tendril,[3] 'catkin' (in those plants that have them). And in all cases there is the seed which belongs to the fruit: by 'fruit' is meant the seed or seeds,[4] together with the seed-vessel. Besides these there are in some cases peculiar parts, such as the gall in the oak, or the tendril in the vine.

In the case of trees we may thus distinguish the annual parts, while it is plain that in annual plants *all* the parts are annual: for the end of their being is attained when the fruit is produced. And with those plants which bear fruit annually, those which take two years (such as celery and certain others[5]) and those which have fruit on them for a longer time —with all these the stem will correspond to the plant's length of life: for plants develop a stem at whatever time they are about to bear seed, seeing that the stem exists for the sake of the seed.

Let this suffice for the definition of these parts: and now we must endeavour to say what each of the parts just mentioned is, giving a general and typical description.

The sap is obvious: some call it simply in all cases 'juice,' as does Menestor[6] among others: others, in

[5] *cf.* 7.1.2 and 3. [6] A Pythagorean philosopher of Sybaris.

δ' ἐν μὲν τοῖς ἄλλοις ἀνωνύμως ἐν δέ τισιν ὀπὸν
καὶ ἐν ἄλλοις δάκρυον. ἶνες δὲ καὶ φλέβες καθ'
αὑτὰ μὲν ἀνώνυμα τῇ δὲ ὁμοιότητι μεταλαμβά-
νουσι τῶν ἐν τοῖς ζῴοις μορίων. ἔχει δὲ ἴσως
καὶ ἄλλας διαφορὰς καὶ ταῦτα καὶ ὅλως τὸ τῶν
φυτῶν γένος· πολύχουν γὰρ ὥσπερ εἰρήκαμεν.
ἀλλ' ἐπεὶ διὰ τῶν γνωριμωτέρων μεταδιώκειν δεῖ
τὰ ἀγνώριστα, γνωριμώτερα δὲ τὰ μείζω καὶ ἐμ-
φανῆ τῇ αἰσθήσει, δῆλον ὅτι καθάπερ ὑφήγηται
4 περὶ τούτων λεκτέον· ἐπαναφορὰν γὰρ ἕξομεν
τῶν ἄλλων πρὸς ταῦτα μέχρι πόσου καὶ πῶς
ἕκαστα μετέχει τῆς ὁμοιότητος. εἰλημμένων δὲ
τῶν μερῶν μετὰ ταῦτα ληπτέον τὰς τούτων
διαφοράς· οὕτως γὰρ ἅμα καὶ ἡ οὐσία φανερὰ
καὶ ἡ ὅλη τῶν γενῶν πρὸς ἄλληλα διάστασις.

Ἡ μὲν οὖν τῶν μεγίστων σχεδὸν εἴρηται· λέγω
δ' οἷον ῥίζης καυλοῦ τῶν ἄλλων· αἱ γὰρ δυνάμεις
καὶ ὧν χάριν ἕκαστον ὕστερον ῥηθήσονται. ἐξ
ὧν γὰρ καὶ ταῦτα καὶ τὰ ἄλλα σύγκειται
πειρατέον εἰπεῖν ἀρξαμένους ἀπὸ τῶν πρώτων.

Πρῶτα δέ ἐστι τὸ ὑγρὸν καὶ θερμόν· ἅπαν γὰρ
φυτὸν ἔχει τινὰ ὑγρότητα καὶ θερμότητα σύμφυτον ὥσπερ καὶ ζῷον, ὧν ὑπολειπόντων γίνεται
γῆρας καὶ φθίσις, τελείως δὲ ὑπολιπόντων θάνα-
5 τος καὶ αὔανσις. ἐν μὲν οὖν τοῖς πλείστοις ἀνώ-

---

[1] Lit. 'muscles and veins.'

[2] *i.e.* the analogy with animals is probably imperfect, but is useful so far as it goes.

[3] 1. 1. 10. [4] *e.g.* the root, as such.

[5] *e.g.* the different forms which roots assume.

the case of some plants give it no special name, while in some they call it 'juice,' and in others 'gum.' Fibre and 'veins'[1] have no special names in relation to plants, but, because of the resemblance, borrow the names of the corresponding parts of animals. [2] It may be however that, not only these things, but the world of plants generally, exhibits also other differences as compared with animals: for, as we have said,[3] the world of plants is manifold. However, since it is by the help of the better known that we must pursue the unknown, and better known are the things which are larger and plainer to our senses, it is clear that it is right to speak of these things in the way indicated: for then in dealing with the less known things we shall be making these better known things our standard, and shall ask how far and in what manner comparison is possible in each case. And when we have taken the parts,[4] we must next take the differences which they exhibit,[5] for thus will their essential nature become plain, and at the same time the general differences between one kind of plant and another.

Now the nature of the most important parts has been indicated already, that is, such parts as the root, the stem, and the rest: their functions and the reasons for which each of them exists will be set forth presently. For we must endeavour to state of what these, as well as the rest, are composed, starting from their elementary constituents.

First come moisture and warmth: for every plant, like every animal, has a certain amount of moisture and warmth which essentially belong to it; and, if these fall short, age and decay, while, if they fail altogether, death and withering ensue. Now in

νυμος ἡ ὑγρότης, ἐν ἐνίοις δὲ ὠνομασμένη καθάπερ εἴρηται. τὸ αὐτὸ δὲ καὶ ἐπὶ τῶν ζώων ὑπάρχει· μόνη γὰρ ἡ τῶν ἐναίμων ὑγρότης ὠνόμασται, δι' ὃ καὶ διῄρηται πρὸς τοῦτο στερήσει· τὰ μὲν γὰρ ἄναιμα τὰ δ' ἔναιμα λέγεται. ἕν τι μὲν οὖν τοῦτο τὸ μέρος καὶ τὸ τούτῳ συνηρτημένον θερμόν.

Ἄλλα δ' ἤδη ἕτερα τῶν ἐντός, ἃ καθ' ἑαυτὰ μέν ἐστιν ἀνώνυμα, διὰ δὲ τὴν ὁμοιότητα ἀπεικάζεται τοῖς τῶν ζώων μορίοις. ἔχουσι γὰρ ὥσπερ ἶνας· ὅ ἐστι συνεχὲς καὶ σχιστὸν καὶ ἐπίμηκες, ἀπαρά-
6 βλαστον δὲ καὶ ἄβλαστον. ἔτι δὲ φλέβας. αὗται δὲ τὰ μὲν ἄλλα εἰσὶν ὅμοιαι τῇ ἰνί, μείζους δὲ καὶ παχύτεραι καὶ παραβλάστας ἔχουσαι καὶ ὑγρότητα. ἔτι ξύλον καὶ σάρξ. τὰ μὲν γὰρ ἔχει σάρκα τὰ δὲ ξύλον. ἔστι δὲ τὸ μὲν ξύλον σχιστόν, ἡ δὲ σὰρξ πάντη διαιρεῖται ὥσπερ γῆ καὶ ὅσα γῆς· μεταξὺ δὲ γίνεται ἰνὸς καὶ φλεβός· φανερὰ δὲ ἡ φύσις αὐτῆς ἐν ἄλλοις τε καὶ ἐν τοῖς τῶν περικαρπίων δέρμασι. φλοιὸς δὲ καὶ μήτρα κυρίως μὲν λέγεται, δεῖ δὲ αὐτὰ καὶ τῷ λόγῳ διορίσαι. φλοιὸς μὲν οὖν ἐστι τὸ ἔσχατον καὶ χωριστὸν τοῦ ὑποκειμένου σώματος. μήτρα δὲ τὸ μεταξὺ τοῦ ξύλου, τρίτον ἀπὸ τοῦ φλοιοῦ οἷον ἐν τοῖς ὀστοῖς μυελός. καλοῦσι δέ τινες τοῦτο

---

[1] πλείστοις conj. Mold.; πρώτοις Ald.H. [2] 1. 1. 3.
[3] ἀπαράβλαστον conj. R.Const.; ἀπαράβλητον UMVAld.
[4] ἔτι δὲ conj. W.; ἔχον Ald. [5] Fibre.
[6] *i.e.* can be split in one direction.
[7] *e.g.* an unripe walnut.

most[1] plants the moisture has no special name, but in some it has such a name, as has been said[2]: and this also holds good of animals: for it is only the moisture of those which have blood which has received a name; wherefore we distinguish animals by the presence or absence of blood, calling some 'animals with blood,' others 'bloodless.' Moisture then is one essential 'part,' and so is warmth, which is closely connected with it.

There are also other internal characters, which in themselves have no special name, but, because of their resemblance, have names analogous to those of the parts of animals. Thus plants have what corresponds to muscle; and this quasi-muscle is continuous, fissile, long: moreover no other growth starts from it either branching from the side[3] or in continuation of it. Again[4] plants have veins: these in other respects resemble the 'muscle,'[5] but they are longer and thicker, and have side-growths and contain moisture. Then there are wood and flesh: for some plants have flesh, some wood. Wood is fissile,[6] while flesh can be broken up in any direction, like earth and things made of earth: it is intermediate between fibre and veins, its nature being clearly seen especially in the outer covering[7] of seed-vessels. Bark and core are properly so called,[8] yet they too must be defined. Bark then is the outside, and is separable from the substance which it covers. Core is that which forms the middle of the wood, being third[9] in order from the bark, and corresponding to the marrow in bones. Some call this part the 'heart,' others call it 'heart-wood': some

[8] *i.e.* not by analogy with animals, like 'muscle,' 'veins,' 'flesh.' [9] Reckoning inclusively.

καρδίαν, οἱ δ' ἐντεριώνην· ἔνιοι δὲ τὸ ἐντὸς τῆς μήτρας αὐτῆς καρδίαν, οἱ δὲ μυελόν.

Τὰ μὲν οὖν μόρια σχεδόν ἐστι τοσαῦτα. σύγκειται δὲ τὰ ὕστερον ἐκ τῶν προτέρων· ξύλον μὲν ἐξ ἰνὸς καὶ ὑγροῦ, καὶ ἔνια σαρκός· ξυλοῦται γὰρ σκληρυνομένη, οἷον ἐν τοῖς φοίνιξι καὶ νάρθηξι καὶ εἴ τι ἄλλο ἐκξυλοῦται, ὥσπερ αἱ τῶν ῥαφανίδων ῥίζαι· μήτρα δὲ ἐξ ὑγροῦ καὶ σαρκός· φλοιὸς δὲ ὁ μέν τις ἐκ πάντων τῶν τριῶν, οἷον ὁ τῆς δρυὸς καὶ αἰγείρου καὶ ἀπίου· ὁ δὲ τῆς ἀμπέλου ἐξ ὑγροῦ καὶ ἰνός· ὁ δὲ τοῦ φελλοῦ[1] ἐκ σαρκὸς καὶ ὑγροῦ. πάλιν δὲ ἐκ τούτων σύνθετα τὰ μέγιστα καὶ πρῶτα ῥηθέντα καθαπερανεὶ μέλη,[2] πλὴν οὐκ ἐκ τῶν αὐτῶν πάντα οὐδὲ ὡσαύτως ἀλλὰ διαφόρως.

Εἰλημμένων δὲ πάντων τῶν μορίων ὡς εἰπεῖν τὰς τούτων διαφορὰς πειρατέον ἀποδιδόναι καὶ τὰς ὅλων τῶν δένδρων καὶ φυτῶν οὐσίας.

III. Ἐπεὶ δὲ συμβαίνει σαφεστέραν[3] εἶναι τὴν μάθησιν διαιρουμένων κατὰ εἴδη,[4] καλῶς ἔχει τοῦτο ποιεῖν ἐφ' ὧν ἐνδέχεται. πρῶτα δέ ἐστι καὶ μέγιστα καὶ σχεδὸν ὑφ' ὧν πάντ' ἢ[5] τὰ πλεῖστα περιέχεται τάδε, δένδρον θάμνος φρύγανον πόα.

Δένδρον μὲν οὖν ἐστι τὸ ἀπὸ ῥίζης μονοστέλεχες

---

[1] φελλοῦ conj. H.; φύλλον UVP_2P_3Ald.; φυλλοῦ M.
[2] *i.e.* root, stem, branch, twig: *cf.* 1. 1. 9.
[3] σαφεστέραν conj. W.; σαφέστερον Ald.
[4] εἴδη here = γένη; *cf.* 6. 1. 2. n.
[5] πάντ' ἢ conj. Sch. after G; πάντη UMVAld.

again call only the inner part of the core itself the 'heart,' while others distinguish this as the 'marrow.'

Here then we have a fairly complete list of the 'parts,' and those last named are composed of the first 'parts'; wood is made of fibre and sap, and in some cases of flesh also; for the flesh hardens and turns to wood, for instance in palms ferula and in other plants in which a turning to wood takes place, as in the roots of radishes. Core is made of moisture and flesh: bark in some cases of all three constituents, as in the oak black poplar and pear; while the bark of the vine is made of sap and fibre, and that of the cork-oak[1] of flesh and sap. Moreover out of these constituents are made the most important parts,[2] those which I mentioned first, and which may be called 'members': however not all of them are made of the same constituents, nor in the same proportion, but the constituents are combined in various ways.

Having now, we may say, taken all the parts, we must endeavour to give the differences between them and the essential characters of trees and plants taken as wholes.

Definitions of the various classes into which plants may be divided.

III. Now since our study becomes more illuminating[3] if we distinguish different kinds,[4] it is well to follow this plan where it is possible. The first and most important classes, those which comprise all or nearly all[5] plants, are tree, shrub, under-shrub, herb.

A tree is a thing which springs from the root with

πολύκλαδον ὀζωτὸν οὐκ εὐαπόλυτον, οἷον ἐλάα
συκῆ ἄμπελος· θάμνος δὲ τὸ ἀπὸ ῥίζης πολύκλαδον, οἷον βάτος παλίουρος. φρύγανον δὲ τὸ
ἀπὸ ῥίζης πολυστέλεχες καὶ πολύκλαδον οἷον
καὶ θύμβρα καὶ πήγανον. πόα δὲ τὸ ἀπὸ ῥίζης
φυλλοφόρον προϊὸν ἀστέλεχες, οὗ ὁ καυλὸς σπερ-
μοφόρος, οἷον ὁ σῖτος καὶ τὰ λάχανα.

2 Δεῖ δὲ τοὺς ὅρους οὕτως ἀποδέχεσθαι καὶ λαμβάνειν ὡς τύπῳ καὶ ἐπὶ τὸ πᾶν λεγομένους· ἔνια
γὰρ ἴσως ἐπαλλάττειν δόξειε, τὰ δὲ καὶ παρὰ τὴν
ἀγωγὴν ἀλλοιότερα γίνεσθαι καὶ ἐκβαίνειν τῆς
φύσεως, οἷον μαλάχη τε εἰς ὕψος ἀναγομένη
καὶ ἀποδενδρουμένη· συμβαίνει γὰρ τοῦτο καὶ
οὐκ ἐν πολλῷ χρόνῳ ἀλλ' ἐν ἓξ ἢ ἑπτὰ μησίν,
ὥστε μῆκος καὶ πάχος δορατιαῖον γίνεσθαι, δι' ὃ
καὶ βακτηρίαις αὐταῖς χρῶνται, πλείονος δὲ χρόνου γινομένου κατὰ λόγον ἡ ἀπόδοσις· ὁμοίως
δὲ καὶ ἐπὶ τῶν τεύτλων· καὶ γὰρ ταῦτα λαμβάνει
μέγεθος· ἔτι δὲ μᾶλλον ἄγνοι καὶ ὁ παλίουρος
καὶ ὁ κιττός, ὥσθ' ὁμολογουμένως ταῦτα γίνεται
3 δένδρα· καί τοι θαμνώδη γέ ἐστιν. ὁ δὲ μύρρινος
μὴ ἀνακαθαιρόμενος ἐκθαμνοῦται καὶ ἡ ἡρακλεωτικὴ καρύα. δοκεῖ δὲ αὕτη γε καὶ τὸν καρπὸν
βελτίω καὶ πλείω φέρειν ἐὰν ῥάβδους τις ἐᾷ

[1] θάμνος . . . πήγανον. W.'s text transposes, without alteration, the definitions of θάμνος and φρύγανον as given in U. φρύγανον δὲ τὸ ἀπὸ ῥίζης καὶ πολυστέλεχες καὶ πολύκλαδον οἷον βάτος παλίουρος, Ald. So also M, but with a lacuna marked before φρύγανον and a note that the definition of θάμνος is wanting. φρύγανον δὲ τὸ ἀπὸ ῥίζης καὶ πολυστέλεχες καὶ πολύκλαδον οἷον καὶ γάμβοη καὶ πήγανον. θάμνος δὲ ἀπὸ ῥίζης πολύκλαδον οἷον βάτος παλίουρος U. So also very nearly P_1P_2. G gives to θάμνος (*frutex*) the definition assigned in U to φρύγανον (*suffrutex*) and the other definition is wanting.

a single stem, having knots and several branches, and it cannot easily be uprooted; for instance, olive fig vine. [1]A shrub is a thing which rises from the root with many branches; for instance, bramble Christ's thorn. An under-shrub is a thing which rises from the root with many stems as well as many branches; for instance, savory [2] rue. A herb is a thing which comes up from the root with its leaves and has no main stem, and the seed is borne on the stem; for instance, corn and pot-herbs.

These definitions however must be taken and accepted as applying generally and on the whole. For in the case of some plants it might seem that our definitions overlap; and some under cultivation appear to become different and depart from their essential nature, for instance, mallow [3] when it grows tall and becomes tree-like. For this comes to pass in no long time, not more than six or seven months, so that in length and thickness the plant becomes as great as a spear, and men accordingly use it as a walking-stick, and after a longer period the result of cultivation is proportionately greater. So too is it with the beets; they also increase in stature under cultivation, and so still more do chaste-tree Christ's thorn ivy, so that, as is generally admitted, these become trees, and yet they belong to the class of shrubs. On the other hand the myrtle, unless it is pruned, turns into a shrub, and so does filbert [4]: indeed this last appears to bear better and more abundant fruit, if one leaves

Note that W.'s transposition gives καὶ . . . καὶ the proper force; § 4 shews that the *typical* φρύγανον in T.'s view was πολυστέλεχες.

[2] θύμβρα conj. W.; γάμβρη MSS. But the first καὶ being meaningless, W. also suggests σισύμβριον for καὶ γάμβρη.

[3] *cf.* Plin. 19. 62. [4] *cf.* 3. 15. 1.

πλείους ὡς τῆς φύσεως θαμνώδους οὔσης. οὐ μονοστέλεχες δ' ἂν δόξειεν οὐδ' ἡ μηλέα οὐδ' ἡ ῥοιὰ οὐδ' ἡ ἄπιος εἶναι, οὐδ' ὅλως ὅσα παραβλαστητικὰ ἀπὸ τῶν ῥιζῶν ἀλλὰ τῇ ἀγωγῇ τοιαῦτα παραιρουμένων τῶν ἄλλων. ἔνια δὲ καὶ ἐῶσι πολυστελέχη διὰ λεπτότητα, καθάπερ ῥόαν μηλέαν· ἐῶσι δὲ καὶ τὰς ἐλάας κοπάδας καὶ τὰς συκᾶς.

4 Τάχα δ' ἄν τις φαίη καὶ ὅλως μεγέθει καὶ μικρότητι διαιρετέον εἶναι, τὰ δὲ ἰσχύϊ καὶ ἀσθενείᾳ καὶ πολυχρονιότητι καὶ ὀλιγοχρονιότητι. τῶν τε γὰρ φρυγανωδῶν καὶ λαχανωδῶν ἔνια μονοστελέχη καὶ οἷον δένδρου φύσιν ἔχοντα γίνεται, καθάπερ ῥάφανος πήγανον, ὅθεν καὶ καλοῦσί τινες τὰ τοιαῦτα δενδρολάχανα, τά τε λαχανώδη πάντα ἢ τὰ πλεῖστα ὅταν ἐγκαταμείνῃ λαμβάνει τινὰς ὥσπερ ἀκρεμόνας καὶ γίνεται τὸ ὅλον ἐν σχήματι δενδρώδει πλὴν ὀλιγοχρονιώτερα.

5 Διὰ δὴ ταῦτα ὥσπερ λέγομεν οὐκ ἀκριβολογητέον τῷ ὅρῳ ἀλλὰ τῷ τύπῳ ληπτέον τοὺς ἀφορισμούς· ἐπεὶ καὶ τὰς διαιρέσεις ὁμοίως, οἷον ἡμέρων ἀγρίων, καρποφόρων ἀκάρπων, ἀνθοφόρων ἀνανθῶν, ἀειφύλλων φυλλοβόλων. τὰ μὲν γὰρ ἄγρια καὶ ἥμερα παρὰ τὴν ἀγωγὴν εἶναι δοκεῖ· πᾶν γὰρ καὶ ἄγριον καὶ ἥμερόν φησιν Ἵππων γίνεσθαι τυγχάνον ἢ μὴ τυγχάνον θεραπείας.

---

[1] *i.e.* so that the tree comes to look like a shrub from the growth of fresh shoots after cutting. *cf.* 2. 6. 12; 2. 7. 2.

[2] ῥάφανος conj. Bod. from G; ῥαφανὶς Ald.

[3] *cf.* 3. 2. 2. The Ionian philosopher. See Zeller, *Pre-Socratic Philosophy* (Eng. trans.), 1. 281 f.

[4] καὶ add. W.; so G.

[5] ἢ conj. Sch.; καὶ UAld.Cam.Bas.H.

a good many of its branches untouched, since it is by nature like a shrub. Again neither the apple nor the pomegranate nor the pear would seem to be a tree of a single stem, nor indeed any of the trees which have side stems from the roots, but they acquire the character of a tree when the other stems are removed. However some trees men even leave with their numerous stems because of their slenderness, for instance, the pomegranate and the apple, and they leave the stems of the olive and the fig cut short.[1]

Exact classification impracticable: other possible bases of classification.

Indeed it might be suggested that we should classify in some cases simply by size, and in some cases by comparative robustness or length of life. For of under-shrubs and those of the pot-herb class some have only one stem and come as it were to have the character of a tree, such as cabbage[2] and rue: wherefore some call these 'tree-herbs'; and in fact all or most of the pot-herb class, when they have been long in the ground, acquire a sort of branches, and the whole plant comes to have a tree-like shape, though it is shorter lived than a tree.

For these reasons then, as we are saying, one must not make a too precise definition; we should make our definitions typical. For we must make our distinctions too on the same principle, as those between wild and cultivated plants, fruit-bearing and fruitless, flowering and flowerless, evergreen and deciduous. Thus the distinction between wild and cultivated seems to be due simply to cultivation, since, as Hippon[3] remarks, any plant may be either[4] wild or cultivated according as it receives or[5] does not receive attention.

ἄκαρπα δὲ καὶ κάρπιμα καὶ ἀνθοφόρα καὶ ἀνανθῆ παρὰ τοὺς τόπους καὶ τὸν ἀέρα τὸν περιέχοντα· τὸν αὐτὸν δὲ τρόπον καὶ φυλλοβόλα καὶ ἀείφυλλα. περὶ γὰρ Ἐλεφαντίνην οὐδὲ τὰς ἀμπέλους οὐδὲ τὰς συκᾶς φασι φυλλοβολεῖν.

6 Ἀλλ' ὅμως τοιαῦτα διαιρετέον· ἔχει γάρ τι τῆς φύσεως κοινὸν ὁμοίως ἐν δένδροις καὶ θάμνοις καὶ τοῖς φρυγανικοῖς καὶ ποιώδεσιν· ὑπὲρ ὧν καὶ τὰς αἰτίας ὅταν τις λέγῃ περὶ πάντων κοινῇ δῆλον ὅτι λεκτέον οὐχ ὁρίζοντα καθ' ἕκαστον· εὔλογον δὲ καὶ ταύτας κοινὰς εἶναι πάντων. ἅμα δὲ καὶ φαίνεταί τινα ἔχειν φυσικὴν διαφορὰν εὐθὺς ἐπὶ τῶν ἀγρίων καὶ τῶν ἡμέρων, εἴπερ ἔνια μὴ δύναται ζῆν ὥσπερ τὰ γεωργούμενα μηδ' ὅλως δέχεται θεραπείαν ἀλλὰ χείρω γίνεται, καθάπερ ἐλάτη πεύκη κήλαστρον καὶ ἁπλῶς ὅσα ψυχροὺς τόπους φιλεῖ καὶ χιονώδεις, ὡσαύτως δὲ καὶ τῶν φρυγανικῶν καὶ ποιωδῶν, οἷον κάππαρις καὶ θέρμος. ἥμερον δὲ καὶ ἄγριον δίκαιον καλεῖν ἀναφέροντα πρός τε ταῦτα καὶ ὅλως πρὸς τὸ ἡμερώτατον· [ὁ δ' ἄνθρωπος ἢ μόνον ἢ μάλιστα ἥμερον.]

IV. Φανεραὶ δὲ καὶ κατ' αὐτὰς τὰς μορφὰς αἱ διαφοραὶ τῶν ὅλων τε καὶ μορίων, οἷον λέγω

---

[1] ἀνθόφορα καὶ ἀνανθῆ conj. Sch. from G; καρπόφορα ἄνθη P_2Ald. [2] *cf.* 1. 9. 5; Plin. 16. 81.

[3] τοιαῦτα conj. W.; διαιρετέον conj. Sch.; τοῖς αὐτοῖς αἱρετέον Ald. The sense seems to be: Though these 'secondary' distinctions are not entirely satisfactory, yet (if we look to the *causes* of different characters), they are indispensable, since they are due to causes which affect all the four classes of our 'primary' distinction.

[4] *i.e.* we must take the *extreme* cases.

[5] *i.e.* plants which entirely refuse cultivation.

Again the distinctions between fruitless and fruit-bearing,[1] flowering and flowerless, seem to be due to position and the climate of the district. And so too with the distinction between deciduous and evergreen. [2] Thus they say that in the district of Elephantine neither vines nor figs lose their leaves.

Nevertheless we are bound to use such distinctions.[3] For there is a certain common character alike in trees, shrubs, under-shrubs, and herbs. Wherefore, when one mentions the causes also, one must take account of all alike, not giving separate definitions for each class, it being reasonable to suppose that the causes too are common to all. And in fact there seems to be some natural difference from the first in the case of wild and cultivated, seeing that some plants cannot live under the conditions of those grown in cultivated ground, and do not submit to cultivation at all, but deteriorate under it; for instance, silver-fir fir holly, and in general those which affect cold snowy country; and the same is also true of some of the under-shrubs and herbs, such as caper and lupin. Now in using the terms 'cultivated' and 'wild' [4] we must make these[5] on the one hand our standard, and on the other that which is in the truest sense[6] 'cultivated.' [7] Now Man, if he is not the only thing to which this name is strictly appropriate, is at least that to which it most applies.

Differences as to appearance and habitat.

IV. Again the differences, both between the plants as wholes and between their parts, may be seen in

[6] ὅλως πρὸς τὸ. ? πρὸς τὸ ὅλως conj. St.

[7] ὁ δ' ἄνθρωπος . . . ἥμερον. I have bracketed this clause, which seems to be an irrelevant gloss.

μέγεθος καὶ μικρότης, σκληρότης μαλακότης, λειότης τραχύτης, φλοιοῦ φύλλων τῶν ἄλλων, ἁπλῶς εὐμορφία καὶ δυσμορφία τις, ἔτι δὲ καὶ καλλικαρπία καὶ κακοκαρπία. πλείω μὲν γὰρ δοκεῖ τὰ ἄγρια φέρειν, ὥσπερ ἀχρὰς κότινος, καλλίω δὲ τὰ ἥμερα καὶ τοὺς χυλοὺς δὲ αὐτοὺς γλυκυτέρους καὶ ἡδίους καὶ τὸ ὅλον ὡς εἰπεῖν εὐκράτους μᾶλλον.

2 Αὗταί τε δὴ φυσικαί τινες ὥσπερ εἴρηται διαφοραί, καὶ ἔτι δὴ μᾶλλον τῶν ἀκάρπων καὶ καρποφόρων καὶ φυλλοβόλων καὶ ἀειφύλλων καὶ ὅσα ἄλλα τοιαῦτα. πάντων δὲ ληπτέον ἀεὶ καὶ τὰς κατὰ τοὺς τόπους· οὐ γὰρ οὐδ' οἷόν τε ἴσως ἄλλως. αἱ δὲ τοιαῦται δόξαιεν ἂν γενικόν τινα ποιεῖν χωρισμόν, οἷον ἐνύδρων καὶ χερσαίων, ὥσπερ ἐπὶ τῶν ζώων. ἔστι γὰρ ἔνια τῶν φυτῶν ἃ οὐ δύναται μὴ ἐν ὑγρῷ ζῆν· διήρηται δὲ ἄλλο κατ' ἄλλο γένος τῶν ὑγρῶν, ὥστε τὰ μὲν ἐν τέλμασι τὰ δὲ ἐν λίμναις τὰ δ' ἐν ποταμοῖς τὰ δὲ καὶ ἐν αὐτῇ τῇ θαλάττῃ φύεσθαι, τὰ μὲν ἐλάττω καὶ ἐν τῇ παρ' ἡμῖν τὰ δὲ μείζω περὶ τὴν ἐρυθράν. ἔνια δὲ ὡσπερεὶ κάθυγρα καὶ ἕλεια, καθάπερ ἰτέα καὶ πλάτανος, τὰ δὲ οὐκ ἐν ὕδατι δυνάμενα ζῆν οὐδ' ὅλως ἀλλὰ διώκοντα τοὺς ξηροὺς τόπους· τῶν δ' ἐλαττόνων ἔστιν ἃ καὶ τοὺς αἰγιαλούς.

---

[1] κατ' αὐτὰς τὰς conj. Sch.; καὶ τά (κα) τ' αὐτὰς τὰς U; κατὸ ταύτας τὰς MVAld.

[2] πάντων . . . τόπους, text perhaps defective.

[3] *i.e.* as to locality. [4] *cf.* 4. 7. 1.

the appearance itself[1] of the plant. I mean differences such as those in size, hardness, smoothness or their opposites, as seen in bark, leaves, and the other parts; also, in general, differences as to comeliness or its opposite and as to the production of good or of inferior fruit. For the wild kinds appear to bear more fruit, for instance, the wild pear and wild olive, but the cultivated plants better fruit, having even flavours which are sweeter and pleasanter and in general better blended, if one may so say.

These then as has been said, are differences of natural character, as it were, and still more so are those between fruitless and fruitful, deciduous and evergreen plants, and the like. But with all the differences in all these cases we must take into account the locality,[2] and indeed it is hardly possible to do otherwise. Such [3] differences would seem to give us a kind of division into classes, for instance, between that of aquatic plants and that of plants of the dry land, corresponding to the division which we make in the case of animals. For there are some plants which cannot live except in wet; and again these are distinguished from one another by their fondness for different kinds of wetness; so that some grow in marshes, others in lakes, others in rivers, others even in the sea, smaller ones in our own sea, larger ones in the Red Sea.[4] Some again, one may say, are lovers of very wet places,[5] or plants of the marshes, such as the willow and the plane. Others again cannot live at all[6] in water, but seek out dry places; and of the smaller sorts there are some that prefer the shore.

[5] *i.e.* though not actually living *in* water.
[6] οὐδ᾽ ὅλως conj. W.; ἐν τούτοις Ald.H. *Minime* G.

3 Οὐ μὴν ἀλλὰ καὶ τουτων εἴ τις ἀκριβολο-
γεῖσθαι θέλοι, τὰ μὲν ἂν εὕροι κοινὰ καὶ ὥσπερ
ἀμφίβια, καθάπερ μυρίκην ἰτέαν κλήθραν, τὰ δὲ
καὶ τῶν ὁμολογουμένων χερσαίων πεφυκότα ποτὲ
ἐν τῇ θαλάττῃ βιοῦν, φοίνικα σκίλλαν ἀνθέρικον.
ἀλλὰ τὰ τοιαῦτα καὶ ὅλως τὸ οὕτω σκοπεῖν οὐκ
οἰκείως ἐστὶ σκοπεῖν· οὐδὲ γὰρ οὐδ' ἡ φύσις οὕτως οὐδ' ἐν τοῖς τοιούτοις ἔχει τὸ ἀναγκαῖον. τὰς
μὲν οὖν διαιρέσεις καὶ ὅλως τὴν ἱστορίαν τῶν φυτῶν οὕτω ληπτέον. [ἅπαντα δ' οὖν καὶ ταῦτα καὶ
τὰ ἄλλα διοισει καθάπερ εἴρηται ταῖς τε τῶν
ὅλων μορφαῖς καὶ ταῖς τῶν μορίων διαφοραῖς, ἢ
τῷ ἔχειν τὰ δὲ μὴ ἔχειν, ἢ τῷ πλείω τὰ δ'
ἐλάττω, ἢ τῷ ἀνομοίως ἢ ὅσοι τρόποι διῄρηνται
4 πρότερον. οἰκεῖον δὲ ἴσως καὶ τοὺς τόπους συμ-
παραλαμβάνειν ἐν οἷς ἕκαστα πέφυκεν ἢ μὴ
πέφυκε γίνεσθαι. μεγάλη γὰρ καὶ αὕτη διαφορὰ
καὶ οὐχ ἥκιστα οἰκεία τῶν φυτῶν διὰ τὸ συνηρ-
τῆσθαι τῇ γῇ καὶ μὴ ἀπολελύσθαι καθάπερ
τὰ ζῷα.]

V. Πειρατέον δ' εἰπεῖν τὰς κατὰ μέρος διαφορὰς ὡς ἂν καθόλου λέγοντας πρῶτον καὶ κοινῶς,

---

1 θέλοι conj. Sch.; θέλει Ald. H.

2 εὕροι conj. Sch.; εὕρῃ Ald.; εὕρῃ H.

3 Presumably as being sometimes found on the shore below high-water mark.

4 ἅπαντα . . . ζῷα. This passage seems not to belong here (W.).

5 τρόποι conj. Sch.; τόποι UMVAld.

However, if one should wish [1] to be precise, one would find [2] that even of these some are impartial and as it were amphibious, such as tamarisk willow alder, and that others even of those which are admitted to be plants of the dry land sometimes live in the sea,[3] as palm squill asphodel. But to consider all these exceptions and, in general, to consider in such a manner is not the right way to proceed. For in such matters too nature certainly does not thus go by any hard and fast law. Our distinctions therefore and the study of plants in general must be understood accordingly. [4] To return—these plants as well as all others will be found to differ, as has been said, both in the shape of the whole and in the differences between the parts, either as to having or not having certain parts, or as to having a greater or less number of parts, or as to having them differently arranged, or because of other differences [5] such as we have already mentioned. And it is perhaps also proper to take into account the situation in which each plant naturally grows or does not grow. For this is an important distinction, and specially characteristic of plants, because they are united to the ground and not free from it like animals.

Characteristic differences in the parts of plants, whether general, special, or seen in qualities and properties.

V. Next we must try to give the differences as to particular parts, in the first instance speaking broadly of those of a general character,[6] and then

[6] *i.e.* those which divide plants into large classes (*e.g.* evergreen and deciduous).

εἶτα καθ' ἕκαστον, ὕστερον ἐπὶ πλεῖον ὥσπερ ἀναθεωροῦντας.

Ἔστι δὲ τὰ μὲν ὀρθοφυῆ καὶ μακροστελέχη καθάπερ ἐλάτη πεύκη κυπάριττος, τὰ δὲ σκολιώτερα καὶ βραχυστελέχη οἷον ἰτέα συκῆ ῥοιά, καὶ κατὰ πάχος δὲ καὶ λεπτότητα ὁμοίως. καὶ πάλιν τὰ μὲν μονοστελέχη τὰ δὲ πολυστελέχη· τοῦτο δὲ ταὐτὸ τρόπον τινὰ καὶ τῷ παραβλαστητικὰ ἢ ἀπαράβλαστα εἶναι· καὶ πολυκλαδῆ καὶ ὀλιγόκλαδα καθάπερ ὁ φοῖνιξ, καὶ ἐν αὐτοῖς τούτοις ἔτι κατὰ ἰσχὺν ἢ πάχος ἢ
2 τὰς τοιαύτας διαφοράς. πάλιν τὰ μὲν λεπτόφλοια, καθάπερ δάφνη φίλυρα, τὰ δὲ παχύφλοια, καθάπερ δρῦς. ἔτι τὰ μὲν λειόφλοια, καθάπερ μηλέα συκῆ, τὰ δὲ τραχύφλοια, καθάπερ ἀγρία δρῦς φελλὸς φοῖνιξ. πάντα δὲ νέα μὲν ὄντα λειοφλοιότερα, ἀπογηράσκοντα δὲ τραχυφλοιότερα, ἔνια δὲ καὶ ῥηξίφλοια, καθάπερ ἄμπελος, τὰ δὲ καὶ ὡς περιπίπτειν, οἷον ἀνδράχλη μηλέα κόμαρος. ἔστι δὲ καὶ τῶν μὲν σαρκώδης ὁ φλοιός, οἷον φελλοῦ δρυὸς αἰγείρου· τῶν δὲ ἰνώδης καὶ ἄσαρκος ὁμοίως δένδρων καὶ θάμνων καὶ ἐπετείων, οἷον ἀμπέλου καλάμου πυροῦ. καὶ τῶν μὲν πολύλοπος, οἷον φιλύρας ἐλάτης ἀμπέλου λινοσπάρτου κρομύων, τῶν δὲ μονόλοπος, οἷον συκῆς

[1] *i.e.* taking account of differences in qualities, etc. See § 4, but the order in which the three kinds of 'differences' are discussed is not that which is here given; the second is taken first and resumed at 6. 1, the third begins at 5. 4, the first at 14. 4.

[2] ταὐτὸ conj. Sch.; αὐτὸ UMVPAld.

[3] τραχυφλοιότερα conj. H. from G; παχυφ. UMAld *cf.* Plin. 16. 126.

of special differences between individual kinds; and after that we must take a wider range, making as it were a fresh survey.[1]

Some plants grow straight up and have tall stems, as silver-fir fir cypress; some are by comparison crooked and have short stems, as willow fig pomegranate; and there are like differences as to degree of thickness. Again some have a single stem, others many stems; and this difference corresponds[2] more or less to that between those which have side-growths and those which have none, or that between those which have many branches and those which have few, such as the date-palm. And in these very instances we have also differences in strength thickness and the like. Again some have thin bark, such as bay and lime; others have a thick bark, such as the oak. And again some have smooth bark, as apple and fig; others rough bark, as 'wild oak' (Valonia oak) cork-oak and date-palm. However all plants when young have smoother bark, which gets rougher[3] as they get older; and some have cracked bark,[4] as the vine; and in some cases it readily drops off, as in andrachne apple[5] and arbutus. And again of some the bark is fleshy, as in cork-oak oak poplar; while in others it is fibrous and not fleshy; and this applies alike to trees shrubs and annual plants, for instance to vines reeds and wheat. Again in some the bark has more than one layer, as in lime silver-fir vine Spanish broom[6] onions[7]; while in some it consists of only

[4] ῥηξίφλοια conj. St.; ῥιζίφοια (?) U; ῥιζίφλοια P.; ῥιζόφλοια P_2Ald. *cf.* 4. 15. 2, Plin. *l.c.*

[5] μηλέα conj. H. Steph., etc.; νηλεία UMPAld.; νήλεια P_2V. *cf.* Plin. *l.c.*

[6] G appears to have read λίνου, σπάρτου.

[7] *cf.* 5. 1 6.

καλάμου αἶρας. κατὰ μὲν δὴ τοὺς φλοιοὺς ἐν τούτοις αἱ διαφοραί.

3 Τῶν δὲ ξύλων αὐτῶν καὶ ὅλως τῶν καυλῶν οἱ μέν εἰσι σαρκώδεις, οἷον δρυὸς συκῆς, καὶ τῶν ἐλαττόνων ῥάμνου τεύτλου κωνείου· οἱ δὲ ἄσαρκοι, καθάπερ κέδρου λωτοῦ κυπαρίττου. καὶ οἱ μὲν ἰνώδεις· τὰ γὰρ τῆς ἐλάτης καὶ τοῦ φοίνικος ξύλα τοιαῦτα· τὰ δὲ ἄϊνα, καθάπερ τῆς συκῆς. ὡσαύτως δὲ καὶ τὰ μὲν φλεβώδη τὰ δ' ἄφλεβα. περὶ δὲ τὰ φρυγανικὰ καὶ θαμνώδη καὶ ὅλως τὰ ὑλήματα καὶ ἄλλας τις ἂν λάβοι διαφοράς· ὁ μὲν γὰρ κάλαμος γονατώδες, ὁ δὲ βάτος καὶ ὁ παλίουρος ἀκανθώδη. ἡ δὲ τύφη καὶ ἔνια τῶν ἑλείων ἢ λιμναίων ὁμοίως ἀδιάφρακτα καὶ ὁμαλῆ, καθάπερ σχοῖνος. ὁ δὲ τοῦ κυπείρου καὶ βουτόμου καυλὸς ὁμαλότητά τινα ἔχει παρὰ τούτους· ἔτι δὲ μᾶλλον ἴσως ὁ τοῦ μύκητος.

4 Αὗται μὲν δὴ δόξαιεν ἂν ἐξ ὧν ἡ σύνθεσις. αἱ δὲ κατὰ τὰ πάθη καὶ τὰς δυνάμεις οἷον σκληρότης μαλακότης γλισχρότης κραυρότης <πυκνότης> μανότης κουφότης βαρύτης καὶ ὅσα ἄλλα τοιαῦτα· ἡ μὲν γὰρ ἰτέα καὶ χλωρὸν εὐθὺ κοῦφον, ὥσπερ ὁ φελλός, ἡ δὲ πύξος καὶ ἡ ἔβενος οὐδὲ αὐανθέντα. καὶ τὰ μὲν σχίζεται, καθάπερ τὰ τῆς

---

[1] ῥάμνου conj. W.; θάμνου P_2; βαλάνου Ald. H.

[2] κωνείου conj. Sch.; κωνίου Ald. U (corrected to κωνείου). *cf.* 7. 6. 4.

[3] δὲ ἄϊνα conj. Sch. from G.; δὲ βῖνα U; δὲ μανά Ald.; δὲ . . . να M.

[4] ὑλήματα conj. Sch. (a general term including shrubs, under-shrubs, etc. *cf.* 1. 6. 7; 1. 10. 6); κλήματα, Ald.

one coat, as in fig reed darnel. Such are the respects in which bark differs.

Next of the woods themselves and of stems generally some are fleshy, as in oak and fig, and, among lesser plants, in buckthorn[1] beet hemlock[2]; while some are not fleshy, for instance, prickly cedar nettle-tree cypress. Again some are fibrous, for of this character is the wood of the silver-fir and the date-palm; while some are not fibrous,[3] as in the fig. In like manner some are full of 'veins,' others veinless. Further in shrubby plants and under-shrubs and in woody plants[4] in general one might find other differences: thus the reed is jointed, while the bramble and Christ's thorn have thorns on the wood. Bulrush and some of the marsh or pond plants are in like manner[5] without joints and smooth, like the rush; and the stem of galingale and sedge has a certain smoothness beyond those just mentioned; and still more perhaps has that of the mushroom.

Differences as to qualities and properties.

These then would seem to be the differences in the parts which make up the plant. Those which belong to the qualities[6] and properties are such as hardness or softness, toughness or brittleness, closeness or openness of texture, lightness or heaviness, and the like. For willow-wood is light from the first, even when it is green, and so is that of the cork-oak; but box and ebony are not light even when dried. Some woods again can be split,[7] such

[5] ὁμοίως, sense doubtful; ὁμωνύμων conj. W.

[6] πάθη, *cf.* 1. 1. 1 n.

[7] σχίζεται conj. W.; σχισθέντα UMVAld.; σχιστά H.; *fissiles* G.

ἐλάτης, τὰ δὲ εὔθραυστα μᾶλλον, οἷον τὰ τῆς ἐλάας. καὶ τὰ μὲν ἄοζα, οἷον τὰ τῆς ἀκτῆς, τὰ δὲ ὀζώδη, οἷον τὰ τῆς πεύκης καὶ ἐλάτης.

5 Δεῖ δὲ καὶ τὰς τοιαύτας ὑπολαμβάνειν τῆς φύσεως. εὔσχιστον μὲν γὰρ ἡ ἐλάτη τῷ εὐθυπορεῖν, εὔθραυστον δὲ ἡ ἐλάα διὰ τὸ σκολιὸν καὶ σκληρόν. εὔκαμπτον δὲ ἡ φίλυρα καὶ ὅσα ἄλλα διὰ τὸ γλίσχραν ἔχειν τὴν ὑγρότητα. βαρὺ δὲ ἡ μὲν πύξος καὶ ἡ ἔβενος ὅτι πυκνά, ἡ δὲ δρῦς ὅτι γεῶδες. ὡσαύτως δὲ καὶ τὰ ἄλλα πάντα πρὸς τὴν φύσιν πως ἀνάγεται.

VI. Διαφέρουσι δὲ καὶ ταῖς μήτραις· πρῶτον μὲν εἰ ἔνια ἔχει ἢ μὴ ἔχει, καθάπερ τινές φασιν ἄλλα τε καὶ τὴν ἀκτήν· ἔπειτα καὶ ἐν αὐτοῖς τοῖς ἔχουσι· τῶν μὲν γάρ ἐστι σαρκώδης τῶν δὲ ξυλώδης τῶν δὲ ὑμενώδης. καὶ σαρκώδης μὲν οἷον ἀμπέλου συκῆς μηλέας ῥοιᾶς ἀκτῆς νάρθηκος. ξυλώδης δὲ πίτυος ἐλάτης πεύκης, καὶ μάλιστα αὕτη διὰ τὸ ἔνδᾳδος εἶναι. τούτων δ' ἔτι σκληρότεραι καὶ πυκνότεραι κρανείας πρίνου δρυὸς κυτίσου συκαμίνου ἐβένου λωτοῦ.

2 Διαφέρουσι δὲ αὐταὶ καὶ τοῖς χρώμασι· μέλαιναι γὰρ τῆς ἐβένου καὶ τῆς δρυός, ἣν καλοῦσι μελάνδρυον. ἅπασαι δὲ σκληρότεραι καὶ κραυρό-

---

[1] *i.e.* break across the grain. εὔθραυστα mP; ἄθραυστα UPAld.; *fragilis* G. *cf.* 5. 5, Plin 16. 186.
[2] ἄοζα conj. Palm. from G; λοξά UPAld
[3] *i.e.* across the grain. [4] *cf.* 5 6. 2. [5] *cf.* 5. 1. 4.
[6] T. appears not to agree as to elder: see below.

as that of the silver-fir, while others are rather breakable,[1] such as the wood of the olive. Again some are without knots,[2] as the stems of elder, others have knots, as those of fir and silver-fir.

Now such differences also must be ascribed to the essential character of the plant: for the reason why the wood of silver-fir is easily split is that the grain is straight, while the reason why olive-wood is easily broken[3] is that it is crooked and hard. Lime-wood and some other woods on the other hand are easily bent because their sap is viscid.[4] Boxwood and ebony are heavy because the grain is close, and oak because it contains mineral matter.[5] In like manner the other peculiarities too can in some way be referred to the essential character.

Further 'special' differences.

VI. Again there are differences in the 'core': in the first place according as plants have any or have none, as some say[6] is the case with elder among other things; and in the second place there are differences between those which have it, since in different plants it is respectively fleshy, woody, or membranous; flesby, as in vine fig apple pomegranate elder ferula; woody, as in Aleppo pine silver-fir fir; in the last-named[7] especially so, because it is resinous. Harder again and closer than these is the core of dog-wood kermes-oak oak laburnum mulberry ebony nettle-tree.

The cores in themselves also differ in colour; for that of ebony and oak is black, and in fact in the oak it is called 'oak-black'; and in all these the core is harder and more brittle than the ordinary

7 αὕτη conj. Sch.; αὐτὴ UAld.; αὐτῇ MV; αὐτῆς P_2.

τεραι τῶν ξύλων· δι' ὃ καὶ οὐχ ὑπομένουσι καμπήν. μανότεραι δὲ αἱ μὲν αἱ δ' οὔ. ὑμενώδεις δ' ἐν μὲν τοῖς δένδροις οὐκ εἰσὶν ἢ σπάνιοι, ἐν δὲ τοῖς θαμνώδεσι καὶ ὅλως τοῖς ὑλήμασιν οἷον καλάμῳ τε καὶ νάρθηκι καὶ τοῖς τοιούτοις εἰσίν. ἔχει δὲ τὴν μήτραν τὰ μὲν μεγάλην καὶ φανεράν, ὡς πρῖνος δρῦς καὶ τἆλλα προειρημένα, τὰ δ' ἀφανεστέραν, οἷον ἐλάα πύξος· οὐ γὰρ ἔστιν ἀφωρισμένην οὕτω λαβεῖν, ἀλλὰ καί φασί τινες οὐ κατὰ τὸ μέσον ἀλλὰ κατὰ τὸ πᾶν ἔχειν· ὥστε μὴ εἶναι τόπον ὡρισμένον· δι' ὃ καὶ ἔνια οὐδ' ἂν δόξειεν ὅλως ἔχειν· ἐπεὶ καὶ τοῦ φοίνικος οὐδεμία φαίνεται διαφορὰ κατ' οὐδέν.

3 Διαφέρουσι δὲ καὶ ταῖς ῥίζαις. τὰ μὲν γὰρ πολύρριζα καὶ μακρόρριζα, καθάπερ συκῆ δρῦς πλάτανος· ἐὰν γὰρ ἔχωσι τόπον, ἐφ' ὁσονοῦν προέρχονται. τὰ δὲ ὀλιγόρριζα, καθάπερ ῥοιὰ μηλέα· τὰ δὲ μονόρριζα, καθάπερ ἐλάτη πεύκη· μονόρριζα δὲ οὕτως, ὅτι μίαν μεγάλην τὴν εἰς βάθος ἔχει μικρὰς δὲ ἀπὸ ταύτης πλείους. ἔχουσι δὲ καὶ τῶν μὴ μονορρίζων ἔνια τὴν ἐκ τοῦ μέσου μεγίστην καὶ κατὰ βάθους, ὥσπερ ἀμυγδαλῆ· ἐλάα δὲ μικρὰν ταύτην τὰς δὲ ἄλλας μείζους καὶ ὡς κεκαρκινωμένας. ἔτι δὲ τῶν μὲν παχεῖαι μᾶλλον τῶν δὲ ἀνωμαλεῖς, καθάπερ δάφνης ἐλάας·
4 τῶν δὲ πᾶσαι λεπταί, καθάπερ ἀμπέλου. διαφέρουσι δὲ καὶ λειότητι καὶ τραχύτητι καὶ πυκνότητι. πάντων γὰρ αἱ ῥίζαι μανότεραι τῶν ἄνω,

---

1 μανότεραι . . . οὔ: text can hardly be sound, but sense is clear. 2 *i.e.* homogeneous. 3 Plin. 16. 127.

4 3. 6. 4 seems to give a different account.

5 *cf.* *C.P.* 3. 23. 5, and καρκινώδης *C.P.* 1. 12. 3; 3. 21. 5.

wood; and for this reason the core of these trees can not be bent. Again the core differs in closeness of texture.[1] A membranous core is not common in trees, if indeed it is found at all; but it is found in shrubby plants and woody plants generally, as in reed ferula and the like. Again in some the core is large and conspicuous, as in kermes-oak oak and the other trees mentioned above; while in others it is less conspicuous, as in olive and box. For in these trees one cannot find it isolated, but, as some say, it is not found in the middle of the stem, being diffused throughout, so that it has no separate place; and for this reason some trees might be thought to have no core at all; in fact in the date-palm the wood is alike throughout.[2]

Differences in root.

3 Again plants differ in their roots, some having many long roots, as fig oak plane; for the roots of these, if they have room, run to any length. Others again have few roots, as pomegranate and apple, others a single root, as silver-fir and fir; these have a single root in the sense that they have one long one 4 which runs deep, and a number of small ones branching from this. Even in some of those which have more than a single root the middle root is the largest and goes deep, for instance, in the almond; in the olive this central root is small, while the others are larger and, as it were, spread out crab-wise.[5] Again the roots of some are mostly stout, of some of various degrees of stoutness, as those of bay and olive; and of some they are all slender, as those of the vine. Roots also differ in degree of smoothness and in density. For the roots of all

πυκνότεραι δὲ ἄλλαι ἄλλων καὶ ξυλωδέστεραι· καὶ αἱ μὲν ἰνώδεις, ὡς αἱ τῆς ἐλάτης, αἱ δὲ σαρκώδεις μᾶλλον, ὥσπερ αἱ τῆς δρυός, αἱ δὲ οἷον ὀζώδεις καὶ θυσανώδεις, ὥσπερ αἱ τῆς ἐλάας· τοῦτο δὲ ὅτι τὰς λεπτὰς καὶ μικρὰς πολλὰς ἔχουσι καὶ ἀθρόας· ἐπεὶ πᾶσαί γε καὶ ταύτας ἀποφύουσιν ἀπὸ τῶν μεγάλων ἀλλ' οὐχ ὁμοίως ἀθρόας καὶ πολλάς.

Ἔστι δὲ καὶ τὰ μὲν βαθύρριζα, καθάπερ δρῦς, τὰ δ' ἐπιπολαιόρριζα, καθάπερ ἐλάα ῥοιὰ μηλέα κυπάριττος. ἔτι δὲ αἱ μὲν εὐθεῖαι καὶ ὁμαλεῖς, αἱ δὲ σκολιαὶ καὶ παραλλάττουσαι· τοῦτο γὰρ οὐ μόνον συμβαίνει διὰ τοὺς τόπους τῷ μὴ εὐοδεῖν ἀλλὰ καὶ τῆς φύσεως αὐτῆς ἐστιν, ὥσπερ ἐπὶ τῆς δάφνης καὶ τῆς ἐλάας· ἡ δὲ συκῆ καὶ τὰ τοιαῦτα σκολιοῦται διὰ τὸ μὴ εὐοδεῖν.

5 Ἅπασαι δ' ἔμμητροι καθάπερ καὶ τὰ στελέχη καὶ οἱ ἀκρεμόνες· καὶ εὔλογον ἀπὸ τῆς ἀρχῆς. εἰσὶ δὲ καὶ αἱ μὲν παραβλαστητικαὶ εἰς τὸ ἄνω, καθάπερ ἀμπέλου ῥόας, αἱ δὲ ἀπαράβλαστοι, καθάπερ ἐλάτης κυπαρίττου πεύκης. αἱ αὐταὶ δὲ διαφοραὶ καὶ τῶν φρυγανικῶν καὶ τῶν ποιωδῶν καὶ τῶν ἄλλων· πλὴν εἰ ὅλως ἔνια μὴ ἔχει, καθάπερ ὕδνον μύκης πέζις κεραύνιον. τὰ μὲν πολύρριζα καθάπερ πυρὸς τίφη κριθή, πᾶν τὸ τοιοῦτο, καθάπερ εἰκαζούσαις· τὰ δ' ὀλιγόρριζα

6 καθάπερ τὰ χεδροπά. σχεδὸν δὲ καὶ τῶν λαχανωδῶν τὰ πλεῖστα μονόρριζα, οἷον ῥάφανος

---

[1] πέζις κεραύνιον: πύξος κράνιον UMVAld.; πέζις conj. Sch. from Athen. 2. 59; κεραύνιον conj. W. *cf.* Plin. 3. 36 and 37, Juv. 5. 117. [2] εἰκαζούσαις: word corrupt; so UMVAld.
[3] Plin. 19. 98.

plants are less dense than the parts above ground, but the density varies in different kinds, as also does the woodiness. Some are fibrous, as those of the silver-fir, some fleshier, as those of the oak, some are as it were branched and tassel-like, as those of the olive; and this is because they have a large number of fine small roots close together; for all in fact produce these from their large roots, but they are not so closely matted nor so numerous in some cases as in others.

Again some plants are deep-rooting, as the oak, and some have surface roots, as olive pomegranate apple cypress. Again some roots are straight and uniform, others crooked and crossing one another. For this comes to pass not merely on account of the situation because they cannot find a straight course; it may also belong to the natural character of the plant, as in the bay and the olive; while the fig and such like become crooked because they can not find a straight course.

All roots have core, just as the stems and branches do, which is to be expected, as all these parts are made of the same materials. Some roots again have side-growths shooting upwards, as those of the vine and pomegranate, while some have no side-growth, as those of silver-fir cypress and fir. The same differences are found in under-shrubs and herbaceous plants and the rest, except that some have no roots at all, as truffle mushroom bullfist[1] 'thunder-truffle.' Others have numerous roots, as wheat one-seeded wheat barley and all plants of like nature, for instance,[2] . . . . Some have few roots, as leguminous plants. [3] And in general most of the pot-herbs have single roots, as cabbage beet celery

τεῦτλον σέλινον λάπαθος· πλὴν ἔνια καὶ ἀποφυάδας ἔχει μεγάλας, οἷον τὸ σέλινον καὶ τὸ τεῦτλον· καὶ ὡς ἂν κατὰ λόγον ταῦτα βαθυρριζότερα τῶν δένδρων. εἰσὶ δὲ τῶν μὲν σαρκώδεις, καθάπερ ῥαφανίδος γογγυλίδος ἄρου κρόκου· τῶν δὲ ξυλώδεις, οἷον εὐζώμου ὠκίμου· καὶ τῶν ἀγρίων δὲ τῶν πλείστων, ὅσων μὴ εὐθὺς πλείους καὶ σχιζόμεναι, καθάπερ πυροῦ κριθῆς καὶ τῆς καλουμένης πόας. αὕτη γὰρ ἐν τοῖς ἐπετείοις καὶ ἐν τοῖς ποιώδεσιν ἡ διαφορὰ τῶν ῥιζῶν ὥστε τὰς μὲν εὐθὺς σχίζεσθαι πλείους οὔσας καὶ ὁμαλεῖς, τῶν δὲ ἄλλων μίαν ἢ δύο τὰς μεγίστας καὶ ἄλλας ἀπὸ τούτων.

7 Ὅλως δὲ πλείους αἱ διαφοραὶ τῶν ῥιζῶν ἐν τοῖς ὑλήμασι καὶ λαχανώδεσιν· εἰσὶ γὰρ αἱ μὲν ξυλώδεις, ὥσπερ αἱ τοῦ ὠκίμου· αἱ δὲ σαρκώδεις, ὥσπερ αἱ τοῦ τεύτλου καὶ ἔτι δὴ μᾶλλον τοῦ ἄρου καὶ ἀσφοδέλου καὶ κρόκου· αἱ δὲ ὥσπερ ἐκ φλοιοῦ καὶ σαρκός, ὥσπερ αἱ τῶν ῥαφανίδων καὶ γογγυλίδων· αἱ δὲ γονατώδεις, ὥσπερ αἱ τῶν καλάμων καὶ ἀγρώστεων καὶ εἴ τι καλαμῶδες, καὶ μόναι δὴ αὗται ἢ μάλισθ' ὅμοιαι τοῖς ὑπὲρ γῆς· ὥσπερ γὰρ κάλαμοί εἰσιν ἐρριζωμένοι ταῖς λεπταῖς. αἱ δὲ λεπυρώδεις ἢ φλοιώδεις, οἷον αἵ τε τῆς σκίλλης καὶ τοῦ βολβοῦ καὶ ἔτι κρομύου καὶ τῶν τούτοις ὁμοίων. αἰεὶ γὰρ ἔστι περιαιρεῖν αὐτῶν.

8 Πάντα δὲ τὰ τοιαῦτα δοκεῖ καθάπερ δύο γένη ῥιζῶν ἔχειν· τοῖς δὲ καὶ ὅλως τὰ κεφαλοβαρῆ καὶ κατάρριζα πάντα· τήν τε σαρκώδη ταύτην

---

[1] The same term being applied to 'herbaceous' plants in general. [2] Plin. 19. 98.

monk's rhubarb; but some have large side-roots, as celery and beet, and in proportion to their size these root deeper than trees. Again of some the roots are fleshy, as in radish turnip cuckoo-pint crocus; of some they are woody, as in rocket and basil. And so with most wild plants, except those whose roots are to start with numerous and much divided, as those of wheat barley and the plant specially [1] called 'grass.' For in annual and herbaceous plants this is the difference between the roots:—Some are more numerous and uniform and much divided to start with, but the others have one or two specially large roots and others springing from them.

To speak generally, the differences in roots are more numerous in shrubby plants and pot-herbs; [2] for some are woody, as those of basil, some fleshy, as those of beet, and still more those of cuckoo-pint asphodel and crocus; some again are made, as it were, of bark and flesh, as those of radishes and turnips; some have joints, as those of reeds and dog's tooth grass and of anything of a reedy character; and these roots alone, or more than any others, resemble the parts above ground; they are in fact like [3] reeds fastened in the ground by their fine roots. Some again have scales or a kind of bark, as those of squill and purse-tassels, and also of onion and things like these. In all these it is possible to strip off a coat.

Now all such plants, seem, as it were, to have two kinds of root; and so, in the opinion of some, this is true generally of all plants which have a solid 'head' [4] and send out roots from it downwards. These have,

[3] *i.e.* the main root is a sort of repetition of the part above ground. [4] *i.e.* bulb, corm, rhizome, etc.

καὶ φλοιώδη, καθάπερ ἡ σκίλλα, καὶ τὰς ἀπὸ ταύτης ἀποπεφυκυίας· οὐ γὰρ λεπτότητι καὶ παχύτητι διαφέρουσι μόνον, ὥσπερ αἱ τῶν δένδρων καὶ τῶν λαχάνων, ἀλλ' ἀλλοῖον ἔχουσι τὸ γένος. ἐκφανεστάτη δ' ἤδη ἥ τε τοῦ ἄρου καὶ ἡ τοῦ κυπείρου· ἡ μὲν γὰρ παχεῖα καὶ λεία καὶ σαρκώδης, ἡ δὲ λεπτὴ καὶ ἰνώδης. διόπερ ἀπορήσειεν ἄν τις εἰ ῥίζας τὰς τοιαύτας θετέον· ᾗ μὲν γὰρ κατὰ γῆς δόξαιεν ἄν, ᾗ δὲ ὑπεναντίως ἔχουσι ταῖς ἄλλαις οὐκ ἂν δόξαιεν. ἡ μὲν γὰρ ῥίζα λεπτοτέρα πρὸς τὸ πόρρω καὶ ἀεὶ σύνοξυς· ἡ δὲ τῶν σκιλλῶν καὶ τῶν βολβῶν καὶ τῶν ἄρων ἀνάπαλιν.

9 Ἔτι δ' αἱ μὲν ἄλλαι κατὰ τὸ πλάγιον ἀφιᾶσι ῥίζας, αἱ δὲ τῶν σκιλλῶν καὶ τῶν βολβῶν οὐκ ἀφιᾶσιν· οὐδὲ τῶν σκορόδων καὶ τῶν κρομύων. ὅλως δέ γε ἐν ταύταις αἱ κατὰ μέσον ἐκ τῆς κεφαλῆς ἠρτημέναι φαίνονται ῥίζαι καὶ τρέφονται. τοῦτο δ' ὥσπερ κῦμα ἢ καρπός, ὅθεν καὶ οἱ ἐγγεοτόκα λέγοντες οὐ κακῶς· ἐπὶ δὲ τῶν ἄλλων τοιοῦτο μὲν οὐδέν ἐστιν· ἐπεὶ δὲ πλεῖον ἡ φύσις ἡ κατὰ ῥίζαν ταύτῃ ἀπορίαν ἔχει. τὸ γὰρ δὴ πᾶν λέγειν τὸ κατὰ γῆς ῥίζαν οὐκ ὀρθόν· καὶ γὰρ ἂν ὁ καυλὸς τοῦ βολβοῦ καὶ ὁ τοῦ γηθύου καὶ

---

[1] τὰς conj. Sch.; τῆς Ald.H.; τὴν . . . ἀποπεφυκυῖαν P.

[2] ἀλλ' ἀλλοῖον ἔχουσι conj. St.; ἀλλὰ λεῖον ἔχοντες PMV Ald.; ἀλλοῖον ἔχ. mBas.mP from G; ἀλλ' ἀλλοῖον ἔχουσαι conj. Scal. [3] *cf.* 4. 10. 5.

[4] καὶ ἀεὶ Ald.; ἀεὶ καὶ conj. W. [5] Plin. 19. 99.

[6] *cf.* the definition of 'root,' 1. 1. 9.

[7] ἐγγεότοκα λέγοντες conj. W.; *cf.* ἡ τῶν ἐγγειοτόκων τούτων γένεσις in Athenaeus' citation of this passage (2. 60);

that is to say, this fleshy or bark-like root, like squill, as well as the[1] roots which grow from this. For these roots not only differ in degree of stoutness, like those of trees and pot-herbs; they are of quite distinct classes.[2] This is at once quite evident in cuckoo-pint and galingale,[3] the root being in the one case thick smooth and fleshy, in the other thin and fibrous. Wherefore we might question if such roots should be called 'roots'; inasmuch as they are under ground they would seem to be roots, but, inasmuch as they are of opposite character to other roots, they would not. For your root gets slenderer as it gets longer and tapers continuously[4] to a point; but the so-called root of squill purse-tassels and cuckoo-pint does just the opposite.

Again, while the others send out roots at the sides, this is not the case[5] with squill and purse-tassels, nor yet with garlic and onion. In general in these plants the roots which are attached to the 'head' in the middle appear to be real roots and receive nourishment,[6] and this 'head' is, as it were, an embryo or fruit; wherefore those who call such plants 'plants which reproduce themselves underground'[7] give a fair account of them. In other kinds of plants there is nothing of this sort.[8] But a difficult question is raised, since here the 'root' has a character which goes beyond what one associates with roots. For it is not right to call all that which is underground 'root,' since in that case the stalk[9] of purse-tassels and that of long onion and in general any part which is under-

ευτεοσ οισαλεγουτες U; ἔν τε τοῖς ὀστοῖς ἀλέγοντες MV (omitting τε) Ald. (omitting τοῖς).

[8] τοιοῦτο μὲν οὐδέν conj. W.; τοῦτο μὲν MSS

[9] ἂν ὁ καυλός conj. St.; ἀνάκαυλος Ald.

ὅλως ὅσα κατὰ βάθους ἐστὶν εἴησαν ἂν ῥίζαι, καὶ τὸ ὕδνον δὲ καὶ ὃ καλοῦσί τινες ἀσχίον καὶ τὸ οὔϊγγον καὶ εἴ τι ἄλλο ὑπόγειόν ἐστιν· ὧν οὐδέν ἐστι ῥίζα· δυνάμει γὰρ δεῖ φυσικῇ διαιρεῖν καὶ οὐ τόπῳ.

10 Τάχα δὲ τοῦτο μὲν ὀρθῶς λέγεται, ῥίζα δὲ οὐδὲν ἧττόν ἐστιν· ἀλλὰ διαφορά τις αὕτη τῶν ῥιζῶν, ὥστε τὴν μέν τινα τοιαύτην εἶναι τὴν δὲ τοιαύτην καὶ τρέφεσθαι τὴν ἑτέραν ὑπὸ τῆς ἑτέρας. καίτοι καὶ αὐταὶ αἱ σαρκώδεις ἐοίκασιν ἕλκειν. τὰς γοῦν τῶν ἄρων πρὸ τοῦ βλαστάνειν στρέφουσι καὶ γίγνονται μείζους κωλυόμεναι διαβῆναι πρὸς τὴν βλάστησιν. ἐπεὶ ὅτι γε πάντων τῶν τοιούτων ἡ φύσις ἐπὶ τὸ κάτω μᾶλλον ῥέπει φανερόν· οἱ μὲν γὰρ καυλοὶ καὶ ὅλως τὰ ἄνω βραχέα καὶ ἀσθενῆ, τὰ δὲ κάτω μεγάλα καὶ πολλὰ καὶ ἰσχυρὰ οὐ μόνον ἐπὶ τῶν εἰρημένων ἀλλὰ καὶ ἐπὶ καλάμου καὶ ἀγρώστιδος καὶ ὅλως ὅσα καλαμώδη καὶ τούτοις ὅμοια. καὶ ὅσα δὴ ναρθηκώδη, καὶ τούτων ῥίζαι μεγάλαι καὶ σαρκώδεις.

11 Πολλὰ δὲ καὶ τῶν ποιωδῶν ἔχει τοιαύτας ῥίζας, οἷον σπάλαξ κρόκος καὶ τὸ περδίκιον καλούμενον· καὶ γὰρ τοῦτο παχείας τε καὶ πλείους ἔχει τὰς ῥίζας ἢ φύλλα· καλεῖται δὲ περδίκιον διὰ τὸ τοὺς πέρδικας ἐγκυλιεσθαι καὶ ὀρύττειν. ὁμοίως δὲ

---

[1] *βάθους* conj. Sch; *βάθος* Ald.

[2] *καὶ ὃ* W. after U; *καὶ* om. Ald.; G omits also *τὸ* before *οὔϊγγον*, making the three plants synonymous. The passage is cited by Athen., *l.c.*, with considerable variation.

[3] *τοιαύτην* conj. St.; *τοσαύτην* MSS.

[4] *i.e.* the fleshy root (tuber, etc.).

[5] *i.e.* the fibrous root (root proper).

ground[1] would be a root, and so would the truffle, the plant which[2] some call puff-ball, the *uingon*, and all other underground plants. Whereas none of these is a root; for we must base our definition on natural function and not on position.

However it may be that this is a true account and yet that such things are roots no less; but in that case we distinguish two different kinds of root, one being of this character[3] and the other of the other, and the one[4] getting its nourishment from the other[5]; though the fleshy roots too themselves seem to draw nourishment. At all events men invert[6] the roots of cuckoo-pint before it shoots, and so they become larger by being prevented from pushing[7] through to make a shoot. For it is evident that the nature of all such plants is to turn downwards for choice; for the stems and the upper parts generally are short and weak, while the underground parts are large numerous and strong, and that, not only in the instances given, but in reeds dog's-tooth grass and in general in all plants of a reedy character and those like them. Those too which resemble ferula[8] have large fleshy roots.

[9]Many herbaceous plants likewise have such roots, as colchicum[10] crocus and the plant called 'partridge-plant'; for this too has thick roots which are more numerous than its leaves. [11](It is called the 'partridge-plant' because partridges roll in it and grub it up.) So too with the plant called in Egypt

[6] στρέφουσι conj. Sch.; τρεφουσι MVAld.; *cf.* 7. 12. 2.

[7] διαβῆναι conj. W.; διαθεῖναι UMV.

[8] *i.e.* have a hollow stem (umbelliferous plants, more or less).

[9] Plin. 19. 99.

[10] σπάλαξ UMV; ἀσπάλαξ mBas.: perhaps corrupt.

[11] Plin. 21. 102.

καὶ τὸ ἐν Αἰγύπτῳ καλούμενον οὔϊγγον· τὰ μὲν γὰρ φύλλα μεγάλα καὶ ὁ βλαστὸς αὐτοῦ βραχύς, ἡ δὲ ῥίζα μακρὰ καί ἐστιν ὥσπερ ὁ καρπός. διαφέρει τε καὶ ἐσθίεται, καὶ συλλέγουσι δὲ ὅταν
12 ὁ ποταμὸς ἀποβῇ στρέφοντες τὰς βώλους. φανερώτατα δὲ καὶ πλείστην ἔχοντα πρὸς τὰ ἄλλα διαφορὰν τὸ σίλφιον καὶ ἡ καλουμένη μαγύδαρις· ἀμφοτέρων γὰρ τούτων καὶ ἁπάντων τῶν τοιούτων ἐν ταῖς ῥίζαις μᾶλλον ἡ φύσις. ταῦτα μὲν οὖν ταύτῃ ληπτέα.

Ἔνιαι δὲ τῶν ῥιζῶν πλείω δόξαιεν ἂν ἔχειν διαφορὰν παρὰ τὰς εἰρημένας· οἷον αἵ τε τῆς ἀραχίδνης καὶ τοῦ ὁμοίου τῷ ἀράκῳ· φέρουσι γὰρ ἀμφότεραι καρπὸν οὐκ ἐλάττω τοῦ ἄνω· καὶ μίαν μὲν ῥίζαν τὸ ἀρακῶδες τοῦτο παχεῖαν ἔχει τὴν κατὰ βάθους, τὰς δ' ἄλλας ἐφ' ὧν ὁ καρπὸς λεπτοτέρας καὶ ἐπ' ἄκρῳ [καὶ] σχιζομένας πολλαχῇ· φιλεῖ δὲ μάλιστα χωρία τὰ ὕφαμμα· φύλλον δὲ οὐδέτερον ἔχει τούτων οὐδ' ὅμοια τοῖς φύλλοις, ἀλλ' ὥσπερ ἀμφίκαρπα μᾶλλόν ἐστιν· ὃ καὶ φαίνεται θαυμάσιον. αἱ μὲν οὖν φύσεις καὶ δυνάμεις τοσαύτας ἔχουσι διαφοράς.

VII. Αὐξάνεσθαι δὲ πάντων δοκοῦσιν αἱ ῥίζαι πρότερον τῶν ἄνω· καὶ γὰρ φύεται εἰς βάθος· οὐδεμία δὲ καθήκει πλέον ἢ ὅσον ὁ ἥλιος ἐφικνεῖται· τὸ γὰρ θερμὸν τὸ γεννῶν· οὐ μὴν ἀλλὰ

[1] οὔϊγγον mBas.H.; οὔϊτον MV; ουϊτον Ald.; *cf.* 1. 1. 7; Plin. 21. 88 (*oetum*).

[2] μεγάλα: text doubtful (W.).

[3] διαφέρει: text doubtful (Sch.).

[4] στρέφοντες τὰς βώλους conj. Coraës; στέφοντες βωμούς UMVAld. [5] ἐν ins. Sch.

ningon[1]; for its leaves are large[2] and its shoots short, while the root is long and is, as it were, the fruit. It is an excellent thing[3] and is eaten; men gather it when the river goes down by turning the clods.[4] But the plants which afford the most conspicuous instances and shew the greatest difference as compared with others are silphium and the plant called *magydaris*; the character of both of these and of all such plants is especially shewn in[5] their roots. Such is the account to be given of these plants.

Again some roots would seem to shew a greater difference[6] than those mentioned, for instance, those of *arakhidna*,[7] and of a plant[8] which resembles *arakos*. For both of these bear a fruit underground which is as large as the fruit above ground, and this *arakos*-like[9] plant has one thick root, namely, the one which runs deep, while the others which bear the 'fruit' are slenderer and branch[10] in many directions at the tip. It is specially fond of sandy ground. Neither of these plants has a leaf nor anything resembling a leaf, but they bear, as it were, two kinds of fruit instead, which seems surprising. So many then are the differences shewn in the characters and functions of roots.

VII. The roots of all plants seem to grow earlier than the parts above ground (for growth does take place downwards[11]). But no root goes down further than the sun reaches, since it is the heat which induces growth. Nevertheless the nature of the soil,

[6] *i.e.* to be even more abnormal: διαφορὰν conj. Sch.; διαφοραὶ Ald. [7] Plin. 21. 89.
[8] tine-tare. See Index, App. (1).
[9] ἀρακῶδες conj. Sch.; σαρκῶδες Ald.G.
[10] καὶ before σχιζ. om. Sch. from G.
[11] *cf.* *C.P.* 1. 12. 7. (cited by Varro, 1. 45. 3); 3. 3. 1.

ταῦτα μεγάλα συμβάλλεται πρὸς βαθυρριζίαν καὶ ἔτι μᾶλλον πρὸς μακρορριζίαν, ἡ τῆς χώρας φύσις ἐὰν ᾖ κούφη καὶ μανὴ καὶ εὐδίοδος· ἐν γὰρ ταῖς τοιαύταις πορρωτέρω καὶ μείζους αἱ αὐξήσεις. φανερὸν δὲ ἐπὶ τῶν ἡμερωμάτων· ἔχοντα γὰρ ὕδωρ ὁπουοῦν δίεισιν ὡς εἰπεῖν, ἐπειδὰν ὁ τόπος ᾖ κενὸς καὶ μηδὲν τὸ ἀντιστατοῦν. ἤγουν ἐν τῷ Λυκείῳ ἡ πλάτανος ἡ κατὰ τὸν ὀχετὸν ἔτι νέα οὖσα ἐπὶ τρεῖς καὶ τριάκοντα πήχεις ἀφῆκεν ἔχουσα τόπον τε ἅμα καὶ τροφήν.

Δόξειε δὲ ὡς εἰπεῖν ἡ συκῆ μακρορριζότατον εἶναι καὶ ὅλως δὲ μᾶλλον τὰ μανὰ καὶ εὐθύρριζα. πάντα δὲ τὰ νεώτερα τῶν παλαιῶν, ἐὰν εἰς ἀκμὴν ἥκωσιν, ἤδη βαθυρριζότερα καὶ μακρορριζότερα. συμφθίνουσι γὰρ καὶ αἱ ῥίζαι τῷ ἄλλῳ σώματι. πάντων δὲ ὁμοίως οἱ χυλοὶ τοῖς φυτοῖς δεινότεροι, τοῖς δὲ ὡς ἐπίπαν· δι' ὃ καὶ ἐνίων πικραὶ ὧν οἱ καρποὶ γλυκεῖς· αἱ δὲ καὶ φαρμακώδεις· ἔνιαι δ' εὐώδεις, ὥσπερ αἱ τῆς ἴριδος.

Ἰδία δὲ ῥίζης φύσις καὶ δύναμις ἡ τῆς Ἰνδικῆς συκῆς· ἀπὸ γὰρ τῶν βλαστῶν ἀφίησι, μέχρι οὗ ἂν συνάψῃ τῇ γῇ καὶ ῥιζωθῇ, καὶ γίνεται περὶ τὸ δένδρον κύκλῳ συνεχὲς τὸ τῶν ῥιζῶν οὐχ ἁπτόμενον τοῦ στελέχους ἀλλ' ἀφεστηκός.

---

1 ταῦτα before μέγαλα om. W.
2 ἡμερωμάτων conj. Sch.; ἡμερωτάτων UP_2Ald.: *cf.* *C.P.* 5. 6. 8.
3 ὁπουοῦν MSS.; ὁποσονοῦν conj. W. from G, *in quantum libeat.* 4 ἐπειδὰν conj. Sch.; ἐπεὶ κἂν UMVPAld.
5 Quoted by Varro, 1. 37. 5.
6 ἐπὶ conj. Sch.; παρὰ P_2; περὶ Ald.
7 συμφθίνουσι: συμφωνοῦσι conj. St.

if it is light open and porous, contributes greatly[1] to deep rooting, and still more to the formation of long roots; for in such soils growth goes further and is more vigorous. This is evident in cultivated plants.[2] For, provided that they have water, they run on, one may say, wherever it may be,[3] whenever[4] the ground is unoccupied and there is no obstacle. [5] For instance the plane-tree by the watercourse in the Lyceum when it was still young sent out its roots a distance of[6] thirty-three cubits, having both room and nourishment.

The fig would seem, one may say, to have the longest roots, and in general plants which have wood of loose texture and straight roots would seem to have these longer. Also young plants, provided that they have reached their prime, root deeper and have longer roots than old ones; for the roots decay along with[7] the rest of the plant's body. And in all cases alike the juices of plants[8] are more powerful in the roots than in other parts, while in some cases they are extremely powerful; wherefore the roots are bitter in some plants whose fruits are sweet; some roots again are medicinal, and some are fragrant, as those of the iris.

The character and function of the roots of the 'Indian fig' (banyan) are peculiar, for this plant sends out roots from the shoots till it has a hold on the ground[9] and roots again; and so there comes to be a continuous circle of roots round the tree, not connected with the main stem but at a distance from it.

[8] τοῖς φυτοῖς Ald.; ταῖς ῥίζαις conj. W. from G: text probably defective.

[9] τῇ γῇ conj. Scal. from G; συκῇ U; τῇ συκῇ P_2Ald.

Παραπλήσιον δὲ τούτῳ μᾶλλον δὲ τρόπον τινὰ θαυμασιώτερον εἴ τι ἐκ τῶν φύλλων ἀφίησι ῥίζαν, οἷόν φασι περὶ Ὀποῦντα ποιάριον εἶναι, ὃ και ἐσθίεσθαί ἐστιν ἡδύ. τὸ γὰρ αὖ τῶν θέρμων θαυμαστὸν ἧττον, ὅτι ἂν ἐν ὕλῃ βαθείᾳ σπαρῇ διείρει τὴν ῥίζαν πρὸς τὴν γῆν καὶ βλαστάνει διὰ τὴν ἰσχύν. ἀλλὰ δὴ τὰς μὲν τῶν ῥιζῶν διαφορὰς ἐκ τούτων θεωρητέον.

VIII. Τῶν δένδρων τὰς τοιαύτας ἄν τις λάβοι διαφοράς. ἔστι γὰρ τὰ μὲν ὀζώδη τὰ δ' ἄνοζα καὶ φύσει καὶ τόπῳ κατὰ τὸ μᾶλλον καὶ ἧττον. ἄνοζα δὲ λέγω οὐχ ὥστε μὴ ἔχειν ὅλως—οὐδὲν γὰρ τοιοῦτο δένδρον, ἀλλ' εἴπερ, ἐπὶ τῶν ἄλλων οἷον σχοῖνος τύφη κύπειρος ὅλως ἐπὶ τῶν λιμνωδῶν—ἀλλ' ὥστε ὀλίγους ἔχειν. φύσει μὲν οἷον ἀκτὴ δάφνη συκῆ ὅλως πάντα τὰ λειόφλοια καὶ ὅσα κοῖλα καὶ μανά. ὀζῶδες δὲ ἐλάα πεύκη κότινος· τούτων δὲ τὰ μὲν ἐν παλισκίοις καὶ νηνέμοις καὶ ἐφύδροις, τὰ δὲ ἐν εὐηλίοις καὶ χειμερίοις καὶ πνευματώδεσι καὶ λεπτοῖς καὶ ξηροῖς· τὰ μὲν γὰρ ἀνοζότερα, τὰ δὲ ὀζωδέστερα τῶν

---

[1] τι conj. W.; τις MSS. [2] Plin. 21. 104.
[3] *cf.* 8. 11. 8; Plin. 18. 133 and 134.
[4] διείρει conj. Sch.; διαιρεῖ P_2Ald.; *cf.* *C.P.* 2. 17. 7.
[5] ὄζος is the knot and the bough starting from it: *cf.* Arist. *de iuv. et sen.* 3.
[6] ἐπὶ τῶν conj. Coraës; ἢ τῶν UM; ἧττον (erased) P (ἐκ τῶν marg.) ἧττον Ald.

Something similar to this, but even more surprising, occurs in those plants which[1] emit roots from their leaves, as they say does a certain herb[2] which grows about Opus, which is also sweet to taste. The peculiarity again of lupins[3] is less surprising, namely that, if the seed is dropped where the ground is thickly overgrown, it pushes[4] its root through to the earth and germinates because of its vigour. But we have said enough for study of the differences between roots.

Of trees (principally) and their characteristic special differences: as to knots.

VIII. One may take it that the following are the differences between trees:—Some have knots,[5] more or less, others are more or less without them, whether from their natural character or because of their position. But, when I say 'without knots,' I do not mean that they have no knots at all (there is no tree like that, but, if it is true of any plants, it is only of[6] other kinds, such as rush bulrush[7] galingale and plants of the lake side[8] generally) but that they have few knots. Now this is the natural character of elder bay fig and all smooth-barked trees, and in general of those whose wood is hollow or of a loose texture. Olive fir and wild olive have knots; and some of these grow in thickly shaded windless and wet places, some in sunny positions exposed to storms and winds,[9] where the soil is light and dry; for the number of knots varies between trees of the

[7] τύφη conj. Bod.; τίφη UAld.H.; *cf.* 1. 5. 3.

[8] ἐπὶ τῶν conj. W.; εἴ τι ἐπὶ τῶν Ald.

[9] πνευματώδεσι conj. Scal.; πυματώδεσι U; πυγματώδεσι MVAld.

ὁμογενῶν. ὅλως δὲ ὀζωδέστερα τὰ ὀρεινὰ τῶν πεδεινῶν καὶ τὰ ξηρὰ τῶν ἑλείων.

2 Ἔτι δὲ κατὰ τὴν φυτείαν τὰ μὲν πυκνὰ ἄνοζα καὶ ὀρθά, τὰ δὲ μανὰ ὀζωδέστερα καὶ σκολιώτερα· συμβαίνει γὰρ ὥστε τὰ μὲν ἐν παλισκίῳ εἶναι τὰ δὲ ἐν εὐηλίῳ. καὶ τὰ ἄρρενα δὲ τῶν θηλειῶν ὀζωδέστερα ἐν οἷς ἐστιν ἄμφω, οἷον κυπάριττος ἐλάτη ὀστρυὶς κρανεία· καλοῦσι γὰρ γένος τι θηλυκρανείαν· καὶ τὰ ἄγρια δὲ τῶν ἡμέρων, καὶ ἁπλῶς καὶ τὰ ὑπὸ ταὐτὸ γένος, οἷον κότινος ἐλάας καὶ ἐρινεὸς συκῆς καὶ ἀχρὰς ἀπίου. πάντα γὰρ ταῦτα ὀζωδέστερα· καὶ ὡς ἐπὶ τὸ πολὺ πάντα τὰ πυκνὰ τῶν μανῶν· καὶ γὰρ τὰ ἄρρενα πυκνότερα καὶ τὰ ἄγρια· πλὴν εἴ τι διὰ πυκνότητα παντελῶς ἄνοζον ἢ ὀλίγοζον, οἷον πύξος λωτός.

3 Εἰσὶ δὲ τῶν μὲν ἄτακτοι καὶ ὡς ἔτυχεν οἱ ὄζοι, τῶν δὲ τεταγμένοι καὶ τῷ διαστήματι καὶ τῷ πλήθει καθάπερ εἴρηται· δι' ὃ καὶ ταξιόζωτα ταῦτα καλοῦσιν. τῶν μὲν γὰρ οἷον δι' ἴσου τῶν δὲ μεῖζον αἰεὶ τὸ πρὸς τῷ πάχει. καὶ τοῦτο κατὰ λόγον. ὅπερ μάλιστα ἔνδηλον καὶ ἐν τοῖς κοτίνοις καὶ ἐν τοῖς καλάμοις· τὸ γὰρ γόνυ καθάπερ ὄζος. καὶ οἱ μὲν κατ' ἀλλήλους, ὥσπερ οἱ τῶν

---

[1] Plin. 16. 125. [2] 1. 8. 1.
[3] ταξιόζωτα conj. W.; ἀξιολογώτατα Ald.; *cf.* ταξίφυλλος, 1. 10. 8. [4] Plin. 16. 122.

same kind. And in general mountain trees have more knots than those of the plain, and those that grow in dry spots than those that grow in marshes.

Again the way in which they are planted makes a difference in this respect; those trees that grow close together are knotless and erect, those that grow far apart have more knots and a more crooked growth; for it happens that the one class are in shade, the others in full sun. Again the 'male' trees have more knots than the 'female' in those trees in which both forms are found, as cypress silver-fir hop-hornbeam cornelian cherry—for there is a kind called 'female cornelian cherry' (cornel)—and wild trees have more knots than trees in cultivation: this is true both in general and when we compare those of the same kind, as the wild and cultivated forms of olive fig and pear. All these have more knots in the wild state; and in general those of closer growth have this character more than those of open growth; for in fact the 'male' plants are of closer growth, and so are the wild ones; except that in some cases, as in box and nettle-tree, owing to the closer growth there are no knots at all, or only a few.

[1] Again the knots of some trees are irregular and set at haphazard, while those of others are regular, alike in their distance apart and in their number, as has been said[2]; wherefore also they are called 'trees with regular knots.'[3] [4] For of some the knots are, as it were, at even distances, while in others the distance between them is greater at the thick end of the stem. And this proportion holds throughout. This is especially evident in the wild olive and in reeds—in which the joint corresponds to the knot in trees. Again some knots are opposite one another,

κοτίνων, οἱ δ' ὡς ἔτυχεν. ἔστι δὲ τὰ μὲν δίοζα, τὰ δὲ τρίοζα, τὰ δὲ πλείους ἔχοντα· ἔνια δὲ πεντάοζά ἐστι. καὶ τῆς μὲν ἐλάτης ὀρθοὶ καὶ οἱ ὄζοι καὶ οἱ
4 κλάδοι ὥσπερ ἐμπεπηγότες, τῶν δὲ ἄλλων οὔ. δι' ὃ καὶ ἰσχυρὸν ἡ ἐλάτη. ἰδιώτατοι δὲ οἱ τῆς μηλέας· ὅμοιοι γὰρ θηρίων προσώποις, εἷς μὲν ὁ μέγιστος ἄλλοι δὲ περὶ αὐτὸν μικροὶ πλείους. εἰσὶ δὲ τῶν ὄζων οἱ μὲν τυφλοί, οἱ δὲ γόνιμοι. λέγω δὲ τυφλοὺς ἀφ' ὧν μηδεὶς βλαστός. οὗτοι δὲ καὶ φύσει καὶ πηρώσει γίνονται, ὅταν ἢ μὴ λυθῇ καὶ ἐκβιάζηται ἢ καὶ ἀποκοπῇ καὶ οἷον ἐπικαυθεὶς πηρωθῇ· γίνονται δὲ μᾶλλον ἐν τοῖς παχέσι τῶν ἀκρεμόνων, ἐνίων δὲ καὶ ἐν τοῖς στελέχεσιν. ὅλως δὲ καὶ τοῦ στελέχους καὶ τοῦ κλάδου καθ' ὃ ἂν ἐπικόψῃ ἢ ἐπιτέμῃ τις, ὄζος γίνεται καθαπερανεὶ διαιρῶν τὸ ἓν καὶ ποιῶν ἑτέραν ἀρχήν, εἴτε διὰ τὴν πήρωσιν εἴτε δι' ἄλλην αἰτίαν· οὐ γὰρ δὴ κατὰ φύσιν τὸ ὑπὸ τῆς πληγῆς.

5 Αἰεὶ δὲ ἐν ἅπασιν οἱ κλάδοι φαίνονται πολυοζότεροι διὰ τὸ μήπω τἀνὰ μέσον προσηυξῆσθαι, καθάπερ καὶ τῆς συκῆς οἱ νεόβλαστοι τραχύτατοι καὶ τῆς ἀμπέλου τὰ ἄκρα τῶν κλημάτων. ὡς γὰρ ὄζος ἐν τοῖς ἄλλοις οὕτω καὶ ὀφθαλμὸς

---

1 *cf.* 4. 4. 12. 2 Plin. 16. 122.
3 *i.e.* primary and secondary branches.
4 *cf.* 5. 2. 2. 5 Plin. 16. 124.
6 *cf.* Arist. *de iuv. et sen.* 3; Plin. 16. 125.
7 *ὅταν . . . πηρωθῇ* conj. W.; *ἢ ὅταν ἡ μὴ λυθῇ καὶ ἐκβιάζηται καὶ ἡ ἀποκοπὴ καὶ* U; *ὅταν ἢ μὴ λυθῇ καὶ ἐκβιάζηται ἢ ἀποκοπῇ* P; *ἢ ὅταν λυθῇ καὶ ἐκβιάζηται ἢ ἀποκοπὴ καὶ οἱ οὐ* P_2; *ὅταν ἢ μὴ λυθῇ καὶ ἐκβιάζηται καὶ ἢ ἀποκοπῇ καὶ* Ald.H.; G differs widely.

as those of the wild olive, while others are set at random. Again some trees have double knots, some treble,[1] some more at the same point; some have as many as five. [2] In the silver-fir both the knots and the smaller branches[3] are set at right angles, as if they were stuck in, but in other trees they are not so. And that is why the silver-fir is such a strong tree.[4] Most peculiar[5] are the knots of the apple, for they are like the faces of wild animals; there is one large knot, and a number of small ones round it. Again some knots are blind,[6] others productive; by 'blind' I mean those from which there is no growth. These come to be so either by nature or by mutilation, according as either the knot[7] is not free and so the shoot does not make its way out, or, a bough having been cut off, the place is mutilated, for example by burning. Such knots occur more commonly in the thicker boughs, and in some cases in the stem also. And in general, wherever one chops or cuts part of the stem or bough, a knot is formed, as though one thing were made thereby into two and a fresh growing point produced, the cause being the mutilation or some other such reason; for the effect of such a blow cannot of course be ascribed to nature.

Again in all trees the branches always seem to have more knots, because the intermediate parts[8] have not yet developed, just as the newly formed branches of the fig are the roughest,[9] and in the vine the highest[10] shoots. [11] (For to the knot in other

[8] *i.e.* the internodes; till the branch is fully grown its knots are closer together, and so seem more numerous: *μήπω τὰνὰ μέσον προσηυξῆσθαι* conj. Sch.; *μήπω τὰνὰ μέσον προσκυζῆθαι* U; *μήτ' ἀνὰ μέσον προσκυζεῖσθαι* MAld.; *μήποτ' ἀνάμεσον προσηυξῆσθαι* P_2.

[9] *i.e.* have most knots.

[10] *i.e.* youngest.

[11] Plin. 16. 125.

ἐν ἀμπέλῳ και ἐν καλάμῳ γονυ . . . ἐνίοις δὲ καὶ οἷον κράδαι γινονται, καθάπερ πτελέᾳ καὶ δρυῒ καὶ μάλιστα ἐν πλατάνῳ· ἐὰν δὲ ἐν τραχέσι καὶ ἀνύδροις καὶ πνευματώδεσι καὶ παντελῶς. πάντως δὲ πρὸς τῇ γῇ καὶ οἷον τῇ κεφαλῇ τοῦ στελέχους ἀπογηρασκόντων τὸ πάθος τοῦτο γίνεται.

6 Ἔνια δὲ καὶ ἴσχει τοὺς καλουμένους ὑπο τινων ἢ γόγγρους ἢ τὸ ἀνάλογον, οἷον ἡ ἐλάα· κυριώτατον γὰρ ἐπὶ ταύτης τοῦτο τοὔνομα καὶ πάσχειν δοκεῖ μάλιστα τὸ εἰρημένον· καλοῦσι δ' ἔνιοι τοῦτο πρέμνον οἱ δὲ κροτώνην οἱ δὲ ἄλλο ὄνομα. τοῖς δὲ εὐθέσι καὶ μονορρίζοις καὶ ἀπαραβλάστοις οὐ γίνεται τοῦθ' ὅλως ἢ ἧττον· [φοῖνιξ δὲ παραβλαστητικόν·] ἡ δὲ ἐλάα καὶ ὁ κότινος καὶ τὰς οὐλότητας ἰδίας ἔχουσι τὰς ἐν τοῖς στελέχεσι.

IX. Ἔστι μὲν οὖν τὰ μὲν ὡς εἰς μῆκος αὐξητικὰ μάλιστ' ἢ μόνον, οἷον ἐλάτη φοῖνιξ κυπάριττος καὶ ὅλως τὰ μονοστελέχη καὶ ὅσα μὴ πολύρριζα μηδὲ πολύκλαδα· <ἡ δὲ φοῖνιξ ἀπαραβλαστητικόν·> τὰ δὲ ὁμοῖα τούτοις ἀνὰ λόγον καὶ εἰς βάθος. ἔνια δ' εὐθὺς σχίζεται, οἷον ἡ

---

[1] The opening of the description of the diseases of trees seems to have been lost. [2] κράδαι; *cf*. *C.P.* 5. 1. 3.

[3] πάντως . . . γίνεται conj. W.; πάντως δὲ ὁ πρὸς τῇ γῇ καὶ οἷον τ. κ. στ. ἀπογηράσκων τῶν παχυτέρων γίνεται Ald.; so U except παχύτερον, and M except παχύτερος.

[4] γόγγρους: *cf*. Hesych., *s.vv.* γόγγρος, κροτώνη.

[5] The word is otherwise unknown.

[6] ἧττον· ἡ δὲ ἐλάα conj. W.; ἧττον· ἡ δὲ φοῖνιξ πάραβλαστητικόν· ἡ δὲ ἐλάα U; so Ald. except παραβλαστικόν. The

trees correspond the 'eye' in the vine, the joint in the reed). . . . .[1] In some trees again there occurs, as it were, a diseased formation of small shoots,[2] as in elm oak and especially in the plane; and this is universal if they grow in rough waterless or windy spots. Apart from any such cause[3] this affection occurs near the ground in what one may call the 'head' of the trunk, when the tree is getting old.

Some trees again have what are called by some 'excrescences'[4] (or something corresponding), as the olive; for this name belongs most properly to that tree, and it seems most liable to the affection; and some call it 'stump,' some *krotone*,[5] others have a different name for it. It does not occur, or only occurs to a less extent, in straight young trees, which have a single root and no side-growths. To the olive[6] also, both wild and cultivated, are peculiar certain thickenings[7] in the stem.

As to habit.

IX. [8] Now those trees which grow chiefly or only[9] in the direction of their height are such as silver-fir date-palm cypress, and in general those which have a single stem and not many roots or branches (the date-palm, it may be added, has no side-growths at all[10]). And trees like[11] these have also similar growth downwards. Some however divide from the first,

note about the palm (φοῖνιξ δὲ παραβλαστητικόν) I have omitted as untrue as well as irrelevant; possibly with ἀπαραβα. for παραβα. it belongs to the next section.

7 οὐλότητας conj. W.; κοιλότητας MSS. (?) Ald.

8 Plin. 16. 125.

9 μάλιστ' ἢ μόνον conj. W.; μάλιστα μανὰ Ald. H.

10 See 3. 8. 6. n.

11 ὅμοια conj. Sch.; ὁμοίως MSS. Sense hardly satisfactory.

μηλέα· τὰ δὲ πολύκλαδα καὶ μείζω τὸν ὄγκον ἔχει τὸν ἄνω, καθάπερ ῥόα· οὐ μὴν ἀλλ' οὖν[1] μέγιστά γε συμβάλλεται πρὸς ἕκαστον ἡ ἀγωγὴ καὶ ὁ τόπος καὶ ἡ τροφή. σημεῖον δ' ὅτι ταὐτὰ πυκνὰ μὲν ὄντα μακρὰ καὶ λεπτὰ γίνεται, μανὰ δὲ παχύτερα καὶ βραχύτερα· καὶ ἐὰν μὲν εὐθύς τις ἀφιῇ τοὺς ὄζους βραχέα, ἐὰν δὲ ἀνακαθαίρῃ μακρά, καθάπερ ἡ ἄμπελος.

2 Ἱκανὸν δὲ κἀκεῖνο πρὸς πίστιν ὅτι καὶ τῶν λαχάνων ἔνια λαμβάνει δένδρου σχῆμα, καθάπερ εἴπομεν[2] τὴν μαλάχην καὶ τὸ τεῦτλον· ἅπαντα δ' ἐν τοῖς οἰκείοις τόποις εὐαυξῆ . . . καὶ τὸ αὐτὸ κάλλιστον.[3] ἐπεὶ καὶ τῶν ὁμογενῶν ἀνοζότερα καὶ μείζω καὶ καλλίω τὰ ἐν τοῖς οἰκείοις, οἷον ἐλάτη ἡ Μακεδονικὴ τῆς Παρνασίας καὶ τῶν ἄλλων. ἅπαντα[4] δὲ ταῦτα καὶ ὅλως ἡ ὕλη ἡ ἀγρία καλλίων καὶ πλείων τοῦ ὄρους ἐν τοῖς προσβορείοις ἢ ἐν τοῖς πρὸς μεσημβρίαν.

3 Ἔστι δὲ τὰ μὲν ἀείφυλλα τὰ δὲ φυλλοβόλα.[5] τῶν μὲν ἡμέρων ἀείφυλλα ἐλάα φοῖνιξ δάφνη μύρρινος πεύκης τι γένος κυπάριττος· τῶν δ' ἀγρίων ἐλάτη πεύκη ἄρκευθος μίλος[6] θυία καὶ ἣν Ἀρκάδες καλοῦσι φελλόδρυν φιλυρέα κέδρος πίτυς ἀγρία μυρίκη πύξος πρῖνος κήλαστρον φιλύκη ὀξυάκανθος ἀφάρκη, ταῦτα δὲ φύεται περὶ τὸν Ὄλυμπον, ἀνδράχλη κόμαρος τέρμινθος

[1] οὖν marked as doubtful in U. [2] 1. 3. 2.

[3] καὶ τὸ αὐτὸ κάλλιστον. The first part of the sentence to which these words belong is apparently lost (W.).

[4] *i.e.* the fir and other trees mentioned in the lost words.

[5] Plin. 16. 80.

[6] μίλος conj. Sch.; σμίλαξ P_2Ald.; *cf.* 3. 3. 3.

such as apple; some have many branches, and their greater mass of growth high up, as the pomegranate: however[1] training position and cultivation chiefly contribute to all of these characters. In proof of which we have the fact that the same trees which, when growing close together, are tall and slender, when grown farther apart become stouter and shorter; and if we from the first let the branches grow freely, the tree becomes short, whereas, if we prune them, it becomes tall,—for instance, the vine.

This too is enough for proof that even some pot-herbs acquire the form of a tree, as we said[2] of mallow and beet. Indeed all things grow well in congenial places. . . .[3] For even among those of the same kind those which grow in congenial places have less knots, and are taller and more comely: thus the silver-fir in Macedon is superior to other silver-firs, such as that of Parnassus. Not only is this true of all these,[4] but in general the wild woodland is more beautiful and vigorous on the north side of the mountain than on the south.

As to shedding of leaves.

Again some[5] trees are evergreen, some deciduous. Of cultivated trees, olive date-palm bay myrtle a kind of fir and cypress are evergreen, and among wild trees silver-fir fir Phoenician cedar yew[6] odorous cedar the tree which the Arcadians call 'cork-oak' (holm-oak) mock-privet prickly cedar 'wild[7] pine' tamarisk box kermes-oak holly alaternus cotoneaster hybrid arbutus[8] (all of which grow about Olympus)

[7] ἀγρία after πίτυς conj. Sch.; after πρῖνος UPAld.: *cf.* 3. 3. 3.
[8] κόμαρος conj. Bod.; σίναρος UMV; οἴναρος Ald.; σύναρος P_2.

ἀγρία δάφνη. δοκεῖ δ' ἡ ἀνδράχλη καὶ ὁ κόμαρος τὰ μὲν κάτω φυλλοβολεῖν τὰ δὲ ἔσχατα τῶν ἀκρεμόνων ἀείφυλλα ἔχειν, ἐπιφύειν δὲ ἀεὶ τοὺς ἀκρεμόνας.

4 Τῶν μὲν οὖν δένδρων ταῦτα. τῶν δὲ θαμνωδῶν κιττὸς βάτος ῥάμνος κάλαμος κεδρίς· ἔστι γάρ τι μικρὸν ὃ οὐ δενδροῦται. τῶν δὲ φρυγανικῶν καὶ ποιωδῶν πήγανον ῥάφανος ῥοδωνία ἰωνία ἀβρότονον ἀμάρακον ἕρπυλλος ὀρίγανον σέλινον ἱπποσέλινον μήκων καὶ τῶν ἀγρίων εἴδη πλείω. διαμένει δὲ καὶ τούτων ἔνια τοῖς ἄκροις τὰ δὲ ἄλλα ἀποβάλλει οἷον ὀρίγανον σέλινον . . . ἐπεὶ καὶ τὸ πήγανον κακοῦται καὶ ἀλλάττεται.

5 Πάντα δὲ καὶ τῶν ἄλλων τὰ ἀείφυλλα στενοφυλλότερα καὶ ἔχοντά τινα λιπαρότητα καὶ εὐωδίαν. ἔνια δ' οὐκ ὄντα τῇ φύσει παρὰ τὸν τόπον ἐστὶν ἀείφυλλα, καθάπερ ἐλέχθη περὶ τῶν ἐν Ἐλεφαντίνῃ καὶ Μέμφει· κατωτέρω δ' ἐν τῷ Δελτα μικρὸν πάνυ χρόνον διαλείπει τοῦ μὴ ἀεὶ βλαστάνειν. ἐν Κρήτῃ δὲ λέγεται πλάτανόν τινα εἶναι ἐν τῇ Γορτυναίᾳ πρὸς πηγῇ τινι ἣ οὐ φυλλοβολεῖ· μυθολογοῦσι δὲ ὡς ὑπὸ ταύτῃ ἐμίγη τῇ Εὐρώπῃ ὁ Ζεύς· τὰς δὲ πλησίας πάσας φυλλοβολεῖν. ἐν δὲ Συβάρει δρῦς ἐστιν εὐσύνοπτος ἐκ τῆς πόλεως ἣ οὐ φυλλοβολεῖ· φασὶ

[1] Plin. 16. 80.
[2] Some words probably missing (W.) which would explain the next two clauses. [3] Plin. 16. 82. [4] 1. 3. 5.
[5] Plin. 12. 11 ; Varro, 1. 7.

andrachne arbutus terebinth 'wild bay' (oleander). Andrachne and arbutus seem to cast their lower leaves, but to keep those at the end of the twigs perennially, and to be always adding leafy twigs. These are the trees which are evergreen.

[1] Of shrubby plants these are evergreen:—ivy bramble buckthorn reed *kedris* (juniper)—for there is a small kind of *kedros* so called which does not grow into a tree. Among under-shrubs and herbaceous plants there are rue cabbage rose gilliflower southernwood sweet marjoram tufted thyme marjoram celery alexanders poppy, and a good many more kinds of wild plants However some of these too, while evergreen as to their top growths, shed their other leaves, as marjoram and celery . . . . . [2] for rue too is injuriously affected and changes its character.

[3] And all the evergreen plants in the other classes too have narrower leaves and a certain glossiness and fragrance. Some moreover which are not evergreen by nature become so because of their position, as was said[4] about the plants at Elephantine and Memphis, while lower down the Nile in the Delta there is but a very short period in which they are not making new leaves. It is said that in Crete[5] in the district of Gortyna there is a plane near a certain spring[6] which does not lose its leaves; (indeed the story is that it was under[7] this tree that Zeus lay with Europa), while all the other plants in the neighbourhood shed their leaves. [8] At Sybaris there is an oak within sight of the city which does not shed

[6] πηγῇ conj. H. from G; σκηνῇ UMVAld.; κηνῇ P_2; κρηνῇ mBas.
[7] ὑπὸ conj. Hemsterhuis; ἐπὶ Ald. [8] Plin. 16. 81.

δὲ οὐ βλαστάνειν αὐτὴν ἅμα ταῖς ἄλλαις ἀλλὰ μετὰ Κύνα. λέγεται δὲ καὶ ἐν Κύπρῳ πλάτανος εἶναι τοιαύτη.

6 Φυλλοβολεῖ δὲ πάντα τοῦ μετοπώρου καὶ μετὰ τὸ μετόπωρον, πλὴν τὸ μὲν θᾶττον τὸ δὲ βραδύτερον ὥστε καὶ τοῦ χειμῶνος ἐπιλαμβάνειν. οὐκ ἀνάλογοι δὲ αἱ φυλλοβολίαι ταῖς βλαστήσεσιν, ὥστε τὰ πρότερον βλαστήσαντα πρότερον φυλλοβολεῖν, ἀλλ' ἔνια πρωϊβλαστεῖ μὲν οὐδὲν δὲ προτερεῖ τῶν ἄλλων, ἀλλά τινων καὶ ὑστερεῖ, καθάπερ ἡ ἀμυγδαλῆ.

7 Τὰ δὲ ὀψιβλαστεῖ μὲν οὐδὲν δὲ ὡς εἰπεῖν ὑστερεῖ τῶν ἄλλων, ὥσπερ ἡ συκάμινος. δοκεῖ δὲ καὶ ἡ χώρα συμβάλλεσθαι καὶ ὁ τόπος ὁ ἔνικμος πρὸς τὸ διαμένειν. τὰ γὰρ ἐν τοῖς ξηροῖς καὶ ὅλως λεπτογείοις πρότερα φυλλοβολεῖ καὶ τὰ πρεσβύτερα δὲ τῶν νέων. ἔνια δὲ καὶ πρὸ τοῦ πεπᾶναι τὸν καρπὸν ἀποβάλλει τὰ φύλλα, καθάπερ αἱ ὄψιαι συκαῖ καὶ ἀχράδες.

Τῶν δ' ἀειφύλλων ἡ ἀποβολὴ καὶ ἡ αὔανσις κατὰ μέρος· οὐ γὰρ δὴ ταὐτὰ αἰεὶ διαμένει, ἀλλὰ τὰ μὲν ἐπιβλαστάνει τὰ δ' ἀφαυαίνεται. τοῦτο δὲ περὶ τροπὰς μάλιστα γίνεται θερινάς. εἰ δέ τινων καὶ μετ' Ἀρκτοῦρον ἢ καὶ κατ' ἄλλην ὥραν ἐπισκεπτέον. καὶ τὰ μὲν περὶ τὴν φυλλοβολίαν οὕτως ἔχει.

---

[1] Plin. 16. 82 and 83.

its leaves, and they say that it does not come into leaf along with the others, but only after the rising of the dog-star. It is said that in Cyprus too there is a plane which has the same peculiarity.

[1] The fall of the leaves in all cases takes place in autumn or later, but it occurs later in some trees than in others, and even extends into the winter. However the fall of the leaf does not correspond to the growth of new leaves (in which case those that come into leaf earlier would lose their leaves earlier), but some (such as the almond) which are early in coming into leaf are not earlier than the rest in losing their leaves, but are even comparatively late.[2]

[3] Others again, such as the mulberry, come into leaf late, but are hardly at all later than the others in shedding their leaves. It appears also that position and a moist situation conduce to keeping the leaves late; for those which grow in dry places, and in general where the soil is light, shed their leaves earlier, and the older trees earlier than young ones. Some even cast their leaves before the fruit is ripe, as the late kinds of fig and pear.

In those which are evergreen the shedding and withering of leaves take place by degrees; for it is not the same[4] leaves which always persist, but fresh ones are growing while the old ones wither away. This happens chiefly about the summer solstice. Whether in some cases it occurs even after the rising of Arcturus or at a quite different season is matter for enquiry. So much for the shedding of leaves.

[2] ὑστερεῖ conj. H.; ὕστερον UMVPAld.
[3] Plin. 16. 84.
[4] ταὐτὰ conj. Sch.; ταῦτα Ald.

X. Τὰ δὲ φύλλα τῶν μὲν ἄλλων δένδρων ὅμοια πάντων αὐτὰ ἑαυτοῖς, τῆς δὲ λεύκης καὶ τοῦ κιττοῦ καὶ τοῦ καλουμένου κρότωνος ἀνόμοια καὶ ἑτεροσχήμονα· τὰ μὲν γὰρ νέα περιφερῆ τὰ δὲ παλαιότερα γωνοειδῆ, καὶ εἰς τοῦτο ἡ μετάστασις πάντων. τοῦ δὲ κιττοῦ ἀνάπαλιν νέου μὲν ὄντος ἐγγωνιώτερα πρεσβυτέρου δὲ περιφερέστερα· μεταβάλλει γὰρ καὶ οὗτος. ἴδιον δὲ καὶ τὸ τῇ ἐλάᾳ καὶ τῇ φιλύρᾳ καὶ τῇ πτελέᾳ καὶ τῇ λεύκῃ συμβαῖνον· στρέφειν γὰρ δοκοῦσιν τὰ ὕπτια μετὰ τροπὰς θερινάς, καὶ τούτῳ γνωρίζουσιν ὅτι γεγένηνται τροπαί.
2 πάντα δὲ τὰ φύλλα διαφέρει κατὰ τὰ ὕπτια καὶ τὰ πρανῆ. καὶ τῶν μὲν ἄλλων τὰ ὕπτια ποιωδέστερα καὶ λειότερα· τὰς γὰρ ἶνας καὶ τὰς φλέβας ἐν τοῖς πρανέσιν ἔχουσιν, ὥσπερ ἡ χεὶρ <τὰ ἄρθρα>· τῆς δ' ἐλάας λευκότερα καὶ ἧττον λεῖα ἐνίοτε καὶ τὰ ὕπτια. πάντα δὴ ἢ τά γε πλεῖστα ἐκφανῆ ἔχει τὰ ὕπτια καὶ ταῦτα γίνεται τῷ ἡλίῳ φανερά. καὶ στρέφεται τὰ πολλὰ πρὸς τὸν ἥλιον· δι' ὃ καὶ οὐ ῥᾴδιον εἰπεῖν ὁπότερον πρὸς τῷ κλῶνι μᾶλλόν ἐστιν· ἡ μὲν γὰρ ὑπτιότης μᾶλλον δοκεῖ ποιεῖν τὸ πρανές, ἡ δὲ φύσις οὐχ ἧττον βούλεται τὸ ὕπτιον, ἄλλως τε καὶ ἡ ἀνάκλασις διὰ τὸν ἥλιον· ἴδοι δ'

---

1 Plin. 16. 85.

2 καὶ τοῦ κιττοῦ καὶ τοῦ MSS· *cf.* Plin. *l.c*; Diosc. 4. 164. καὶ τοῦ κικίου τοῦ καὶ conj. W.; Galen, *Lex. Hipp.*, gives κίκιον as a name for the root of κρότων. *cf. C.P.* 2. 16. 4.

3 *i.e.* not 'entire.' 'Young leaves' = leaves of the young tree.

4 This seems to contradict what has just been said.

5 τὰ ἄρθρα add. Sch. from Plin. 16. 88, *incisuras*. *cf.* Arist. *H.A.* 1. 15, where Plin. (11. 274) renders ἄρθρα *incisuras*.

Differences in leaves.

X. [1] Now, while the leaves of all other trees are all alike in each tree, those of the abele ivy [2] and of the plant called *kroton* (castor-oil plant) are unlike one another and of different forms. The young leaves in these are round, the old ones angular,[3] and eventually all the leaves assume that form. On the other hand [4] in the ivy, when it is young, the leaves are somewhat angular, but when it is older, they become rounder: for in this plant too a change of form takes place. There is a peculiarity special to the olive lime elm and abele: their leaves appear to invert the upper surface after the summer solstice, and by this men know that the solstice is past. Now all leaves differ as to their upper and under surfaces; and in most trees the upper surfaces are greener and smoother, as they have the fibres and veins in the under surfaces, even as the human hand has its 'lines,'[5] but even the upper surface of the leaf of the olive is sometimes whiter and less smooth.[6] So all or most leaves display their upper surfaces, and it is these surfaces which are exposed to the light.[7] Again most leaves turn towards the sun; wherefore also it is not easy to say which surface is next to the twig [8]; for, while the way in which the upper surface is presented seems rather to make the under surface closer to it, yet nature desires equally that the upper surface should be the nearer, and this is specially seen in the turning back [9] of the leaf towards the sun. One

[6] ἐνίοτε καὶ τὰ ὕπτια conj. W.; λεῖα δὲ καὶ τὰ τοῦ κιττοῦ MSS. A makeshift correction of an obscure passage.

[7] *cf.* Plin. *l.c.* [8] *i.e.* is the under one.

[9] Whereby the under surface is exposed to it: see above.

ἄν τις ὅσα πυκνὰ καὶ κατ' ἄλληλα, καθάπερ τὰ τῶν μυρρίνων.

3 Οἴονται δέ τινες καὶ τὴν τροφὴν τῷ ὑπτίῳ διὰ τοῦ πρανοῦς εἶναι, διὰ τὸ ἔνικμον ἀεὶ τοῦτο καὶ χνοῶδες εἶναι, οὐ καλῶς λέγοντες. ἀλλὰ τοῦτο μὲν ἴσως συμβαίνει χωρὶς τῆς ἰδίας φύσεως καὶ διὰ τὸ μὴ ὁμοίως ἡλιοῦσθαι, ἡ δὲ τροφὴ διὰ τῶν φλεβῶν ἢ ἰνῶν ὁμοίως ἀμφοτέροις· ἐκ θατέρου δ' εἰς θάτερον οὐκ εὔλογον μὴ ἔχουσι πόρους μηδὲ βάθος δι' οὗ· ἀλλὰ περὶ μὲν τροφῆς διὰ τίνων ἕτερος λόγος.

4 Διαφέρουσι δὲ καὶ τὰ φύλλα πλείοσι διαφοραῖς· τὰ μὲν γάρ ἐστι πλατύφυλλα, καθάπερ ἄμπελος συκῆ πλάτανος, τὰ δὲ στενόφυλλα, καθάπερ ἐλάα ῥόα μύρρινος· τὰ δ' ὥσπερ ἀκανθόφυλλα, καθάπερ πεύκη πίτυς κέδρος· τὰ δ' οἷον σαρκόφυλλα· τοῦτο δ' ὅτι σαρκῶδες ἔχουσι τὸ φύλλον, οἷον κυπάριττος μυρίκη μηλέα, τῶν δὲ φρυγανικῶν κνέωρος στοιβὴ καὶ ποιωδῶν ἀείζωον πόλιον· [τοῦτο δὲ καὶ πρὸς τοὺς σῆτας τοὺς ἐν τοῖς ἱματίοις ἀγαθόν·] τὰ γὰρ αὖ τῶν τευτλίων ἢ ῥαφάνων ἄλλον τρόπον σαρκώδη καὶ τὰ τῶν πηγανίων καλουμένων· ἐν πλάτει γὰρ καὶ οὐκ ἐν στρογγυλότητι τὸ σαρκῶδες. καὶ τῶν θαμνωδῶν

5 δὲ ἡ μυρίκη σαρκῶδες τὸ φύλλον ἔχει. ἔνια δὲ

---

[1] *cf.* 1. 8. 3; 1. 10. 8; Plin. 16. 92.

[2] ἐκ θατέρου δ' εἰς conj. Sch. from G; δὲ ἐκ θατέρου εἰς with stop at ἰνῶν Ald. [3] δι' οὗ I conj.; δι' ὧν U.

[4] ἀκανθόφυλλα conj. W.; σπανόφυλλα UMAld.; ἀνόφυλλα P_2; *cf.* 3. 9. 5, whence Sch. conj. τριχόφυλλα: Plin. *l.c.* has *capillata pino cedro.*

[5] μηλέα probably corrupt; omitted by Plin. *l.c.*

may observe this in trees whose leaves are crowded and opposite,[1] such as those of myrtle.

Some think that the nourishment too is conveyed to the upper surface through the under surface, because this surface always contains moisture and is downy, but they are mistaken. It may be that this is not due to the trees' special character, but to their not getting an equal amount of sunshine, though the nourishment conveyed through the veins or fibres is the same in both cases. That it should be conveyed from one side to the other[2] is improbable, when there are no passages for it nor thickness for it to pass through.[3] However it belongs to another part of the enquiry to discuss the means by which nourishment is conveyed.

Again there are various other differences between leaves; some trees are broad-leaved, as vine fig and plane, some narrow-leaved, as olive pomegranate myrtle. Some have, as it were, spinous[4] leaves, as fir Aleppo pine prickly cedar; some, as it were, fleshy leaves; and this is because their leaves are of fleshy substance, as cypress tamarisk apple,[5] among under-shrubs *kneoros* and *stoibe*, and among herbaceous plants house-leek and hulwort. [6] This plant is good against moth in clothes. For the leaves of beet and cabbage are fleshy in another way, as are those of the various plants called rue; for their fleshy character is seen in the flat instead of in the round.[7] Among shrubby plants the tamarisk[8] has fleshy

[6] Probably a gloss.

[7] Or 'solid,' such leaves being regarded as having, so to speak, three, and not two dimensions. στρόγγυλος = 'thick-set' in Arist. *H.A.* 9. 44.

[8] μυρίκη probably corrupt; μ. was mentioned just above, among *trees*; ἐρείκη conj. Dalec.

καὶ καλαμόφυλλα, καθάπερ ὁ φοῖνιξ καὶ ὁ κόϊξ
καὶ ὅσα τοιαῦτα· ταῦτα δὲ ὡς καθ' ὅλου εἰπεῖν
γωνιόφυλλα· καὶ γὰρ ὁ κάλαμος καὶ ὁ κύπειρος
καὶ ὁ βούτομος καὶ τἆλλα δὲ τῶν λιμνωδῶν
τοιαῦτα· πάντα δὲ ὥσπερ ἐκ δυοῖν σύνθετα καὶ
τὸ μέσον οἷον τρόπις, οὗ ἐν τοῖς ἄλλοις μέγας
πόρος ὁ μέσος. διαφέρουσι δὲ καὶ τοῖς σχήμασι·
τὰ μὲν γὰρ περιφερῆ, καθάπερ τὰ τῆς ἀπίου, τὰ
δὲ προμηκέστερα, καθάπερ τὰ τῆς μηλέας· τὰ δὲ
εἰς ὀξὺ προήκοντα καὶ παρακανθίζοντα, καθάπερ
τὰ τοῦ μίλακος. καὶ ταῦτα μὲν ἄσχιστα· <τὰ δὲ
σχιστὰ> καὶ οἷον πριονώδη, καθάπερ τὰ τῆς
ἐλάτης καὶ τὰ τῆς πτερίδος· τρόπον δέ τινα
σχιστὰ καὶ τὰ τῆς ἀμπέλου, καὶ τὰ τῆς συκῆς
6 δὲ ὥσπερ ἂν εἴποι τις κορωνοποδώδη. ἔνια δὲ
καὶ ἐντομὰς ἔχοντα, καθάπερ τὰ τῆς πτελέας καὶ
τὰ τῆς Ἡρακλεωτικῆς καὶ τὰ τῆς δρυός. τὰ δὲ
καὶ παρακανθίζοντα καὶ ἐκ τοῦ ἄκρου καὶ ἐκ τῶν
πλαγίων, οἷον τὰ τῆς πρίνου καὶ τὰ τῆς δρυὸς
καὶ μίλακος καὶ βάτου καὶ παλιούρου καὶ τὰ τῶν
ἄλλων. ἀκανθῶδες δὲ ἐκ τῶν ἄκρων καὶ τὸ τῆς
πεύκης καὶ πίτυος καὶ ἐλάτης ἔτι δὲ κέδρου καὶ
κεδρίδος. φυλλάκανθον δὲ ὅλως ἐν μὲν τοῖς
δένδροις οὐκ ἔστιν οὐδὲν ὧν ἡμεῖς ἴσμεν, ἐν δὲ
τοῖς ἄλλοις ὑλήμασίν ἐστιν, οἷον ἥ τε ἄκορνα καὶ
ἡ δρυπὶς καὶ ὁ ἄκανος καὶ σχεδὸν ἅπαν τὸ τῶν
ἀκανωδῶν γένος· ὥσπερ γὰρ φύλλον ἐστὶν ἡ
ἄκανθα πᾶσιν· εἰ δὲ μὴ φύλλα τις ταῦτα θήσει,

---

1 Plin. *l.c.* and 13. 30. 2 οὗ ἐν conj. W.; ὅθεν Ald. H.
3 παρακανθίζοντα conj. Sch.; παραγωνίζοντα UMVAld.
4 τὰ δὲ σχιστὰ add. W.

leaves. Some again have reedy leaves, as date-palm doum-palm and such like. But, generally speaking, the leaves of these end in a point; for reeds galingale sedge and the leaves of other marsh plants are of this character. [1] The leaves of all these are compounded of two parts, and the middle is like a keel, placed where in [2] other leaves is a large passage dividing the two halves. Leaves differ also in their shapes; some are round, as those of pear, some rather oblong, as those of the apple; some come to a sharp point and have spinous projections [3] at the side, as those of smilax. So far I have spoken of undivided leaves; but some are divided [4] and like a saw, as those of silver-fir and of fern. To a certain extent those of the vine are also divided, while those of the fig one might compare to a crow's foot.[5] [6] Some leaves again have notches, as those of elm filbert and oak, others have spinous projections both at the tip and at the edges, as those of kermes-oak oak smilax bramble Christ's thorn and others. The leaf of fir Aleppo pine silver-fir and also of prickly cedar and *kedris* (juniper) [7] has a spinous point at the tip. Among other trees there is none that we know which has spines for leaves altogether, but it is so with other woody plants, as *akorna drypis* pine-thistle and almost all the plants which belong to that class.[8] For in all these spines, as it were, take the place of leaves, and, if one is not to reckon these

[5] κορωνοποδώδη conj. Gesner. The fig-leaf is compared to a crow's foot, Plut. *de defect. orac.* 3; σκολοπώδη Ald., which word is applied to thorns by Diosc. [6] Plin. 16. 90.

[7] κεδρίδος conj. Dalec.; κεδρίας MSS. *cf.* Plin. *l.c.*, who seems to have read ἀγρίας.

[8] ἀκανωδῶν conj. W., *cf.* 1. 13. 3; ἀκανθωδῶν MSS.; ἀκανθῶν P_2.

συμβαίνοι ἂν ὅλως ἄφυλλα εἶναι, ἐνίοις δὲ ἄκανθαν μὲν εἶναι φύλλον δὲ ὅλως οὐκ ἔχειν, καθάπερ ὁ ἀσφάραγος.

7 Πάλιν δ' ὅτι τὰ μὲν ἄμισχα, καθάπερ τὰ τῆς σκίλλης καὶ τοῦ βολβοῦ, τὰ δ' ἔχοντα μίσχον. καὶ τὰ μὲν μακρόν, οἷον ἡ ἄμπελος καὶ ὁ κιττός, τὰ δὲ βραχὺν καὶ οἷον ἐμπεφυκότα, καθάπερ ἐλάα καὶ οὐχ ὥσπερ ἐπὶ τῆς πλατάνου καὶ ἀμπέλου προσηρτημένον. διαφορὰ δὲ καὶ τὸ μὴ ἐκ τῶν αὐτῶν εἶναι τὴν πρόσφυσιν, ἀλλὰ τοῖς μὲν πλείστοις ἐκ τῶν κλάδων τοῖς δὲ καὶ ἐκ τῶν ἀκρεμόνων, τῆς δρυὸς δὲ καὶ ἐκ τοῦ στελέχους, τῶν δὲ λαχανωδῶν τοῖς πολλοῖς εὐθὺς ἐκ τῆς ῥίζης, οἷον κρομύου σκόρδου κιχορίου, ἔτι δὲ ἀσφοδέλου σκίλλης βολβοῦ σισυριγχίου καὶ ὅλως τῶν βολβωδῶν· καὶ τούτων δὲ οὐχ ἡ πρώτη μόνον ἔκφυσις ἀλλὰ καὶ ὅλος ὁ καυλὸς ἄφυλλον. ἐνίων δ' ὅταν γένηται, φύλλα εἰκός, οἷον θριδακίνης ὠκίμου σελίνου καὶ τῶν σιτηρῶν ὁμοίως. ἔχει δ' ἔνια τούτων καὶ τὸν καυλὸν εἶτ' ἀκανθίζοντα, ὡς ἡ θριδακίνη καὶ τὰ φυλλάκανθα πάντα καὶ τῶν θαμνωδῶν δὲ καὶ ἔτι μᾶλλον, οἷον βάτος παλίουρος.

8 Κοινὴ δὲ διαφορὰ πάντων ὁμοίως δένδρων καὶ τῶν ἄλλων ὅτι τὰ μὲν πολύφυλλα τὰ δ' ὀλιγόφυλλα. ὡς δ' ἐπὶ τὸ πᾶν τὰ πλατύφυλλα ταξίφυλλα, καθάπερ μύρρινος, τὰ δ' ἄτακτα καὶ ὡς ἔτυχε, καθάπερ σχεδὸν τὰ πλεῖστα τῶν ἄλλων

---

[1] Plin. 16. 91. [2] ἐπὶ conj. W.; ἡ Ald.H.

[3] ἐνίων . . . εἰκός. So Sch. explains: text probably defective.

as leaves, they would be entirely leafless, and some would have spines but no leaves at all, as asparagus.

[1] Again there is the difference that some leaves have no leaf-stalk, as those of squill and purse-tassels, while others have a leaf-stalk. And some of the latter have a long leaf-stalk, as vine and ivy, some, as olive, a short one which grows, as it were, into the stem and is not simply attached to it, as it is in [2] plane and vine. Another difference is that the leaves do not in all cases grow from the same part, but, whereas in most trees they grow from the branches, in some they grow also from the twigs, and in the oak from the stem as well; in most pot-herbs they grow directly from the root, as in onion garlic chicory, and also in asphodel squill purse-tassels Barbary-nut, and generally in plants of the same class as purse-tassels; and in these not merely the original growth but the whole stalk is leafless. In some, when the stalk is produced, the leaves may be expected to grow,[3] as in lettuce basil celery, and in like manner in cereals. In some of these the stalk presently becomes spinous, as in lettuce and the whole class of plants with spinous leaves, and still more in shrubby plants, as bramble and Christ's thorn.

[4] Another difference which is found in all trees alike and in other plants as well is that some have many, some few leaves. And in general those that have flat leaves [5] have them in a regular series, as myrtle, while in other instances the leaves are in no particular order, but set at random, as in most other

[4] Plin. 16. 92.

[5] πλατύφυλλα UVP; πολύφυλλα conj. W.; but πλατύτης is one of the 'differences' given in the summary below.

[ἦν]. ἴδιον δὲ ἐπὶ τῶν λαχανωδῶν, οἷον κρομύου γητείου, τὸ κοιλόφυλλον.

Ἁπλῶς δ' αἱ διαφοραὶ τῶν φύλλων ἢ μεγέθει ἢ πλήθει ἢ σχήματι ἢ πλατύτητι ἢ στενότητι ἢ κοιλότητι ἢ τραχύτητι ἢ λειότητι καὶ τῷ παρακανθίζειν ἢ μή. ἔτι δὲ κατὰ τὴν πρόσφυσιν ὅθεν ἢ δι' οὗ· τὸ μὲν ὅθεν, ἀπὸ ῥίζης ἢ κλάδου ἢ καυλοῦ ἢ ἀκρεμόνος· τὸ δὲ δι' οὗ, ἢ διὰ μίσχου ἢ δι' αὑτοῦ καὶ εἰ δὴ πολλὰ ἐκ τοῦ αὐτοῦ. καὶ ἔνια καρποφόρα, μεταξὺ περιειληφότα τὸν καρπόν, ὥσπερ ἡ Ἀλεξανδρεία δάφνη ἐπιφυλλόκαρπος.

Αἱ μὲν οὖν διαφοραὶ τῶν φύλλων κοινοτέρως πᾶσαι εἴρηνται καὶ σχεδόν εἰσιν ἐν τούτοις.

(Σύγκειται δὲ τὰ μὲν ἐξ ἰνὸς καὶ φλοιοῦ καὶ σαρκός, οἷον τὰ τῆς συκῆς καὶ τῆς ἀμπέλου, τὰ δὲ ὥσπερ ἐξ ἰνὸς μόνον, οἷον τοῦ καλάμου καὶ σίτου.
9 τὸ δὲ ὑγρὸν ἁπάντων κοινόν· ἅπασι γὰρ ἐνυπάρχει καὶ τούτοις καὶ τοῖς ἄλλοις τοῖς ἐπετείοις [μίσχος ἄνθος καρπὸς εἴ τι ἄλλο]· μᾶλλον δὲ καὶ τοῖς μὴ ἐπετείοις· οὐδὲν γὰρ ἄνευ τούτου. δοκεῖ δὲ καὶ τῶν μίσχων τὰ μὲν ἐξ ἰνῶν μόνον συγκεῖσθαι, καθάπερ τὰ τοῦ σίτου καὶ τοῦ καλάμου, τὰ δ' ἐκ τῶν αὐτῶν, ὥσπερ οἱ καυλοί.

---

[1] τῶν ἄλλων ἦν MSS.; τῶν ποιωδῶν conj. W. ἦν, at all events, cannot be right. [2] Plin. 19. 100.
[3] ἢ στενότητι ἢ κοιλότητι: so G; ἢ κοιλότητι ἢ στενότητι MSS. [4] *i.e.* petiolate. [5] *i.e.* sessile.
[6] *i.e.* compound: εἰ δὴ conj. W.; εἴδη UMVAld.
[7] The passage from here to the end of the chapter is a digression.

plants.[1] [2] It is peculiar to pot-herbs to have hollow leaves, as in onion and horn-onion.

To sum up, the differences between leaves are shewn in size, number, shape, hollowness, in breadth,[3] roughness and their opposites, and in the presence or absence of spinous projections; also as to their attachment, according to the part from which they spring or the means by which they are attached; the part from which they spring being the root or a branch or the stalk or a twig, while the means by which they are attached may be a leaf-stalk,[4] or they may be attached directly;[5] and there may be[6] several leaves attached by the same leaf-stalk. Further some leaves are fruit-bearing, enclosing the fruit between them, as the Alexandrian laurel, which has its fruit attached to the leaves.

These are all the differences in leaves stated somewhat generally, and this is a fairly complete list of examples.

Composition of the various parts of a plant.

[7] (Leaves are composed some of fibre bark and flesh, as those of the fig and vine, some, as it were, of fibre alone, as those of reeds and corn. But moisture is common to all, for it is found both in leaves and in the other annual parts,[8] leaf-stalk, flower, fruit and so forth but more especially in the parts which are not annual[9]; in fact no part is without it. Again it appears that some leaf-stalks are composed only of fibre, as those of corn and reeds, some of the same materials as the stalks.

[8] μίσχος . . . ἄλλο has no construction; probably a (correct) gloss, taken from 1, 2, 1.

[9] *i.e.* while these are young, W.

10 Τῶν δ' ἀνθῶν τὰ μὲν ἐκ φλοιοῦ καὶ φλεβὸς καὶ σαρκός, <τὰ δ' ἐκ σαρκὸς> μόνον, οἷον τὰ ἐν μέσῳ τῶν ἄρων.

Ὁμοίως δὲ καὶ ἐπὶ τῶν καρπῶν· οἱ μὲν γὰρ ἐκ σαρκὸς καὶ ἰνός, οἱ δὲ ἐκ σαρκὸς μόνον, οἱ δὲ καὶ ἐκ δέρματος σύγκεινται· τὸ δὲ ὑγρὸν ἀκολουθεῖ καὶ τούτοις. ἐκ σαρκὸς μὲν καὶ ἰνὸς ὁ τῶν κοκκυμήλων καὶ σικύων, ἐξ ἰνὸς δὲ καὶ δέρματος ὁ τῶν συκαμίνων καὶ τῆς ῥόας. ἄλλοι δὲ κατ' ἄλλον τρόπον μεμερισμένοι. πάντων δὲ ὡς εἰπεῖν τὸ μὲν ἔξω φλοιὸς τὸ δ' ἐντὸς σὰρξ τῶν δὲ καὶ πυρήν.)

XI. Ἔσχατον δ' ἐν ἅπασι τὸ σπέρμα. τοῦτο δὲ ἔχον ἐν ἑαυτῷ σύμφυτον ὑγρὸν καὶ θερμόν, ὧν ἐκλιπόντων ἄγονα, καθάπερ τὰ ᾠά. καὶ τῶν μὲν εὐθὺ τὸ σπέρμα μετὰ τὸ περιέχον, οἷον φοίνικος καρύου ἀμυγδάλης, πλείω δὲ τούτων τὰ ἐμπεριέχοντα, ὡς τὰ τοῦ φοίνικος. τῶν δὲ μεταξὺ σὰρξ καὶ πυρήν, ὥσπερ ἐλάας καὶ κοκκυμηλέας καὶ ἑτέρων. ἔνια δὲ καὶ ἐν λοβῷ, τὰ δ' ἐν ὑμένι, τὰ δ' ἐν ἀγγείῳ, τὰ δὲ καὶ γυμνόσπερμα τελείως.

2 Ἐν λοβῷ μὲν οὐ μόνον τὰ ἐπέτεια, καθάπερ τὰ χεδροπὰ καὶ ἕτερα πλείω τῶν ἀγρίων, ἀλλὰ καὶ τῶν δένδρων ἔνια, καθάπερ ἥ τε κερωνία, ἥν τινες καλοῦσι συκῆν Αἰγυπτίαν, καὶ ἡ κερκὶς καὶ ἡ κολοιτία περὶ Λιπάραν· ἐν ὑμένι δ' ἔνια τῶν

---

[1] τὰ U; τὸ Ald.
[2] τὰ δ' ἐκ σαρκὸς preserved only in mBas.; om. UMVP$_2$. Sch. reads τό.
[3] ἄρων conj. W.; αἱρῶν MSS.
[4] *i.e.* rind.
[5] Plin. 18. 53.
[6] οὐ conj. Sch.; οὖν Ald.H.

Of flowers some[1] are composed of bark veins and flesh, some of flesh only,[2] as those in the middle of cuckoo-pint.[3]

So too with fruits; some are made of flesh and fibre, some of flesh alone, and some of skin[4] also. And moisture is necessarily found in these also. The fruit of plums and cucumbers is made of flesh and fibre, that of mulberries and pomegranates of fibre and skin. The materials are differently distributed in different fruits, but of nearly all the outside is bark, the inside flesh, and this in some cases includes a stone.)

Differences in seeds.

XI. Last in all plants comes the seed. This possesses in itself natural moisture and warmth, and, if these fail, the seeds are sterile, like eggs in the like case. In some plants the seed comes immediately inside the envelope, as in date filbert almond (however, as in the case of the date, there may be more than one covering). In some cases again there is flesh and a stone between the envelope and the seed, as in olive plum and other fruits. Some seeds again are enclosed in a pod, some in a husk, some in a vessel, and some are completely naked.

5 Enclosed in a pod are not[6] only the seeds of annual plants, as leguminous plants, and of considerable numbers of wild plants, but also those of certain trees, as the carob-tree (which some[7] call the 'Egyptian fig'), Judas-tree,[8] and the *koloitia*[9] of the Liparae islands. In a husk are enclosed the

[7] *ἥν τινες* conj. St. from G; *ἥντινα* Ald.H.
[8] Clearly not the *κερκίς* (aspen) described 3. 14. 2.
[9] *κολοιτία* MSS.; *κολουτέα* conj. St., *cf.* 3. 17. 2 n.

ἐπετείων, ὥσπερ ὁ πυρὸς καὶ ὁ κέγχρος· ὡσαύτως δὲ καὶ ἐναγγειοσπέρματα καὶ γυμνοσπέρματα. ἐναγγειοσπέρματα μὲν οἷον ἥ τε μήκων καὶ ὅσα μηκωνικά· τὸ γὰρ σήσαμον ἰδιωτέρως· γυμνοσπέρματα δὲ τῶν τε λαχάνων πολλά, καθάπερ ἄνηθον κορίαννον ἄννησον κύμινον μάραθον καὶ
3 ἕτερα πλείω. τῶν δὲ δένδρων οὐδὲν γυμνόσπερμον ἀλλ' ἢ σαρξὶ περιεχόμενον ἢ κελύφεσιν, τὰ μὲν δερματικοῖς, ὥσπερ ἡ βάλανος καὶ τὸ Εὐβοϊκόν, τὰ δὲ ξυλώδεσιν, ὥσπερ ἡ ἀμυγδάλη καὶ τὸ κάρυον. οὐδὲν δὲ ἐναγγειόσπερμον, εἰ μή τις τὸν κῶνον ἀγγεῖον θήσει διὰ τὸ χωρίζεσθαι τῶν καρπῶν.

Αὐτὰ δὲ τὰ σπέρματα τῶν μὲν εὐθὺ σαρκώδη, καθάπερ ὅσα καρυηρὰ καὶ βαλανηρά· τῶν δὲ ἐν πυρῆνι τὸ σαρκῶδες ἔχεται, καθάπερ ἐλάας καὶ δαφνίδος καὶ ἄλλων. τῶν δ' ἐμπύρηνα μόνον ἢ πυρηνώδη γε καὶ ὥσπερ ξηρά, καθάπερ τὰ κνηκώδη καὶ κεγχραμιδώδη καὶ πολλὰ τῶν λαχανηρῶν. ἐμφανέστατα δὲ τὰ τοῦ φοίνικος· οὐδὲ γὰρ κοιλότητα ἔχει τοῦτο οὐδεμίαν ἀλλ' ὅλον ξηρόν· οὐ μὴν ἀλλ' ὑγρότης δή τις καὶ θερμότης ὑπάρχει δῆλον ὅτι καὶ τούτῳ, καθάπερ εἴπομεν.

---

[1] μηκωνικὰ . . . τὸ γὰρ conj. W. from G; μήκωνι· κατὰ γὰρ UMVAld.

[2] κορίαννον ἄννησον conj. Sch.; κοριάννησον UMAld.; κοράννησον V; *cf.* Plin. 19. 119.

[3] ἢ κελύφεσιν conj. Sch., *cf. C.P.* 4. 1. 2; ἢ δὲ κύμασιν U; Plin. 15. 112, *crusta teguntur glandes.* [4] Plin. 15. 113.

seeds of some annuals, as wheat and millet; and in like manner some plants have their seeds in a vessel, some have them naked. In a vessel are those of the poppy and plants of the poppy kind;[1] (the case of sesame however is somewhat peculiar), while many pot-herbs have their seeds naked, as dill coriander[2] anise cummin fennel and many others. No tree has naked seeds, but either they are enclosed in flesh or in shells,[3] which are sometimes of leathery nature, as the acorn and the sweet chestnut, sometimes woody, as almond and nut. Moreover no tree has its seeds in a vessel, unless one reckons a cone as a vessel, because it can be separated from the fruits.

The actual seeds are in some cases fleshy in themselves, as all those which resemble nuts or acorns; [4] in some cases the fleshy part is contained in a stone, as in olive bay and others. The seeds in some plants again merely consist of a stone,[5] or at least are of stone-like character, and are, as it were,[6] dry; for instance those of plants like safflower millet and many pot-herbs. Most obviously of this character are those of the date,[7] for they contain no cavity, but are throughout dry [8];—not but what there must be even in them some moisture and warmth, as we have said.[9]

[5] *ἐμπύρηνα μόνον ἢ πυρηνώδη* conj. Sch.; *ἐν πυρῆνι μόνον ἢ πυρηνώδει* Ald. (P has *πυρηνώδη*).

[6] *i.e.* no seed can really be without moisture; *cf.* 1. 11. 1.

[7] *cf. C.P.* 5. 18. 4.

[8] *ξηρὸν* I conj., as required by the next clause; *ἔξορθον* PAld.; *ἔξορρον* W. from Sch. conj. The germ in the date-stone is so small as to be undiscoverable, whence the stone seems to be homogeneous throughout, with no cavity for the germ.

[9] 1. 10. 9.

4 Διαφέρουσι δὲ καὶ τῷ τὰ μὲν ἀθρόα μετ'
ἀλλήλων εἶναι, τὰ δὲ διεστῶτα καὶ στοιχηδόν,[1]
ὥσπερ τὰ τῆς κολοκύντης καὶ σικύας καὶ τῶν
δένδρων, ὡς Περσικῆς μηλέας. καὶ τῶν ἀθρόων
τὰ μὲν ἐνί τινι[2] περιέχεσθαι, καθάπερ τὰ τῆς ῥόας
καὶ τῆς ἀπίου καὶ μηλέας καὶ τῆς ἀμπέλου καὶ
συκῆς· τὰ δὲ μετ' ἀλλήλων μὲν εἶναι, μὴ περιέχεσθαι δὲ ὑφ' ἑνός, ὥσπερ τὰ σταχυηρὰ τῶν
ἐπετείων, εἰ μή τις θείη τὸν στάχυν ὡς περιέχον·
οὕτω δ' ἔσται καὶ ὁ βότρυς καὶ τἆλλα τὰ
βοτρυώδη καὶ ὅσα δὴ φέρει δι' εὐβοσίαν καὶ
χώρας ἀρετὴν ἀθρόους τοὺς καρπούς, ὥσπερ ἐν
Συρίᾳ φασὶ καὶ ἄλλοθι τὰς ἐλάας.[3]
5 Ἀλλὰ καὶ αὕτη[4] δοκεῖ τις εἶναι διαφορὰ τὸ[5] τὰ
μὲν ἀφ' ἑνὸς μίσχου καὶ μιᾶς προσφύσεως
ἀθρόα γίνεσθαι, καθάπερ ἐπί τε τῶν βοτρυηρῶν
καὶ σταχυηρῶν εἴρηται μὴ περιεχόμενα κοινῷ
τινι γίνεσθαι· τὰ δὲ μὴ γίνεσθαι. ἐπεὶ καθ'
ἕκαστόν γε λαμβάνοντι τῶν σπερμάτων ἢ τῶν
περιεχόντων ἰδίαν ἀρχὴν ἔχει τῆς προσφύσεως,
οἷον ἥ τε ῥὰξ καὶ ἡ ῥόα καὶ πάλιν ὁ πυρὸς καὶ ἡ
κριθή. ἥκιστα δ' ἂν δόξειεν τὰ τῶν μήλων καὶ
τὰ τῶν ἀπίων, ὅτι συμψαύει τε καὶ περιείληπται
καθάπερ ὑμένι τινὶ δερματικῷ περὶ ὃν τὸ περι-
6 κάρπιον· ἀλλ' ὅμως καὶ τούτων ἕκαστον ἰδίαν
ἀρχὴν ἔχει καὶ φύσιν· φανερώτατα δὲ τῷ

---

[1] στοιχηδόν conj. W.; σχεδὸν Ald.
[2] ἐνί τινι conj. Sch.; ἔν τινι Ald.
[3] *cf.* Plin. 15. 15.
[4] αὕτη conj. Sch.; αὐτὴ Ald.
[5] τὸ conj. W.; τῷ Ald.

Further seeds differ in that in some cases they are massed together, in others they are separated and arranged in rows,[1] as those of the gourd and bottle-gourd, and of some trees, such as the citron. Again of those that are massed together some differ in being contained in a single[2] case, as those of pomegranate pear apple vine and fig; others in being closely associated together, yet not contained in a single case, as, among annuals, those which are in an ear—unless one regards the ear as a case. In that case the grape-cluster and other clustering fruits will come under the description, as well as all those plants which on account of good feeding or excellence of soil bear their fruits massed together,[3] as they say the olive does in Syria and elsewhere.

But this[4] too seems to be a point of difference, that[5] some grow massed together from a single stalk and a single attachment, as has been said in the case of plants with clusters or ears whose seeds do not grow contained in one common case; while others grow otherwise. For in these instances, if one takes each seed or case separately, it has its own special point of attachment, for instance each grape or pomegranate,[6] or again each grain of wheat or barley. This would seem to be least of all the case with the seeds of apples and pears, since[7] these touch one another[8] and are enclosed in a sort of skin-like membrane, outside which is the fruit-case.[9] However each of these too has its own peculiar point of attachment and character; this is most

[6] ἥ τε . . . ῥόα.: text perhaps defective; ἥ τε ῥὰξ βότρυας καὶ τῆς ῥόας ὁ πυρήν conj. Bod.
[7] ὅτι conj. Sch.; ὅπι U; ὅποι PMAld.
[8] *cf.* 8. 5. 2. [9] *i.e.* pulp.

κεχωρίσθαι τὰ τῆς ῥόας· ὁ γὰρ πυρὴν ἑκάστῳ προσπέφυκεν, οὐχ ὥσπερ τῶν συκῶν ἄδηλα διὰ τὴν ὑγρότητα. καὶ γὰρ τούτῳ ἔχουσι διαφορὰν καίπερ ἀμφότερα περιεχόμενα σαρκώδει τινὶ καὶ τῷ τοῦτο περιειληφότι μετὰ τῶν ἄλλων· τὰ μὲν γὰρ περὶ ἕκαστον ἔχει πυρῆνα τὸ σαρκῶδες τοῦτο τὸ ὑγρόν, αἱ δὲ κεγχραμίδες ὥσπερ κοινόν τι πᾶσαι, καθάπερ καὶ τὸ γίγαρτον καὶ ὅσα τὸν αὐτὸν ἔχει τρόπον. ἀλλὰ τὰς μὲν τοιαύτας διαφορὰς τάχ' ἄν τις λάβοι πλείους· ὧν δεῖ τὰς κυριωτάτας καὶ μάλιστα τῆς φύσεως μὴ ἀγνοεῖν.

XII. Αἱ δὲ κατὰ τοὺς χυλοὺς καὶ τὰ σχήματα καὶ τὰς ὅλας μορφὰς σχεδὸν φανεραὶ πᾶσιν, ὥστε μὴ δεῖσθαι λόγου· πλὴν τοσοῦτόν γ' ὅτι σχῆμα οὐδὲν περικάρπιον εὐθύγραμμον οὐδὲ γωνίας ἔχει. τῶν δὲ χυλῶν οἱ μέν εἰσιν οἰνώδεις, ὥσπερ ἀμπέλου συκαμίνου μύρτου· οἱ δ' ἐλαώδεις, ὥσπερ ἐλάας δάφνης καρύας ἀμυγδαλῆς πεύκης πίτυος ἐλάτης· οἱ δὲ μελιτώδεις, οἷον σύκου φοίνικος διοσβαλάνου· οἱ δὲ δριμεῖς, οἷον ὀριγάνου θύμβρας καρδάμου νάπυος· οἱ δὲ πικροί, ὥσπερ ἀψινθίου κενταυρίου. διαφέρουσι δὲ καὶ ταῖς εὐωδίαις, οἷον ἀννήσου κεδρίδος· ἐνίων δὲ ὑδαρεῖς ἂν δόξαιεν, οἷον οἱ τῶν κοκκυμηλέων· οἱ δὲ ὀξεῖς, ὥσπερ ῥοῶν

---

[1] i.e. of the pulp. [2] τούτῳ conj. Sch.; τοῦτο Ald.

[3] τὸν om. St.: i.e. the seeds are arranged in compartments of the pulp.

obvious in the separation of the pomegranate seeds, for the stone is attached to each, and the connexion is not, as in figs, obscured by the moisture.[1] For here[2] too there is a difference, although in both cases the seeds are enclosed in a sort of fleshy substance, as well as in the case which encloses this and the other parts of the fruit. For in the pomegranate the stones have this moist fleshy substance enclosing each[3] separate stone; but in the case of fig-seeds, as well as in that of grape-stones and other plants which have the same arrangement, the same pulp is common to all.[4] However one might find more such differences, and one should not ignore the most important of them, namely those which specially belong to the plant's natural character.

Differences in taste.

XII. The differences in taste, shape, and form as a whole are tolerably evident to all, so that they do not need explanation; except that it should be stated that[5] the case containing the fruit is never right-lined in shape and never has angles. [6] Of tastes some are like wine, as those of vine mulberry and myrtle; some are like olive-oil, as, besides olive itself, bay hazel almond fir Aleppo pine silver-fir; some like honey, as fig date chestnut; some are pungent, as marjoram savory cress mustard; some are bitter, as wormwood centaury. Some also are remarkably fragrant, as anise and juniper[7]; of some the smell would seem to be insipid,[8] as in plums; of others sharp, as in pomegranates and

[4] *i.e.* the fruit is not divided into compartments.
[5] πλὴν ἢ τοσοῦτον conj. W.; πλὴν τοσοῦτον ἢ UMAld.
[6] Plin. 19. 186; 15. 109. [7] *cf.* 1. 9. 4. [8] Lit. watery.

καὶ ἐνίων μήλων. ἁπάντων δὲ οἰνώδεις καὶ τοὺς ἐν τούτῳ τῷ γένει θετέον· ἄλλοι δὲ ἐν ἄλλοις εἴδεσιν· ὑπὲρ ὧν ἁπάντων ἀκριβέστερον ἐν τοῖς περὶ χυλῶν ῥητέον, αὐτάς τε τὰς ἰδέας διαριθμουμένους ὁπόσαι καὶ τὰς πρὸς ἀλλήλους διαφορὰς καὶ τίς ἡ ἑκάστου φύσις καὶ δύναμις.

2 Ἔχει δὲ καὶ ἡ τῶν δένδρων αὐτῶν ὑγρότης, ὥσπερ ἐλέχθη, διάφορα εἴδη· ἡ μὲν γάρ ἐστιν ὀπώδης, ὥσπερ ἡ τῆς συκῆς καὶ τῆς μήκωνος· ἡ δὲ πιττώδης, οἷον ἐλάτης πεύκης τῶν κωνοφόρων· ἄλλη δ' ὑδαρής, οἷον ἀμπέλου ἀπίου μηλέας, καὶ τῶν λαχανωδῶν δέ, οἷον σικύου κολοκύντης θριδακίνης· αἱ δὲ [ἤδη] δριμύτητά τινα ἔχουσι, καθάπερ ἡ τοῦ θύμου καὶ θύμβρας· αἱ δὲ καὶ εὐωδίαν, ὥσπερ αἱ τοῦ σελίνου ἀνήθου μαράθου καὶ τῶν τοιούτων. ὡς δ' ἁπλῶς εἰπεῖν ἅπασαι κατὰ τὴν ἰδίαν φύσιν ἑκάστου δένδρου καὶ ὡς καθ' ὅλου εἰπεῖν φυτοῦ· πᾶν γὰρ ἔχει κρᾶσίν τινα καὶ μίξιν ἰδίαν, ἥπερ οἰκεία δῆλον ὅτι τυγχάνει τοῖς ὑποκειμένοις καρποῖς· ὧν τοῖς πλείστοις συνεμφαίνεταί τις ὁμοιότης οὐκ ἀκριβὴς οὐδὲ σαφής· ἀλλ' ἐν τοῖς περικαρπίοις· διὸ μᾶλλον κατεργασίαν λαμβάνει καὶ πέψιν καθαρὰν καὶ εἰλικρινῆ ἡ τοῦ

---

[1] cf. C.P. 6. 6. 4.

[2] T. is said to have written a treatise περὶ χυμῶν.

[3] ὀπώδης. ὀπός is used specially of the juice of the fig itself.

[4] μήκωνος probably corrupt: it should be a tree.

some kinds of apples. [1] But the smells even of those in this class must in all cases be called wine-like, though they differ in different kinds, on which matter we must speak more precisely, when we come to speak of flavours,[2] reckoning up the different kinds themselves, and stating what differences there are between them, and what is the natural character and property of each.

Now the sap of the trees themselves assumes different kinds of tastes as was said; sometimes it is milky,[3] as that of the fig and poppy,[4] sometimes like pitch, as in silver-fir fir and the conifers; sometimes it is insipid, as in vine pear and apple, as well as such pot-herbs as cucumber gourd lettuce; while others[5] again have a certain pungency, such as the juice of thyme and savory; others have a fragrance, such as the juices of celery dill fennel and the like. To speak generally, all saps correspond to the special character of the several trees, one might almost add, to that of each plant. For every plant has a certain temperament and composition of its own, which[6] plainly belongs in a special sense to the fruits of each. And in most of these is seen a sort of correspondence with the character of the plant as a whole, which is not however exact nor obvious; it is chiefly[7] in the fruit-cases[8] that it is seen, and that is why it is the character of the flavour which becomes more complete and matures into something separate and

[5] I have bracketed ἤδη: ? a dittography of αἱ δὲ.
[6] ἥπερ mBas. H; εἴπερ MAld.
[7] ἀλλ' ἐν . . . μᾶλλον MSS. (?) Ald. H; γὰρ for διὸ conj. W., omitting stop before it.
[8] *i.e.* the pulp: so G. *cf.* 1. 11. 6.

χυλοῦ φύσις· δεῖ γὰρ ὥσπερ τὸ μὲν ὕλην ὑπολαβεῖν τὸ δὲ εἶδος καὶ μορφήν.

3 Ἔχει δὲ αὐτὰ τὰ σπέρματα καὶ οἱ χιτῶνες οἱ περὶ αὐτὰ διαφορὰν τῶν χυλῶν. ὡς δ' ἁπλῶς εἰπεῖν ἅπαντα τὰ μόρια τῶν δένδρων καὶ φυτῶν, οἷον ῥίζα καυλὸς ἀκρεμὼν φύλλον καρπός, ἔχει τινὰ οἰκειότητα πρὸς τὴν ὅλην φύσιν, εἰ καὶ παραλλάττει κατά τε τὰς ὀσμὰς καὶ τοὺς χυλούς, ὡς τὰ μὲν εὔοσμα καὶ εὐώδη τὰ δ' ἄοσμα καὶ ἄχυλα παντελῶς εἶναι τῶν τοῦ αὐτοῦ μορίων.

4 Ἐνίων γὰρ εὔοσμα τὰ ἄνθη μᾶλλον ἢ τὰ φύλλα, τῶν δὲ ἀνάπαλιν τὰ φύλλα μᾶλλον καὶ οἱ κλῶνες, ὥσπερ τῶν στεφανωματικῶν· τῶν δὲ οἱ καρποί· τῶν δ' οὐδέτερον· ἐνίων δ' αἱ ῥίζαι· τῶν δέ τι μέρος. ὁμοίως δὲ καὶ κατὰ τοὺς χυλούς· τὰ μὲν γὰρ βρωτὰ τὰ δ' ἄβρωτα τυγχάνει καὶ ἐν φύλλοις καὶ περικαρπίοις. ἰδιώτατον δὲ τὸ ἐπὶ τῆς φιλύρας· ταύτης γὰρ τὰ μὲν φύλλα γλυκέα καὶ πολλὰ τῶν ζώων ἐσθίει, ὁ δὲ καρπὸς οὐδενὶ βρωτός· ἐπεὶ τό γε ἀνάπαλιν οὐδὲν θαυμαστόν, ὥστε τὰ μὲν φύλλα μὴ ἐσθίεσθαι τοὺς δὲ καρποὺς οὐ μόνον ὑφ' ἡμῶν ἀλλὰ καὶ ὑπὸ τῶν ἄλλων ζώων. ἀλλὰ καὶ περὶ τούτου καὶ τῶν ἄλλων τῶν τοιούτων ὕστερον πειρατέον θεωρεῖν τὰς αἰτίας·

XIII. Νῦν δὲ τοσοῦτον ἔστω δῆλον, ὅτι κατὰ πάντα τὰ μέρη πλείους εἰσὶ διαφοραὶ πολλαχῶς·

---

[1] *i.e.* the pulp. [2] *i.e.* the flavour.

[3] Sense: Every tree has a characteristic juice of its own, which is however specially recognisable in its fruit; in the tree as a whole its character is not always apparent. Hence the importance of the flavour (which is seen in the fruit-pulp), since it is this which determines the specific character,

distinct; in fact we must consider the one[1] as 'matter,' the other[2] as 'form' or specific character.[3]

Again the seeds themselves and the coats containing them have different flavours. And, to speak generally, all parts of trees and plants, as root stem branch leaf fruit, have a certain relationship to the character of the whole, even if[4] there is variation in scents and tastes, so that of the parts of the same plant some are fragrant and sweet to the taste, while others are entirely scentless and tasteless.

For in some plants the flowers are more fragrant than the leaves, in others on the contrary it is rather the leaves and twigs which are fragrant, as in those used for garlands. In others again it is the fruits; in others it is neither[5] of these parts, but, in some few cases, the root or some part of it. And so too with the flavours. Some leaves and some fruit-pulps are, and some are not good for food. [6] Most peculiar is the case of the lime: the leaves of this are sweet, and many animals eat them, but the fruit no creature eats, (for, as to the contrary case, it would not be at all surprising that the leaves should not be eaten, while the fruits were eaten not only by us but by other animals). But concerning this and other such matters we must endeavour to consider the causes on some other occasion.

Differences in flowers.

XIII. For the present let so much be clear, that in all the parts of plants there are numerous differ-

the pulp of fruit in general being, in Aristotelian language, the 'matter,' while the flavour is 'form.' *cf.* *C.P.* 6. 6. 6.

[4] *εἰ καὶ* conj. Sch.; *ἡ δὲ* U; *εἰ δὲ* MVAld.

[5] *οὐδέτερον* seems inaccurately used, as four parts have been mentioned. [6] *cf.* 3. 10. 5; Plin 16. 65.

ἐπεὶ καὶ τῶν ἀνθῶν τὰ μέν ἐστι χνοώδη, καθάπερ τὸ τῆς ἀμπέλου και συκαμινου καὶ τοῦ κιττοῦ· τὰ δὲ φυλλώδη, καθάπερ ἀμυγδαλῆς μηλέας ἀπίου κοκκυμηλέας. καὶ τὰ μὲν μέγεθος ἔχει, τὸ δὲ τῆς ἐλάας φυλλῶδες ὂν ἀμέγεθες. ὁμοίως δὲ καὶ ἐν τοῖς ἐπετείοις καὶ ποιώδεσι τὰ μὲν φυλλώδη τὰ δὲ χνοώδη. πάντων δὲ τὰ μὲν δίχροα τὰ δὲ μονόχροα. τὰ μὲν τῶν δένδρων τά γε πολλὰ μονόχροα καὶ λευκανθῆ· μόνον γὰρ ὡς εἰπεῖν τὸ τῆς ῥόας φοινικοῦν καὶ ἀμυγδαλῶν τινων ὑπέρυθρον· ἄλλου δὲ οὐδενὸς τῶν ἡμέρων οὔτε ἀνθῶδες οὔτε δίχρουν, ἀλλ' εἴ τινος τῶν ἀγρίων, οἷον τὸ τῆς ἐλάτης· κρόκινον γὰρ τὸ ταύτης ἄνθος· καὶ ὅσα δή φασιν ἐν τῇ ἔξω θαλάττῃ ῥόδων ἔχειν τὴν χρόαν.

2 Ἐν δὲ τοῖς ἐπετείοις σχεδὸν τά γε πλείω τοιαῦτα καὶ δίχροα καὶ διανθῆ. λέγω δὲ διανθὲς ὅτι ἕτερον ἄνθος ἐν τῷ ἄνθει ἔχει κατὰ μέσον, ὥσπερ τὸ ῥόδον καὶ τὸ κρίνον καὶ τὸ ἴον τὸ μέλαν. ἔνια δὲ καὶ μονόφυλλα φύεται διαγραφὴν ἔχοντα μόνον τῶν πλειόνων, ὥσπερ τὸ τῆς ἰασιώνης· οὐ γὰρ κεχώρισται ταύτης ἐν τῷ ἄνθει τὸ φύλλον ἕκαστον· οὐδὲ δὴ τοῦ λειρίου τὸ κάτω μέρος, ἀλλὰ ἐκ τῶν ἄκρων ἀποφύσεις γωνιώδεις. σχεδὸν δὲ καὶ τὸ τῆς ἐλάας τοιοῦτόν ἐστιν.

3 Διαφέρει δὲ καὶ κατὰ τὴν ἔκφυσιν καὶ θέσιν· τὰ μὲν γὰρ ἔχει περὶ αὐτὸν τὸν καρπόν, οἷον ἄμ-

---

[1] *i.e.* petaloid.
[2] ἀγρίων Ald.; αἰτίων U; ἀντιῶν MV; ποντίων conj. W.
[3] *i.e.* corolla and stamens, etc.
[4] *i.e.* are gamopetalous (or gamosepalous).

enees shewn in a variety of ways. Thus of flowers some are downy, as that of the vine mulberry and ivy, some are 'leafy,'[1] as in almond apple pear plum. Again some of these flowers are conspicuous, while that of the olive, though it is 'leafy,' is inconspicuous. Again it is in annual and herbaceous plants alike that we find some leafy, some downy. All plants again have flowers either of two colours or of one; most of the flowers of trees are of one colour and white, that of the pomegranate being almost the only one which is red, while that of some almonds is reddish. The flower of no other cultivated trees is gay nor of two colours, though it may be so with some uncultivated[2] trees, as with the flower of silver-fir, for its flower is of saffron colour; and so with the flowers of those trees by the ocean which have, they say, the colour of roses.

However, among annuals, most are of this character—their flowers are two-coloured and twofold.[3] I mean by 'twofold' that the plant has another flower inside the flower, in the middle, as with rose lily violet. Some flowers again consist of a single 'leaf,'[4] having merely an indication of more, as that of bindweed.[5] For in the flower of this the separate 'leaves' are not distinct; nor is it so in the lower part of the narcissus,[6] but there are angular projections[7] from the edges. And the flower of the olive is nearly of the same character.

But there are also differences in the way of growth and the position of the flower; some plants have it

[5] *cf. C.P.* 2. 18. 2 and 3; Plin. 21. 65.
[6] λειρίου conj. Sch., *i.e.* narcissus, *cf.* 6. 6. 9; χειρίου MSS.
[7] *i.e.* something resembling separate 'leaves' (petals or sepals).

πελος ἐλάα· ἧς καὶ ἀποπίπτοντα διατετρημενα φαίνεται, καὶ τοῦτο σημεῖον λαμβάνουσιν εἰ καλῶς ἀπήνθηκεν· ἐὰν γὰρ συγκαυθῇ ἢ βρεχθῇ, συναποβάλλει τὸν καρπὸν καὶ οὐ τετρημένον γίγνεται· σχεδὸν δὲ καὶ τὰ πολλὰ τῶν <ἀνθῶν> ἐν μέσῳ τὸ περικάρπιον ἔχει, τάχα δὲ καὶ ἐπ' αὐτοῦ τοῦ περικαρπίου, καθάπερ ῥόα μελέα ἄπιος κοκκυμηλέα μύρρινος, καὶ τῶν γε φρυγανικῶν ῥοδωνία καὶ τὰ πολλὰ τῶν στεφανωτικῶν· κάτω γὰρ ὑπὸ τὸ ἄνθος ἔχει τὰ σπέρματα· φανερώτατον δὲ ἐπὶ τοῦ ῥόδου διὰ τὸν ὄγκον. ἔνια δὲ καὶ ἐπ' αὐτῶν τῶν σπερμάτων, ὥσπερ ὁ ἄκανος καὶ ὁ κνῆκος καὶ πάντα τὰ ἀκανώδη· καθ' ἕκαστον γὰρ ἔχει τὸ ἄνθος. ὁμοίως δὲ καὶ τῶν ποιωδῶν ἔνια, καθάπερ τὸ ἄνθεμον· ἐν δὲ τοῖς λαχανηροῖς ὅ τε σίκυος καὶ ἡ κολοκύντη καὶ ἡ σικύα· πάντα γὰρ ἐπὶ τῶν καρπῶν ἔχει καὶ προσαυξανομένων ἐπιμένει τὰ ἄνθη πολὺν χρόνον.

Ἄλλα δὲ ἰδιωτέρως, οἷον ὁ κιττὸς καὶ ἡ συκάμινος· ἐν αὐτοῖς μὲν γὰρ ἔχει τοῖς ὅλοις περικαρπίοις, οὐ μὴν οὔτε ἐπ' ἄκροις οὔτ' ἐπὶ περιειληφόσι καθ' ἕκαστον, ἀλλ' ἐν τοῖς ἀνὰ μέσον· εἰ μὴ ἄρα οὐ σύνδηλα διὰ τὸ χνοῶδες.

Ἔστι δὲ καὶ ἄγονα τῶν ἀνθῶν ἔνια, καθάπερ ἐπὶ τῶν σικύων ἃ ἐκ τῶν ἄκρων φύεται τοῦ κλή-

---

[1] *cf.* 3. 16. 4. [2] Lacuna in text; ἀνθῶν I conj.
[3] τάχα Ald.; τινα W. after Sch. conj.
[4] ἄπιος conj. Bod.; ἄγνος Ald.H.
[5] *i.e.* composites.
[6] σπερμάτων conj. Dalec. from G; στομάτων Ald.
[7] ἄκανος conj. W.; ἄκαρος UV.
[8] ἀκανώδη conj. W.; ἀνθώδη Ald.H. *cf.* 1. 10. 6; 6. 4. 4.

close above the fruit, as vine and olive; in the latter, when the flowers drop off, they are seen to have a hole through them,[1] and this men take for a sign whether the tree has blossomed well; for if the flower is burnt up or sodden, it sheds the fruit along with itself, and so there is no hole through it. The majority of flowers[2] have the fruit-case in the middle of them, or, it may be,[3] the flower is on the top of the fruit-case, as in pomegranate apple pear[4] plum and myrtle, and among under-shrubs, in the rose and in many of the coronary plants. For these have their seeds below, beneath the flower, and this is most obvious in the rose because of the size of the seed-vessel. In some cases[5] again the flower is on top of the actual seeds,[6] as in pine-thistle[7] safflower and all thistle-like[8] plants; for these have a flower attached to each seed. So too with some herbaceous plants, as *anthemon*, and among pot-herbs, with cucumber[9] gourd and bottle-gourd; all these have their flowers attached on top of the fruits,[10] and the flowers persist for a long time while the fruits are developing.

In some other plants the attachment is peculiar, as in ivy and mulberry; in these the flower is closely attached to the whole[11] fruit-case; it is not however set above it, nor in a seed-vessel that envelops each[12] separately, but it occurs in the middle part of the structure—except that in some cases it is not easily recognised because it is downy.

[13] Again some flowers are sterile, as in cucumbers those which grow at the ends of the shoot, and that

[9] ὅ τε σίκυος conj. W.; ὅπερ σίκυος UM; ὁ περσίκυος Ald.
[10] καρπῶν conj. Sch.; ἄκρων Ald.H.
[11] *i.e.* compound. [12] οὔτ' ἐπὶ I conj. for οὔτε.
[13] *cf.* Arist. *Probl.* 20. 3.

ματος, δι' ὃ καὶ ἀφαιροῦσιν αὐτά· κωλύει γὰρ τὴν τοῦ σικύου βλάστησιν. φασὶ δὲ καὶ τῆς μηλέας τῆς Μηδικῆς ὅσα μὲν ἔχει τῶν ἀνθῶν ὥσπερ ἠλακάτην τινὰ πεφυκυῖαν ἐκ μέσου ταῦτ' εἶναι γόνιμα, ὅσα δὲ μὴ ἔχει ταῦτ' ἄγονα. εἰ δὲ καὶ ἐπ' ἄλλου τινὸς ταῦτα συμβαίνει τῶν ἀνθοφόρων ὥστε ἄγονον ἄνθος φύειν εἴτε κεχωρισμένον εἴτε μή, σκεπτέον. ἐπεὶ γένη γε ἔνια καὶ ἀμπέλου καὶ ῥόας ἀδυνατεῖ τελεοκαρπεῖν, ἀλλὰ μέχρι τοῦ ἄνθους ἡ γένεσις.

5 (Γίνεται δὲ καὶ τό γε τῆς ῥόας ἄνθος πολὺ καὶ πυκνὸν καὶ ὅλως ὁ ὄγκος πλατὺς ὥσπερ ὁ τῶν ῥόδων· κάτωθεν δ' ἑτεροῖος· οἷος δίωτος μικρὸς ὥσπερ ἐκτετραμμένος ὁ κύτινος ἔχων τὰ χείλη μυχώδη.)

Φασὶ δέ τινες καὶ τῶν ὁμογενῶν τὰ μὲν ἀνθεῖν τὰ δ' οὔ, καθάπερ τῶν φοινίκων τὸν μὲν ἄρρενα ἀνθεῖν τὸν δὲ θῆλυν οὐκ ἀνθεῖν ἀλλ' εὐθὺ προφαίνειν τὸν καρπόν.

Τὰ μὲν οὖν τῷ γένει ταὐτὰ τοιαύτην τὴν δια-

---

[1] *i.e.* the pistil.

[2] *i.e.* as seen from above: *καὶ ὅλων . . . ῥόδων* describes the corolla, *κάτωθεν . . . μυχώδη* the undeveloped ovary, including the adherent calyx.

[3] *ῥόδων* conj. Bod.; *ῥοῶν* Ald.

[4] *κάτωθεν . . . μυχώδη* I conj.; *δ' ἕτεροι δι' ὧν ὡς μικρὸν ὥσπερ ἐκτετραμμένος κότινος ἔχων τὰ χείλη μυχώδη* UMVAld. (except that Ald. has *ἄνω* for *χείλη* and *ἐκτετραμμένον*: so also P, but *ἐκτετραμμένος*). The sentence explains incidentally why the pomegranate flower was called *κύτινος* (*cf.* 2. 6. 12; *C.P.* 1. 14. 4; 2. 9. 3; 2. 9. 9; Diosc. 1. 110; Plin. 23. 110

is why men pluck them off, for they hinder the growth of the cucumber. And they say that in the citron those flowers which have a kind of distaff[1] growing in the middle are fruitful, but those that have it not are sterile. And we must consider whether it occurs also in any other flowering plants that they produce sterile flowers, whether apart from the fertile flowers or not. For some kinds of vine and pomegranate certainly are unable to mature their fruit, and do not produce anything beyond the flower.

(The flower of the pomegranate is produced abundantly and is solid[2]: in general appearance it is a substantial structure with a flat top, like the flower of the rose[3]; but,[4] as seen from below, the inferior part of the flower is different-looking, being like a little two-eared jar turned on one side and having its rim indented.)

Some say that even of plants of the same kind 5
some specimens flower while others do not; for instance that the 'male' date-palm flowers but the 'female' does not, but exhibits its fruit without any antecedent flower.

Such[6] is the difference which we find between

and 111), *i.e.* because it resembled a κύτος (see LS. *s.v.*). T. chooses the particular form of jar called δίωτος, because the indentations between the sepals suggest this: . This is called ἐκτετραμμένος, because the weight of the developing fruit causes it to take up at one stage a horizontal position, like a jar lying on its side; χείλη refers to the jar (for the plural *cf.* the use of ἄντυγες), μυχώδη to the indentations in the calyx (a jar having ordinarily an unindented rim).

[5] ὁμογενῶν conj. Sch.; ὁμοιογενῶν Ald.

[6] ταὐτὰ τοιαύτην I conj. from G; τοιαῦτα τὴν UM; τοιαύτην P.

φορὰν ἔχει, καθάπερ ὅλως ὅσα μὴ δύναται τελεοκαρπεῖν. ἡ δὲ τοῦ ἄνθους φύσις ὅτι πλείους ἔχει διαφορὰς φανερὸν ἐκ τῶν προειρημένων.

XIV. Διαφέρει δὲ τὰ δένδρα καὶ τοῖς τοιούτοις κατὰ τὴν καρποτοκίαν· τὰ μὲν γὰρ ἐκ τῶν νέων βλαστῶν φέρει τὰ δ' ἐκ τῶν ἔνων τὰ δ' ἐξ ἀμφοτέρων. ἐκ μὲν τῶν νέων συκῆ ἄμπελος· ἐκ δὲ τῶν ἔνων ἐλάα ῥόα μηλέα ἀμυγδαλῆ ἄπιος μύρρινος καὶ σχεδὸν τὰ τοιαῦτα πάντα· ἐκ δὲ τῶν νέων ἐὰν ἄρα τι συμβῇ κυῆσαι καὶ ἀνθῆσαι (γίνεται γὰρ καὶ ταῦτ' ἐνίοις, ὥσπερ καὶ τῷ μυρρίνῳ καὶ μάλισθ' ὡς εἰπεῖν περὶ τὰς βλαστήσεις τὰς μετ' Ἀρκτοῦρον) οὐ δύναται τελεοῦν ἀλλ' ἡμιγενῆ φθείρεται· ἐξ ἀμφοτέρων δὲ καὶ τῶν ἔνων καὶ τῶν νέων εἴ τινες ἄρα μηλέαι τῶν διφόρων ἢ εἴ τι ἄλλο κάρπιμον· ἔτι δὲ ὁ ὄλυνθος ἐκπέττων καὶ σῦκα φέρων ἐκ τῶν νέων.

2 Ἰδιωτάτη δὲ ἡ ἐκ τοῦ στελέχους ἔκφυσις, ὥσπερ τῆς ἐν Αἰγύπτῳ συκαμίνου· ταύτην γάρ φασι φέρειν ἐκ τοῦ στελέχους· οἱ δὲ ταύτῃ τε καὶ ἐκ τῶν ἀκρεμόνων, ὥσπερ τὴν κερωνίαν· αὕτη γὰρ καὶ ἐκ τούτων φέρει πλὴν οὐ πολύν· καλοῦσι δὲ κερωνίαν ἀφ' ἧς τὰ σῦκα τὰ Αἰγύπτια καλούμενα.

---

[1] ? *i.e.* that, like the 'female' date-palm, they have no flower.

[2] τοιαῦτα πάντα· ἐκ δὲ τῶν νέων ἐὰν ἄρα τι conj. W.; τοιαῦτα· πάντα γὰρ ἐκ τῶν ἔνων· ἐὰν δὲ ἄρα τι MSS.

[3] *cf.* 3. 5. 4.

[4] διφόρων conj. Sch. from G; διαφόρων UAld.

plants of the same kind; and the like may be said[1] in general of those which cannot mature their fruit. And it is plain from what has been said that flowers shew many differences of character.

Differences in fruits.

XIV. Again as to the production of fruit trees differ in the following respects. Some bear on their new shoots, some on last year's wood, some on both. Fig and vine bear on their new shoots; on last year's wood olive pomegranate apple almond pear myrtle and almost all such trees. And, if any of these does[2] happen to conceive and to produce flowers on its new shoots, (for this does occur in some cases, as with myrtle, and especially, one may say, in the growth which is made after the rising of Arcturus)[3] it can not bring them to perfection, but they perish half-formed. Some apples again of the twice-bearing[4] kinds and certain other fruit-trees bear both on last year's wood and on the new shoots; and so does the *olynthos*,[5] which ripens its fruit as well as bearing figs on the new shoots.

Most peculiar is the growth of fruit direct from the stem, as in the sycamore; for this, they say, bears fruit on the stem. Others say that it bears both in this way and[6] also on the branches, like the carob; for the latter bears on the branches too, though not abundantly: (the name carob is given to the tree which produces what are called 'Egyptian

[5] ὄλυνθος is not elsewhere used for a kind of fig: ἔτι δὲ συκῇ τοὺς ὀλύνθους ἐκπέττουσα καὶ σῦκα φέρουσα conj. Sch. somewhat drastically.

[6] ταύτῃ τε καὶ ἐκ conj. W.; ταύτης μὲν ἐκ UMVAld. *cf.* 4. 2. 4.

ἔστι δὲ καὶ τὰ μὲν ἀκρόκαρπα τῶν δένδρων και ὅλως τῶν φυτῶν τὰ δὲ πλαγιόκαρπα τὰ δ' ἀμφοτέρως. πλείω δ' ἀκρόκαρπα τῶν ἄλλων ἢ τῶν δένδρων, οἷον τῶν τε σιτηρῶν τὰ σταχυώδη καὶ τῶν θαμνωδῶν ἐρείκη καὶ σπειραία καὶ ἄγνος καὶ ἄλλ' ἄττα καὶ τῶν λαχανωδῶν τὰ κεφαλόρριζα. ἐξ ἀμφοτέρων δὲ καὶ τῶν δένδρων ἔνια καὶ τῶν λαχανωδῶν, οἷον βλίτον ἀδράφαξυς ῥάφανος· ἐπεὶ καὶ ἐλάα ποιεῖ πως τοῦτο, καί φασιν ὅταν ἄκρον ἐνέγκῃ σημεῖον εὐφορίας εἶναι. ἀκρόκαρπος δέ πως καὶ ὁ φοῖνιξ· πλὴν τοῦτό γε καὶ ἀκρόφυλλον καὶ ἀκρόβλαστον· ὅλως γὰρ ἐν τῷ ἄνω πᾶν τὸ ζωτικόν. τὰς μὲν οὖν κατὰ <τὰ> μέρη διαφορὰς πειρατέον ἐκ τούτων θεωρεῖν.

8 Αἱ δὲ τοιαῦται τῆς ὅλης οὐσίας φαίνονται· δῆλον ὅτι τὰ μὲν ἥμερα τὰ δ' ἄγρια· καὶ τὰ μὲν κάρπιμα τὰ δ' ἄκαρπα· καὶ ἀείφυλλα καὶ φυλλοβολα, καθάπερ ἐλέχθη, τὰ δ' ὅλως ἄφυλλα· καὶ τὰ μὲν ἀνθητικὰ τὰ δ' ἀνανθῆ· καὶ πρωϊβλαστῆ δὲ καὶ πρωϊκαρπα τὰ δὲ ὀψιβλαστῆ καὶ ὀψίκαρπα· ὡσαύτως δὲ καὶ ὅσα παραπλήσια τούτοις. και πως τά γε τοιαῦτα ἐν τοῖς μέρεσιν ἢ οὐκ ἄνευ τῶν μερῶν ἐστιν. ἀλλ' ἐκείνη ἰδιωτάτη καὶ τροπόν τινα μεγίστη διάστασις, ἥπερ καὶ ἐπὶ τῶν ζώων, ὅτι τὰ μὲν ἔνυδρα τὰ δὲ χερσαῖα· καὶ γὰρ τῶν φυτῶν

---

[1] Plin 16. 112.
[2] τοῦτο conj. Sch.; τούτου UAld.; τοῦτον M.
[3] τὰ add. W.; *cf.* 1. 13. 1.

figs'). [1] Again some trees, and some plants in general, produce fruit at the top, others at the sides, others in both ways. But bearing fruit at the top is less common in trees than in other plants, as among grains in those which have an ear, among shrubby plants in heath privet chaste tree and certain others, and among pot-herbs in those with a bulbous root. Among plants which bear both on the top and at the sides are certain trees and certain pot-herbs, as blite orach cabbage. I say trees, since the olive does this too in a way, and they say that, when it bears at the top, it is a sign of fruitfulness. The date-palm too bears at the top, in a sense, but this[2] tree also has its leaves and shoots at the top; indeed it is in the top that its whole activity is seen. Thus we must endeavour to study in the light of the instances mentioned the differences seen in the[3] various parts of the plant.

General differences (affecting the whole plant).

But there appear to be the following differences which affect the plant's whole being: some are cultivated, some wild; some fruitful, some barren; some evergreen, some deciduous, as was said, while some again have no leaves at all; some are flowering plants, some flowerless; some are early, some late in producing their shoots and fruits; and there are other differences similar to these. Now it may be said that[4] such differences are seen in the parts, or at least that particular parts are concerned in them. But the special, and in a way the most important distinction is one which may be seen in animals too, namely, that some are of the water, some of the land. For

[4] καί πως τά γε τοιαῦτα conj. Sch.; καὶ πῶν τά γε ταῦτα U; καὶ τά γε τοιαῦτα Ald.

ἔστι τι τοιοῦτον γένος ὃ οὐ δύναται φύεσθαι <μὴ> ἐν ὑγρῷ· τὰ δὲ φύεται μέν, οὐχ ὅμοια δὲ ἀλλὰ χείρω. πάντων δὲ τῶν δένδρων ὡς ἁπλῶς εἰπεῖν καὶ τῶν φυτῶν εἴδη πλείω τυγχάνει καθ' ἕκαστον
4 γένος· σχεδὸν γὰρ οὐδέν ἐστιν ἁπλοῦν· ἀλλ' ὅσα μὲν ἥμερα καὶ ἄγρια λέγεται ταύτην ἐμφανεστάτην καὶ μεγίστην ἔχει διαφοράν, οἷον συκῆ ἐρινεός, ἐλάα κότινος, ἄπιος ἀχράς· ὅσα δ' ἐν ἑκατέρῳ τούτων τοῖς καρποῖς τε καὶ φύλλοις καὶ ταῖς ἄλλαις μορφαῖς τε καὶ τοῖς μορίοις. ἀλλὰ τῶν μὲν ἀγρίων ἀνώνυμα τὰ πλεῖστα καὶ ἔμπειροι ὀλίγοι· τῶν δὲ ἡμέρων καὶ ὠνομασμένα τὰ πλείω καὶ ἡ αἴσθησις κοινοτέρα· λέγω δ' οἷον ἀμπέλου συκῆς ῥόας μηλέας ἀπίου δάφνης μυρρίνης τῶν ἄλλων· ἡ γὰρ χρῆσις οὖσα κοινὴ συνθεωρεῖν ποιεῖ τὰς διαφοράς.

5 Ἴδιον δὲ καὶ τοῦτ' ἐφ' ἑκατέρων· τὰ μὲν γὰρ ἄγρια τῷ ἄρρενι καὶ τῷ θήλει ἢ μόνοις ἢ μάλιστα διαιροῦσι, τὰ δὲ ἥμερα πλείοσιν ἰδέαις. ἔστι δὲ τῶν μὲν ῥᾷον λαβεῖν καὶ διαριθμῆσαι τὰ εἴδη, τῶν δὲ χαλεπώτερον διὰ τὴν πολυχοΐαν.

Ἀλλὰ δὴ τὰς μὲν τῶν μορίων διαφορὰς καὶ τῶν ἄλλων οὐσιῶν ἐκ τούτων πειρατέον θεωρεῖν. περὶ δὲ τῶν γενέσεων μετὰ ταῦτα λεκτέον· τοῦτο γὰρ ὥσπερ ἐφεξῆς τοῖς εἰρημένοις ἐστίν.

of plants too there is a class which cannot grow except[1] in moisture, while others will indeed grow on dry land, but they lose their character and are inferior. Again of all trees, one might almost say, and of all plants there are several forms to each kind; for hardly any kind contains but a single form. But the plants which are called respectively cultivated and wild shew this difference in the clearest and most emphatic way, for instance the cultivated and wild forms of fig olive and pear. In each of these pairs there are differences in fruit and leaves, and in their forms and parts generally. But most of the wild kinds have no names and few know about them, while most of the cultivated kinds have received names[2] and they are more commonly observed; I mean such plants as vine fig pomegranate apple pear bay myrtle and so forth; for, as many people make use of them, they are led also to study the differences.

But there is this peculiarity as to the two classes respectively; in the wild kinds men find only or chiefly the distinction of 'male' and 'female,' while in the cultivated sorts they recognise a number of distinguishing features. In the former case it is easy to mark and count up the different forms, in the latter it is harder because the points of difference are numerous.

However we have said enough for study of the differences between parts and between general characters. We must now speak of the methods of growth, for this subject comes naturally after what has been said.

[1] μὴ add. W.
[2] ὠνομασμένα τὰ πλείω conj. Sch.; ὠνομασμένων πλείω Ald.

BOOK II

B

I. Αἱ γενέσεις τῶν δένδρων καὶ ὅλως τῶν φυτῶν ἢ αὐτόμαται ἢ ἀπὸ σπέρματος ἢ ἀπὸ ῥίζης ἢ ἀπὸ παρασπάδος ἢ ἀπὸ ἀκρεμόνος ἢ ἀπὸ κλωνὸς ἢ ἀπ' αὐτοῦ τοῦ στελέχους εἰσίν, ἢ ἔτι τοῦ ξύλου κατακοπέντος εἰς μικρά· καὶ γὰρ οὕτως ἔνια φύεται. τούτων δὲ ἡ μὲν αὐτόματος πρώτη τις, αἱ δὲ ἀπὸ σπέρματος καὶ ῥίζης φυσικώταται δόξαιεν ἄν· ὥσπερ γὰρ αὐτόμαται καὶ αὗταί· δι' ὃ καὶ τοῖς ἀγρίοις ὑπάρχουσιν· αἱ δὲ ἄλλαι τέχνης ἢ δὴ προαιρέσεως.

2 Ἅπαντα δὲ βλαστάνει κατά τινα τῶν τρόπων τούτων, τὰ δὲ πολλὰ κατὰ πλείους· ἐλάα μὲν γὰρ πάντως φύεται πλὴν ἀπὸ τοῦ κλωνός· οὐ γὰρ δύναται καταπηγνυμένη, καθάπερ ἡ συκῆ τῆς κράδης καὶ ἡ ῥόα τῆς ῥάβδου. καίτοι φασί γέ τινες ἤδη καὶ χάρακος παγείσης καὶ πρὸς τὸν κιττὸν συμβιῶσαι καὶ γενέσθαι δένδρον· ἀλλὰ σπάνιόν τι τὸ τοιοῦτον· θάτερα δὲ τὰ πολλὰ τῆς φύσεως. συκῆ δὲ τοὺς μὲν ἄλλους τρόπους

[1] ἔνια φύεται conj. Sch.; ἀναφύεται Ald.

BOOK II

Of Propagation, especially of Trees.

Of the ways in which trees and plants originate. Instances of degeneration from seed.

I. The ways in which trees and plants in general originate are these:—spontaneous growth, growth from seed, from a root, from a piece torn off, from a branch or twig, from the trunk itself; or again from small pieces into which the wood is cut up (for some trees can be produced[1] even in this manner). Of these methods spontaneous growth comes first, one may say, but growth from seed or root would seem most natural; indeed these methods too may be called spontaneous; wherefore they are found even in wild kinds, while the remaining methods depend on human skill or at least on human choice.

However all plants start in one or other of these ways, and most of them in more than one. Thus the olive is grown in all the ways mentioned, except from a twig; for an olive-twig will not grow if it is set in the ground, as a fig or pomegranate will grow from their young shoots. Not but what some say that cases have been known in which, when a stake of olive-wood was planted to support ivy, it actually lived along with it and became a tree; but such an instance is a rare exception, while the other methods of growth are in most cases the natural ones. The fig grows in all the ways mentioned,

φύεται πάντας, ἀπὸ δὲ τῶν πρέμνων καὶ τῶν ξύλων οὐ φύεται· μηλέα δὲ καὶ ἄπιος καὶ ἀπὸ τῶν ἀκρεμόνων σπανίως. οὐ μὴν ἀλλὰ τά γε πολλὰ ἢ πάνθ' ὡς εἰπεῖν ἐνδέχεσθαι δοκεῖ καὶ ἀπὸ τούτων, ἐὰν λεῖοι καὶ νέοι καὶ εὐαυξεῖς ὦσιν. ἀλλὰ φυσικώτεραί πως ἐκεῖναι· τὸ δὲ ἐνδεχόμενον ὡς δυνατὸν ληπτέον.

3 Ὅλως γὰρ ὀλίγα τὰ ἀπὸ τῶν ἄνω μᾶλλον βλαστάνοντα καὶ γεννώμενα, καθάπερ ἄμπελος ἀπὸ τῶν κλημάτων· αὕτη γὰρ οὐκ ἀπὸ τῆς πρώρας ἀλλ' ἀπὸ τοῦ κλήματος φύεται, καὶ εἰ δή τι τοιοῦτον ἕτερον ἢ δένδρον ἢ φρυγανῶδες, ὥσπερ δοκεῖ τό τε πήγανον καὶ ἡ ἰωνία καὶ τὸ σισύμβριον καὶ ὁ ἕρπυλλος καὶ τὸ ἑλένιον. κοινοτάτη μὲν οὖν ἐστὶ πᾶσιν ἥ τε ἀπὸ τῆς παρασπάδος καὶ ἀπὸ σπέρματος. ἅπαντα γὰρ ὅσα ἔχει σπέρματα καὶ ἀπὸ σπέρματος γίνεται· ἀπὸ δὲ παρασπάδος καὶ τὴν δάφνην φασίν, ἐάν τις τὰ ἔρνη παρελὼν φυτεύσῃ. δεῖ δὲ ὑπόρριζον εἶναι μάλιστά γε τὸ παρασπώμενον ἢ ὑπόπρεμνον. οὐ μὴν ἀλλὰ καὶ ἄνευ τούτου θέλει βλαστάνειν καὶ ῥόα καὶ μηλέα ἐαρινή· βλαστάνει δὲ καὶ ἀμυγδαλῆ φυτευομένη.

4 κατὰ πλείστους δὲ τρόπους ὡς εἰπεῖν ἡ ἐλάα βλαστάνει· καὶ γὰρ ἀπὸ τοῦ στελέχους καὶ ἀπὸ τοῦ πρέμνου κατακοπτομένου καὶ ἀπὸ τῆς ῥίζης [καὶ ἀπὸ τοῦ ξύλου] καὶ ἀπὸ ῥάβδου καὶ χάρακος ὥσπερ εἴρηται. τῶν δ' ἄλλων ὁ μύρρινος· καὶ γὰρ οὗτος ἀπὸ τῶν ξύλων καὶ τῶν πρέμνων

---

[1] τά γε πολλὰ πάνθ' conj. Sch.; ἢ before πάνθ' ins. St.; τά τε πολλὰ πάνθ' Ald.

[2] εὐαυξεῖς conj. H; αὐξεῖς UMVAld.

[3] οὐκ I conj.; οὐδ' MSS.

except from root-stock and cleft wood; apple and pear grow also from branches, but rarely. However it appears that most, if not practically all,[1] trees may grow from branches, if these are smooth young and vigorous.[2] But the other methods, one may say, are more natural, and we must reckon what may occasionally occur as a mere possibility.

In fact there are quite few plants which grow and are brought into being more easily from the upper parts, as the vine is grown from branches; for this, though it cannot[3] be grown from the 'head,'[4] yet can be grown from the branch, as can all similar trees and under-shrubs, for instance, as it appears, rue gilliflower bergamot-mint tufted thyme calamint. So the commonest ways of growth with all plants are from a piece torn off or from seed; for all plants that have seeds grow also from seed. And they say that the bay too grows[5] from a piece torn off, if one takes off the young shoots and plants them; but it is necessary that the piece torn off should have part of the root or stock[6] attached to it. However the pomegranate and 'spring apple'[7] will grow even without this, and a slip of almond[8] grows if it is planted. The olive grows, one may say, in more ways than any other plant; it grows from a piece of the trunk or of the stock,[9] from the root, from a twig, and from a stake, as has been said.[10] Of other plants the myrtle also can be propagated in several ways; for this too grows from pieces of wood

[4] πρώρας, *cf.* Col. 3. 10. 1, *caput vitis vocat πρώραν.* Sch. restores the word, *C.P.* 3. 14. 7.

[5] *cf.* *C.P.* 1. 3. 2.

[6] *i.e.* a 'heel' (Lat. *perna*).

[7] *cf.* *C.P.* 2. 11. 6; Athen. 3. 23.

[8] *cf.* *Geop.* 10. 3. 9.

[9] καὶ ἀπὸ τοῦ ξύλου om. Julius Pontedeva on Varro 1. 39. 3: a gloss on ἀπὸ τοῦ πρέμνου κατακ.

[10] 2. 1. 2.

φύεται. δεῖ δὲ καὶ τούτου καὶ τῆς ἐλάας τὰ ξύλα διαιρεῖν μὴ ἔλαττω σπιθαμιαίων καὶ τὸν φλοιὸν μὴ περιαιρεῖν.

Τὰ μὲν οὖν δένδρα βλαστάνει καὶ γίνεται κατὰ τοὺς εἰρημένους τρόπους· αἱ γὰρ ἐμφυτεῖαι καὶ οἱ ἐνοφθαλμισμοὶ καθάπερ μίξεις τινές εἰσιν ἢ κατ' ἄλλον τρόπον γενέσεις, περὶ ὧν ὕστερον λεκτέον.

II. Τῶν δὲ φρυγανωδῶν και ποιωδῶν τὰ μὲν πλεῖστα ἀπὸ σπέρματος ἢ ῥίζης τὰ δὲ και ἀμφοτέρως· ἔνια δὲ καὶ ἀπὸ των βλαστῶν, ὥσπερ εἴρηται. ῥοδωνία δὲ καὶ κρινωνία κατακοπέντων τῶν καυλῶν, ὥσπερ καὶ ἡ ἄγρωστις. φύεται δὲ ἡ κρινωνία καὶ ἡ ῥοδωνία καὶ ὅλου τοῦ καυλοῦ τεθέντος. ἰδιωτάτη δὲ ἡ ἀπὸ δακρύου· καὶ γὰρ οὕτω δοκεῖ τὸ κρίνον φύεσθαι, ὅταν ξηρανθῇ τὸ ἀπορρυέν. φασὶ δὲ καὶ ἐπὶ τοῦ ἱπποσελίνου· καὶ γὰρ τοῦτο ἀφίησι δάκρυον. φύεται δέ τις καὶ κάλαμος, ἐάν τις διατέμνων τὰς ἠλακάτας πλαγίας τιθῇ καὶ κατακρύψῃ κόπρῳ καὶ γῇ. ἰδίως δὲ ἀπὸ ῥίζης [τῷ] φύεσθαι καὶ τὰ κεφαλόρριζα.

2 Τοσαυταχῶς δὲ οὔσης τῆς δυνάμεως τὰ μὲν πολλὰ τῶν δένδρων, ὥσπερ ἐλέχθη πρότερον, ἐν πλείοσι τρόποις φύεται· ἔνια δὲ ἀπὸ σπέρματος

---

[1] ἐμφυτεῖαι conj. R. Const.; ἐμφυλέαι (with erasures) U; ἐμφυλείαι V; ἐμφυλεῖαι Ald.
[2] 2. 1. 3; *cf.* *C.P.* 1. 4. 4 and 6.
[3] *i.e.* bulbil. *cf.* 6. 6. 8; 9. 1. 4; *C.P.* 1. 4. 6; Plin. 21. 24.
[4] ἐπὶ conj. W.; ἀπὸ P_2Ald.
[5] δέ τις καὶ Ald,; τις om. W. after Sch.

and also from pieces of the stock. It is necessary however with this, as with the olive, to cut up the wood into pieces not less than a span long and not to strip off the bark.

Trees then grow and come into being in the above-mentioned ways; for as to methods of grafting[1] and inoculation, these are, as it were, combinations of different kinds of trees; or at all events these are methods of growth of a quite different class and must be treated of at a later stage.

II. Of under-shrubs and herbaceous plants the greater part grow from seed or a root, and some in both ways; some of them also grow from cuttings, as has been said,[2] while roses and lilies grow from pieces of the stems, as also does dog's-tooth grass. Lilies and roses also grow when the whole stem is set. Most peculiar is the method of growth from an exudation[3]; for it appears that the lily grows in this way too, when the exudation that has been produced has dried up. They say the same of[4] alexanders, for this too produces an exudation. There is a certain[5] reed also which grows if one cuts it in lengths from joint to joint and sets them[6] sideways, burying it in dung and soil. Again they say that plants which have a bulbous root are peculiar in their way of growing[7] from the root.

The capacity for growth being shewn in so many ways, most trees, as was said before,[8] originate in several ways; but some come[9] only from seed, as silver-

[6] *cf.* 1. 4. 4; Plin. 17. 145; Col. 4. 32. 2; τιθῇ conj. Sch.; ᾗ Ald.; ? θῇ.

[7] *i.e.* by offset bulbs. Text probably defective; *cf. C.P.* 1. 4. 1. τῷ U; τὸ UMV. [8] 2. 1. 1.

[9] φύεται I conj.; φησὶν ἔστιν or φασὶν ἔστιν MSS.; ὡς φασίν ἔστιν Ald.; παραγίνεται conj. W.

φύεται μονον, οἷον ἐλάτη πεύκη πίτυς ὅλως πᾶν τὸ κωνοφόρον· ἔτι δὲ καὶ φοῖνιξ, πλὴν εἰ ἄρα ἐν Βαβυλῶνι καὶ ἀπὸ τῶν ῥάβδων [ὥς] φασί τινες μολεύειν. κυπάριττος δὲ παρὰ μὲν τοῖς ἄλλοις ἀπὸ σπέρματος, ἐν Κρήτῃ δὲ καὶ ἀπὸ τοῦ στελέχους, οἷον ἐπὶ τῆς ὀρείας ἐν Τάρρᾳ· παρὰ τούτοις γάρ ἐστιν ἡ κουριζομένη κυπάριττος· αὕτη δὲ ἀπὸ τῆς τομῆς βλαστάνει πάντα τρόπον τεμνομένη καὶ ἀπὸ γῆς καὶ ἀπὸ τοῦ μέσου καὶ ἀπὸ τοῦ ἀνωτέρω· βλαστάνει δὲ ἐνιαχοῦ καὶ ἀπὸ τῶν ῥιζῶν σπανίως δέ.

3 Περὶ δὲ δρυὸς ἀμφισβητοῦσιν· οἱ μὲν γὰρ ἀπὸ σπέρματός φασι μόνον, οἱ δὲ καὶ ἀπὸ ῥίζης γλίσχρως· οἱ δὲ καὶ ἀπ' αὐτοῦ τοῦ στελέχους κοπέντος. ἀπὸ παρασπάδος δὲ καὶ ῥίζης οὐδὲν φύεται τῶν μὴ παραβλαστανόντων.

4 Ἁπάντων δὲ ὅσων πλείους αἱ γενέσεις, ἡ ἀπο παρασπάδος καὶ ἔτι μᾶλλον ἡ ἀπὸ παραφυάδος ταχίστη καὶ εὐαυξής, ἐὰν ἀπὸ ῥίζης ἡ παραφυὰς ᾖ. καὶ τὰ μὲν οὕτως ἢ ὅλως ἀπὸ φυτευτηρίων πεφυτευμένα πάντα δοκεῖ τοὺς καρποὺς ἐξομοιοῦν. ὅσα δ' ἀπὸ τοῦ καρποῦ τῶν δυναμενων καὶ οὕτως βλαστάνειν, ἅπανθ' ὡς εἰπεῖν χείρω, τὰ δὲ καὶ ὅλως ἐξίσταται τοῦ γένους, οἷον ἄμπελος μηλέα συκῆ ῥοιὰ ἄπιος· ἔκ τε γὰρ τῆς κεγχραμίδος οὐδὲν γίνεται γένος ὅλως ἥμερον, ἀλλ' ἢ ἐρινεὸς ἢ ἀγρία συκῆ, διαφέρουσα πολλάκις καὶ τῇ χροίᾳ· καὶ γὰρ ἐκ μελαίνης λευκὴ καὶ ἐκ λευκῆς μέλαινα

---

[1] μολεύειν conj. Sch.; μωλύειν MSS.; μοσχεύειν conj. R. Const. (*cf.* *C.P.* 1. 2 1). But *cf.* Hesych. *s.v.* μολεύειν.
[2] Plin. 16. 141. [3] ἐπὶ conj. W.; τὸ UMVAld.

fir fir Aleppo pine, and in general all those that bear cones: also the date-palm, except that in Babylon it may be that, as some say, they take cuttings[1] from it. The cypress in most regions grows from seed, but in Crete[2] from the trunk also, for instance in[3] the hill country about Tarra; for there grows the cypress which they clip, and when cut it shoots in every possible way, from the part which has been cut, from the ground, from the middle, and from the upper parts; and occasionally, but rarely, it shoots from the roots also.

About the oak accounts differ; some say it only grows from seed, some from the root also, but not vigorously, others again that it grows from the trunk itself, when this is cut. But no tree grows from a piece torn off or from a root except those which make side-growths.

However in all the trees which have several methods of originating the quickest method and that which promotes the most vigorous growth is from a piece torn off, or still better from a sucker, if this is taken from the root. And, while all the trees which are propagated thus or by some kind of slip[4] seem to be alike in their fruits to the original tree, those raised from the fruit, where this method of growing is also possible, are nearly all inferior, while some quite lose the character of their kind, as vine apple fig pomegranate pear. As for the fig,[5] no cultivated kind is raised from its seed, but either the ordinary wild fig or some wild kind is the result, and this often differs in colour from the parent; a black fig gives a

[4] φυτευτήριον: a general term including παραφυάς and παρασπάς.
[5] *cf. C.P.* 1. 9.

γίνεται· ἔκ τε τῆς ἀμπέλου τῆς γενναίας ἀγεννής· καὶ πολλάκις ἕτερον γένος· ὁτὲ δὲ ὅλως οὐδὲν ἥμερον ἀλλ' ἄγριον ἐνίοτε καὶ τοιοῦτον ὥστε μὴ ἐκπέττειν τὸν καρπόν· αἱ δ' ὥστε μηδὲ ἁδρύνειν ἀλλὰ μέχρι τοῦ ἀνθῆσαι μόνον ἀφικνεῖσθαι.

Φύονται δὲ καὶ ἐκ τῶν τῆς ἐλάας πυρήνων ἀγριέλαιος, καὶ ἐκ τῶν τῆς ῥόας κόκκων τῶν γλυκέων ἀγεννεῖς, καὶ ἐκ τῶν ἀπυρήνων σκληραί, πολλάκις δὲ καὶ ὀξεῖαι. τὸν αὐτὸν δὲ τρόπον καὶ ἐκ τῶν ἀπίων καὶ ἐκ τῶν μηλέων· ἐκ μὲν γὰρ τῶν ἀπίων μοχθηρὰ ἡ ἀχράς, ἐκ δὲ τῶν μηλέων χείρων τε τῷ γένει καὶ ἐκ γλυκείας ὀξεῖα, καὶ ἐκ στρουθίου Κυδώνιος. χείρων δὲ καὶ ἡ ἀμυγδαλῆ καὶ τῷ χυλῷ καὶ τῷ σκληρὰ ἐκ μαλακῆς· δι' ὃ καὶ αὐξηθεῖσαν ἐγκεντρίζειν κελεύουσιν, εἰ δὲ μὴ τὸ μόσχευμα μεταφυτεύειν πολλάκις.

Χείρων δὲ καὶ ἡ δρῦς· ἀπὸ γοῦν τῆς ἐν Πύρρᾳ πολλοὶ φυτεύσαντες οὐκ ἐδύνανθ' ὁμοίαν ποιεῖν. δάφνην δὲ καὶ μυρρίνην διαφέρειν ποτέ φασιν, ὡς ἐπὶ τὸ πολὺ δ' ἐξίστασθαι καὶ οὐδὲ τὸ χρῶμα διασώζειν, ἀλλ' ἐξ ἐρυθροῦ καρποῦ γίνεσθαι μέλαιναν, ὥσπερ καὶ τὴν ἐν Ἀντάνδρῳ· πολλάκις δὲ καὶ τὴν κυπάριττον ἐκ θηλείας ἄρρενα. μάλιστα δὲ τούτων ὁ φοῖνιξ δοκεῖ διαμένειν ὥσπερ εἰπεῖν τελείως τῶν ἀπὸ σπέρματος, καὶ πεύκη ἡ κωνοφόρος καὶ πίτυς ἡ φθειροποιός. ταῦτα μὲν οὖν ἐν τοῖς ἡμερωμένοις. ἐν δὲ τοῖς

---

[1] φύονται conj. W.; φυτευονται Ald.H.; φύεται Vo.cod.Cas.
[2] γλυκέων conj. St.; γλαυκίων UMVAld.
[3] *cf.* Athen. 3. 20 and 23.
[4] *cf. C.P.* 1. 9. 1.
[5] In Lesbos: *cf.* 3. 9. 5
[6] *cf. C.P.* 1. 9. 2.

white, and conversely. Again the seed of an excellent vine produces a degenerate result, which is often of quite a different kind; and at times this is not a cultivated kind at all, but a wild one of such a character that it does not ripen its fruit; with others again the result is that the seedlings do not even mature fruit, but only get as far as flowering.

Again the stones of the olive give[1] a wild olive, and the seeds of a sweet pomegranate[2] give a degenerate kind, while the stoneless kind gives a hard sort and often an acid fruit. So also is it with seedlings of pears and apples; pears give a poor sort of wild pears, apples produce an inferior kind which is acid instead of sweet; quince produces wild quince.[3] Almond again raised from seed is inferior in taste and in being hard instead of soft; and this is why men[4] bid us graft on to the almond, even when it is fully grown, or, failing that, frequently plant the offsets.

The oak also deteriorates from seed; at least many persons having raised trees from acorns of the oak at Pyrrha[5] could not produce one like the parent tree. On the other hand they say that bay and myrtle sometimes improve by seeding, though usually they degenerate and do not even keep their colour, but red fruit gives black—as happened with the tree in Antandros; and frequently seed of a 'female' cypress produces a 'male' tree. The date-palm seems to be about the most constant of these trees, when raised from seed, and also the 'cone-bearing pine'[6] (stone-pine) and the 'lice-bearing pine.'[7] So much for degeneration in cultivated trees; among wild kinds it is plain that more in proportion

[7] Plin. 16. 49. The 'lice' are the seeds which were eaten. *cf.* Hdt. 4. 109, φθειροτραγέουσι; Theocr. 5. 49.

ἀγρίοις δῆλον ὅτι πλείω κατὰ λόγον ὡς ἰσχυροτέροις· ἐπεὶ θάτερόν γε καὶ ἄτοπον, εἰ δὴ χείρω καὶ ἐν ἐκείνοις καὶ ὅλως ἐν τοῖς ἀπὸ σπέρματος μόνον· εἰ μή τι τῇ θεραπείᾳ δύνανται μεταβάλλειν.

7 Διαφέρουσι δὲ καὶ τόποι τόπων καὶ ἀὴρ ἀέρος· ἐνιαχοῦ γὰρ ἐκφέρειν ἡ χώρα δοκεῖ τὰ ὅμοια, καθάπερ καὶ ἐν Φιλίπποις· ἀνάπαλιν ὀλίγα καὶ ὀλιγαχοῦ λαμβάνειν μεταβολήν, ὥστε ἐκ σπέρματος ἀγρίου ποιεῖν ἥμερον ἢ ἐκ χείρονος ἁπλῶς βέλτιον· τοῦτο γὰρ ἐπὶ τῆς ῥόας μόνον ἀκηκόαμεν ἐν Αἰγύπτῳ καὶ ἐν Κιλικίᾳ συμβαίνειν· ἐν Αἰγύπτῳ μὲν γὰρ τὴν ὀξεῖαν καὶ σπαρεῖσαν καὶ φυτευθεῖσαν γλυκεῖαν γίνεσθαί πως ἢ οἰνώδη· περὶ δὲ Σόλους τῆς Κιλικίας περὶ ποταμὸν τὸν Πίναρον, οὗ ἡ μάχη πρὸς Δαρεῖον ἐγένετο, πᾶσαι γίνονται ἀπύρηνοι.

8 Εὔλογον δὲ καὶ εἴ τις τὸν παρ' ἡμῶν φοίνικα φυτεύοι ἐν Βαβυλῶνι, κάρπιμόν τε γίνεσθαι καὶ ἐξομοιοῦσθαι τοῖς ἐκεῖ. τὸν αὐτὸν δὲ τρόπον καὶ εἴ τις ἑτέρα προσάλληλον ἔχει καρπὸν τόπῳ· κρείττων γὰρ οὗτος τῆς ἐργασίας καὶ τῆς θεραπείας. σημεῖον δ' ὅτι μεταφερόμενα τἀκεῖθεν ἄκαρπα τὰ δὲ καὶ ὅλως ἀβλαστῆ γίνεται.

9 Μεταβάλλει δὲ καὶ τῇ τροφῇ καὶ διὰ τὴν

---

[1] *i.e.* that they should improve from seed.
[2] Whereas wild trees are produced *only* from seed.
[3] *i.e.* improve a degenerate seedling.
[4] ἁπλῶς : ? om. Sch.
[5] *cf. C.P.* 1. 9. 2.

degenerate from seed, since the parent trees are stronger. For the contrary[1] would be very strange, seeing that degenerate forms are found even in cultivated trees,[2] and among these only in those which are raised from seed. (As a general rule these are degenerate, though men may in some cases effect a change[3] by cultivation).

Effects of situation, climate, tendance.

Again differences in situation and climate affect the result. In some places, as at Philippi, the soil seems to produce plants which resemble their parent; on the other hand a few kinds in some few places seem to undergo a change, so that wild seed gives a cultivated form, or a poor form one actually better.[4] We have heard that this occurs, but only with the pomegranate, in Egypt[5] and Cilicia; in Egypt a tree of the acid kind both from seeds and from cuttings produces one whose fruit has a sort of sweet taste,[6] while about Soli in Cilicia near the river Pinaros (where the battle with Darius was fought) all those pomegranates raised from seed are without stones.

If anyone were to plant our palm at Babylon, it is reasonable to expect that it would become fruitful and like the palms of that country. And so would it be with any other country which has fruits that are congenial to that particular locality; for the locality[7] is more important than cultivation and tendance. A proof of this is the fact that things transplanted thence become unfruitful, and in some cases refuse to grow altogether.

There are also modifications due to feeding[8] and

[6] Or 'wine-like.' Cited by Apollon. *Hist. Mir.* 43.
[7] οὗτος conj. W.; αὐτὸς Ald.
[8] τῇ τροφῇ conj. W.; τῆς τροφῆς UMVAld.

ἄλλην ἐπιμέλειαν, οἷς καὶ τὸ ἄγριον ἐξημεροῦται καὶ αὐτῶν δὲ τῶν ἡμερων ἔνια ἀπαγριοῦται, οἷον ῥόα καὶ ἀμυγδαλῆ. ἤδη δέ τινες καὶ ἐκ κριθῶν ἀναφῦναί φασι πυροὺς καὶ ἐκ πυρῶν κριθὰς καὶ ἐπὶ τοῦ αὐτοῦ πυθμένος ἄμφω. ταῦτα μὲν οὖν ὡς μυθωδέστερα δεῖ δέχεσθαι. μεταβάλλει δ' οὖν τὰ μεταβάλλοντα τὸν τρόπον τοῦτον αὐτομάτως· ἐξαλλαγῇ δὲ χώρας, ὥσπερ ἐν Αἰγύπτῳ καὶ Κιλικίᾳ περὶ τῶν ῥοῶν εἴπομεν, οὐδὲ διὰ μίαν θεραπείαν.

Ὡσαύτως δὲ καὶ ὅπου τὰ κάρπιμα ἄκαρπα γίνεται, καθάπερ τὸ πέρσιον τὸ ἐξ Αἰγύπτου καὶ ὁ φοῖνιξ ἐν τῇ Ἑλλάδι καὶ εἰ δή„ τις κομίσειε τὴν ἐν Κρήτῃ λεγομένην αἴγειρον. ενιοι δέ φασι καὶ τὴν ὄην ἐὰν εἰς ἀλεεινὸν ἔλθῃ σφόδρα τόπον ἄκαρπον γίνεσθαι· φύσει γὰρ ψυχρόν. εὔλογον δὲ ἀμφότερα συμβαίνειν κατὰ τὰς ἐναντιώσεις, εἴπερ μηδ' ὅλως ἔνια φύεσθαι θέλει μεταβάλλοντα τοὺς τόπους. καὶ κατὰ μὲν τὰς χώρας αἱ τοιαῦται μεταβολαί.

11 Κατὰ δὲ τὴν φυτείαν τὰ ἀπὸ τῶν σπερμάτων φυτευόμενα, καθάπερ ἐλέχθη· παντοῖαι γὰρ αἱ ἐξαλλαγαὶ καὶ τούτων. τῇ θεραπείᾳ δὲ μεταβάλλει ῥόα καὶ ἀμυγδαλῆ· ῥόα μὲν κόπρον ὑείαν λαβοῦσα καὶ ὕδατος πλῆθος ῥυτοῦ· ἀμυγδαλῆ δὲ ὅταν πάτταλόν τις ἐνθῇ, καὶ τὸ δάκρυον ἀφαιρῇ τὸ ἐπιρρέον πλείω χρόνον καὶ τὴν ἄλλην ἀποδιδῷ

[1] ἔνια ἀπαγριοῦται οἷον conj. W.; ἔνια καὶ ἀπορῇ τε ῥόα UV; ἔ. καὶ ἀπορῇ τὰ ῥόα M; ἔ. καὶ ἀπορρεῖ τὰ ῥόα Ald.

[2] *i.e.* cultivation has nothing to do with it.

[3] 2. 2. 7. [4] *cf.* 3. 3. 4. [5] Plin. 17. 242.

[6] *i.e.* improve. *cf.* 2. 2. 6 *ad fin.*

attention of other kinds, which cause the wild to become cultivated, or again cause some cultivated kinds to go wild,[1] such as pomegranate and almond. Some say that wheat has been known to be produced from barley, and barley from wheat, or again both growing on the same stool; but these accounts should be taken as fabulous. Anyhow those things which do change in this manner do so spontaneously,[2] and the alteration is due to a change of position (as we said[3] happens with pomegranates in Egypt and Cilicia), and not to any particular method of cultivation.

So too is it when fruit-bearing trees become unfruitful, for instance the *persion* when moved from Egypt, the date-palm when planted in Hellas, or the tree which is called 'poplar' in Crete,[4] if anyone should transplant it. [5] Some again say that the sorb becomes unfruitful if it comes into a very warm position, since it is by nature cold-loving. It is reasonable to suppose that both results follow because the natural circumstances are reversed, seeing that some things entirely refuse to grow when their place is changed. Such are the modifications due to position.

As to those due to method of culture, the changes which occur in things grown from seed are as was said; (for with things so grown also the changes are of all kinds). Under cultivation the pomegranate and the almond change character,[6] the pomegranate if it receives pig-manure[7] and a great deal of river water, the almond if one inserts a peg and[8] removes for some time the gum which exudes and gives the other

[7] *cf.* *C.P.* 2. 14. 2; 3. 9. 3; Plin. 17. 259; Col. 5. 10. 15 and 16.
[8] *cf.* 2. 7. 6; *C.P.* 1. 17. 10; 2. 14. 1; Plin. 17. 252.

12 θεραπείαν. ὡσαύτως δὲ δῆλον ὅτι καὶ ὅσα ἐξημεροῦται τῶν ἀγρίων ἢ ἀπαγριοῦται τῶν ἡμέρων· τὰ μὲν γὰρ θεραπείᾳ τὰ δ' ἀθεραπευσίᾳ μεταβάλλει· πλὴν εἴ τις λέγοι μηδὲ μεταβολὴν ἀλλ' ἐπίδοσιν εἰς τὸ βέλτιον εἶναι καὶ χεῖρον· οὐ γὰρ οἷόν τε τὸν κότινον ποιεῖν ἐλάαν οὐδὲ τὴν ἀχράδα ποιεῖν ἄπιον οὐδὲ τὸν ἐρινεὸν συκῆν. ὃ γὰρ ἐπὶ τοῦ κοτινου φασὶ συμβαινειν, ὥστ' ἐὰν περικοπεὶς τὴν θαλιαν ὅλως μεταφυτευθῇ φέρειν φαυλίας, μετακίνησίς τις γίνεται οὐ μεγάλη. ταῦτα μὲν οὖν ὁποτέρως δεῖ λαβεῖν οὐθὲν ἂν διαφέροι.

III. Φασὶ δ' οὖν αὐτομάτην τινὰ γίνεσθαι τῶν τοιούτων μεταβολήν, ὁτὲ μὲν τῶν καρπῶν ὁτὲ δὲ καὶ ὅλως αὐτῶν τῶν δένδρων, ἃ καὶ σημεῖα νομίζουσιν οἱ μάντεις· οἷον ῥόαν ὀξεῖαν γλυκεῖαν ἐξενεγκεῖν καὶ γλυκεῖαν ὀξεῖαν· καὶ πάλιν ἁπλῶς αὐτὰ τὰ δένδρα μεταβάλλειν, ὥστε ἐξ ὀξείας γλυκεῖαν γίνεσθαι καὶ ἐκ γλυκείας ὀξεῖαν· χεῖρον δὲ τὸ εἰς γλυκεῖαν μεταβάλλειν. καὶ ἐξ ἐρινεοῦ συκῆν καὶ ἐκ συκῆς ἐρινεόν· χεῖρον δὲ τὸ ἐκ συκῆς. καὶ ἐξ ἐλάας κότινον καὶ ἐκ κοτίνου ἐλάαν· ἥκιστα δὲ τοῦτο. πάλιν δὲ συκῆν ἐκ

---

[1] περικοπεὶς conj W.; περισκοπτεῖς U; περικόπτης Ald.

[2] φαυλίας conj. Salm.; φαύλους U; θάλος Ald. *cf.* Plin. 16. 244. These olives produced little oil, but were valued for perfumery: see *C.P.* 6. 8. 3 and 5; *de odor.*, 15.

[3] οὐ add. Salm.; om. MSS. (?) Ald. H.

attention required. In like manner plainly some wild things become cultivated and some cultivated things become wild; for the one kind of change is due to cultivation, the other to neglect:—however it might be said that this is not a change but a natural development towards a better or an inferior form; (for that it is not possible to make a wild olive pear or fig into a cultivated olive pear or fig). As to that indeed which is said to occur in the case of the wild olive, that if the tree is transplanted with its top-growth entirely cut off,[1] it produces 'coarse olives,'[2] this is no[3] very great change. However it can make no difference which way[4] one takes this.

Of spontaneous changes in the character of trees, and of certain marvels.

III. [5]Apart from these changes it is said that in such plants there is a spontaneous kind of change, sometimes of the fruit, sometimes of the tree itself as a whole, and soothsayers call such changes portents. For instance, an acid pomegranate, it is said, may produce sweet fruit, and conversely; and again, in general, the tree itself sometimes undergoes a change, so that it becomes sweet[6] instead of acid, or the reverse happens. And the change to sweet is considered a worse portent. Again a wild fig may turn into a cultivated one, or the contrary change take place; and the latter is a worse portent. So again a cultivated olive may turn into a wild one, or conversely, but the latter change is rare. So again a white fig

[4] *i.e.* whether nature or man is said to cause the admitted change. [5] Plin. 17. 242.

[6] *i.e. all* the fruit is now acid instead of sweet, or the reverse. Sch. brackets ἐξ ὀξείας . . . ὀξεῖαν.

λευκῆς μέλαιναν καὶ ἐκ μελαίνης λευκήν. ὁμοίως δὲ τοῦτο καὶ ἐπὶ ἀμπέλου.

2 Καὶ ταῦτα μὲν ὡς τέρατα καὶ παρὰ φύσιν ὑπολαμβάνουσιν· ὅσα δὲ συνήθη τῶν τοιούτων οὐδὲ θαυμάζουσιν ὅλως· οἷον τὸ τὴν κάπνειον ἄμπελον καλουμένην καὶ ἐκ μέλανος βότρυος λευκὸν καὶ ἐκ λευκοῦ μέλανα φέρειν· οὐδὲ γὰρ οἱ μάντεις τὰ τοιαῦτα κρίνουσιν· ἐπεὶ οὐδὲ ἐκεῖνα, παρ' οἷς πέφυκεν ἡ χώρα μεταβάλλειν, ὥσπερ ἐλέχθη περὶ τῆς ῥόας ἐν Αἰγύπτῳ· ἀλλὰ τὸ ἐνταῦθα θαυμαστόν, διὰ τὸ μίαν μόνον ἢ δύο, καὶ ταύτας ἐν τῷ παντὶ χρόνῳ σπανίας. οὐ μὴν ἀλλ' εἴπερ συμβαίνει, μᾶλλον ἐν τοῖς καρποῖς γίνεσθαι τὴν παραλλαγὴν ἢ ἐν ὅλοις τοῖς δένδροις.

3 Ἐπεὶ καὶ τοιαύτη τις ἀταξία γίνεται περὶ τοὺς καρπούς· οἷον ἤδη ποτὲ συκῇ τὰ σῦκα ἔφυσεν ἐκ τοῦ ὄπισθεν τῶν θρίων· καὶ ῥόα δὲ καὶ ἄμπελος ἐκ τῶν στελεχῶν, καὶ ἄμπελος ἄνευ φύλλων καρπὸν ἤνεγκεν. ἐλάα δὲ τὰ μὲν φύλλα ἀπέβαλε τὸν δὲ καρπὸν ἐξήνεγκεν· ὃ καὶ Θετταλῷ τῷ Πεισιστράτου γενέσθαι λέγεται. συμβαίνει δὲ καὶ διὰ χειμῶνας τοῦτο καὶ δι' ἄλλας αἰτίας ἔνια τῶν δοκούντων εἶναι παρὰ λόγον οὐκ ὄντων δέ· οἷον ἐλάα ποτ' ἀποκαυθεῖσα τελέως ἀνεβλάστησεν ὅλη, καὶ αὐτὴ καὶ ἡ θαλία. ἐν δὲ τῇ Βοιωτίᾳ καταβρωθέντων τῶν ἐρνῶν ὑπ' ἀττελέβων πάλιν

---

[1] ἐπὶ conj. Sch.; ἐξ Ald.H.
[2] *cf.* *C.P.* 5. 3. 1 and 2; Arist. *de gen. an.* 4. 4; Hesych. *s.v.* καπνίας; Schol. ad Ar. *Vesp.* 151. [3] 2. 2. 7.
[4] εἰκὸς has perhaps dropped out. Sch.
[5] θρίων conj. R. Const., *cf.* *C.P.* 5. 1. 7 and 8; 5. 2. 2; ἐρινεῶν P_2Ald. *cf.* also Athen. 3. 11.

may change into a black one, and conversely; and similar changes occur in[1] the vine.

Now these changes they interpret as miraculous and contrary to nature; but they do not even feel any surprise at the ordinary changes, for instance, when the 'smoky' vine,[2] as it is called, produces alike white grapes instead of black or black grapes instead of white. Of such changes the soothsayers take no account, any more than they do of those instances in which the soil produces a natural change, as was said[3] of the pomegranate in Egypt. But it is surprising when such a change occurs in our own country, because there are only one or two instances and these separated by wide intervals of time. However, if such changes occur, it is natural[4] that the variation should be rather in the fruit than in the tree as a whole. In fact the following irregularity also occurs in fruits; a fig-tree has been known to produce its figs from behind the leaves,[5] pomegranate and vines from the stem, while the vine has been known to bear fruit without leaves. The olive again has been known to lose its leaves and yet produce its fruit; this is said to have happened to Thettalos, son of Pisistratus. This may be due to inclement weather; and some changes, which seem to be abnormal, but are not really so, are due to other accidental causes; [6]for instance, there was an olive that, after being completely burnt down, sprang up again entire, the tree and all its branches. And in Boeotia an olive whose young shoots[7] had been eaten off by locusts grew again: in this case however[8] the

[6] *cf.* Hdt. 8. 55; Plin. 17. 241.

[7] ἐρνῶν conj. Sch.; ἔργων P_2Ald.; κλάδων mU.

[8] *i.e.* the portent was not so great as in the other case quoted, as the tree itself had not been destroyed.

ἀνεβλάστησε· τὰ δ' οἷον ἀπέπεσεν. ἥκιστα δ' ἴσως τὰ τοιαῦτα ἄτοπα διὰ το φανερὰς ἔχειν τὰς αἰτίας, ἀλλὰ μᾶλλον τὸ μὴ ἐκ τῶν οἰκείων τόπων φέρειν τοὺς καρποὺς ἢ μὴ οἰκείους·[1] καὶ μάλιστα δ' εἰ[2] τῆς ὅλης φύσεως γίνεται μεταβολή, καθάπερ ἐλέχθη.[3] περὶ μὲν οὖν τὰ δένδρα τοιαῦται τινές εἰσι μεταβολαί.

IV. Τῶν δὲ ἄλλων τό τε σισύμβριον εἰς μίνθαν δοκεῖ μεταβάλλειν, ἐὰν μὴ κατέχηται τῇ θεραπείᾳ, δι' ὃ καὶ μεταφυτεύουσι πολλάκις, καὶ ὁ πυρὸς εἰς αἶραν. ταῦτα μὲν οὖν ἐν τοῖς δένδροις αὐτομάτως, εἴπερ γίνεται. τὰ δ' ἐν τοῖς ἐπετείοις διὰ παρασκευῆς· οἷον ἡ τίφη καὶ ἡ ζειὰ μεταβάλλουσιν εἰς πυρὸν ἐὰν πτισθεῖσαι σπείρωνται, καὶ τοῦτ' οὐκ εὐθὺς ἀλλὰ τῷ τρίτῳ ἔτει. σχεδὸν δὲ παραπλήσιον τοῦτό γε τῷ τὰ σπέρματα κατὰ τὰς χώρας μεταβάλλειν· μεταβάλλει γὰρ καὶ ταῦτα καθ' ἑκάστην χώραν καὶ σχεδὸν ἐν τῷ ἴσῳ χρόνῳ καὶ ἡ τίφη. μεταβάλλουσι δὲ καὶ οἱ ἄγριοι πυροὶ καὶ αἱ κριθαὶ θεραπευόμεναι καὶ ἐξημερούμεναι κατὰ τὸν ἴσον χρόνον.

2 Καὶ ταῦτα μὲν ἔοικε χώρας τε μεταβολῇ και θεραπείᾳ γίνεσθαι· καὶ ἔνια ἀμφοτέροις, τὰ δὲ τῇ θεραπείᾳ μόνον· οἷον πρὸς τὸ τὰ ὄσπρια μὴ γίνεσθαι ἀτεράμονα[4] βρέξαντα κελεύουσιν ἐν νίτρῳ[5]

[1] οἰκείους· καὶ I conj.; οἰκειοῦται UMV; οἰκείως Ald.H.; ἐοικότας conj. W. [2] εἰ ins. Sch. [3] 2. 3. 1.
[4] *cf.* 6. 7. 2; Plin. 19. 176.
[5] *i.e.* to prevent the change which cultivated soil induces.

shoots had, so to speak, only been shed. But after all such phenomena are perhaps far from strange, since the cause in each case is obvious; rather is it strange that trees should bear fruit not at the places where it naturally forms, or else fruit which does not belong to the character[1] of the tree. And most surprising of all is it when,[2] as has been said,[3] there is a change in the entire character of the tree. Such are the changes which occur in trees.

Of spontaneous and other changes in other plants.

IV. [4] Of other plants it appears that bergamot-mint turns into cultivated mint, unless it is fixed by special attention; and this is why men frequently transplant[5] it; [6] so too wheat turns into darnel. Now in trees such changes, if they occur, are spontaneous, but in annual plants they are deliberately brought about: for instance, one-seeded wheat and rice-wheat change[7] into wheat, if bruised before they are sown; and this does not happen at once, but in the third year. This change resembles that produced in the seeds by difference of soil[8]; for these grains vary according to the soil, and the change takes about the same time as that which occurs in one-seeded wheat. Again wild wheats and barleys also with tendance and cultivation change in a like period.

These changes appear to be due to change of soil and cultivation, and in some cases the change is due to both, in others to cultivation alone; for instance, in order that pulses may not become uncookable,[9]

[6] But see reff. under *αἶρα* in Index.
[7] *cf. C.P.* 5. 6. 12; Plin. 18. 93.
[8] *χώραν* conj. St.; *ὥραν* Ald.H.
[9] *ἀτεράμονα* conj. W.; *ἀτέραμνα* UAld. *cf.* 8. 8. 6 and 7; *C.P.* 4. 7. 2; 4. 12. 1 and 8; *Geop.* 2. 35. 2; 2. 41.

νύκτα τῇ ὑστεραίᾳ σπείρειν ἐν ξηρᾷ· φακοὺς ὥστε ἁδροὺς γίνεσθαι φυτεύουσιν ἐν βολίτῳ· τοὺς ἐρεβίνθους δέ, ὥστε μεγάλους, αὐτοῖς τοῖς κελύφεσι βρέξαντα σπείρειν. μεταβάλλουσι δὲ καὶ κατὰ τὰς ὥρας τοῦ σπόρου πρὸς κουφότητα καὶ ἀλυπίαν· οἷον ἐάν τις τοὺς ὀρόβους ἐαρινοὺς σπείρῃ τρισάλυποι γίνονται, καὶ οὐχ ὡς οἱ μετοπωρινοὶ βαρεῖς.

3 Γίνεται δὲ καὶ ἐν τοῖς λαχάνοις μεταβολὴ διὰ τὴν θεραπείαν· οἷον τὸ σέλινον, ἐὰν σπαρὲν καταπατηθῇ καὶ κυλινδρωθῇ, ἀναφύεσθαί φασιν οὖλον. μεταβάλλει δὲ καὶ τὴν χώραν ἐξαλλάττοντα, καθάπερ καὶ τἆλλα. καὶ τὰ μὲν τοιαῦτα κοινὰ πάντων ἐστίν. εἰ δὲ κατά τινα πήρωσιν ἢ ἀφαίρεσιν μέρους δένδρον ἄγονον γίνεται, καθάπερ τὰ ζῷα, τοῦτο σκεπτέον· οὐδὲν γοῦν φανερὸν κατά γε τὴν διαίρεσιν εἰς τὸ πλείω καὶ ἐλάττω φέρειν ὥσπερ κακούμενον, ἀλλ' ἢ ἀπόλλυται τὸ ὅλον ἢ διαμένον καρποφορεῖ. τὸ δὲ γῆρας κοινή τις φθορὰ πᾶσιν.

4 Ἄτοπον δ' ἂν δόξειε μᾶλλον εἰ ἐν τοῖς ζῴοις αἱ τοιαῦται μεταβολαὶ φυσικαὶ καὶ πλείους· καὶ γὰρ κατὰ τὰς ὥρας ἔνια δοκεῖ μεταβάλλειν, ὥσπερ ὁ ἱέραξ καὶ ἔποψ καὶ ἄλλα τῶν ὁμοίων ὀρνέων. καὶ κατὰ τὰς τῶν τόπων ἀλλοιώσεις, ὥσπερ ὁ ὕδρος εἰς ἔχιν ξηραινομένων τῶν λιβά-

---

[1] νύκτα I conj.; νυκτὶ MSS.
[2] ἐν βολίτῳ conj. Milas. on *Geop.* 3. 27; ἔμβολον UMV Ald. *cf.* *C.P.* 5 6. 11; Col. 2 10. 15; Plin. 18. 198.
[3] *cf.* *C.P.* 5. 6. 11; *Geop.* 2. 3. 6.
[4] ἀλυπίαν conj. Sch.; δι' ἀλυπίας M; δι' ἀλυπίαν Ald.

men bid one moisten the seed in nitre for a night[1] and sow it in dry ground the next day. To make lentils vigorous they plant the seeds in dung[2]; to make chick-peas large they bid one moisten the seed while still in the pods,[3] before sowing. Also the time of sowing makes differences which conduce to digestibility and harmlessness[4]: thus, if one sows vetches[5] in spring, they become quite harmless and are not indigestible like those sown in autumn.

Again in pot-herbs change is produced by cultivation; for instance, they say that,[6] if celery seed is trodden and rolled in after sowing, it comes up curly; it also varies from change of soil, like other things. Such variations are common to all; we must now consider whether a tree, like animals, becomes unproductive from mutilation or removal of a part. At all events it does not appear that division[7] is an injury, as it were, which affects the amount of fruit produced; either the whole tree perishes, or else, if it survives,[8] it bears fruit. Old age however is a cause which in all plants puts an end to life. . . . .[9]

It would seem more surprising if[10] the following changes occurred in animals naturally and frequently; some animals do indeed seem to change according to the seasons, for instance, the hawk the hoopoe and other similar birds. So also changes in the nature of the ground produce changes in animals, for instance, the water-snake changes into a viper, if the marshes

[5] *cf.* Plin. 18. 139; Col. 2. 10. 34.
[6] *cf.* *C.P.* 5. 6. 7; *Geop.* 12. 23. 2.
[7] γε conj. Sch.; τε Ald.
[8] διάμενον conj. Sch.; διαμένοντα Ald.
[9] Something seems to have been lost at the end of § 3.
[10] εἰ ins. Sch.; τοιαῦται may however mean 'the above-mentioned,' and refer to something which has been lost.

δων. φανερώτατα δὲ καὶ κατὰ τὰς γενέσεις ἔνια, καὶ μεταβάλλει διὰ πλειόνων ζωων· οἷον ἐκ κάμπης γίνεται χρυσαλλὶς εἶτ᾽ ἐκ ταύτης ψυχή· καὶ ἐπ᾽ ἄλλων δ᾽ ἐστὶ τοῦτο πλειόνων, οὐδὲν ἴσως ἄτοπον, οὐδ᾽ ὅμοιον τὸ ζητούμενον. ἀλλ᾽ ἐκεῖνο συμβαίνει περὶ τὰ δένδρα καὶ ὅλως πᾶσαν τὴν ὕλην, ὥσπερ ἐλέχθη καὶ πρότερον, ὥστε αὐτομάτην μεταβλαστάνειν μεταβολῆς τινος γινομένης ἐκ τῶν οὐρανίων τοιαύτης. τὰ μὲν οὖν περὶ τὰς γενέσεις καὶ μεταβολὰς ἐκ τούτων θεωρητέον.

V. Ἐπεὶ δὲ καὶ αἱ ἐργασίαι καὶ αἱ θεραπεῖαι μεγάλα συμβάλλονται, καὶ ἔτι πρότερον αἱ φυτεῖαι καὶ ποιοῦσι μεγάλας διαφοράς, λεκτέον καὶ περὶ τούτων.

Καὶ πρῶτον περὶ τῶν φυτειῶν. αἱ μὲν οὖν ὧραι πρότερον εἴρηνται καθ᾽ ἃς δεῖ. τὰ δὲ φυτὰ λαμβάνειν κελεύουσιν ὡς κάλλιστα καὶ ἐξ ὁμοίας γῆς εἰς ἣν μέλλεις φυτεύειν, ἢ χειρονος· τοὺς δὲ γυροὺς προορύττειν ὡς πλείστου χρόνου καὶ βαθυτέρους αἰεὶ καὶ τοῖς ἐπιπολαιορριζοτέροις.

---

[1] *i.e.* in the instance given the development of an insect exhibits, not one, but a series of changes from one creature to another.

[2] Whereas the metamorphoses mentioned above are independent of climatic conditions.

[3] δὲ conj. W.; τε Ald.

[4] κάλλιστα conj. W., *cf. C.P.* 3. 24. 1; τάχιστα MVAld.; τὰ χίστα U.

dry up. Most obvious are certain changes in regard to the way in which animals are produced, and such changes run through a series of creatures[1]; thus a caterpillar changes into a chrysalis, and this in turn into the perfect insect; and the like occurs in a number of other cases. But there is hardly anything abnormal in this, nor is the change in plants, which is the subject of our enquiry, analogous to it. That kind of change occurs in trees and in all woodland plants generally, as was said before, and its effect is that, when a change of the required character occurs in the climatic conditions, a spontaneous change in the way of growth ensues.[2] These instances must suffice for investigation of the ways in which plants are produced or modified.

Of methods of propagation, with notes on cultivation.

V. Since however methods of cultivation and tendance largely contribute, and, before these, methods of planting, and cause great differences, of these too we must speak.

And first of methods of planting: as to the seasons, we have already stated at what seasons one should plant. Further[3] we are told that the plants chosen should be the best possible,[4] and should be taken from soil resembling that in which you are going to plant them, or else inferior[5]; also the holes should be dug[6] as long as possible beforehand, and should always be deeper than the original holes, even for those whose roots do not run very deep.

[5] *i.e.* the shift should be into better soil, if possible. *cf. C.P.* 3. 5. 2.

[6] γυροὺς προορύττειν conj. R. Const.; πυροὺς προσορύττειν UMVAld. *cf. C.P.* 3. 4. 1.

2 Λέγουσι δέ τινες ὡς οὐδεμία κατωτέρω διϊκνεῖται τριῶν ἡμιποδίων· δι᾽ ὃ καὶ ἐπιτιμῶσι τοῖς ἐν μείζονι βάθει φυτεύουσιν· οὐκ ἐοίκασι δὲ ὀρθῶς λέγειν ἐπὶ πολλῶν· ἀλλ᾽ ἐὰν ἢ χώματος ἐπιλάβηται βαθέος ἢ καὶ χώρας τοιαύτης ἢ καὶ τόπου, πολλῷ μακροτέραν ὠθεῖ τὸ τῇ φύσει βαθύρριζον. πεύκην δέ τις ἔφη μεταφυτεύων μεμοχλευμένην μείζω τὴν ῥίζαν ἔχειν ὀκτάπηχυν καίπερ οὐχ ὅλης ἐξαιρεθείσης ἀλλ᾽ ἀπορραγεισης.

3 Τὰ δὲ φυτευτήρια ἐὰν μὲν ἐνδέχηται ὑπόρριζα, εἰ δὲ μή, δεῖ μᾶλλον ἀπὸ τῶν κάτω ἢ τῶν ἄνω λαμβάνειν, πλὴν ἀμπέλου· καὶ τὰ μὲν ἔχοντα ῥίζας ὀρθὰ ἐμβάλλειν, τὰ δὲ μὴ ἔχοντα ὑποβάλλειν τοῦ φυτευτηρίου ὅσον σπιθαμὴν ἢ μικρῷ πλεῖον. ἔνιοι δὲ κελεύουσι καὶ τῶν ὑπορρίζων ὑποβαλλειν, τιθέναι δὲ καὶ τὴν θέσιν ὁμοίως ἥνπερ εἶχεν ἐπὶ τῶν δένδρων τὰ πρόσβορρα καὶ τὰ πρὸς ἕω καὶ τὰ πρὸς μεσημβρίαν. ὅσα δὲ ἐνδέχεται τῶν φυτῶν καὶ προμοσχεύειν· τὰ μὲν ἐπ᾽ αὐτῶν τῶν δένδρων, οἷον ἐλάας ἀπίου μηλέας συκῆς· τὰ δ᾽ ἀφαιροῦντας, οἷον ἀμπέλου· ταύτην γὰρ οὐχ οἷόν τε ἐπ᾽ αὐτῆς μοσχεύειν.

4 Ἐὰν δὲ μὴ ὑπόρριζα τὰ φυτὰ μηδὲ ὑπόπρεμνα

---

[1] ἀλλ᾽ ἐὰν . . . τοιούτου. ἐὰν ᾖ μὲν σώματος M; so V, but ἢ; ᾖ om. PAld.; χώματος H; κενώματος for σώματος and εὐδιόδου for ᾖ καὶ τόπου conj. W. χώρας refers to exposure, etc., τόπου (sc. τοιούτου) to quality of soil: so G.

[2] Plin. 16. 129; Xen. *Oec.* 19. 3. [3] *cf.* *C.P.* 3. 6.

Some say that no root goes down further than a foot and a half, and accordingly they blame those who plant deeper. However there are many instances in which it appears that what they say does not hold good; a plant which is naturally deep-rooting pushes much deeper if it finds either a deep mass of soil or a position which favours such growth or again the kind of ground which favours it.[1] In fact,[2] a man once said that when he was transplanting a fir which he had uprooted with levers, he found that it had a root more than eight cubits long, though the whole of it had not been removed, but it was broken off.

The slips for planting should be taken, if possible, with roots attached, or, failing that, from the lower[3] rather than from the higher parts of the tree, except in the case of the vine; those that have roots should be set upright,[4] while in the case of those which have none about[5] a handsbreadth or rather more of the slip should be buried. Some say that part even of those which have roots should be buried, and that the position[6] should be the same as that of the tree from which the slip was taken, facing north or east or south, as the case may be. With those plants with which it is possible, shoots from the boughs should also, they say, be planted, some being set on the trees themselves,[7] as with olive pear apple and fig, but in other cases, as in that of the vine, they must be set separately, for that the vine cannot be grafted on itself.

If the slips cannot be taken with root or stock

[4] *cf.* *C.P.* 3. 6. 4; Xen. *Oec.* 19. 9.
[5] ὅσον conj. Sch.; οἷον P_2Ald.
[6] *cf.* *C.P.* 3. 5. 2. [7] *i.e.* grafted.

λαμβάνειν, καθάπερ τῆς ἐλάας, σχίσαντά τε τὸ ξύλον κάτωθεν καὶ λίθον ἐμβαλόντα φυτεύειν· ὁμοίως δὲ καὶ τῆς ἐλάας καὶ συκῆς καὶ τῶν ἄλλων. φυτεύεται δὲ ἡ συκῆ καὶ ἐάν τις κράδην παχεῖαν ἀποξύνας σφύρᾳ παίῃ, ἄχρι οὗ ἂν ἀπολίπῃ μικρὸν ὑπὲρ τῆς γῆς, εἶτ' αὐτῆς ἄμμον βαλὼν ἄνωθεν ἐπιχώσῃ· καὶ γίνεσθαι δή φασι καὶ καλλίω ταῦτα τὰ φυτά, μέχρι οὗ ἂν ᾖ νέα.

5 Παραπλησία καὶ τῶν ἀμπέλων, ὅταν ἀπὸ τοῦ παττάλου· προοδοποιεῖ γὰρ ὁ πάτταλος ἐκείνῳ τῷ κλήματι διὰ τὴν ἀσθένειαν· φυτεύουσιν οὕτω καὶ ῥόαν καὶ ἄλλα τῶν δένδρων. ἡ συκῆ δέ, ἐὰν ἐν σκίλλῃ φυτευθῇ, θᾶττον παραγίνεται καὶ ἧττον ὑπὸ σκωλήκων κατεσθίεται. ὅλως δὲ πᾶν ἐν σκίλλῃ φυτευόμενον εὐβλαστὲς καὶ θᾶττον αὐξάνεται. ὅσα δὲ ἐκ τοῦ στελέχους καὶ διακοπτόμενα φυτεύεται, κάτω τρέποντα τὴν τομὴν δεῖ φυτεύειν, διακόπτειν δὲ μὴ ἔλαττω σπιθαμιαίων, ὥσπερ ἐλέχθη, καὶ τὸν φλοιὸν προσεῖναι· φύεται δ' ἐκ τῶν τοιούτων ἔρνη· βλαστανόντων δ' ἀεὶ προσχωννύειν, ἄχρι οὗ ἂν γένηται ἄρτιον· αὕτη μὲν οὖν τῆς ἐλάας ἰδία καὶ τοῦ μυρρίνου, αἱ δ' ἄλλαι κοινότεραι πᾶσιν.

6 Ἄριστον δὲ καὶ ῥιζώσασθαι καὶ φυτείας μάλιστα τῆς τυχούσης ἡ συκῆ. φυτεύειν δὲ ῥόας μὲν

---

[1] ἡ before τῆς om. W. [2] τε τὸ conj. W.; τό τε MVP.
[3] καὶ τῆς ἐλαίας U; ἐλάας MVP; so W.
[4] Plin. 17, 123. [5] *cf.* *C.P.* 3. 12. 1.
[6] *cf.* 7. 13. 4; *C.P.* 5. 6. 10 (where another bulb, σχῖνος, is mentioned as being put to the same use); Athen. 3. 13; Plin. 17. 87.

attached, as with the olive,[1] they say that one must[2] split the wood at the lower end and plant with a stone on top; and the fig and other trees must be treated in like manner with the olive.[3] The fig[4] is also propagated by sharpening a stout shoot and driving it in with a hammer, till only a small piece of it is left above ground, and then piling sand above so as to earth it up; and they say that the plants thus raised grow finer up to a certain age.

Similar is the method used with vines, when they are propagated by the 'peg'[5] method; for the peg makes a passage for that sort of shoot on account of its weakness; and in the same manner men plant the pomegranate and other trees. The fig progresses more quickly and is less eaten by grubs, if the cutting is set in a squill-bulb[6]; in fact anything so planted is vigorous and grows faster. All those trees which are propagated by pieces cut from the stem should be planted with the cut part downwards,[7] and the pieces cut off should not be less than a handsbreadth in length, as was said,[8] and the bark should be left on. From such pieces new shoots grow, and as they grow, one should keep on heaping up earth about them, till the tree becomes strong.[9] This kind of propagation is peculiar to the olive and myrtle, while the others are more or less common to all trees.

The fig is better than any other tree at striking roots, and will, more than any other tree, grow by any method of propagation. [10] We are told that,

[7] *cf. Geop.* 9. 11. 8.

[8] 2. 5. 3, where however the method of propagation is different.

[9] ἄρτιον Ald.; ἀρτιτελῆ conj. W. (*quoad satis corroboretur* G; *donec robur planta capiat* Plin. 17. 124); ἄρτιτεων U; ἄρτι τέων MV; ἄρτι τεῶν P_2. [10] *cf. C.P.* 3. 7.

καὶ μυρρίνους καὶ δάφνας πυκνὰς κελεύουσι, μὴ πλέον διεστώσας ἢ ἐννέα πόδας, μηλέας δὲ μικρῷ μακρότερον, ἀπίους δὲ καὶ ὄγχνας ἔτι μᾶλλον, ἀμυγδαλᾶς δὲ καὶ συκᾶς πολλῷ πλέον, ὡσαύτως δὲ καὶ τὴν ἐλάαν. ποιεῖσθαι δὲ καὶ πρὸς τὸν τόπον τὰς ἀποστάσεις· ἐν γὰρ τοῖς ὀρεινοῖς ἐλάττους ἢ ἐν τοῖς πεδεινοῖς.

Μέγιστον δὲ ὡς εἰπεῖν τὸ τὴν πρόσφορον ἑκάστῳ χώραν ἀποδιδόναι· τότε γὰρ εὐθενεῖ μάλιστα. ὡς δ' ἁπλῶς εἰπεῖν ἐλάᾳ μὲν καὶ συκῇ καὶ ἀμπέλῳ τὴν πεδεινήν φασιν οἰκειοτάτην εἶναι, τοῖς δὲ ἀκροδρύοις τὰς ὑπωρείας. χρὴ δὲ καὶ ἐν αὐτοῖς τοῖς ὁμογενέσι μὴ ἀγνοεῖν τὰς οἰκείας. ἐν πλείστῃ δὲ ὡς εἰπεῖν διαφορᾷ τὰ τῶν ἀμπέλων ἐστίν· ὅσα γάρ ἐστι γῆς εἴδη, τοσαῦτά τινές φασι καὶ ἀμπέλων εἶναι. φυτευόμενα μὲν οὖν κατὰ φύσιν ἀγαθὰ γίνεσθαι παρὰ φύσιν δὲ ἄκαρπα. ταῦτα μὲν οὖν ὥσπερ κοινὰ πάντων.

VI. Τῶν δὲ φοινίκων ἴδιος ἡ φυτεία παρὰ τἆλλα καὶ ἡ μετὰ ταῦτα θεραπεία. φυτεύουσι γὰρ πλείους εἰς ταὐτὸ τιθέντες δύο κάτω καὶ δύο ἄνωθεν ἐπιδοῦντες, πρανεῖς δὲ πάντας. τὴν γὰρ ἔκφυσιν οὐκ ἐκ τῶν ὑπτίων καὶ κοίλων ποιεῖται, καθάπερ τινές φασιν, ἀλλ' ἐκ τῶν ἄνω, δι' ὃ καὶ ἐν τῇ ἐπιζεύξει τῶν ἐπιτιθεμένων οὐ δεῖ περικαλύπτειν τὰς ἀρχὰς ὅθεν ἡ ἔκφυσις· φανεραὶ δ'

---

[1] ἐλάαν conj. Bod. (*cf.* Plin. 17. 88); ῥοιὰν UAld.H.
[2] ἐλάττονι conj. Sch.; ἔλαττον Ald.
[3] *i.e.* apples pears plums, etc.

in planting the pomegranate myrtle or bay, one should set two trees close together, not further than nine feet apart, apples a little further, pears and wild pears still further, almonds and figs further still, and in like manner the olive.[1] Again the distance apart must be regulated by the nature of the ground, being less [2] in hilly parts than in low ground.

Most important of all, one may say, is it to assign to each the suitable soil; for then is the tree most vigorous. Speaking generally, they say that low ground is most suitable for the olive fig and vine, and the lower slopes of hills for fruit trees.[3] Nor should one fail to note what soil suits each variety even of those closely related. There is the greatest difference, one may say, between the different kinds of vine: for they say that there are as many kinds of vine as there are of soil. If they are planted as their nature requires, they turn out well, if otherwise, they are unfruitful. And these remarks apply almost equally to all trees.

Of the propagation of the date-palm; of palms in general.

VI. [4] The method of propagating date-palms is peculiar and exceptional, as also is their subsequent cultivation. They plant several seeds together, putting two below and two above, which are fastened on; but all face downwards.[5] For germination starts not, as some say, from the 'reverse' or hollow side,[6] but from the part [7] which is uppermost; wherefore in joining on the seeds which are placed above one must not cover up the points from which the growth

[4] Plin. 13. 32.
[5] *i.e.* with the grooved side downwards.
[6] *i.e.* the grooved side. [7] *i.e.* the round side.

εἴσι τοῖς ἐμπείροις. διὰ τοῦτο δ' εἰς τὸ αὐτὸ πλείους τιθέασιν ὅτι ἀπὸ τοῦ ἑνὸς ἀσθενὴς ἡ φυτεία. τούτων δὲ αἵ τε ῥίζαι πρὸς ἀλλήλας συμπλέκονται καὶ εὐθὺς αἱ πρῶται βλαστήσεις, ὥστε ἓν γίνεσθαι τὸ στέλεχος.

2 Ἡ μὲν οὖν ἀπὸ τῶν καρπῶν φυτεία τοιαύτη τις· ἡ δ' ἀφ' αὐτοῦ, ὅταν ἀφέλωσι τὸ ἄνω ἐν ᾧπερ ὁ ἐγκέφαλος[1]· ἀφαιροῦσι δὲ ὅσον δίπηχυ· σχίσαντες δὲ τοῦτο κάτω τιθέασι τὸ ὑγρόν[2]· φιλεῖ δὲ χώραν ἁλμώδη[3]· δι' ὃ καὶ ὅπου μὴ τοιαύτη τυγχάνει περιπάττουσιν ἅλας οἱ γεωργοί· τοῦτο δὲ δεῖ ποιεῖν μὴ περὶ αὐτὰς τὰς ῥίζας ἀλλ' ἄποθεν ἀποστήσαντα περιπάττειν ὅσον ἡμίεκτον· ὅτι δὲ τοιαύτην ζητεῖ χώραν κἀκεῖνο ποιοῦνται σημεῖον· πανταχοῦ γὰρ ὅπου πλῆθος φοινίκων ἁλμώδεις αἱ χῶραι· καὶ γὰρ ἐν Βαβυλῶνί φασιν, ὅπου οἱ φοίνικες πεφύκασι, καὶ ἐν Λιβύῃ δὲ καὶ ἐν Αἰγύπτῳ καὶ Φοινίκῃ καὶ τῆς Συρίας δὲ τῆς κοίλης, ἐν ᾗ γ' οἱ πλεῖστοι τυγχάνουσιν, ἐν τρισὶ μόνοις τόποις ἁλμώδεσιν εἶναι τοὺς δυναμένους θησαυρίζεσθαι· τοὺς δ' ἐν τοῖς ἄλλοις οὐ διαμένειν ἀλλὰ σήπεσθαι, χλωροὺς δ' ἡδεῖς εἶναι καὶ καταναλίσκειν οὕτω.

3 Φιλεῖ δὲ καὶ ὑδρείαν σφόδρα τὸ δένδρον· περὶ δὲ κόπρου διαμφισβητοῦσιν· οἱ μὲν γὰρ οὔ φασι χαίρειν ἀλλ' ἐναντιώτατον εἶναι, οἱ δὲ καὶ χρῆσθαι καὶ ἐπίδοσιν πολλὴν ποιεῖν. δεῖν δ' ὑδρεύειν εὖ μάλα κατὰ τῆς κόπρου, καθάπερ οἱ ἐν

---

[1] *i.e.* 'cabbage.'

[2] τοῦτο . . . ὑγρόν : I have inserted δὲ, otherwise retaining the reading of Ald.; τούτου κάτω· τιθέασι δ' ἔνυγρον conj. W. *cf.* Plin. 13. 36. τὸ ὑγρόν, viz. the cut end.

[3] ἁλμώδη conj. W.; ἀμμώδη P_2Ald.H.

is to come; and these can be recognised by experts. And the reason why they set several together is that a plant that grows from one only is weak. The roots which grow from these seeds become entangled together and so do the first shoots from the very start, so that they combine to make a single stem.

Such is the method of growing from the fruits. But propagation is also possible from the tree itself, by taking off the top, which contains the 'head.'[1] They take off about two cubits' length, and, splitting it, set the moist end.[2] It likes a soil which contains salt[3]; wherefore, where such soil is not available, the growers sprinkle salt about it; and this must not be done about the actual roots: one must keep the salt some way off and sprinkle about a gallon. To shew that it seeks such a soil they offer the following proof; wherever date-palms grow abundantly, the soil is salt,[4] both in Babylon, they say, where the tree is indigenous, in Libya in Egypt and in Phoenicia; while in Coele-Syria, where are[5] most palms, only in three districts, they say, where the soil is salt, are dates produced which can be stored; those that grow in other districts do not keep, but rot, though when fresh they are sweet and men use[6] them at that stage.

[7]The tree is likewise very fond of irrigation; as to dung there is a difference of opinion: some say that the date-palm does not like it, but that it is most injurious, others that it gladly accepts[8] it and makes good growth thereby, but plenty of water should be

[4] ἁλμώδεις conj. W.; ἀμμώδεις Ald. H.

[5] ἐν ᾗ γ' οἱ conj. W.; ἵν' Ἴνδοι U; ἣν Ἴνδοι MVAld.

[6] καταναλίσκειν Ald.; καταναλίσκεσθαι conj. W.

[7] Plin. 13. 28.

[8] καὶ χρῆσθαι conj. Sch.; κεχρῆσθαι Ald.; ? κεχάρησθαι.

Ῥόδῳ. τοῦτο μὲν οὖν ἐπισκεπτέον· ἴσως γὰρ οἱ μὲν οὕτως οἱ δ' ἐκείνως θεραπεύουσιν, καὶ μετὰ μὲν τοῦ ὕδατος ὠφέλιμον ἡ κόπρος ἄνευ δὲ τούτου βλαβερά. ὅταν δὲ ἐνιαύσιος γένηται, μεταφυτεύουσι καὶ τῶν ἁλῶν συμπαραβάλλουσι, καὶ πάλιν ὅταν διετής· χαίρει γὰρ σφόδρα τῇ μεταφυτείᾳ.

4 Μεταφυτεύουσι δὲ οἱ μὲν ἄλλοι τοῦ ἦρος· οἱ δὲ ἐν Βαβυλῶνι περὶ τὸ ἄστρον, ὅτε καὶ ὅλως οἵ γε πολλοὶ φυτεύουσιν, ὡς καὶ παραγινομένου καὶ αὐξανομένου θᾶττον. νέου μὲν ὄντος οὐχ ἅπτονται, πλὴν ἀναδοῦσι τὴν κόμην, ὅπως ὀρθοφυῇ τ' ᾖ καὶ αἱ ῥάβδοι μὴ ἀπαρτῶνται. μετὰ δὲ ταῦτα περιτέμνουσιν, ὁπόταν ἁδρὸς ἤδη γένηται καὶ πάχος ἔχῃ. ἀπολείπουσι δὲ ὅσον σπιθαμὴν τῶν ῥάβδων. φέρει δὲ ἕως μὲν ἂν ᾖ νέος ἀπύρηνον τὸν καρπόν, μετὰ δὲ τοῦτο πυρηνώδη.

5 Ἄλλοι δέ τινες λέγουσιν ὡς οἵ γε κατὰ Συρίαν οὐδεμίαν προσάγουσιν ἐργασίαν ἀλλ' ἢ διακαθαίρουσι καὶ ἐπιβρέχουσιν, ἐπιζητεῖν δὲ μᾶλλον τὸ ναματιαῖον ὕδωρ ἢ τὸ ἐκ τοῦ Διός· εἶναι δὲ πολὺ τοιοῦτον ἐν τῷ αὐλῶνι ἐν ᾧ καὶ τὰ φοινικόφυτα τυγχάνει, τὸν αὐλῶνα δὲ τοῦτον λέγειν τοὺς Σύρους ὅτι διατείνει διὰ τῆς Ἀραβίας μέχρι τῆς ἐρυθρᾶς θαλάσσης καὶ πολλοὺς φάσκειν ἐληλυθέναι· τούτου δὲ ἐν τῷ κοιλοτάτῳ πεφυκέναι τοὺς φοίνικας. ταῦτα μὲν οὖν τάχ' ἀμφοτέρως ἂν εἴη· κατὰ γὰρ τὰς χώρας, ὥσπερ καὶ

---

[1] *cf.* 7. 5. 1. [2] Plin. 13. 37.
[3] συμπαραβάλλουσι conj. Sch. from G; συμπαραλαμβάνουσ. UAld. [4] *cf.* Plin. 13. 38.

given, after manuring, as the Rhodians use. This then is matter for enquiry; it may be that there are two distinct methods of cultivation, and that dung, if accompanied by watering,[1] is beneficial, though without it it is harmful. [2]When the tree is a year old, they transplant it and give plenty[3] of salt, and this treatment is repeated when it is two years old. for it delights greatly in being transplanted.

[4]Most transplant in the spring, but the people of Babylon about the rising of the dog-star, and this is the time when most people propagate it, since it then germinates and grows more quickly. As long as it is young, they do not touch it, except that they tie up the foliage, so that it may grow straight[5] and the slender branches may not hang down.[6] At a later stage they prune it, when it is more vigorous and has become a stout tree, leaving the slender branches only about a handsbreadth long. So long as it is young, it produces its fruit without a stone, but later on the fruit has a stone.

However some say that the people of Syria use no cultivation, except cutting out wood and watering, also that the date-palm requires spring water rather than water from the skies; and that such water is abundant in the valley in which are the palm-groves. And they add that the Syrians say that this valley[7] extends through Arabia to the Red Sea,[8] and that many profess to have visited it,[9] and that it is in the lowest part of it that the date-palms grow. Now both accounts may be true, for it is not strange that

[5] ὀρθοφυῇ τ' ᾖ conj. W.; ὀρθοφύηται P_2Ald.
[6] ἀπαρτῶνται conj. R. Const.; ἀπορθῶνται P_2M Ald.
[7] *cf.* Diod. 3. 41.
[8] *i.e.* the Arabian Gulf.
[9] ἐληλυθέναι Ald.; διεληλυθέναι conj. W.

αὐτὰ τὰ δένδρα, διαφέρειν καὶ τὰς ἐργασίας οὐκ ἄτοπον.

6 Γένη δὲ τῶν φοινικων ἐστὶ πλείω· πρῶτον μὲν καὶ ὥσπερ ἐν μεγίστῃ διαφορᾷ τὸ μὲν κάρπιμον τὸ δὲ ἄκαρπον, ἐξ ὧν οἱ περὶ Βαβυλῶνα τάς τε κλίνας καὶ τἆλλα σκεύη ποιοῦνται. ἔπειτα τῶν καρπίμων οἱ μὲν ἄρρενες αἱ δὲ θήλειαι· διαφέρουσι δὲ ἀλλήλων, καθ' ἃ ὁ μὲν ἄρρην ἄνθος πρῶτον φέρει ἐπὶ τῆς σπάθης, ἡ δὲ θήλεια καρπὸν εὐθὺ μικρόν. αὐτῶν δὲ τῶν καρπῶν διαφοραὶ πλείους· οἱ μὲν γὰρ ἀπύρηνοι οἱ δὲ μαλακοπύρηνοι· τὰς χροιὰς οἱ μὲν λευκοὶ οἱ δὲ μέλανες οἱ δὲ ξανθοί· τὸ δ' ὅλον οὐκ ἐλάττω χρώματά φασιν εἶναι τῶν συκῶν οὐδ' ἁπλῶς τὰ γένη· διαφέρειν δὲ καὶ κατὰ τὰ μεγέθη καὶ κατὰ τὰ σχήματα· καὶ γὰρ σφαιροειδεῖς ἐνίους ὡσανεὶ μῆλα καὶ τὰ μεγέθη τηλικούτους ὡς τέτταρας εἰς τὸν πῆχυν εἶναι, [ἕπτα καὶ εὐπόδους]· ἄλλους δὲ μικροὺς ἡλίκους ἐρεβίνθους. καὶ τοῖς χυλοῖς δὲ πολὺ διαφέροντας.

7 Κράτιστον δὲ καὶ τῶν λευκῶν καὶ τῶν μελάνων τὸ βασιλικὸν καλούμενον γένος ἐν ἑκατέρῳ καὶ μεγέθει καὶ ἀρετῇ· σπάνια δ' εἶναι ταῦτα λέγουσι· σχεδὸν γὰρ ἐν μόνῳ τῷ Βαγῴου κήπῳ τοῦ παλαιοῦ περὶ Βαβυλῶνα. ἐν Κύπρῳ δὲ ἴδιόν τι γένος φοινίκων ἐστὶν ὃ οὐ πεπαίνει τὸν καρπόν, ἀλλ' ὠμὸς ὢν ἡδὺς σφόδρα καὶ γλυκύς ἐστι· τὴν δὲ γλυκύτητα ἰδίαν ἔχει. ἔνιοι δ' οὐ μόνον δια-

---

1 Plin. 13. 39.

2 πρῶτον conj. Sch.; πρῶτος UMVAld.

3 πῆχυν conj. R. Const. from Plin. 13. 45. and G, *cf.* Diod. 2. 53; στάχυν UMVAld.

4 ἕπτα καὶ εὐπόδους UMV: the words perhaps conceal a

in different soils the methods of cultivation should differ, like the trees themselves.

[1] There are several kinds of palm. To begin with, to take first the most important difference;—some are fruitful and some not; and it is from this latter kind that the people of Babylon make their beds and other furniture. Again of the fruitful trees some are 'male,' others 'female'; and these differ from one another in that the 'male' first[2] bears a flower on the spathe, while the 'female' at once bears a small fruit. Again there are various differences in the fruits themselves; some have no stones, others soft stones; as to colour, some are white, some black, some yellow; and in general they say that there is not less variety of colour and even of kind than in figs; also that they differ in size and shape, some being round like apples and of such a size that four of them make up a cubit[3] in length, . . . [4] while others are small,[5] no bigger than chick-peas; and that there is also much difference in flavour.

The best kind alike in size and in quality, whether of the white or black variety, is that which in either form is called 'the royal palm'; but this, they say, is rare; it grows hardly anywhere except in the park of the ancient Bagoas,[6] near Babylon. In Cyprus[7] there is a peculiar kind of palm which does not ripen its fruit, though, when it is unripe, it is very sweet and luscious, and this lusciousness is of a peculiar kind. Some palms again[8] differ not merely

gloss on πῆχυν, *e.g.* εἷς πῆχυς δύο πόδες (Salm.); om. G; ἐνίοτε καὶ ἐπὶ πόδα conj. W. [5] Plin. 13. 42.

[6] Βαγῴου: Βάττου MSS. corr. by R. Const. from Plin. 13. 41. τοῦ παλαίου apparently distinguishes this Bagoas from some more recent wearer of the name.

[7] Plin. 13. 33. [8] Plin. 13. 28.

φέρουσι τοῖς καρποῖς ἀλλὰ καὶ αὐτῷ τῷ δένδρῳ κατά τε τὸ μῆκος καὶ τὴν ἄλλην μορφήν· οὐ γὰρ μεγάλοι καὶ μακροὶ ἀλλὰ βραχεῖς, ἔτι δὲ καρπιμώτεροι τῶν ἄλλων καὶ καρποφοροῦντες εὐθὺς τριετεῖς· πολλοὶ δὲ καὶ οὗτοι περὶ Κύπρον. εἰσὶ δὲ καὶ περὶ Συρίαν καὶ περὶ Αἴγυπτον φοίνικες οἳ φέρουσι τετραετεῖς καὶ πενταετεῖς ἀνδρομήκεις ὄντες.

8 Ἕτερον δ' ἔτι γένος ἐν Κύπρῳ, ὃ καὶ τὸ φύλλον πλατύτερον ἔχει καὶ τὸν καρπὸν μείζω πολλῷ καὶ ἰδιόμορφον· μεγέθει μὲν ἡλίκος ῥόα τῷ σχήματι δὲ προμήκης, οὐκ εὔχυλος δὲ ὥσπερ ἄλλοι ἀλλ' ὅμοιος ταῖς ῥόαις, ὥστε μὴ καταδέχεσθαι ἀλλὰ διαμασησαμένους ἐκβάλλειν. γένη μὲν οὖν, ὥσπερ εἴρηται, πολλά. θησαυρίζεσθαι δὲ μόνους δύνασθαί φασι τῶν ἐν Συρίᾳ τοὺς ἐν τῷ αὐλῶνι, τοὺς δ' ἐν Αἰγύπτῳ καὶ Κύπρῳ καὶ παρὰ τοῖς ἄλλοις χλωροὺς ἀναλίσκεσθαι.

9 Ἔστι δὲ ὁ φοῖνιξ ὡς μὲν ἁπλῶς εἰπεῖν μονοστέλεχες καὶ μονοφυές· οὐ μὴν ἀλλὰ γίνονταί τινες καὶ διφυεῖς, ὥσπερ ἐν Αἰγύπτῳ, καθάπερ δικρόαν ἔχοντες· τὸ δ' ἀνάστημα τοῦ στελέχους ἀφ' οὗ ἡ σχίσις καὶ πεντάπηχυ· πρὸς ἄλληλα δέ πως ἰσάζοντα. φασὶ δὲ καὶ τοὺς ἐν Κρήτῃ πλείους εἶναι τοὺς διφυεῖς, ἐνίους δὲ καὶ τριφυεῖς· ἐν δὲ τῇ Λαπαίᾳ τινὰ καὶ πεντακέφαλον· οὐκ ἄλογον γοῦν ἐν ταῖς εὐτροφωτέραις χώραις πλείω γίνεσθαι τὰ τοιαῦτα καὶ τὸ ὅλον δὲ τὰ εἴδη πλείω καὶ τὰς διαφοράς.

---

[1] ὅμοιος conj. Bod.; ὁμοίως UMVAld. [2] cf. § 5.
[3] Plin. 13. 38; cf. 4. 2. 7, where the name (κουκιόφορον) of this tree is given.

in their fruits but in the character of the tree itself as to stature and general shape; for instead of being large and tall they are low growing; but these are more fruitful than the others, and they begin to bear as soon as they are three years old; this kind too is common in Cyprus. Again in Syria and Egypt there are palms which bear when they are four or five years old, at which age they are the height of a man.

There is yet another kind in Cyprus, which has broader leaves and a much larger fruit of peculiar shape; in size it is as large as a pomegranate, in shape it is long; it is not however juicy like others, but like[1] a pomegranate, so that men do not swallow it, but chew it and then spit it out. Thus, as has been said, there are many kinds. The only dates that will keep, they say, are those which grow in the Valley[2] of Syria, while those that grow in Egypt Cyprus and elsewhere are used when fresh.

The palm, speaking generally, has a single and simple stem; however there are some with two stems, as in Egypt,[3] which make a fork, as it were; the length of the stem up to the point where it divides is as much as five cubits, and the two branches of the fork are about equal in length. They say that the palms in Crete more often than not have this double stem, and that some of them have three stems; and that in Lapaia one with five heads has been known. It is after all not surprising[4] that in more fertile soils such instances should be commoner, and in general that more kinds and more variation should be found under such conditions.

[4] οὐκ ἄλογον γοῦν conj. W. (οὐκ ἄλογον δ' Sch.); οὐ καλῶς γοῦν Ald.MU (marked doubtful).

10 Ἄλλο δέ τι γένος ἐστὶν ὅ φασι γίνεσθαι πλεῖστον περὶ τὴν Αἰθιοπίαν, ὃ καλοῦσι κόϊκας· οὗτοι δὲ θαμνώδεις, οὐχὶ ἓν τὸ στέλεχος ἔχοντες ἀλλὰ πλείω καὶ ἐνίοτε συνηρτημένα μέχρι τινὸς εἰς ἕν, τὰς δὲ ῥάβδους οὐ μακρὰς μὲν ἀλλ' ὅσον πηχυαίας, ἀλλὰ λείας, ἐπὶ δὲ τῶν ἄκρων τὴν κόμην. ἔχουσι δὲ καὶ τὸ φύλλον πλατὺ καὶ ὥσπερ ἐκ δυοῖν συγκείμενον ἐλαχίστοιν. καλοὶ δὲ καὶ τῇ ὄψει φαίνονται· τὸν δὲ καρπὸν καὶ τῷ σχήματι καὶ τῷ μεγέθει καὶ τῷ χυλῷ διάφορον ἔχουσι· στρογγυλώτερον γὰρ καὶ μείζω καὶ εὐστομώτερον ἧττον δὲ γλυκύν. πεπαίνουσι δὲ ἐν τρισὶν ἔτεσιν ὥστ' ἀεὶ καρπὸν ἔχειν, ἐπικαταλαμβάνοντος τοῦ νέου τὸν ἕνον· ποιοῦσι δὲ καὶ ἄρτους ἐξ αὐτῶν· περὶ μὲν οὖν τούτων ἐπισκεπτέον.

11 Οἱ δὲ χαμαιρριφεῖς καλούμενοι τῶν φοινίκων ἕτερόν τι γένος ἐστὶν ὥσπερ ὁμώνυμον· καὶ γὰρ ἐξαιρεθέντος τοῦ ἐγκεφάλου ζῶσι καὶ κοπέντες ἀπὸ τῶν ῥιζῶν παραβλαστάνουσι. διαφέρουσι δὲ καὶ τῷ καρπῷ καὶ τοῖς φύλλοις· πλατὺ γὰρ καὶ μαλακὸν ἔχουσι τὸ φύλλον, δι' ὃ καὶ πλέκουσιν ἐξ αὐτοῦ τάς τε σπυρίδας καὶ τοὺς φορμούς· πολλοὶ δὲ καὶ ἐν τῇ Κρήτῃ γίνονται καὶ ἔτι μᾶλλον ἐν Σικελίᾳ. ταῦτα μὲν οὖν ἐπὶ πλεῖον εἴρηται τῆς ὑποθέσεως.

---

[1] Plin. 13. 47.

[2] *κόικας* conj. Salm. *cf.* 1. 10. 5, and the probable reading in Plin. *l.c.*

[3] *συνηρτημένα μέχρι τινὸς εἰς ἕν* conj. W.; *συνηρτημένας μὲν*

[1] There is another kind which is said to be abundant in Ethiopia, called the doum-palm[2]; this is a shrubby tree, not having a single stem but several, which sometimes are joined together up to a certain point[3]; and the leaf-stalks are not long,[4] only the length of a cubit, but they are plain,[5] and the leafage is borne only at the tip. The leaf is broad and, as it were, made up of at least[6] two leaflets. This tree is fair to look upon, and its fruit in shape size and flavour differs from the date, being rounder larger and pleasanter to the taste, though not so luscious. It ripens in three years, so that there is always fruit on the tree, as the new fruit overtakes that of last year. And they make bread out of it. These reports then call for enquiry.

[7] The dwarf-palm, as it is called, is a distinct kind, having nothing but its name[8] in common with other palms. For if the head is removed, it survives, and, if it is cut down, it shoots again from the roots. It differs too in the fruit and leaves; for the leaf is broad and flexible, and so they weave their baskets and mats out of it. It is common in Crete and still more so in Sicily.[9] However in these matters we have said more than our purpose required.

εἰς ἕν U; *συνηρτημένα μέχρι τινὸς εἰσι* Ald.; *συνηρτημένας μὲν μέχρι τινὸς εἶεν* MV.

[4] *μὲν* ins. W. after Sch. (omitted above).

[5] *i.e.* without leaflets, except at the tip.

[6] *ἐλαχίστοιν* Bas.; *ἐλαχίστων* U. *cf.* Arist. *Eth. N.* 5. 3. 3, *ἐν ἐλαχίστοις δυσίν*.

[7] Plin. 13. 39.

[8] For *ὁμώνυμον cf.* 9. 10. 1 n.

[9] A dwarf palm is now abundant at Selinunte: *cf.* Verg. *Aen.* 3. 705, *palmosa Selinus*.

12 Ἐν δὲ ταῖς τῶν ἄλλων φυτείαις ἀνάπαλιν τίθενται τὰ φυτευτήρια, καθάπερ τῶν κλημάτων. οἱ μὲν οὖν οὐθὲν διαφέρειν φασὶν ἥκιστα δὲ ἐπὶ τῶν ἀμπέλων· ἔνιοι δὲ ῥόαν δασύνεσθαι καὶ σκιάζειν μᾶλλον τὸν καρπόν· ἔτι δὲ ἧττον ἀποβάλλειν τοὺς κυτίνους. συμβαίνειν δὲ τοῦτό φασι καὶ ἐπὶ τῆς συκῆς· οὐ γὰρ ἀποβάλλειν ἀνάπαλιν φυτευθεῖσαν, ἔτι δ' εὐβατωτέραν γίνεσθαι· οὐκ ἀποβάλλειν δὲ οὐδ' ἐάν τις ἀποκλάσῃ φυομένης εὐθὺς τὸ ἄκρον.

Αἱ μὲν οὖν φυτεῖαι καὶ γενέσεις ὃν τρόπον ἔχουσι σχεδὸν ὡς τύπῳ περιλαβεῖν εἴρηνται.

VII. Περὶ δὲ τῆς ἐργασίας καὶ τῆς θεραπείας τὰ μέν ἐστι κοινὰ τὰ δὲ ἴδια καθ' ἕκαστον. κοινὰ μὲν ἥ τε σκαπάνη καὶ ἡ ὑδρεία καὶ ἡ κόπρωσις, ἔτι δὲ ἡ διακάθαρσις καὶ ἀφαίρεσις τῶν αὔων. διαφέρουσι δὲ τῷ μᾶλλον καὶ ἧττον. τὰ μὲν φίλυδρα καὶ φιλόκοπρα τὰ δ' οὐχ ὁμοίως, οἷον ἡ κυπάριττος, ἥπερ οὐ φιλόκοπρον οὐδὲ φίλυδρον, ἀλλὰ καὶ ἀπόλλυσθαί φασιν ἐάν γε νέαν οὖσαν ἐφυδρεύωσι πολλῷ. ῥόα δὲ καὶ ἄμπελος φίλυδρα. συκῆ δὲ εὐβλαστοτέρα μὲν ὑδρευομένη τὸν δὲ καρπὸν ἴσχει χείρω πλὴν τῆς Λακωνικῆς· αὕτη δὲ φίλυδρος.

---

[1] ἀνάπαλιν conj. Sch.; τἀνάπαλιν Ald. *cf.* *C.P.* 2. 9. 4; *Geop.* 10. 45; Plin. 17. 84. [2] οὖν ins. H.

[3] δασύνεσθαι: see LS. reff. *s.v.* δασύς.

[4] *cf.* *C.P.* 2. 9. 3.

[5] εὐβατωτέραν (*i.e.* 'more manageable'). The reference is to a method of keeping the tree dwarf (Bod.). Plin. *l.c.* has

Further notes on the propagation of trees.

To return to the other trees:—in propagating them they set the cuttings upside down,[1] as with vine-shoots. Some however[2] say that that makes no difference, and least of all in propagating the vine; while others contend that the pomegranate thus propagated has a bushier growth[3] and shades the fruit better, and also that it is then[4] less apt to shed the flower. This also occurs, they say, with the fig; when it is set upside down, it does not shed its fruit, and it makes a more accessible[5] tree; and it does not shed its fruit, even if one breaks off the top[6] as it begins to grow.

Thus we have given a general sketch of what we find about methods of propagation, and of the ways in which these trees are reproduced.

Of the cultivation of trees.

VII. [7] As to cultivation and tendance some requirements apply equally to all trees, some are peculiar to one. Those which apply equally to all are spadework watering and manuring, and moreover pruning and removal of dead wood. But different trees differ in the degree. Some love moisture and manure, some not so much, as the cypress,[8] which[9] is fond neither of manure nor of water, but actually dies, they say, if it is overwatered when young. But the pomegranate and vine are water-loving. The fig grows more vigorously if it is watered, but then its fruit is inferior, except in the case of the Laconian variety, which is water-loving.[10]

scansilem (so also G), which seems to be a rendering of εὐβατ. εὐβατοτέραν U.

[6] τὸ ἄκρον conj. R. Const. after G; τὸν καρπὸν UMVP_2Ald.

[7] Plin. 17. 246. [8] Plin. 17. 247.

[9] ᾗπερ conj. W. from G; ὥσπερ Ald. [10] *cf. C.P.* 3. 6. 6.

2 Διακαθαίρεσθαι δὲ πάντα ζητεῖ· βελτίω γὰρ τῶν αὔων ἀφαιρουμένων ὥσπερ ἀλλοτρίων, ἃ καὶ τὰς αὐξήσεις καὶ τὰς τροφὰς ἐμποδίζει. δι' ὃ καὶ ... ὅταν ᾖ γεράνδρυον ὅλως κόπτουσιν· ἡ γὰρ βλάστησις νέα γίνεται τοῦ δένδρου. πλείστης δὲ διακαθάρσεώς φησιν Ἀνδροτίων δεῖσθαι μύρρινον καὶ ἐλάαν· ὅσῳ γὰρ ἂν ἐλάττω καταλίπῃς, ἄμεινον βλαστήσει καὶ τὸν καρπὸν οἴσει πλείω· πλὴν ἀμπέλου δῆλον ὅτι· ταύτῃ γὰρ ἀναγκαιότερον καὶ πρὸς βλάστησιν καὶ πρὸς εὐκαρπίαν. ἁπλῶς δὲ καὶ ταύτην καὶ τὴν ἄλλην θεραπείαν πρὸς τὴν ἰδίαν φύσιν ἑκάστῳ ποιητέον.

3 Δεῖσθαι δέ φησιν Ἀνδροτίων καὶ κόπρου δριμυτάτης καὶ πλείστης ὑδρείας, ὥσπερ καὶ τῆς διακαθάρσεως, ἐλάαν καὶ μύρρινον καὶ ῥόαν· οὐ γὰρ ἔχειν μήτραν οὐδὲ νόσημα κατὰ γῆς οὐδέν· ἀλλ' ἐπειδὰν παλαιὸν ᾖ τὸ δένδρον, ἀποτέμνειν δεῖν τοὺς ἀκρεμόνας ἔπειτα τὸ στέλεχος θεραπεύειν ὡσπερὰν ἐξ ἀρχῆς φυτευθέν· οὕτω δε φασι πολυχρονιώτερα καὶ ἰσχυρότατα μύρρινον εἶναι καὶ ἐλάαν. ταῦτα μὲν οὖν ἐπισκέψαιτ' ἄν τις, εἰ καὶ μὴ πάντα ἀλλὰ περί γε τῆς μήτρας.

4 Ἡ δὲ κόπρος οὔτε πᾶσιν ὁμοίως οὔθ' ἡ αὐτὴ πᾶσιν ἁρμόττει· τὰ μὲν γὰρ δριμείας δεῖται τὰ δ' ἧττον τὰ δὲ παντελῶς κούφης. δριμυτάτη δὲ ἡ τοῦ ἀνθρώπου· καθάπερ καὶ Χαρτόδρας ἀρίστην μὲν ταύτην εἶναί φησι, δευτέραν δὲ τὴν ὑείαν, τρίτην δὲ αἰγός, τετάρτην δὲ προβάτου,

---

[1] Plin. 17. 248.
[2] Name of tree missing. Sch.
[3] *cf.* *C.P.* 3. 10. 4.
[4] ταύτῃ conj. W.; ταύτης Ald.

1 All trees require pruning; for they are improved by removal of the dead wood, which is, as it were, a foreign body, and prevents growth and nourishment. Wherefore when the (tree)[2] becomes old, they cut off all its boughs: for then the tree breaks afresh. Androtion[3] says that the myrtle and olive need more pruning than any other trees; for the smaller you leave them, the better they will grow, and they will bear better fruit. But the vine of course needs pruning even more; for it is in the case of this tree[4] more necessary for promoting both growth and fruitfulness. However, speaking generally, both this and other kinds of tendance must be suited to the particular natural character in each case.

Androtion further says that the olive the myrtle and the pomegranate require the most pungent manure and the heaviest watering, as well as the most thorough pruning, for that then they do not get 'softwood'[5] nor any disease underground; but when the tree is old, he adds, one should cut off the boughs, and then attend to the stem as though it were a tree just planted. Thus[6] treated they say that the myrtle and olive are longer lived and very robust. These statements might be a subject for further enquiry, or, if not all of them, at least what is stated of the 'softwood.'

Manure does not suit all alike, nor is the same manure equally good for all. Some need it pungent, some less so, some need it quite light. The most pungent is human dung: thus Chartodras[7] says that this is the best, pig-manure being second to it, goat-manure third, fourth that of sheep, fifth that of

[5] *i.e.* effete sap-wood. [6] *οὕτω* conj. W.; *οἱ* Ald.
[7] Name perhaps corrupt.

πέμπτην δὲ βοός, ἕκτην δὲ τὴν λοφούρων. ἡ δὲ συρματῖτις ἄλλη καὶ ἄλλως· ἡ μὲν γὰρ ἀσθενεστέρα ταύτης ἡ δὲ κρείττων.

5 Τὴν δὲ σκαπάνην πᾶσιν οἴονται συμφέρειν, ὥσπερ καὶ τὴν ὄσκαλσιν τοῖς ἐλάττοσιν· εὐτραφέστερα γὰρ γίνεσθαι. τρέφειν δὲ δοκεῖ καὶ ὁ κονιορτὸς ἔνια καὶ θάλλειν ποιεῖν, οἷον τὸν βότρυν, δι' ὃ καὶ ὑποκονίουσι πολλάκις· οἱ δὲ καὶ τὰς συκᾶς ὑποσκάπτουσιν ἔνθα τούτου δεῖ. Μεγαροῖ δὲ καὶ τοὺς σικύους καὶ τὰς κολοκύντας, ὅταν οἱ ἐτησίαι πνεύσωσι, σκάλλοντες κονιορτοῦσι καὶ οὕτω γλυκυτέρους καὶ ἁπαλωτέρους ποιοῦσιν οὐχ ὑδρεύοντες. τοῦτο μὲν οὖν ὁμολογούμενον. τὴν δ' ἄμπελον οὔ φασί τινες δεῖν [ἢ] ὑποκονίειν οὐδ' ὅλως ἅπτεσθαι περκάζοντος τοῦ βότρυος, ἀλλ' εἴπερ ὅταν ἀπομελανθῇ. οἱ δὲ τὸ ὅλον μηδὲ τότε πλὴν ὅσον ὑποτῖλαι τὴν βοτάνην· ὑπὲρ μὲν οὖν τούτων ἀμφισβητοῦσιν.

6 Ἐὰν δέ τι μὴ φέρῃ καρπὸν ἀλλ' εἰς βλάστησιν τρέπηται, σχίζουσι τοῦ στελέχους τὸ κατὰ γῆν καὶ λίθον ἐντιθέασιν ὅπως ἂν ῥαγῇ, καί φασι φέρειν. ὁμοίως δὲ καὶ ἐάν τις τῶν ῥιζῶν τινας περιτέμῃ, δι' ὃ καὶ τῶν ἀμπέλων ὅταν τραγῶσι τοῦτο ποιοῦσι τὰς ἐπιπολῆς. τῶν δὲ συκῶν πρὸς τῷ περιτέμνειν καὶ τέφραν περιπάττουσι καὶ κατασχάζουσι τὰ στελέχη καί φασι φέρειν μᾶλλον. ἀμυγδαλῇ δὲ καὶ πάτταλον ἐγκόψαντες

---

1 Lit. 'bushy tails,' *i.e.* horses asses mules.
2 *cf. C.P.* 3. 16. 3.
3 δεῖ ins. H; so apparently G read.
4 δεῖν ὑποκονίειν οὐδ' ὅλως conj. W. (so Sch., but keeping [ἢ] after δεῖν); δεῖν ἢ ὑποκινιεῖν οὐδ' ὅλως UMV; δεῖν ἢ ὑποκονιεῖν ἢ ὅλως Ald.
5 Plin. 17. 253 and 254.

oxen, and sixth that of beasts of burden.[1] Litter manure is of different kinds and is applied in various ways: some kinds are weaker, some stronger.

Spade-work is held to be beneficial to all trees, and also hoeing for the smaller ones, as they then become more vigorous. Even dust[2] is thought to fertilise some things and make them flourish, for instance the grape; wherefore they often put dust to the roots of the vine. Some also dig in dust about the figs in places where it is deficient.[3] In Megara, when the etesian winds are past, they cover the cucumber and gourd plants with dust by raking, and so make the fruits sweeter and tenderer by not watering. On this point there is general agreement. But some say that dust should not be put to the vine,[4] and that it should not be meddled with at all when the grape is turning, or, if at all, only when it has turned black. Some again say that even then nothing should be done except to pluck up the weeds. So on this point there is a difference of opinion.

5 If a tree does not bear fruit but inclines to a leafy growth, they split that part of the stem which is underground and insert a stone corresponding[6] to the crack thus made, and then, they say, it will bear. The same result follows, if one cuts off some of the roots, and accordingly they thus treat the surface roots of the vine when it runs to leaf. In the case of figs, in addition to root-pruning,[7] they also sprinkle ashes about the tree, and make gashes in the stems, and then, they say, it bears better. [8] Into the almond tree they drive an iron peg, and, having thus made

[6] ὅπως ἂν ῥαγῇ Ald.: so G; ? ὅπου; ὅπως ἀνεώγῃ conj. W. *cf.* *Geop.* 5. 35. [7] Plin. *l.c.*
[8] *cf.* 2. 2. 11; *C.P.* 1. 17. 10; 2. 14. 1; Plin. 7. 253.

σιδηροῦν ὅταν τετράνωσιν ἄλλον ἀντεμβάλλουσι δρύϊνον καὶ τῇ γῇ κρύπτουσιν· ὃ καὶ καλοῦσί τινες κολάζειν ὡς ὑβρίζον τὸ δένδρον.

Ταὐτὸν δὲ τοῦτο καὶ ἐπὶ τῆς ἀπίου καὶ ἐπ' ἄλλων τινὲς ποιοῦσιν. ἐν Ἀρκαδίᾳ δὲ καὶ εὐθύνειν καλοῦσι τὴν ὄαν· πολὺ γὰρ τὸ δένδρον τοῦτο παρ' αὐτοῖς ἐστι. καί φασιν, ὅταν πάθῃ τοῦτο, τὰς μὲν μὴ φερούσας φέρειν τὰς δὲ μὴ πεττούσας ἐκπέττειν καλῶς. ἀμυγδαλῆν δὲ καὶ ἐκ πικρᾶς γίγνεσθαι γλυκεῖαν, ἐάν τις περιορύξας τὸ στέλεχος καὶ τιτράνας ὅσον τε παλαιστιαῖον τὸ πανταχόθεν ἀπορρέον δάκρυον ἐπὶ ταὐτὸ ἐᾷ καταρρεῖν. τοῦτο μὲν οὖν ἂν εἴη πρός τε τὸ φέρειν ἅμα καὶ πρὸς τὸ εὐκαρπεῖν.

VIII. Ἀποβάλλει δὲ πρὸ τοῦ πέψαι τὸν καρπὸν ἀμυγδαλῆ μηλέα ῥόα ἄπιος καὶ μάλιστα δὴ πάντων συκῆ καὶ φοῖνιξ, πρὸς ἃ καὶ τὰς βοηθείας ζητοῦσι· ὅθεν καὶ ὁ ἐρινασμός· ἐκ γὰρ τῶν ἐκεῖ κρεμαννυμένων ἐρινῶν ψῆνες ἐκδυόμενοι κατεσθίουσι καὶ πιαίνουσι τὰς κορυφάς. διαφέρουσι δὲ καὶ αἱ χῶραι πρὸς τὰς ἀποβολάς· περὶ γὰρ Ἰταλίαν οὔ φασιν ἀποβάλλειν, δι' ὃ οὐδ' ἐρι-

---

[1] The operation being performed at the base of the tree. *cf.* § 7. [2] ἐκπέττειν conj. R. Const.; εἰσπέττειν UMAld.

[3] Plin. 17. 252.

[4] τὸ πανταχόθεν conj. W.; πανταχόθεν τὸ MSS.; so apparently G. *cf. C.P.* 2. 14. 4.

[5] πέψαι conj. Sch.; πέμψαι Ald.

[6] ἐκεῖ κρεμαννυμένων ἐρινῶν I conj.; ἐκεῖ κρεμαννυμένων Ald.: ἐπικρεμαμένων ἐρινῶν conj. W.: but the present partic. is used *C.P.* 2. 9. 5.

a hole, insert in its place a peg of oak-wood and bury it[1] in the earth, and some call this 'punishing' the tree, since its luxuriance is thus chastened.

Some do the same with the pear and with other trees. In Arcadia they have a similar process which is called 'correcting' the sorb (for that tree is common in that country). And they say that under this treatment those trees that would not bear do so, and those that would not ripen their fruit now ripen[2] them well. [3] It is also said that the almond becomes sweet, instead of bitter, if one digs round the stem and, having bored a hole about a palmsbreadth, allows the gum which exudes from all sides[4] to flow down into it and collect. The object of this would be alike to make the tree bear and to improve the fruit.

Of remedies for the shedding of the fruit: caprification.

VIII. Trees which are apt to shed their fruit before ripening[5] it are almond apple pomegranate pear and, above all, fig and date-palm; and men try to find the suitable remedies for this. This is the reason for the process called 'caprification'; gall-insects come out of the wild figs which are hanging there,[6] eat the tops of the cultivated figs and so make them swell.[7] The shedding of the fruit differs according to the soil: in Italy[8] they say that it does not occur, and so they do not use caprification,[9]

[7] πιαίνουσι MVAld.; διείρουσι conj. W. ? πεπαίνουσι, 'ripen,' which is the word used in the parallel pass. *C.P.* 2. 9. 6, the object of the process being to cause the figs to dry.

[8] Plin. 15. 81. 'Italy' means South Italy. *cf.* 4. 5. 5 and 6; 5. 8. 1.

[9] ἐρινάζουσιν conj. Bod.; ἐρινοῦσιν Ald. H.

νάζουσιν· οὐδ' ἐν τοῖς καταβορείοις καὶ λεπτογείοις, οἷον ἐπὶ Φαλύκῳ τῆς Μεγαρίδος· οὐδὲ τῆς Κορινθίας ἔν τισι τόποις. ὡσαύτως δὲ καὶ ἡ τῶν πνευμάτων κατάστασις· βορείοις γὰρ μᾶλλον ἢ νοτίοις ἀποβάλλουσι, κἂν ψυχρότερα καὶ πλείω γένηται μᾶλλον· ἔτι δ' αὐτῶν τῶν δένδρων ἡ φύσις· τὰ πρωΐα γὰρ ἀποβάλλει, τὰ δ' ὄψια οὐκ ἐκβάλλει, καθάπερ ἡ Λακωνικὴ καὶ αἱ ἄλλαι. δι' ὃ καὶ οὐκ ἐρινάζουσι ταύτας. ταῦτα μὲν οὖν ἔν τε τοῖς τόποις καὶ τοῖς γένεσι καὶ τῇ καταστάσει τοῦ ἀέρος ἔχει τὰς διαφοράς.

2 Οἱ δὲ ψῆνες ἐκδύονται μὲν ἐκ τοῦ ἐρινεοῦ, καθάπερ εἴρηται· γίνονται δ' ἐκ τῶν κεγχραμίδων. σημεῖον δὲ λέγουσιν, ὅτι ἐπειδὰν ἐκδύωσιν οὐκ ἔνεισι κεγχραμίδες. ἐκδύονται δὲ οἱ πολλοὶ ἐγκαταλιπόντες ἢ πόδα ἢ πτερόν. γένος δέ τι καὶ ἕτερόν ἐστι τῶν ψηνῶν, ὃ καλοῦσι κεντρίνας· οὗτοι δ' ἀργοὶ καθάπερ κηφῆνες· καὶ τοὺς εἰσδυομένους τῶν ἑτέρων κτείνουσιν αὐτοὶ δὲ ἐναποθνήσκουσιν. ἐπαινοῦσι δὲ μάλιστα τῶν ἐρινῶν τὰ μέλανα τὰ ἐκ τῶν πετρωδῶν χωρίων· πολλὰς
3 γὰρ ἔχει ταῦτα κεγχραμίδας. γιγνώσκεται δὲ τὸ ἐρινασμένον τῷ ἐρυθρὸν εἶναι καὶ ποικίλον καὶ ἰσχυρόν· τὸ δ' ἀνερίναστον λευκὸν καὶ ἀσθενές· προστιθέασι δὲ τοῖς δεομένοις ὅταν ὕσῃ. ὅπου δὲ πλεῖστος κονιορτός, ἐνταῦθα πλεῖστα καὶ

---

[1] *cf.* 8. 2. 11.
[2] ψυχρότερα καὶ πλείω conj. Sch.; τεχνοτέρα καὶ πλείων MV Ald.; τεχρότερα καὶ πλείω U.
[3] πρωΐα conj. Sch. from G; πρῶτα Ald.H.
[4] Plin. 17. 255 and 256.

nor is it practised in places which face north nor in those with light soils, as at Phalykos[1] in the Megarid, nor in certain parts of the district of Corinth. Also conditions as to wind make a difference; the fruit is shed more with northerly than with southerly winds, and this also happens more if the winds are cold and frequent.[2] Moreover the character of the tree itself makes a difference; for some kinds, such as the Laconian and other such kinds, shed their early[3] figs but not the later ones. Wherefore caprification is not practised with these. Such are the changes to which the fig is subject in respect of locality kind and climatic conditions.

[4] Now the gall-insects come, as has been said, out of the wild fig, and they are engendered from the seeds. The proof given of this is that, when they come out, there are no seeds left in the fruit; and most of them in coming out leave a leg or a wing behind. There is another kind of gall-insect which is called *kentrines*; these insects are sluggish, like drones, they kill those of the other kind who are entering the figs, and they themselves die in the fruit. The black kind of wild fig which grows in rocky places is most commended for caprification, as these figs contain numerous seeds.[5] A fig which has been subject to caprification is known by being red and parti-coloured and stout, while one which has not been so treated is pale and sickly. The treatment is applied to the trees which need it, after rain. The wild figs are most plentiful and most potent

[5] *i.e.* and so should produce more gall-insects: in *C.P.* 2. 9. 6 it is implied that the insect is produced by putrefaction of the seeds of the wild fig.

ἰσχυρότατα τὰ ἐρινὰ γίνεται. φασὶ δὲ ἐρινάζειν καὶ τὸ πόλιον, ὁπόταν αὐτῷ καρπὸς ᾖ πολύς, καὶ τοὺς τῆς πτελέας κωρύκους· ἐγγίνεται γὰρ καὶ ἐν τούτοις θηρίδι' ἄττα. κνῖπες ὅταν ἐν ταῖς συκαῖς γίνωνται κατεσθίουσι τοὺς ψῆνας. ἄκος δὲ τούτου φασὶν εἶναι τοὺς καρκίνους προσπερονᾶν· πρὸς γὰρ τούτους τρέπεσθαι τοὺς κνῖπας. ἀλλὰ γὰρ δὴ ταῖς μὲν συκαῖς αὗται βοήθειαι.

4 Τοῖς δὲ φοίνιξιν αἱ ἀπὸ τῶν ἀρρένων πρὸς τοὺς θήλεις· οὗτοι γάρ εἰσιν οἱ ἐπιμένειν ποιοῦντες καὶ ἐκπέττειν, ὃ καλοῦσί τινες ἐκ τῆς ὁμοιότητος ὀλυνθάζειν. γίνεται δὲ τόνδε τὸν τρόπον. ὅταν ἀνθῇ τὸ ἄρρεν, ἀποτέμνουσι τὴν σπάθην ἐφ' ἧς τὸ ἄνθος εὐθὺς ὥσπερ ἔχει, τόν τε χνοῦν καὶ τὸ ἄνθος καὶ τὸν κονιορτὸν κατασείουσι κατὰ τοῦ καρποῦ τῆς θηλείας· κἂν τοῦτο πάθῃ, διατηρεῖ καὶ οὐκ ἀποβάλλει. φαίνεται δ' ἀμφοῖν ἀπὸ τοῦ ἄρρενος τοῖς θήλεσι βοήθεια γίνεσθαι· θῆλυ γὰρ καλοῦσι τὸ καρποφόρον· ἀλλ' ἡ μὲν οἷον μῖξις· ἡ δὲ κατ' ἄλλον τρόπον.

---

[1] *ὁπότ' ἂν . . . πολύς* conj. W. from G, *cum copiose fructificat*; *ὁπόταν αἰγίπυρος ᾖ πολύς* MSS. U adds *καὶ* before *ὁπόταν*.

[2] *κωρύκους* I conj. In 3. 14. 1. the elm is said to bear *κωουκίδες* which contain gnat-like creatures; these growths are called *κωρυκώδη τινα κοῖλα* 3. 15. 4; and in 3. 7. 3. the

where there is most dust. And they say that hulwort also, when it fruits freely,[1] and the 'gall-bags'[2] of the elm are used for caprification. For certain little creatures are engendered in these also. When the *knips* is found in figs, it eats the gall-insects. It is to prevent this, it is said, that they nail up the crabs; for the *knips* then turns its attention to these. Such are the ways of assisting the fig-trees.

With dates it is helpful to bring the male to the female; for it is the male which causes the fruit to persist and ripen, and this process some call, by analogy, 'the use of the wild fruit.'[3] The process is thus performed: when the male palm is in flower, they at once cut off the spathe on which the flower is, just as it is, and shake the bloom with the flower and the dust over the fruit of the female, and, if this is done to it, it retains the fruit and does not shed it. In the case both of the fig and of the date it appears that the 'male' renders aid to the 'female,'—for the fruit-bearing tree is called 'female'—but while in the latter case there is a union of the two sexes, in the former the result is brought about somewhat differently.

same thing is referred to as *τὸ θυλακῶδες τοῦτο*, where *τοῦτο* = 'the well-known'; *cf.* also 9. 1. 2, where Sch. restores *κωρύκους*; *cf.* Pall. 4. 10. 28. *κυπαίρους* (?) U; *κυπέρους* MV; *κύπεριν* Ald.; *κυττάρους* conj. W.

[3] *ὀλυνθάζειν*, from *ὄλυνθος*, a kind of wild fig, as *ἐρινάζειν*, from *ἐρινός*, the wild fig used for caprification. *cf.* *C.P.* 3. 18. 1.

Γ'

I. Ἐπεὶ δὲ περὶ τῶν ἡμέρων δένδρων εἴρηται, λεκτέον ὁμοίως καὶ περὶ τῶν ἀγρίων, εἴ τέ τι ταὐτὸν καὶ ἕτερον ἔχουσι τοῖς ἡμέροις εἴ θ' ὅλως ἴδιον τῆς φύσεως.

Αἱ μὲν οὖν γενέσεις ἁπλαῖ τινες αὐτῶν εἰσι· πάντα γὰρ ἢ ἀπὸ σπέρματος ἢ ἀπὸ ῥίζης φύεται. τοῦτο δ' οὐχ ὡς οὐκ ἐνδεχόμενον καὶ ἄλλως, ἀλλ' ἴσως διὰ τὸ μὴ πειρᾶσθαι μηδένα μηδὲ φυτεύειν· ἐκφύοιτο[1] δ' ἂν εἰ λαμβάνοιεν τόπους ἐπιτηδείους καὶ θεραπείαν τὴν ἁρμόττουσαν· ὥσπερ καὶ νῦν τὰ ἀλσώδη καὶ φίλυδρα, λέγω δ' οἷον πλάτανον ἰτέαν λεύκην αἴγειρον πτελέαν· ἅπαντα γὰρ ταῦτα καὶ τὰ τοιαῦτα φυτευόμενα βλαστάνει καὶ τάχιστα καὶ κάλλιστα ἀπὸ τῶν παρασπάδων, ὥστε καὶ μεγάλας οὔσας ἤδη καὶ ἰσοδένδρους ἄν τις μεταθῇ διαμένειν· φυτεύεται δὲ τὰ πολλὰ αὐτῶν καὶ καταπηγνύμενα, καθάπερ ἡ λεύκη καὶ ἡ αἴγειρος.

2 Τούτων μὲν οὖν πρὸς τῇ σπερματικῇ καὶ τῇ ἀπὸ τῶν ῥιζῶν καὶ αὕτη γένεσίς ἐστι· τῶν δὲ

---

[1] ἐκφύοιτο conj. W.; ἐπιφύοιτο UMVAld.

BOOK III

Of Wild Trees.

Of the ways in which wild trees originate.

I. Now that we have spoken of cultivated trees, we must in like manner speak of wild ones, noting in what respects they agree with or differ from cultivated trees, and whether in any respects their character is altogether peculiar to themselves.

Now the ways in which they come into being are fairly simple; they all grow either from seed or from a root. But the reason of this is not that they could not possibly grow in any other way, but merely perhaps that no one even tries to plant them otherwise; whereas they might grow[1] from slips, if they were provided with a suitable position and received the fitting kind of tendance, as may be said even now of the trees of woodland and marsh, such as plane willow abele black poplar and elm; all these and other similar trees grow very quickly and well when they are planted from pieces torn off, so that[2] they survive, even if at the time of shifting they are already tall and as big as trees. Most of these are simply planted by being set firmly, for instance, the abele and the black poplar.

Such is the way in which these originate as well as from seed or from roots; the others grow only

[2] ὥστε καὶ μεγ. conj. Sch.; καὶ ὥστε καὶ μεγ. UM; καὶ ὥστε μεγ. PAld.

ἄλλων ἐκεῖναι· πλὴν ὅσα μόνον ἀπὸ σπέρματος φύεται, καθάπερ ἐλάτη πεύκη πίτυς. ὅσα δὲ ἔχει σπέρμα καὶ καρπόν, κἂν ἀπὸ ῥίζης γίνηται, καὶ ἀπὸ τούτων· ἐπεὶ καὶ τὰ δοκοῦντα ἄκαρπα εἶναι γεννᾶν φασιν, οἷον πτελέαν ἰτέαν. σημεῖον δὲ λέγουσιν οὐ μόνον ὅτι φύεται πολλὰ τῶν ῥιζῶν ἀπηρτημένα καθ' οὓς ἂν ᾖ τόπους, ἀλλὰ καὶ τὰ συμβαίνοντα θεωροῦντες, οἷον ἐν Φενεῷ τῆς Ἀρκαδίας, ὡς ἐξερράγη τὸ συναθροισθὲν ὕδωρ ἐν τῷ πεδίῳ φραχθέντων τῶν βερέθρων· ὅπου μὲν ἐγγὺς ἦσαν ἰτέαι πεφυκυῖαι τοῦ καταποθέντος τόπου, τῷ ὑστέρῳ ἔτει μετὰ τὴν ἀναξήρανσιν ἐνταῦθα αὖθις ἀναφῦναί φασιν ἰτέαν· ὅπου δὲ πτελέαι αὖθις πτελέας, καθάπερ καὶ ὅπου πεῦκαι καὶ ἔλαται πεύκας καὶ ἐλάτας, ὥσπερ μιμουμένων κἀκείνων.

3 Ἀλλὰ τὴν ἰτέαν ταχὺ προκαταβάλλειν πρὸ τοῦ τελείως ἁδρῦναι καὶ πέψαι τὸν καρπόν· δι' ὃ καὶ τὸν ποιητὴν οὐ κακῶς προσαγορεύειν αὐτὴν ὠλεσίκαρπον.

Τῆς δὲ πτελέας κἀκεῖνο σημεῖον ὑπολαμβάνουσιν· ὅταν γὰρ ἀπὸ τῶν πνευμάτων εἰς τοὺς ἐχομένους τόπους ὁ καρπὸς ἀπενεχθῇ, φύεσθαί φασι. παραπλήσιον δὲ ἔοικεν εἶναι τὸ συμβαῖνον ὃ καὶ ἐπὶ τῶν φρυγανικῶν καὶ ποιωδῶν τινών ἐστιν· οὐκ ἐχόντων γὰρ σπέρμα φανερόν, ἀλλὰ

---

[1] *cf.* 5. 4. 6.

[2] 'Katavothra' (now called 'the devil's holes,' see Lawson, cited below); *cf.* Paus. 8. 14; Catull. 68. 109; Plut. *de sera numinis vindicta*, 557 c: Plin. 31. 36; Frazer, *Pausanias and other Greek Sketches*, pp. 315 foll.; Lawson, *Modern Greek Folklore and Ancient Greek Religion*, p. 85.

in these two ways—while some of them, such as silver-fir fir and Aleppo pine grow *only* from seed. All those that have seed and fruit, even if they grow from a root, will grow from seed too; for they say that even those which, like elm and willow, appear to have no fruit reproduce themselves. For proof they give the fact that many such trees come up at a distance from the roots of the original tree, whatever the position may be; and further, they have observed a thing which occasionally happens; for instance, when at Pheneos[1] in Arcadia the water which had collected in the plain since the underground channels[2] were blocked burst forth, where there were willows growing near the inundated region, the next year after it had dried up they say that willows grew again; and where there had been elms, elms[3] grew, even as, where there had been firs and silver-firs, these trees reappeared—as if the former trees followed the example[4] of the latter.

But the willow is said to shed its fruit early, before it is completely matured and ripened; and so the poet[5] not unfittingly calls it "the willow which loses its fruit."

That the elm also reproduces itself the following is taken to be a proof: when the fruit is carried by the winds to neighbouring spots, they say that young trees grow from it. Something similar to this appears to be what happens in the case of certain under-shrubs and herbaceous plants; though they have no visible seed, but some of them only a sort of

[3] πτελέας αὖθις πτελέας conj. St.; πτελέας ἀντὶ πελέας U; πτελέας ἀντὶ πτελέας MV; πτελέας αὖθις πτελέας P; πτελέα αὖθις πτελέας Ald.

[4] *i.e.* by growing from seed, as conifers normally do.

[5] Homer, *Od.* 10. 510; *cf.* Plin. 16. 110.

τῶν μὲν οἷον χνοῦν τῶν δ' ἄνθος, ὥσπερ τὸ θύμον, ὅμως ἀπὸ τούτων βλαστάνουσιν. ἐπεὶ ἥ γε πλάτανος ἔχει φανερῶς καὶ ἀπὸ τούτων φύεται. τοῦτο δ' ἐξ ἄλλων τε δῆλον κἀκεῖνο μέγιστον σημεῖον· ὤφθη γὰρ ἤδη ποτὲ πεφυκυῖα πλάτανος ἐν τρίποδι χαλκῷ.

4 Ταύτας τε δὴ τὰς γενέσεις ὑποληπτέον εἶναι τῶν ἀγρίων καὶ ἔτι τὰς αὐτομάτους, ἃς καὶ οἱ φυσιολόγοι λέγουσιν· Ἀναξαγόρας μὲν τὸν ἀέρα πάντων φάσκων ἔχειν σπέρματα καὶ ταῦτα συγκαταφερόμενα τῷ ὕδατι γεννᾶν τὰ φυτά· Διογένης δὲ σηπομένου τοῦ ὕδατος καὶ μίξιν τινὰ λαμβάνοντος πρὸς τὴν γῆν· Κλείδημος δὲ συνεστάναι μὲν ἐκ τῶν αὐτῶν τοῖς ζώοις, ὅσῳ δὲ θολερωτέρων καὶ ψυχροτέρων τοσοῦτον ἀπέχειν τοῦ ζῶα εἶναι. [λέγουσι δέ τινες καὶ ἄλλοι περὶ τῆς γενέσεως.]

5 Ἀλλ' αὕτη μὲν ἀπηρτημένη πώς ἐστι τῆς αἰσθήσεως. ἄλλαι δὲ ὁμολογούμεναι καὶ ἐμφανεῖς, οἷον ὅταν ἔφοδος γένηται ποταμοῦ παρεκβάντος τὸ ῥεῖθρον ἢ καὶ ὅλως ἑτέρωθι ποιησαμένου, καθάπερ ὁ Νέσος ἐν τῇ Ἀβδηρίτιδι πολλάκις μεταβαίνει, καὶ ἅμα τῇ μεταβάσει τοσαύτην ὕλην συγγεννᾷ τοῖς τόποις, ὥστε τῷ τρίτῳ ἔτει συνηρεφεῖν. καὶ πάλιν ὅταν ἐπομβρίαι κατάσχωσι πλείω χρόνον· καὶ γὰρ ἐν ταύταις βλαστήσεις γίνονται φυτῶν. ἔοικε δὲ ἡ μὲν τῶν ποταμῶν ἔφοδος ἐπάγειν σπέρματα καὶ καρπούς, καὶ τοὺς ὀχετούς φασι τὰ τῶν ποιωδῶν· ἡ δ' ἐπομβρία

---

1 *cf.* *C.P.* 1. 5. 2.

2 Sc. of Apollonia, the 'Ionian' philosopher.

3 *cf.* *C.P.* 1. 10. 3; 3. 23. 1; Arist. *Meteor.* 2. 9.

down, and others only a flower, such as thyme, young plants nevertheless grow from these. As for the plane, it obviously has seeds, and seedlings grow from them. This is evident in various ways, and here is a very strong proof—a plane-tree has before now been seen which came up in a brass pot.

Such we must suppose are the ways in which wild trees originate, apart from the spontaneous ways of which natural philosophers tell. [1]Anaxagoras says that the air contains the seeds of all things, and that these, carried down by the rain, produce the plants; while Diogenes[2] says that this happens when water decomposes and mixes in some sort with earth. [3]Kleidemos maintains that plants are made of the same elements as animals, but that they fall short of being animals in proportion as their composition is less pure and as they are colder. [4] And there are other philosophers also who speak of spontaneous generation.

But this kind of generation is somehow beyond the ken of our senses. There are other admitted and observable kinds, as when a river in flood gets over its banks or has altogether changed its course, even as the Nesos in the district of Abdera often alters its course, and in so doing causes such a growth of forest in that region that by the third year it casts a thick shade. The same result ensues when heavy rains prevail for a long time; during these too many plants shoot up. Now, as the flooding of a river, it would appear, conveys seeds and fruits of trees, and, as they say, irrigation channels convey the[5] seeds of herbaceous plants, so heavy

[4] λέγουσι . . . γενεσέως apparently a gloss (W.).

[5] τὰ conj. W.; τὴν MAld.

τοῦτο ποιεῖ ταὐτό· συγκαταφερει γὰρ πολλὰ τῶν σπερμάτων, καὶ ἅμα σῆψίν τινα τῆς γῆς καὶ τοῦ ὕδατος· ἐπεὶ καὶ ἡ μίξις αὐτὴ τῆς Αἰγυπτίας
6 γῆς δοκεῖ τινα γεννᾶν ὕλην. ἐνιαχοῦ δε, ἂν μόνον ὑπεργάσωνται καὶ κινήσωσιν, εὐθὺς ἀναβλαστάνει τὰ οἰκεῖα τῆς χώρας, ὥσπερ ἐν Κρήτῃ κυπάριττοι. γίνεται δὲ παραπλήσιόν τι τούτῳ καὶ ἐν τοῖς ἐλάττοσιν· ἅμα γὰρ κινουμένης ἀναβλαστάνει πόα τις ἐν ἑκάστοις. ἐν δὲ τοῖς ἡμιβρόχοις ἐὰν ὑπονεάσῃς φαίνεσθαί φασι τρίβολον. αὗται μὲν οὖν ἐν τῇ μεταβολῇ τῆς χώρας εἰσίν, εἴτε καὶ ἐνυπαρχόντων σπερμάτων εἴτε καὶ αὐτῆς πως διατιθεμένης· ὅπερ ἴσως οὐκ ἄτοπον ἐγκατακλειομένων ἅμα τῶν ὑγρῶν· ἐνιαχοῦ δὲ καὶ ὑδάτων ἐπιγινομένων ἰδιώτερον ἀνατεῖλαι ὕλης πλῆθος, ὥσπερ ἐν Κυρήνῃ πιττώδους τινὸς γενομένου καὶ παχέος· οὕτως γὰρ ἀνεβλάστησεν ἡ πλησίον ὕλη πρότερον οὐκ οὖσα. φασὶ δὲ καὶ τό γε σίλφιον οὐκ ὂν πρότερον ἐκ τοιαύτης τινὸς αἰτίας φανῆναι. τρόποι μὲν οὖν τοιοῦτοι τῶν τοιουτων γενέσεων.

II. Πάντα δὲ κάρπιμα ἢ ἄκαρπα, καὶ ἀείφυλλα ἢ φυλλοβόλα, καὶ ἀνθοῦντα ἢ ἀνανθῆ· κοιναὶ

---

[1] ἡ δ' . . . ταὐτὸ conj. W.; ἡ δ' ἐπ. τοῦτ' αὖ ἐποίει ταὐτό UMV (δ' αὖ marked doubtful in U); ἡ δ' ἐπ. τοῦτ' αὐτὸ ἐποίει Ald. [2] Plin. 16. 142.

[3] *i.e.* and is released by working the ground.

[4] *cf.* *C.P.* 1. 5. 1; Plin. 16. 143, who gives the date A.U.C. 130; *cf.* 19. 41.

rain acts in the same way[1]; for it brings down many of the seeds with it, and at the same time causes a sort of decomposition of the earth and of the water. In fact, the mere mixture of earth with water in Egypt seems to produce a kind of vegetation. And in some places, if the ground is merely lightly worked and stirred, the plants native to the district immediately spring up; [2] for instance, the cypress in Crete. And something similar to this occurs even in smaller plants; as soon as the earth is stirred, wherever it may be, a sort of vegetation comes up. And in partly saturated soil, if you break up the ground, they say that caltrop appears. Now these ways of origination are due to the change which takes place in the soil, whether there were seeds in it already, or whether the soil itself somehow produces the result. And the latter explanation is perhaps not strange, seeing that the moist element is also locked up in the soil.[3] Again, in some places they say that after rain a more singular abundance of vegetation has been known to spring up; for instance, at Cyrene, after a heavy pitchy shower had fallen: for it was under these circumstances that there sprang up the wood[4] which is near the town, though till then it did not exist. They say also that silphium[5] has been known to appear from some such cause, where there was none before. [6] Such are the ways in which these kinds of generation come about.

Of the differences between wild and cultivated trees.

II. All trees are either fruit-bearing or without fruit, either evergreen or deciduous, either flowering

[5] *cf.* 6. 3. [6] τοιοῦτοι MSS.; τοσοῦτοι conj. W.

γάρ τινες διαιρέσεις ἐπὶ πάντων εἰσὶν ὁμοίως ἡμέρων τε καὶ ἀγρίων. ἴδια δὲ πρὸς τὰ ἥμερα τῶν ἀγρίων ὀψικαρπία τε καὶ ἰσχὺς καὶ πολυκαρπία τῷ προφαίνειν· πεπαίνει τε γὰρ ὀψιαίτερον καὶ τὸ ὅλον ἀνθεῖ καὶ βλαστάνει ὡς ἐπὶ τὸ πᾶν· καὶ ἰσχυρότερα τῇ φύσει· καὶ προφαίνει μὲν πλείω καρπὸν ἐκπέττει δ' ἧττον, εἰ μὴ καὶ πάντα ἀλλά γε τὰ ὁμογενῆ, οἷον ἐλάας καὶ ἀπίου κότινος καὶ ἀχράς. ἅπαντα γὰρ οὕτως, πλὴν εἴ τι σπάνιον, ὥσπερ ἐπὶ τῶν κρανείων καὶ τῶν οὔων· ταῦτα γὰρ δή φασι πεπαίτερα καὶ ἡδύτερα τὰ ἄγρια τῶν ἡμέρων εἶναι· καὶ εἰ δή τι ἄλλο μὴ προσδέχεται γεωργίαν ἢ δένδρον ἢ καί τι τῶν ἐλαττόνων, οἷον τὸ σίλφιον καὶ ἡ κάππαρις καὶ τῶν χεδροπῶν ὁ θέρμος, ἃ καὶ μάλιστ' ἄν τις
2 ἄγρια τὴν φύσιν εἴποι. τὸ γὰρ μὴ προσδεχόμενον ἡμέρωσιν, ὥσπερ ἐν τοῖς ζώοις, τοῦτο ἄγριον τῇ φύσει. καίτοι φησὶν Ἵππων ἅπαν καὶ ἥμερον καὶ ἄγριον εἶναι, καὶ θεραπευόμενον μὲν ἥμερον μὴ θεραπευόμενον δὲ ἄγριον, τῇ μὲν ὀρθῶς λέγων τῇ δὲ οὐκ ὀρθῶς. ἐξαμελούμενον γὰρ ἅπαν χεῖρον γίνεται καὶ ἀπαγριοῦται, θεραπευόμενον δὲ οὐχ ἅπαν βέλτιον, ὥσπερ εἴρηται. ὃ δὴ χωριστέον καὶ τὰ μὲν ἄγρια τὰ δ' ἥμερα λεκτέον,

---

[1] εἰ μὴ . . . ὁμογενῆ conj. W.; εἰ μὴ καὶ πάντα τὰ ἄλλα καὶ τὰ ὁμοιογενῆ UMVAld.H.

[2] *cf.* *C.P.* 3. 1. 4. [3] *cf.* 1. 3. 5 n.

[4] *i.e.* the terms 'cultivated' and 'wild' do not denote distinct 'kinds.'

or flowerless; for certain distinctions apply to all trees alike, whether cultivated or wild. To wild trees, as compared with cultivated ones, belong the special properties of fruiting late, of greater vigour, of abundance of fruit, produced if not matured; for they ripen their fruit later, and in general their time of flowering and making growth is later; also they are more vigorous in growth, and so, though they produce more fruit, they ripen it less; if[1] this is not universally true, at least it holds good of the wild olive and pear as compared with the cultivated forms of these trees. This is generally true with few exceptions, as in the cornelian cherry and sorb; for the wild forms of these, they say, ripen their fruit better, and it is sweeter than in the cultivated forms. 2 And the rule also does not hold good of anything which does not admit of cultivation, whether it be a tree or one of the smaller plants, as silphium caper and, among leguminous plants, the lupin; these one might say are specially wild in their character. For, as with animals which do not submit to domestication, so a plant which does not submit to cultivation may be called wild in its essential character. However Hippon[3] declares that of every plant there exists both a cultivated and a wild form, and that 'cultivated' simply means[4] that the plant has received attention, while 'wild' means that it has not; but though he is partly right, he is partly wrong. It is true that any plant deteriorates by neglect and so becomes wild; but it is not true that every plant may be improved by attention,[5] as has been said. Wherefore[6] we must make our distinction and call some things wild, others culti-

[5] *i.e.* and so become 'cultivated.'
[6] ὃ δὴ MSS.; διὸ conj. Sch. from G.

ὥσπερ τῶν ζώων τὰ συνανθρωπευόμενα καὶ τὰ δεχόμενα τιθασείαν.[1]

3 Ἀλλὰ τοῦτο μὲν οὐδὲν ἴσως διαφέρει ποτέρως ῥητέον. ἅπαν δὲ τὸ ἐξαγριούμενον τοῖς τε καρποῖς χεῖρον γίνεται καὶ αὐτὸ βραχύτερον καὶ φύλλοις καὶ κλωσὶ καὶ φλοιῷ καὶ τῇ ὅλῃ μορφῇ· καὶ γὰρ πυκνότερα καὶ οὐλότερα καὶ σκληρότερα καὶ ταῦτα καὶ ὅλη ἡ φύσις γίνεται, ὡς ἐν τούτοις μάλιστα τῆς διαφορᾶς τῶν ἡμέρων καὶ τῶν ἀγρίων γινομένης. δι' ὃ καὶ ὅσα τῶν ἡμερουμένων τοιαῦτα τυγχάνει, ταῦτα ἄγριά φασιν εἶναι, καθάπερ τὴν πεύκην καὶ τὴν κυπάριττον, ἢ ὅλως ἢ τὴν ἄρρενα, καὶ τὴν καρύαν δὲ καὶ τὴν διοσβάλανον.

4 Ἔτι τε τῷ φιλόψυχρα καὶ ὀρεινὰ μᾶλλον εἶναι· καὶ γὰρ τοῦτο λαμβάνεται πρὸς τὴν ἀγριότητα τῶν δένδρων καὶ ὅλως τῶν φυτῶν, εἴτ' οὖν καθ' αὑτὸ λαμβανόμενον εἴτε κατὰ συμβεβηκός.

Ὁ μὲν οὖν τῶν ἀγρίων ἀφορισμὸς εἴθ' οὕτως ἢ καὶ ἄλλως ληπτέος, οὐδὲν ἂν ἴσως διενέγκοι πρὸς τὰ νῦν· ἐκεῖνο δὲ ἀληθές, ὥς γε τῷ τύπῳ καὶ ἁπλῶς εἰπεῖν, ὅτι μᾶλλον ὀρεινὰ τὰ ἄγρια καὶ εὐθενεῖ τὰ πλείω καὶ μᾶλλον ἐν τούτοις τοῖς τόποις, ἐὰν μή τις λαμβάνῃ τὰ φίλυδρα καὶ παραποτάμια καὶ ἀλσώδη. ταῦτα γὰρ καὶ τὰ
5 τοιαῦτα τυγχάνει πεδεινὰ μᾶλλον. οὐ μὴν ἀλλ' ἔν γε τοῖς μεγάλοις ὄρεσιν, οἷον Παρνησῷ τε καὶ Κυλλήνῃ καὶ Ὀλύμπῳ τῷ Πιερικῷ τε καὶ τῷ Μυσίῳ καὶ εἴ που τοιοῦτον ἕτερον, ἅπαντα

---

[1] τιθασείαν conj. W., *cf.* Plat. *Pol.* 264 C; τιθάσιον UMAld.

vated—the latter class corresponding to those animals which live with man and can be tamed.[1]

But perhaps it does not matter which way this should be put. Any tree which runs wild deteriorates in its fruits, and itself becomes dwarfed in leaves branches bark and appearance generally; for under cultivation these parts, as well as the whole growth of the tree, become closer, more compact[2] and harder; which indicates that the difference between cultivated and wild is chiefly shown in these respects. And so those trees which show these characteristics under cultivation they say are really wild, for instance fir cypress, or at least the 'male' kind, hazel and chestnut.

Moreover these wild forms are distinguished by having greater liking for cold and for hilly country: for that too is regarded as a means of recognising wild trees and wild plants generally, whether it is so regarded in itself or as being only incidentally a distinguishing mark.

So the definition of wild kinds, whether it should be thus made or otherwise, perhaps makes no difference for our present purpose. But it is certainly true, speaking[3] broadly and generally, that the wild trees are more to be found in hilly country, and that the greater part of them flourish more in such regions, with the exception of those which love water or grow by river sides or in woods; these and such-like trees are rather trees of the plain. However on great mountains, such as Parnassus Cyllene the Pierian and the Mysian Olympus, and such regions anywhere

[2] οὐλότερα conj. W. from G, *spissiora*; ὀρθότερα MSS. *cf.* *C.P.* 6. 11. 8.

[3] ὥς γε conj. Sch.; ὥστε UM; ὡς ἐν Ald.H.

φύεται διὰ τὴν πολυειδίαν τῶν τόπων· ἔχουσι γὰρ καὶ λιμνώδεις καὶ ἐνύγρους καὶ ξηροὺς καὶ γεώδεις καὶ πετρώδεις καὶ τοὺς ἀνὰ μέσον λειμῶνας καὶ σχεδὸν ὅσαι διαφοραὶ τῆς γῆς· ἔτι δὲ τοὺς μὲν κοίλους καὶ εὐδιεινοὺς τοὺς δὲ μετεώρους καὶ προσηνέμους· ὥστε δύνασθαι παντοῖα καὶ τὰ ἐν τοῖς πεδίοις φέρειν.

6 Οὐδὲν δ' ἄτοπον οὐδ' εἰ ἔνια μὴ οὕτω πάμφορα τῶν ὀρῶν, ἀλλ' ἰδιωτέρας τινὸς ὕλης ἢ πάσης ἢ τῆς πλείστης, οἷον ἐν τῇ Κρήτῃ τὰ 'Ιδαῖα· κυπάριττος γὰρ ἐκεῖ· καὶ τὰ περὶ Κιλικίαν καὶ Συρίαν, ἐν οἷς κέδρος· ἐνιαχοῦ δὲ τῆς Συρίας τέρμινθος. αἱ γὰρ διαφοραὶ τῆς χώρας τὴν ἰδιότητα ποιοῦσιν. ἀλλ' εἴρηται τὸ ἴδιον ὡς ἐπὶ πᾶν.

III. Ἴδια δὲ τὰ τοιάδε τῶν ὀρεινῶν, ἃ ἐν τοῖς πεδίοις οὐ φύεται, [περὶ τὴν Μακεδονίαν] ἐλάτη πεύκη πίτυς ἀγρία φίλυρα ζυγία φηγὸς πύξος ἀνδράχλη μίλος ἄρκευθος τέρμινθος ἐρινεὸς φιλύκη ἀφάρκη καρύα διοσβάλανος πρῖνος. τὰ δὲ καὶ ἐν τοῖς πεδίοις μυρίκη πτελέα λεύκη ἰτέα αἴγειρος κρανεία θηλυκρανεία κλήθρα δρῦς λακάρη ἀχρὰς μηλέα ὀστρύα κήλαστρον μελία παλίουρος ὀξυάκανθος <σφένδαμνος,> ἣν ἐν μὲν τῷ

---

[1] ἐν . . . 'Ιδαῖα conj. W. (after Sch., who conj. τὰ ἐν); τὰ ἐν κρήτῃ τῇ 'Ιδαίᾳ UAld.

[2] *i.e.* it is not meant that a tree which is 'special' to Mount Ida (*e.g.*) occurs *only* there.

[3] περὶ τὴν Μακ.? a gloss; περί τε τὴν Μακ. MP$_2$Ald.; τε om. P.

else, all kinds grow, because of the diversity of positions afforded them. For such mountains offer positions which are marshy, wet, dry, deep-soiled or rocky; they have also their meadow land here and there, and in fact almost every variety of soil; again they present positions which lie low and are sheltered, as well as others which are lofty and exposed to wind; so that they can bear all sorts, even those which belong to the plains.

Yet it is not strange that there should be some mountains which do not thus bear all things, but have a more special kind of vegetation to a great extent if not entirely; for instance the range of Ida in Crete[1]; for there the cypress grows; or the hills of Cilicia and Syria, on which the Syrian cedar grows, or certain parts of Syria, where the terebinth grows. For it is the differences of soil which give a special character to the vegetation. [2] (However the word 'special' is used here in a somewhat extended sense.)

Of mountain trees: of the differences found in wild trees.

III. The following trees are peculiar to mountain country and do not grow in the plains; [3] let us take Macedonia as an example. Silver-fir fir 'wild pine' lime *zygia* Valonia oak box andrachne yew Phoenician cedar terebinth wild fig alaternus hybrid arbutus hazel chestnut kermes-oak. The following grow also in the plain: tamarisk elm abele willow black poplar cornelian cherry cornel alder oak *lakare* (bird-cherry) wild pear apple hop-hornbeam holly manna-ash Christ's thorn cotoneaster maple,[4] which

[4] σφένδαμνος add. Palm. in view of what follows; ὀξυάκαρτα ἄκανθος UPAld.Bas.; ἄκανθος P_2.

ὄρει πεφυκυῖαν ζυγίαν καλοῦσιν, ἐν δὲ τῷ πεδίῳ
γλεῖνον. οἱ δ' ἄλλως διαιροῦσι καὶ ἕτερον ποιοῦσιν εἶδος σφενδάμνου καὶ ζυγίας.
2 Ἅπαντα δὲ ὅσα κοινὰ τῶν ὀρῶν καὶ τῶν
πεδίων, μείζω μὲν καὶ καλλίω τῇ ὄψει τὰ ἐν τοῖς
πεδίοις γίνεται, κρείττω δὲ τῇ χρείᾳ τῇ τε τῶν
ξύλων καὶ τῇ τῶν καρπῶν τὰ ὀρεινά· πλὴν
ἀχράδος καὶ ἀπίου καὶ μηλέας· αὗται δ' ἐν τοῖς
πεδίοις κρείττους οὐ μόνον τοῖς καρποῖς ἀλλὰ καὶ
τοῖς ξύλοις· ἐν γὰρ τοῖς ὄρεσι μικραὶ καὶ ὀζώδεις
καὶ ἀκανθώδεις γίνονται· πάντα δὲ καὶ ἐν τοῖς
ὄρεσιν, ὅταν ἐπιλάβωνται τῶν οἰκείων τόπων, καὶ
καλλίω φύεται καὶ εὐθενεῖ μᾶλλον· ὡς δὲ ἁπλῶς
εἰπεῖν τὰ ἐν τοῖς ὁμαλέσι τῶν ὀρῶν καὶ μάλιστα,
τῶν δὲ ἄλλων τὰ ἐν τοῖς κάτω καὶ κοίλοις· τὰ δ'
ἐπὶ τῶν ἄκρων χείριστα, πλὴν εἴ τι τῇ φύσει
3 φιλόψυχρον· ἔχει δὲ καὶ ταῦτ' αὖ τινα διαφορὰν
ἐν τοῖς ἀνομοίοις τῶν τόπων, ὑπὲρ ὧν ὕστερον
λεκτέον· νῦν δὲ διαιρετέον ἕκαστον κατὰ τὰς διαφορὰς τὰς εἰρημένας.
Ἀείφυλλα μὲν οὖν ἐστι τῶν ἀγρίων ἃ καὶ
πρότερον ἐλέχθη, ἐλάτη πεύκη πίτυς ἀγρία πύξος
ἀνδράχλη μίλος ἄρκευθος τέρμινθος φιλύκη
ἀφάρκη δάφνη φελλόδρυς κήλαστρον ὀξυάκανθος
πρῖνος μυρίκη· τὰ δὲ ἄλλα πάντα φυλλοβολεῖ·
πλὴν εἴ τι περιττὸν ἐνιαχοῦ, καθάπερ ἐλέχθη περὶ
τῆς ἐν τῇ Κρήτῃ πλατάνου καὶ δρυὸς καὶ εἴ που
τόπος τις ὅλως εὔτροφος.

---

[1] δ' ἄλλως conj. Sch. from G; δ' αὖ Ald. [2] Plin. 16. 77.
[3] i.e. are not always of the poorest quality. ταῦτ' αὖ τινα conj. W.; ταῦτα αὐτῶν Ald.H. [4] 1. 9. 3.

when it grows in the mountains, is called *zygia*, when in the plain, *gleinos*: others, however,[1] classify differently and make maple and *zygia* distinct trees.

2 All those trees which are common to both hill and plain are taller and finer in appearance when they grow in the plain; but the mountain forms are better as to producing serviceable timber and fruits, with the exception of wild pear pear and apple; these are in the plain better in fruit and also in timber; for in the hills they grow small with many knots and much spinous wood. But even on the mountains all trees grow fairer and are more vigorous when they have secured a suitable position; and, to speak generally, those which grow on the level parts of the mountains are specially fair and vigorous; next to these come those which grow on the lower parts and in the hollows; while those that grow on the heights are of the poorest quality, except any that are naturally cold-loving. But even these shew some variation[3] in different positions, of which we must speak later; for the present we must in our distinctions in each case take account only of the differences already mentioned.

Now among wild trees those are evergreen which were mentioned before,[4] silver-fir fir 'wild pine' box andrachne yew Phoenician cedar terebinth alaternus hybrid arbutus bay *phellodrys*[5] (holm-oak) holly cotoneaster kermes-oak tamarisk; but all the others shed their leaves, unless it be that in certain places they keep them exceptionally, as was said[6] of the plane and oak in Crete and in any other place which is altogether favourable to luxuriant growth.

[5] φελλόδρυς conj. Bod., *cf.* 1. 9. 3; φελλὸς δρῦς UMV(?)Ald.
[6] 1. 9. 5.

4 Κάρπιμα δὲ τὰ μὲν ἄλλα πάντα· περὶ δὲ ἰτέας καὶ αἰγείρου καὶ πτελέας, ὥσπερ ἐλέχθη,[1] διαμφισβητοῦσιν. ἔνιοι δὲ τὴν αἴγειρον μόνην ἀκαρπεῖν φασιν, ὥσπερ καὶ οἱ ἐν Ἀρκαδίᾳ, τὰ δὲ ἄλλα πάντα τὰ ἐν τοῖς ὄρεσι καρποφορεῖν. ἐν Κρήτῃ δὲ καὶ αἴγειροι κάρπιμοι πλείους εἰσί·[2] μία μὲν ἐν τῷ στομίῳ τοῦ ἄντρου τοῦ ἐν τῇ Ἴδῃ,[3] ἐν ᾧ τὰ ἀναθήματα ἀνάκειται, ἄλλη δὲ μικρὰ πλησίον· ἀπωτέρω δὲ μάλιστα δώδεκα σταδίους περί τινα κρήνην Σαύρου καλουμένην πολλαί. εἰσὶ δὲ καὶ ἐν τῷ πλησίον ὄρει τῆς Ἴδης ἐν τῷ Κινδρίῳ καλουμένῳ καὶ περὶ Πραισίαν δὲ ἐν τοῖς ὄρεσιν. οἱ δὲ μόνον τῶν τοιούτων τὴν πτελέαν κάρπιμον εἶναί φασι, καθάπερ οἱ περὶ Μακεδονίαν.

5 Μεγάλη δὲ διαφορὰ πρὸς καρπὸν καὶ ἀκαρπίαν καὶ ἡ τῶν τόπων φύσις, ὥσπερ ἐπί τε τῆς περσέας ἔχει καὶ τῶν φοινίκων· ἡ μὲν ἐν Αἰγύπτῳ καρποφορεῖ καὶ εἴ που τῶν πλησίον τόπων, ἐν Ῥόδῳ δὲ μέχρι τοῦ ἀνθεῖν μόνον ἀφικνεῖται. ὁ δὲ φοῖνιξ περὶ μὲν Βαβυλῶνα θαυμαστός, ἐν τῇ Ἑλλάδι δὲ οὐδὲ πεπαίνει, παρ᾽ ἐνίοις δὲ ὅλως οὐδὲ προφαίνει καρπόν.

6 Ὁμοίως δὲ καὶ ἕτερα πλείω τοιαῦτ᾽ ἐστίν· ἐπεὶ καὶ τῶν ἐλαττόνων ποαρίων καὶ ὑλημάτων ἐν τῇ

---

[1] 2. 2. 10.

[2] *cf.* 2. 2. 10. It appears that the buds of the poplar were mistaken for fruit (Sch.); *cf.* Diosc. 1. 81. Later writers perpetuated the error by calling them κόκκοι.

[3] τοῦ ἐν τῇ Ἴδῃ conj. Sch.; τοῦ ἐν τῷ Ἴδῃ U; τοῦ ἐν τῷ Ἴδης MV; ἐν τῇ Ἴδῃ Ald.H.

Most trees are fruit-bearing, but about willow black poplar and elm men hold different opinions, as was said[1]; and some, as the Arcadians, say that only the black poplar is without fruit, but that all the other mountain trees bear fruit. However in Crete there are a number of black poplars which bear fruit[2]; there is one at the mouth of the cave on mount Ida,[3] in which the dedicatory offerings are hung, and there is another small one not far off, and there are quite a number about a spring called the Lizard's Spring about twelve furlongs off. There are also some in the hill-country of Ida in the same neighbourhood, in the district called Kindria and in the mountains about Praisia.[4] Others again, as the Macedonians, say that the elm is the only tree of this class which bears fruit.

Again the character of the position makes a great difference as to fruit-bearing, as in the case of the *persea*[5] and the date-palm. The *persea* of Egypt bears fruit, and so it does wherever it grows in the neighbouring districts, but in Rhodes[6] it only gets as far as flowering. The date-palm in the neighbourhood of Babylon is marvellously fruitful; in Hellas it does not even ripen its fruit, and in some places it does not even produce any.

The same may be said of various other trees: in fact even[7] of smaller herbaceous plants and bushes some are fruitful, others not, although the latter are

[4] Πραισίαν conj. Meurs. *Creta*; τιρασίαν UMVAld.
[5] *cf.* 4. 2. 5. περσέαι conj. R. Const.; περσείας U; περσίας Ald.
[6] Ῥόδῳ conj. R. Const. from G, so too Plin. 16. 111; ῥόα Ald. *cf.* 1. 13. 5. for a similar corruption.
[7] ἐπεὶ καὶ conj. Sch. from G; ἐπεὶ δὲ καὶ Ald.

αὐτῇ χώρᾳ καὶ συνόρῳ χώρᾳ τὰ μὲν κάρπιμα τὰ δ' ἄκαρπα γίνεται· καθάπερ καὶ τὸ κενταύριον ἐν τῇ Ἠλείᾳ, τὸ μὲν ἐν τῇ ὀρεινῇ κάρπιμον, τὸ δ' ἐν τῷ πεδίῳ ἄκαρπον ἀλλὰ μονον ἀνθεῖ, τὸ δ' ἐν τοῖς κοίλοις τόποις οὐδ' ἀνθεῖ πλὴν κακῶς. δοκεῖ δ' οὖν καὶ τῶν ἄλλων τῶν ὁμογενῶν καὶ ἐν μιᾷ προσηγορίᾳ τὸ μὲν ἄκαρπον εἶναι τὸ δὲ κάρπιμον, οἷον πρῖνος ὁ μὲν κάρπιμος ὁ δ' ἄκαρπος· καὶ
7 κλήθρα δὲ ὡσαύτως· ἀνθεῖ δ' ἄμφω. σχεδὸν δὲ ὅσα καλοῦσιν ἄρρενα τῶν ὁμογενῶν ἄκαρπα· καὶ τούτων τὰ μὲν πολλὰ ἀνθεῖν φασι τὰ δ' ὀλίγον τὰ δ' ὅλως οὐδ' ἀνθεῖν· τὰ δὲ ἀνάπαλιν, τὰ μὲν ἄρρενα μόνα καρποφορεῖν, οὐ μὴν ἀλλ' ἀπό γε τῶν ἀνθῶν φύεσθαι τὰ δένδρα, καθάπερ καὶ ἀπὸ τῶν καρπῶν ὅσα κάρπιμα· καὶ ἐν ἀμφοῖν οὕτως ἐνίοτε πυκνὴν εἶναι τὴν ἔκφυσιν ὥστε τοὺς ὀρεοτύπους οὐ δύνασθαι διϊέναι μὴ ὁδοποιησαμένους.

8 Ἀμφισβητεῖται δὲ καὶ περὶ τῶν ἀνθῶν ἐνίων, ὥσπερ εἴπομεν. οἱ μὲν γὰρ καὶ δρῦν ἀνθεῖν οἴονται καὶ τὴν Ἡρακλεῶτιν καρύαν καὶ διοσβάλανον, ἔτι δὲ πεύκην καὶ πίτυν· οἱ δ' οὐδὲν τούτων, ἀλλὰ τὸν ἴουλον τὸν ἐν ταῖς καρύαις καὶ τὸ βρύον τὸ δρύϊνον καὶ τὸν κύτταρον τὸν πιτύ-

---

[1] χώρᾳ καὶ Ald.; ᾗ καὶ conj. St.

[2] *i.e.* the 'males' are sterile whether they flower or not. καὶ τούτων τὰ μὲν πολλὰ I conj.; τούτων τὰ πολλὰ τὰ μὲν Ald.

[3] ? *i.e.* the flowers of the 'female' tree.

[4] *i.e.* (*a*) in those trees whose 'male' form is sterile, whether it bears flowers or not; (*b*) in those whose 'male'

growing in the same place as the former, or[1] quite near it. Take for instance the centaury in Elea; where it grows in hill-country, it is fruitful; where it grows in the plain, it bears no fruit, but only flowers; and where it grows in deep valleys, it does not even flower, unless it be scantily. Any way it appears that, even of other plants which are of the same kind and all go by the same name, one will be without fruit, while another bears fruit; for instance, one kermes-oak will be fruitful, another not; and the same is true of the alder, though both produce flowers. And, generally speaking, all those of any given kind which are called 'male' trees are without fruit, and that though[2] some of these, they say, produce many flowers, some few, some none at all. On the other hand they say that in some cases it is only the 'males' that bear fruit, but that, in spite of this, the trees grow from the flowers,[3] (just as in the case of fruit-bearing trees they grow from the fruit). And they add that in both cases,[4] the crop of seedlings[5] which comes up is sometimes so thick that the woodmen cannot get through except by clearing a way.

There is also a doubt about the flower of some trees, as we said. Some think that the oak bears flowers, and also the filbert the chestnut and even the fir and Aleppo pine; some however think that none of these has a flower, but that,—resembling[6] and corresponding to the wild figs which drop off prematurely, we have in the nuts the catkin,[7] in the

form alone bears fruit, but the fruit is infertile. The passage is obscure: W. gives up the text.

[5] ἔκφυσιν. *cf.* 7. 4. 3.

[6] ὅμοιον conj. W.; ὁμοίαν UAld. *cf.* 3. 7. 3.

[7] *cf.* 3. 5. 5.

ἴνον ὅμοιον καὶ ἀνάλογον εἶναι τοῖς προαποπτώτοις ἐρινοῖς. οἱ δὲ περὶ Μακεδονίαν οὐδὲ ταῦτά φασιν ἀνθεῖν ἄρκευθον ὀξύην ἀρίαν σφένδαμνον. ἔνιοι δὲ τὰς ἀρκεύθους δύο εἶναι, καὶ τὴν μὲν ἑτέραν ἀνθεῖν μὲν ἄκαρπον δ' εἶναι, τὴν δὲ ἑτέραν οὐκ ἀνθεῖν μὲν καρπὸν δὲ φέρειν εὐθὺς προφαινόμενον, ὥσπερ καὶ τὰς συκᾶς τὰ ἐρινά. συμβαίνει δ' οὖν ὥστε ἐπὶ δύο ἔτη τὸν καρπὸν ἔχειν μόνον τοῦτο τῶν δένδρων. ταῦτα μὲν οὖν ἐπισκεπτέον.

IV. Ἡ δὲ βλάστησις τῶν μὲν ἅμα γίνεται καὶ τῶν ἡμέρων, τῶν δὲ μικρὸν ἐπιλειπομένη, τῶν δ' ἤδη πλέον, ἁπάντων δὲ κατὰ τὴν ἠρινὴν ὥραν. ἀλλὰ τῶν καρπῶν ἡ παραλλαγὴ πλείων· ὥσπερ δὲ καὶ πρότερον εἴπομεν, οὐ κατὰ τὰς βλαστήσεις αἱ πεπάνσεις ἀλλὰ πολὺ διαφέρουσιν· ἐπεὶ καὶ τῶν ὀψικαρποτέρων, ἃ δή τινές φασιν ἐνιαυτοφορεῖν, οἷον ἄρκευθον καὶ πρῖνον, ὅμως αἱ βλαστήσεις τοῦ ἦρος. αὐτὰ δ' αὐτῶν τὰ ὁμογενῆ τῷ πρότερον καὶ ὕστερον διαφέρει κατὰ τοὺς τόπους· πρῶτα μὲν γὰρ βλαστάνει τὰ ἐν τοῖς ἕλεσιν, ὡς οἱ περὶ Μακεδονίαν λέγουσι, δεύτερα δὲ τὰ ἐν τοῖς πεδίοις, ἔσχατα δὲ τὰ ἐν τοῖς ὄρεσιν.

2 Αὐτῶν δὲ τῶν καθ' ἕκαστα δένδρων τὰ μὲν

---

[1] *i.e.* the male flower, *cf.* Schol. on Ar. *Vesp.* 1111. Θεόφραστος κυρίως λέγει κύτταρον τὴν προάνθησιν τῆς πίτυος: but no explanation of such a use of the word suggests itself. *cf.* 3. 3. 8; 4 8. 7.

[2] ἀρίαν conj. Sch., *cf.* 3. 4. 2; 3. 16. 3; 3. 17. 1; ὀξύνην ἀγρίαν Ald.

oak the oak-moss, in the pine the 'flowering tuft.'[1] The people of Macedonia say that these trees also produce no flowers—Phoenician cedar beech *aria*[2] (holm-oak) maple. Others distinguish two kinds of Phoenician cedar, of which one bears flowers but bears no fruit, while the other, though it has no flower, bears a fruit which shows itself at once[3]—just as wild figs produce their abortive fruit. However that may be,[4] it is a fact that this is the only tree which keeps its fruit for two years. These matters then need enquiry.

Of the times of budding and fruiting of wild, as compared with cultivated, trees.

IV. Now the budding of wild trees occurs in some cases at the same time as that of the cultivated forms, but in some cases somewhat, and in some a good deal later; but in all cases it is during the spring season. But there is greater diversity in the time of fruiting; as we said before, the times of ripening do not correspond to those of budding, but there are wide differences. For even in the case of those trees which are somewhat late in fruiting,—which some say take a year to ripen their fruit—such as Phoenician cedar and kermes-oak, the budding nevertheless takes place in the spring. Again there are differences of time between individual trees of the same kind, according to the locality; those in the marshes bud earliest, as the Macedonians say, second to them those in the plains, and latest those in the mountains.

Again of particular trees some wild ones bud

[3] *i.e.* without antecedent flower.
[4] δ' οὖν conj. W.; σχεδὸν UMVAld.

συναναβλαστάνει τοῖς ἡμέροις, οἷον ἀνδράχλη
ἀφάρκη· ἀχρὰς δὲ μικρῷ ὕστερον τῆς ἀπίου. τὰ
δὲ καὶ πρὸ ζεφύρου καὶ μετὰ πνοὰς εὐθὺ ζεφύρου.
καὶ πρὸ ζεφύρου μὲν κρανεία καὶ θηλυκρανεία,
μετὰ ζέφυρον δὲ δάφνη κλήθρα, πρὸ ἰσημερίας δὲ
μικρὸν φίλυρα ζυγία φηγὸς συκῆ· πρωίβλαστα
δὲ καὶ καρύα καὶ δρῦς καὶ ἀκτέος· ἔτι δὲ μᾶλλον
τὰ ἄκαρπα δοκοῦντα καὶ ἀλσώδη, λεύκη πτελέα
ἰτέα αἴγειρος· πλάτανος δὲ μικρῷ ὀψιαίτερον
τούτων. τὰ δὲ ἄλλα ὥσπερ ἐνισταμένου τοῦ
ἦρος, οἷον ἐρινεὸς φιλύκη ὀξυάκανθος παλίουρος
τέρμινθος καρύα διοσβάλανος· μηλέα δ' ὀψί-
βλαστος· ὀψιβλαστότατον δὲ σχεδὸν ἴψος ἀρία
τετραγωνία θύεια μίλος. αἱ μὲν οὖν βλαστήσεις
οὕτως ἔχουσιν.

3 Αἱ δὲ ἀνθήσεις ἀκολουθοῦσι μὲν ὡς εἰπεῖν κατὰ
λόγον, οὐ μὴν ἀλλὰ παραλλάττουσι, μᾶλλον δὲ
καὶ ἐπὶ πλέον ἡ τῶν καρπῶν τελείωσις. κρανεία
μὲν γὰρ ἀποδίδωσι περὶ τροπὰς θερινὰς ἡ πρώϊος
σχεδὸν ὥσπερ πρῶτον· ἡ δ' ὄψιος, ἣν δή τινες
καλοῦσι θηλυκρανείαν, μετ' αὐτὸ τὸ μετόπωρον·
ἔστι δὲ ὁ ταύτης καρπὸς ἄβρωτος καὶ τὸ ξύλον
ἀσθενὲς καὶ χαῦνον· τοσαυτη δὴ διαφορὰ περὶ
4 ἄμφω. τέρμινθος δὲ περὶ πυροῦ ἀμητὸν ἢ μικρῷ

---

[1] See below, n. 4.

[2] τὰ ἀκ. δοκ. καὶ ἀλσ. conj. W.; τὰ ἀκ. καὶ δοκ. καὶ ἀλσ. U MP; τὰ ἀκ. τὰ δοκ. ἀλσ. Ald.

[3] ὥσπερ apologises for the unusual sense given to ἐνιστ.

along with the cultivated forms, as andrachne and hybrid arbutus; and the wild pear is a little later than the cultivated. Some again bud both before zephyr begins to blow, and immediately after it has been blowing. Before it come cornelian cherry and cornel, after it bay and alder; a little before the spring equinox come lime *zygia* Valonia oak fig. Hazel[1] oak and elder are also early in budding, and still more those trees which seem to have no fruit and to grow in groves,[2] abele elm willow black poplar; and the plane is a little later than these. The others which bud when the spring is, as it were, becoming established,[3] are such as wild fig alaternus cotoneaster Christ's thorn terebinth hazel[4] chestnut. The apple is late in budding, latest of all generally are *ipsos*[5] (cork-oak) *aria* (holm-oak) *tetragonia* odorous cedar yew. Such are the times of budding.

The flowering times in general follow in proportion; but they present some irregularity, and so in still more cases and to a greater extent do the times at which the fruit is matured. The cornelian cherry produces its fruit about the summer solstice; the early kind, that is to say, and this tree is about the earliest of all.[6] The late form, which some call 'female cornelian cherry' (cornel), fruits quite at the end of autumn. The fruit of this kind is inedible and its wood is weak and spongy; that is what the difference between the two kinds amounts to. The terebinth produces its fruit about the time of wheat-harvest or

(usually 'beginning'). τὰ δ' ἄλλα ὥσπερ ἐνιστ. conj. W.; τὰ δ' ἄλλως περ' U; τὰ δὲ ἄλλως περιενισταμένου MAld. H.

[4] καρύα can hardly be right both here and above.

[5] See Index.

[6] σχεδὸν ὥσπερ πρῶτον not in G, nor in Plin. (16. 105); text perhaps defective.

ὀψιαίτερον ἀποδίδωσι καὶ μελία καὶ σφένδαμνος τοῦ θέρους τὸν καρπόν· κλήθρα δὲ καὶ καρύα καὶ ἀχράδων τι γένος μετοπώρου· δρῦς δὲ καὶ διοσβάλανος ὀψιαίτερον ἔτι περὶ Πλειάδος δύσιν, ὡσαύτως δὲ καὶ φιλύκη καὶ πρῖνος καὶ παλίουρος καὶ ὀξυάκανθος μετὰ Πλειάδος δύσιν· ἡ δ' ἀρία χειμῶνος ἀρχομένου· καὶ ἡ μηλέα μὲν τοῖς πρώτοις ψύχεσιν, ἀχρὰς δὲ ὀψία χειμῶνος· ἀνδράχλη δὲ καὶ ἀφάρκη τὸ μὲν πρῶτον πεπαίνουσιν ἅμα τῷ βότρυϊ περκάζοντι, τὸ δὲ ὕστερον, δοκεῖ γὰρ ταῦτα δίκαρπα, ἀρχομένου τοῦ χειμῶνος, ἐλάτη δὲ καὶ
5 μίλος ἀνθοῦσι μικρὸν πρὸ ἡλίου τροπῶν· [καὶ τῆς γε ἐλάτης τὸ ἄνθος κρόκινον καὶ ἄλλως καλόν·] τὸν δὲ καρπὸν ἀφιᾶσι μετὰ δύσιν Πλειάδος. πεύκη δὲ καὶ πίτυς προτεροῦσι τῇ βλαστήσει μικρόν, ὅσον πεντεκαίδεκα ἡμέραις, τοὺς δὲ καρποὺς ἀποδιδόασι μετὰ Πλειάδα κατὰ λόγον.

Ταῦτα μὲν οὖν μετριωτέραν μὲν ἔχει παραλλαγήν· πάντων δὲ πλείστην ἡ ἄρκευθος καὶ ἡ κήλαστρος καὶ ἡ πρῖνος· ἡ μὲν γὰρ ἄρκευθος ἐνιαύσιον ἔχειν δοκεῖ· περικαταλαμβάνει γὰρ ὁ νέος τὸν περυσινόν. ὡς δέ τινές φασιν, οὐδὲ πεπαίνει, δι' ὃ καὶ προαφαιροῦσι καὶ χρόνον τινὰ τηροῦσιν· ἐὰν δὲ ἐᾷ
6 ἐπὶ τοῦ δένδρου τις, ἀποξηραίνεται. φασὶ δὲ καὶ τὴν πρῖνον οἱ περὶ Ἀρκαδίαν ἐνιαυτῷ τελειοῦν· ἅμα γὰρ τὸν ἔνον πεπαίνει καὶ τὸν νέον ὑποφαίνει· ὥστε τοῖς τοιούτοις συμβαίνει συνεχῶς τὸν καρπὸν ἔχειν. φασὶ δέ γε καὶ τὴν κήλαστρον ὑπὸ τοῦ

---

[1] ἀποδ. καὶ μελία U; ἀποδίδωσι· μελία Ald. Some confusion in text, but sense clear.

[2] ὀψία: ? ἡ ὀψία W.

a little later, manna-ash[1] and maple in summer; alder hazel and a certain kind of wild pear in autumn; oak and chestnut later still, about the setting of the Pleiad; and in like manner alaternus kermes-oak Christ's-thorn cotoneaster after the setting of the Pleiad; *aria* (holm-oak) when winter is beginning, apple with the first cold weather, wild pear late[2] in winter. Andrachne and hybrid arbutus first ripen their fruit when the grape is turning, and again[3] when winter is beginning; for these trees appear to bear twice. As for[4] silver-fir and yew, they flower a little before the solstice; [5](the flower of the silver-fir is yellow and otherwise pretty); they bear their fruit after the setting of the Pleiad. Fir and Aleppo pine are a little earlier in budding, about fifteen days, but produce their fruit after the setting of the Pleiad, though proportionately earlier than silver-fir and yew.

In these trees then the difference of time is not considerable; the greatest difference is shewn in Phoenician cedar holly and kermes-oak; for Phoenician cedar appears to keep its fruit for a year, the new fruit overtaking that of last year; and, according to some, it does not ripen it at all; wherefore men gather it unripe and keep it, whereas if it is left on the tree, it shrivels up. The Arcadians say that the kermes-oak also takes a year to perfect its fruit; for it ripens last year's fruit at the same time that the new fruit appears on it; the result of which is that such trees always have fruit on them. They say also

[3] After ὕστερον Ald. adds ἀνθοῦντι (so also H and G); Plin. 13. 121. omits it; om. W. after Sch.

[4] γὰρ Ald.; δὲ conj. W.

[5] Probably an early gloss, W. *cf.* Plin. 16. 106.

χειμῶνος ἀποβάλλειν. ὀψίκαρπα δὲ σφόδρα καὶ φίλυρα[1] καὶ πύξος. [τὸν[2] δὲ καρπὸν ἄβρωτον ἔχει παντὶ ζώῳ φίλυρα θηλυκρανεία πύξος. ὀψίκαρπα δὲ καὶ κιττὸς καὶ ἄρκευθος καὶ πεύκη καὶ ἀνδράχλη.] ὡς δὲ οἱ περὶ Ἀρκαδίαν φασίν, ἔτι τούτων ὀψικαρπότερα σχεδὸν δὲ πάντων ὀψιαίτερα τετραγωνία[3] θύεια μίλος. αἱ μὲν οὖν τῶν καρπῶν ἀποβολαὶ καὶ πεπάνσεις τῶν ἀγρίων[4] τοιαύτας ἔχουσι διαφορὰς οὐ μόνον πρὸς τὰ ἥμερα ἀλλὰ καὶ πρὸς ἑαυτά.

V. Συμβαίνει δ' ὅταν ἄρξωνται βλαστάνειν τὰ μὲν ἄλλα συνεχῆ τήν τε βλάστησιν καὶ τὴν αὔξησιν ποιεῖσθαι, πεύκην δὲ καὶ ἐλάτην καὶ δρῦν διαλείπειν, καὶ τρεῖς ὁρμὰς εἶναι καὶ τρεῖς ἀφιέναι βλαστούς, δι' ὃ καὶ τρίσλοποι· πᾶν γὰρ δὴ δένδρον ὅταν βλαστάνῃ λοπᾷ· πρῶτον μὲν ἄκρου ἔαρος εὐθὺς ἱσταμένου τοῦ Θαργηλιῶνος, ἐν δὲ τῇ Ἴδῃ περὶ πεντεκαίδεκα μάλιστα ἡμέρας· μετὰ δὲ ταῦτα διαλιπόντα περὶ τριάκοντα ἢ μικρῷ πλείους ἐπιβάλλεται πάλιν ἄλλους βλαστοὺς ἀπ' ἄκρας τῆς κορυνήσεως τῆς ἐπὶ τῷ προτέρῳ βλαστῷ· καὶ τὰ μὲν ἄνω τὰ δ' εἰς τὰ πλάγια κύκλῳ ποιεῖται τὴν βλάστησιν, οἷον γόνυ

[1] φίλυρα Ald.; φιλυρέα conj. Sch.
[2] τὸν δὲ . . . . ἀνδράχλη. Apparently a gloss, W.
[3] τετραγωνία conj. Sch. (τετρα- omitted after -τερα): *cf.* § 2; γωνία MV; γωνίεια U.
[4] τῶν ἀγρίων after πεπάνσεις conj. Sch.; after ἥμερα Ald.
[5] Plin. 16. 100.

that holly loses its fruit owing to the winter. Lime[1] and box are very late in fruiting, (lime has a fruit which no animal can eat, and so have cornel and box. Ivy Phoenician cedar fir and andrachne are late fruiting[2]) though, according to the Arcadians, still later than these and almost latest of all are *tetragonia*[3] odorous cedar and yew. Such then are the differences as to the time of shedding and ripening their fruit between wild[4] as compared with cultivated trees, and likewise as compared with one another.

Of the seasons of budding.

V. [5] Now most trees, when they have once begun to bud, make their budding and their growth continuously, but with fir silver-fir and oak there are intervals. They make three fresh starts in growth and produce three separate sets of buds; wherefore also they lose their bark thrice[6] a year. For every tree loses its bark when it is budding. This first happens in mid-spring[7] at the very beginning of the month Thargelion,[8] on Mount Ida within about fifteen days of that time; later, after an interval of about thirty days or rather more, the tree[9] puts on fresh buds which start from the head of the knobby growth[10] which formed at the first budding-time; and it makes its budding partly on the top of this,[11] partly all round it laterally,[12] using the knob formed at the

[6] *τρίσλοποι* conj. Sch.; *τρίσλοιποι* UM_2V; *τρίσλεποι* M_1Ald. *cf.* 4. 15. 3; 5. 1. 1.

[7] *ἔαρος* conj. R. Const.; *ἀέρος* VAld. *cf.* Plin. *l.c.*

[8] About May.

[9] What follows evidently applies only to the oak.

[10] *κορυνησέως* conj. Sch.; *κορύνης ἕως* UMV; *κορυφῆς ἕως* Ald.

[11] *cf.* 3. 6. 2.

[12] *τὰ* add. Sch.

ποιησάμενα τὴν τοῦ πρώτου βλαστοῦ κορύνην, ὥσπερ καὶ ἡ πρώτη βλάστησις ἔχει. γίνεται δὲ τοῦτο περὶ τὸν Σκιρροφοριῶνα λήγοντα.[1]

2 Κατὰ δὲ ταύτην τὴν βλάστησιν καὶ ἡ κηκὶς φύεται πᾶσα, καὶ ἡ λευκὴ καὶ ἡ μέλαινα· φύεται δὲ ὡς ἐπὶ τὸ πολὺ νυκτὸς ἀθρόος· ἐφ' ἡμέραν δὲ μίαν αὐξηθεῖσα, πλὴν τῆς πιττοειδοῦς, ἐὰν ὑπὸ τοῦ καύματος ληφθῇ ξηραίνεται, καὶ ἀναυξὴς ἐπὶ τὸ μεῖζον, ἐγίνετο γὰρ ἂν μείζων τῷ μεγέθει. διόπερ τινὲς αὐτῶν οὐ μεῖζον ἔχουσι κυάμου τὸ μέγεθος.[2] ἡ δὲ μέλαινα καὶ ἐπὶ πλείους ἡμέρας ἔγχλωρός[3] ἐστι, καὶ αὐξάνονται καὶ λαμβάνουσιν ἔνιαι μέγεθος μήλου.

Διαλείποντα[4] δὲ μετὰ τοῦτο περὶ πεντεκαίδεκα ἡμέρας πάλιν τὸ τρίτον ἐπιβάλλεται βλαστοὺς Ἑκατομβαιῶνος, ἐλαχίστας ἡμέρας τῶν πρότερον· ἴσως γὰρ ἓξ ἢ ἑπτὰ τὸ πλεῖστον· ἡ δὲ βλάστησις ὁμοία καὶ τὸν αὐτὸν τρόπον. παρελθουσῶν δὲ τούτων οὐκέτι εἰς μῆκος ἀλλ' εἰς πάχος ἡ αὔξησις τρέπεται.

3 Πᾶσι μὲν οὖν τοῖς δένδροις αἱ βλαστήσεις φανεραί, μάλιστα δὲ τῇ ἐλάτῃ καὶ τῇ πεύκῃ διὰ τὸ στοιχεῖν τὰ γόνατα καὶ ἐξ ἴσου τοὺς ὄζους ἔχειν. ὥρα δὲ καὶ πρὸς τὸ τέμνεσθαι τὰ ξύλα τότε διὰ τὸ λοπᾶν· ἐν γὰρ τοῖς ἄλλοις καιροῖς οὐκ εὐπεριαίρετος ὁ φλοιός, ἀλλὰ καὶ περιαιρεθέντος μέλαν τὸ ξύλον γίνεται καὶ τῇ ὄψει χεῖρον· ἐπεὶ καὶ πρός γε τὴν χρείαν οὐδέν, ἀλλὰ καὶ

[1] About June.
[2] *cf.* 3. 7. 4; 3. 8. 6; Plin. 16. 27.
[3] ἔγχλωρος conj. Coraës; εὔχλωρος Ald.
[4] διαλείποντα conj. St.; διαλείπουσαι Ald.H.

first budding as a sort of joint, just as in the case of the first budding. This happens about the end of the month Skirrophorion.[1]

[2] (It is only at the time of this second budding that the galls also are produced, both the white and the black; the liquid forming them is mostly produced in quantity at night, and, after swelling for one day—except the part which is of resinous character—it hardens if it is caught by the heat, and so cannot grow any more; otherwise it would have grown greater in bulk; wherefore in some trees the formation is not larger than a bean. The black gall is for several days of a pale green[3] colour; then it swells and sometimes attains the size of an apple.)

Then, after an interval[4] of about fifteen days, the tree for the third time puts on buds in the month Hekatombaion[5]; but this growth continues for fewer days than on either of the previous occasions, perhaps for six or seven at most. However the formation of the buds is as before and takes place in the same manner. After this period there is no increase in length, but the only increase is in thickness.

The periods of budding can be seen in all trees, but especially in fir and silver-fir, because the joints of these are in a regular series and have the knots at even distances. It is then the season also for cutting the timber, because the bark is being shed[6]; for at other times the bark is not easy to strip off, and moreover, if it is stripped off, the wood turns black[7] and is inferior in appearance; for as to its utility[8] this makes no difference, though the wood

[5] About July.
[6] λοπᾶν conj. Sch.; λοιπᾶν UMV; λιπᾶν Ald.
[7] *cf.* Plin. 16. 74.
[8] γε conj. Sch.; τε Ald.

ἰσχυρότερον, ἐὰν μετὰ τὴν πέπαυσιν τῶν καρπῶν τμηθῇ.

4 Ταῦτα μὲν οὖν ἴδια τῶν προειρημένων δένδρων. αἱ δὲ βλαστήσεις αἱ ἐπὶ Κυνὶ καὶ Ἀρκτούρῳ γινόμεναι μετὰ την ἐαρινὴν σχεδὸν κοιναὶ πάντων· ἔνδηλοι δὲ μᾶλλον ἐν τοῖς ἡμέροις καὶ τούτων μάλιστα συκῇ καὶ ἀμπέλῳ καὶ ῥοιᾷ καὶ ὅλως ὅσα εὔτραφῆ καὶ ὅπου χώρα τοιαύτη· δι' ὃ καὶ τὴν ἐπ' Ἀρκτούρῳ πλείστην φασὶ γίνεσθαι περὶ Θετταλίαν καὶ Μακεδονίαν· ἅμα γὰρ συμβαίνει καὶ τὸ μετόπωρον καλὸν γίνεσθαι καὶ μακρόν, ὥστε καὶ τὴν μαλακότητα συμβάλλεσθαι τοῦ ἀέρος. ἐπεὶ καὶ ἐν Αἰγύπτῳ διὰ τοῦθ' ὡς εἰπεῖν αἰεὶ βλαστάνει τὰ δένδρα, ἢ καὶ μικρόν τινα διαλείπει χρόνον.

5 Ἀλλὰ τὰ μὲν περὶ τὰς ἐπιβλαστήσεις, ὥσπερ εἴρηται, κοινά, τὰ δὲ περὶ τὰς διαλείψεις ἀπὸ τῆς πρώτης ἴδια τῶν λεχθέντων. ἴδιον δ' ἐνίοις ὑπάρχει καὶ τὸ τῆς καλουμένης κάχρυος, οἷον τοῖς [τε] προειρημένοις· ἔχει γὰρ καὶ ἐλάτη καὶ πεύκη καὶ δρῦς, καὶ ἔτι φίλυρα καὶ καρύα καὶ διοσβάλανος καὶ πίτυς. αὗται δὲ γίνονται δρυὶ μὲν πρὸ τῆς βλαστήσεως ὑποφαινούσης τῆς ἠρινῆς ὥρας. ἔστι δ' ὡσπερεὶ κύησις φυλλικὴ μεταξὺ πίπτουσα τῆς ἐξ ἀρχῆς ἐποιδήσεως καὶ τῆς φυλλικῆς βλαστήσεως· τῇ δ' ὄῃ ἐστὶ τοῦ

[1] δένδρων conj. R. Const.; καρπῶν Ald.H.
[2] *cf. C.P.* 1. 10. 6; 1. 12. 4; 1. 13. 3; 1. 13. 5; 1. 13. 10; Plin. 16. 98. [3] *cf. C.P.* 1. 14. 11. [4] *cf.* 5. 1. 4; Plin. 16. 30.

is stronger if it is cut after the ripening of the fruit.

Now what has been said is peculiar to the above-mentioned trees.[1] [2] But the buddings which take place at the rising of the dog-star and at that of Arcturus after the spring budding are common to nearly all, though they may be most clearly seen in cultivated trees, and, among these, especially in fig vine pomegranate, and in general in all those that are luxuriant in growth or are growing in rich soil. Accordingly they say that the budding at the rising of Arcturus is most considerable in Thessaly and Macedonia[3]; for it also happens that the autumn in these countries is a fair and a long season; so that the mildness of the climate also contributes. Indeed it is for this reason, one may say, that in Egypt too the trees are always budding, or at least that the process is only suspended for quite a short time.

Now the facts as to the later buddings apply, as has been said, to all trees alike; but those which belong to the intervals after the first period of budding are peculiar to those mentioned above. Peculiar to some also is the growth of what are called 'winter buds,'[4] for instance in the above-mentioned trees; silver-fir fir and oak have them, and also lime hazel chestnut and Aleppo pine. These are found in the oak before the leaf-buds grow, when the spring season is just beginning. This growth consists of a sort of leaf-like formation,[5] which occurs between the first swelling of the leaf-buds and the time when they burst into leaf. In the sorb[6] it

[5] ἐστι . . . φυλλικὴ: ἐστι conj. R. Const.; ὡσπερεὶ conj. Sch.; ἔτι δὲ ὥσπερ ἡ κύησις φυλακὴ UAld.H.; φυλλικὴ mBas. etc.

[6] τῇ δ' ὄῃ ἐστὶ conj. W. (*cf.* the description of ὄη, 3. 12. 8); τῇ δ' ἰδιότητι Ald.

μετοπώρου μετὰ τὴν φυλλοβολίαν εὐθὺς λιπαρά τις καὶ ὥσπερ ἐπῳδηκυῖα, καθαπερανεὶ μέλλουσα βλαστάνειν, καὶ διαμένει τὸν χειμῶνα μέχρι τοῦ ἦρος. ἡ δὲ Ἡρακλεωτικὴ μετὰ τὴν ἀποβολὴν τοῦ καρποῦ φύει τὸ βοτρυῶδες ἡλίκον σκώληξ εὐμεγέθης, ἐξ ἑνὸς μίσχου πλείω δή, ἃ καλοῦσί τινες
6 ἰούλους. τούτων ἕκαστον ἐκ μικρῶν σύγκειται μορίων φολιδωτῶν τῇ τάξει, καθάπερ οἱ στρόβιλοι τῆς πεύκης, ὥστε μὴ ἀνομοίαν εἶναι τὴν ὄψιν στροβίλῳ νέῳ καὶ χλωρῷ πλὴν προμηκέστερον καὶ σχεδὸν ἰσόπαχες διόλου. τοῦτο δὲ αὔξεται τὸν χειμῶνα· (καὶ ἅμα τῷ ἦρι χάσκει τὰ φολιδωτὰ καὶ ξανθὰ γίνεται), καὶ τὸ μῆκος λαμβάνει καὶ τριδάκτυλον· ὅταν δὲ τοῦ ἦρος τὸ φύλλον βλαστάνῃ, ταῦτ' ἀποπίπτει καὶ τὰ τοῦ καρύου καλυκώδη περικάρπια γίνεται συμμεμυκότα κατὰ τοῦ μίσχου, τοσαῦτα ὅσα καὶ ἦν τὰ ἄνθη· τούτων δ' ἐν ἑκάστῳ κάρυον ἕν. περὶ δὲ τῆς φιλύρας ἐπισκεπτέον, καὶ εἴ τι ἄλλο καχρυοφόρον.

VI. Ἔστι δὲ καὶ τὰ μὲν εὐαυξῆ τὰ δὲ δυσαυξῆ. εὐαυξῆ μὲν τά τε πάρυδρα, οἷον πτελέα πλάτανος λεύκη αἴγειρος ἰτέα· καί τοι περὶ ταύτης ἀμφισβητοῦσί τινες ὡς δυσαυξοῦς· καὶ τῶν καρποφόρων δὲ ἐλάτη πεύκη δρῦς. εὐαυξέστατον δὲ . . . μίλος

[1] εὐθὺς λιπαρὰ conj. Sch.; τις add. W.; εὐθὺς αἱ παρὰ τῆς U.
[2] φύει conj. W.; φύεται Ald. [3] *i.e.* catkins. *cf.* 3. 3. 8.
[4] πλείω δὴ conj. Sch.; πιώδη UMVAld.; πλείονα U ?.
[5] *cf.* 3. 10. 4.
[6] συμμεμυκότα κατὰ τοῦ μ.: G evidently had a different text; ? συμπεφυκότα W.

occurs in the autumn after the shedding of the leaves, and has from the first a glistening look,[1] as though swelling had taken place, just as if it were about to burst into leaves; and it persists through the winter till the spring. The filbert after casting its fruit produces[2] its clustering growth,[3] which is as large as a good-sized grub: several[4] of these grow from one stalk, and some call them catkins. Each of these is made up of small processes arranged like scales, and resembles the cone of the fir, so that its appearance is not unlike that of a young green fir-cone, except that it is longer and almost of the same thickness throughout. This grows through the winter (when spring comes, the scale-like processes open and turn yellow); it grows to the length of three fingers, but, when in spring the leaves are shooting, it falls off, and the cup-like[5] fruit-cases of the nut are formed, closed all down[6] the stalk and corresponding[7] in number to the flowers; and in each of these is a single nut. The case of the lime and of any other tree that produces winter-buds needs further consideration.

Of the comparative rate of growth in trees, and of the length of their roots.

VI. Some trees are quick-growing, some slow. Quick-growing are those which grow by the waterside, as elm plane abele black poplar willow; (however some dispute about the last-named, and consider it a slow grower:) and of fruit-bearing trees, silver-fir fir oak. Quickest growing of all are . . .[8] yew *lakara*

[7] ὅσα καὶ ᾖν τὰ ἄνθη conj. W.; ὅσα καὶ κατὰ ἄνθη Ald.

[8] Lacuna in text (Sch. W.). The following list of trees also appears to be in confusion, and includes some of both classes.

καὶ λάκαρα φηγὸς ἄρκευθος σφένδαμνος ὀστρυα ζυγία μελία κλήθρα πίτυς ἀνδράχλη κρανεία πύξος ἀχράς. καρποφορεῖ δ' εὐθὺς ἐλάτη πεύκη πίτυς, κἂν ὁπηλικονοῦν μέγεθος λάβωσιν.

2 Ἡ δὲ αὔξησις καὶ ἡ βλάστησις τῶν μὲν ἄλλων ἄτακτος κατὰ τοὺς τόπους τῶν βλαστῶν, τῆς δ' ἐλάτης ὡρισμένη καὶ συνεχὴς καὶ ὕστερον. ὅταν γὰρ ἐκ τοῦ στελέχους τὰ πρῶτα σχισθῇ, πάλιν ἐξ ἐκείνου ἡ ἑτέρα σχίσις γίνεται κατὰ τὸν αὐτὸν τρόπον, καὶ τοῦτ' ἀεὶ ποιεῖ κατὰ πάσας τὰς ἐπιβλαστήσεις. ἐν δὲ τοῖς ἄλλοις οὐδ' οἱ ὄζοι κατ' ἀλλήλους πλὴν ἐπί τινων ὀλίγων, οἷον κοτίνου καὶ ἄλλων· ἔχει δὲ καὶ τῇδε διαφορὰν ἡ αὔξησις κοινῇ πάντων ὁμοίως ἡμέρων τε καὶ ἀγρίων· τὰ μὲν γὰρ καὶ ἐκ τοῦ ἄκρου τῶν βλαστῶν καὶ ἐκ τῶν πλαγίων φύεται, καθάπερ ἄπιος ῥόα συκῆ μύρρινος σχεδὸν τὰ πλεῖστα· τὰ δ' ἐκ τοῦ ἄκρου μὲν οὐκ ἀνίησιν ἐκ δὲ τῶν πλαγίων, καὶ αὐτὸ προωθεῖται τὸ ὑπάρχον, ὥσπερ καὶ τὸ ὅλον στέλεχος καὶ οἱ ἀκρεμόνες. συμβαίνει δὲ τοῦτο ἐπὶ τῆς Περσικῆς καρύας καὶ τῆς Ἡρακλεωτικῆς καὶ

3 ἄλλων. ἁπάντων δὲ τῶν τοιούτων εἰς ἓν φύλλον ἀποτελευτῶσιν οἱ βλαστοί, δι' ὃ καὶ εὐλόγως οὐκ ἐπιβλαστάνει καὶ αὐξάνεται μὴ ἔχοντα ἀρχήν. (ὁμοία δὲ τρόπον τινὰ ἡ αὔξησις καὶ τοῦ σίτου·

---

[1] κατὰ . . . βλαστῶν conj. W.; κατὰ τοὺς τρόπους (corrected to τόπους) καὶ βλαστούς U; MVP insert τοὺς before βλαστούς.
[2] ἐκείνου . . . κατὰ conj. W.; ἐκείνου ἡ ἑτέρα σχίζεται τὰ ἴσα καὶ UAld.
[3] ἄλλων: ? ἐλάας W.; I suggest ἄλλων ἐλαῶν.

(bird-cherry) Valonia oak Phoenician cedar maple hop-hornbeam *zygia* manna-ash alder Aleppo pine andrachne cornelian cherry box wild pear. But silver-fir fir and Aleppo pine bear fruit from the very first, whatever size they have attained.

While the growth and budding of most trees are irregular as regards the position in which the buds appear,[1] the growth and budding of the silver-fir follow a regular rule, and its development afterwards is also in a regular sequence. For, when the trunk first divides, then again from the divided trunk the second division[2] takes place in like manner, and so the tree goes on with each fresh formation of buds. In other trees not even the knots are opposite to one another, except in some few cases, as wild olive and others.[3] Here too we find a difference in the manner of growth which belongs to all trees alike, both cultivated and wild: in some cases the growth is from the top of the shoots and also from the side-buds,[4] as in pear pomegranate fig myrtle and the majority of trees, one may say: in some cases the growth is not from the top, but only from the side-buds, and the already existing part is pushed out[5] further, as is the whole trunk with the upper branches. This occurs in the walnut and in the filbert as well as in other trees. In all such trees the buds end in a single leaf[6]; wherefore it is reasonable that they should not make fresh buds and growth from this point, as they have no point of departure. (To a certain extent the growth of corn is similar; for it

[4] ἐκ τοῦ . . . πλαγίων: ? ἐκ τοῦ ἄκρου καὶ ἐκ τῶν πλαγίων βλαστῶν. *cf.* 3. 5. 1.

[5] *i.e.* grows without dividing. *cf.* Plin. 16. 100. (of different trees).

[6] φύλλον perhaps conceals some other word.

καὶ γὰρ οὗτος ἀεὶ τῇ προώσει τοῦ ὑπάρχοντος αὐξάνεται, κἂν κολοβωθῇ τὰ φύλλα, καθάπερ ἐν τοῖς ἐπιβοσκομένοις· πλὴν οὗτός γε οὐκ ἐκ τοῦ πλαγίου παραφύει, καθάπερ ἔνια τῶν χεδροπῶν.) αὕτη μὲν οὖν διαφορά τις ἂν εἴη βλαστήσεως ἅμα καὶ αὐξήσεως.

4 Βαθύρριζα δὲ οὔ φασί τινες εἶναι τὰ ἄγρια διὰ τὸ φύεσθαι πάντα ἀπὸ σπέρματος, οὐκ ἄγαν ὀρθῶς λέγοντες. ἐνδέχεται γὰρ ὅταν ἐμβιώσῃ πόρρω καθιέναι τὰς ῥίζας· ἐπεὶ καὶ τῶν λαχάνων τὰ πολλὰ τοῦτο ποιεῖ, καίπερ ἀσθενέστερα ὄντα καὶ ἐναργῶς φυόμενα <ἐν> τῇ γῇ. βαθυρριζότατον δ᾽ οὖν δοκεῖ τῶν ἀγρίων εἶναι ἡ πρῖνος· ἐλάτη δὲ καὶ πεύκη μετρίως, ἐπιπολαιότατον δὲ θραύπαλος καὶ κοκκυμηλέα καὶ σποδιάς· αὕτη δ᾽ ἐστὶν ὥσπερ ἀγρία κοκκυμηλέα. ταῦτα μὲν οὖν καὶ ὀλιγόρριζα· ὁ δὲ θραύπαλος πολύρριζον. συμβαίνει δὲ τοῖς ἄλλοις τοῖς μὴ κατὰ βάθους ἔχουσι, καὶ οὐχ ἥκιστα ἐλάτῃ καὶ πεύκῃ, προρρίζοις ὑπὸ τῶν πνευμάτων ἐκπίπτειν.

5 Οἱ μὲν οὖν περὶ Ἀρκαδίαν οὕτω λέγουσιν. οἱ δ᾽ ἐκ τῆς Ἴδης βαθυρριζότερον ἐλάτην δρυὸς ἀλλ᾽ ἐλάττους ἔχειν καὶ εὐθυρριζοτέραν εἶναι· βαθυρριζότατον δὲ καὶ τὴν κοκκυμηλέαν καὶ τὴν Ἡρακλεωτικήν, τὰς δὲ ῥίζας λεπτὰς καὶ ἰσχυρὰς τὴν Ἡρακλεωτικήν, τὴν δὲ κοκκυμηλέαν πολύρριζον, ἄμφω δ᾽ ἐμβιῶναι δεῖν· δυσώλεθρον δὲ τὴν κοκκυμηλέαν. ἐπιπολῆς δὲ σφένδαμνον καὶ

---

[1] τοῦ ὑπάρχοντος conj. Sch. from G; τῇ ὑπαρχούσῃ Ald.
[2] οὐδ᾽: ᾽ οὐκ W. [3] Plin. 16. 127.
[4] ἐμβιώσῃ: cf. 3. 6. 5; C.P. 1. 2. 1.

also regularly increases by pushing forward of the already existing part,[1] even if the leaves are mutilated, as in corn which is bitten down by animals. Corn however does not[2] make side-growths, as some leguminous plants do.) Here then we may find a difference which occurs both in the making of buds and in the making of fresh growth.

[3] Some say that wild trees are not deep rooting, because they all grow from seed; but this is not a very accurate statement. For it is possible that, when they are well established,[4] they may send their roots down far; in fact even most pot-herbs do this, though these are not so strong as trees, and are undoubtedly grown from seed planted in the ground.[5] The kermes-oak however seems to be the deepest rooting of wild trees; silver-fir and fir are only moderately so, and shallowest are joint-fir plum bullace (which is a sort of wild plum). The last two also have few roots, while joint-fir has many. Trees which do not root deep,[6] and especially silver-fir and fir, are liable to be rooted up by winds.

So the Arcadians say. But the people who live near Mount Ida say that the silver fir is deeper rooting[7] than the oak,[8] and has straighter roots, though they are fewer. Also that those which have the deepest roots are plum and filbert, the latter having strong slender roots, the former having many: but they add that both trees must be well established to acquire these characters; also that plum is very tenacious of life. Maple, they say,

[5] ἐναργῶς . . . γῇ: so G; ἐν add. W.
[6] βάθους conj. Sch.; βάθος Ald.
[7] βαθορριζότερον conj. W.; βαθυρριζότατον UMVAld.
[8] Proverbial for its hold on the ground; *cf.* Verg. *Aen.* 4 441 foll.

ὀλίγας· τὴν δὲ μελίαν πλείους καὶ εἶναι πυκνόρριζον καὶ βαθύρριζον. ἐπιπολῆς δὲ καὶ ἄρκευθον καὶ κέδρον· καὶ κλήθρας λεπτὰς καὶ ὁμαλεῖς· ἔτι δ' ὀξύην· καὶ γὰρ τοῦτ' ἐπιπολαιόρριζον καὶ ὀλιγόρριζον. τὴν δὲ οὔαν ἐπιπολαίους μὲν ἰσχυρὰς δὲ καὶ παχείας καὶ δυσωλέθρους πλήθει δὲ μετρίας. βαθύρριζα μὲν οὖν καὶ οὐ βαθύρριζα τὰ τοιαῦτ' ἐστίν.

VII. Ἀποκοπέντος δὲ τοῦ στελέχους τὰ μὲν ἄλλα πάνθ' ὡς εἰπεῖν παραβλαστάνει, πλὴν ἐὰν αἱ ῥίζαι πρότερον τύχωσι πεπονηκυῖαι· πεύκη δὲ καὶ ἐλάτη τελέως ἐκ ῥιζῶν αὐτοετεῖς αὐαίνονται καὶ ἐὰν τὸ ἄκρον ἐπικοπῇ. συμβαίνει δὲ ἴδιόν τι περὶ τὴν ἐλάτην· ὅταν γὰρ κοπῇ ἢ κολουσθῇ ὑπὸ πνεύματος ἢ καὶ ἄλλου τινὸς περὶ τὸ λεῖον τοῦ στελέχους—ἔχει γὰρ μέχρι τινὸς λεῖον καὶ ἄοζον καὶ ὁμαλὸν ἱκανὸν ἱστῷ πλοίου—περιφύεται μικρόν, ὑποδεέστερον εἰς ὕψος, καὶ καλοῦσιν οἱ μὲν ἄμφαυξιν οἱ δὲ ἀμφίφυαν, τῷ μὲν χρώματι μέλαν τῇ δὲ σκληρότητι ὑπερβάλλον, ἐξ οὗ τοὺς κρατῆρας ποιοῦσιν οἱ περὶ Ἀρκαδίαν·
2 τὸ δὲ πάχος οἷον ἂν τύχῃ τὸ δένδρον, ὅσῳπερ ἂν ἰσχυρότερον καὶ ἐγχυλότερον ᾖ παχύτερον. συμβαίνει δὲ κἀκεῖνο ἴδιον ἐν ταὐτῷ τούτῳ περὶ

---

[1] σφ. καὶ ὀλίγας conj. W.; σφ. κατ' ὀλίγον UMVAld.
[2] *i.e.* not very fibrous.
[3] *cf.* Hdt. 6. 37, and the proverb πίτυος τρόπον ἐκτρίβεσθαι.
[4] ὁμαλὸν conj. Scal.; ὅμοιον Ald.
[5] ἱκανὸν ἱστῷ πλοίου conj. W.; ἢ καὶ ἡλίκον πλεῖον Ald.; so UH, but with πλοῖον.

has shallow roots and few of them[1]; but manna-ash has more and they are thickly matted and run deep; Phoenician cedar and prickly cedar, they say, have shallow roots, those of alder are slender and 'plain,'[2] as also are those of beech; for this too has few roots, and they are near the surface. Sorb, they say, has its roots near the surface, but they are strong and thick and hard to kill, though not very numerous. Such are the trees which are or are not deep-rooting.

Of the effects of cutting down the whole or part of a tree.

VII. Almost all trees shoot from the side if the trunk is cut down, unless the roots have previously been injured; but fir and silver-fir wither away[3] completely from the roots within the year, if merely the top has been cut off. And there is a peculiar thing about the silver-fir; when it is topped or broken off short by wind or some other cause affecting the smooth part of the trunk—for up to a certain height the trunk is smooth knotless and plain[4] (and so suitable for making a ship's mast[5]),—a certain amount of new growth forms round it, which does not however grow much vertically; and this is called by some *amphauxis*[6] and by others *amphiphya*[6]; it is black in colour and exceedingly hard, and the Arcadians make their mixing-bowls out of it; the thickness is in proportion[7] to the tree, according as that is more or less vigorous and sappy, or again according to its thickness. There[8] is this peculiarity too in the silver-fir in the same connexion;

[6] Two words meaning 'growth about,' *i.e. callus*.
[7] οἷον ἂν conj. W.; οἷον ἐὰν Ald.; ὅσον ἂν conj. Scal.
[8] Plin. 16. 123.

τὴν ἐλάτην· ὅταν μὲν γάρ τις τοὺς ὄζους ἅπαντας ἀφελὼν ἀποκόψῃ τὸ ἄκρον, ἀποθνήσκει ταχέως· ὅταν δὲ τὰ κατωτέρω τὰ κατὰ τὸ λεῖον ἀφέλῃ, ζῇ τὸ κατάλοιπον, περὶ ὃ δὴ καὶ ἡ ἄμφαυξις φύεται. ζῇ δὲ δῆλον ὅτι τῷ ἔγχυλον εἶναι καὶ χλωρόν, εἴπερ ἀπαράβλαστον. ἀλλὰ γὰρ τοῦτο μὲν ἴδιον τῆς ἐλάτης.

3 Φέρει δὲ τὰ μὲν ἄλλα τόν τε καρπὸν τὸν ἑαυτῶν καὶ τὰ κατ' ἐνιαυτὸν ἐπιγινόμενα ταῦτα, φύλλον ἄνθος βλαστόν· τὰ δὲ καὶ βρύον ἢ ἕλικα· τὰ δὲ πλείω, καθάπερ ἥ τε πτελέα τόν τε βότρυν καὶ τὸ θυλακῶδες τοῦτο, καὶ συκῆ καὶ τὰ ἐρινὰ τὰ προαποπίπτοντα καὶ εἴ τινες ἄρα τῶν συκῶν ὀλυνθοφοροῦσιν· ἴσως δὲ τρόπον τινὰ καρπὸς οὗτος. ἀλλ' ἡ Ἡρακλεωτικὴ καρύα τὸν ἴουλον καὶ ἡ πρῖνος τὸν φοινικοῦν κόκκον ἡ δὲ δάφνη τὸ βότρυον. φέρει μὲν καὶ ἡ καρποφόρος, εἰ μὴ καὶ πᾶσα ἀλλά τοι γένος τι αὐτῆς, οὐ μὴν ἀλλὰ πλέον ἡ ἄκαρπος, ἣν δὴ καὶ ἄρρενά τινες καλοῦσιν. ἀλλ' ἡ πεύκη τὸν προαποπίπτοντα κύτταρον.

4 Πλεῖστα δὲ πάντων ἡ δρῦς παρὰ τὸν καρπόν, οἷον τήν τε κηκίδα τὴν μικρὰν καὶ τὴν ἑτέραν

[1] *i.e.* and so does not, like other trees under like treatment, put its strength into these. *cf. C.P.* 5. 17. 4.
[2] ἑαυτῶν conj. Sch. from G; αὐτὸν Ald.
[3] The leaf-gall, *cf.* 2. 8. 3; 3. 14. 1. For τοῦτο *cf.* 3. 18. 11; 4. 7. 1.
[4] Lat. *grossi*. *cf. C.P.* 5. 1. 8.
[5] τινὰ καρπὸς conj. Sch.; τινὰ ἄκαρπος UAld.

when, after taking off all the branches, one cuts off the top, it soon dies; yet, when one takes off the lower parts, those about the smooth portion of the trunk, what is left survives, and it is on this part that the *amphauxis* forms. And plainly the reason why the tree survives is that it is sappy and green because it has no side-growths.[1] Now this is peculiar to the silver-fir.

Of other things borne by trees besides their leaves flowers and fruit.

Now, while other trees bear merely their own 2
fruit and the obvious parts which form annually, to wit, leaf flower and bud, some bear also catkins or tendrils, and some produce other things as well, for instance the elm its 'cluster' and the familiar bag-like thing,[3] the fig both the immature figs which drop off and (in some kinds) the untimely figs[4]—though perhaps in a sense[5] these should be reckoned as fruit. Again filbert produces its catkin,[6] kermes-oak its scarlet 'berry,'[7] and bay its 'cluster.'[8] The fruit-bearing sort of bay also produces this, or at all events[9] one kind certainly does so; however the sterile kind, which some call the 'male,' produces it in greater quantity. The fir again bears its 'tuft,'[10] which drops off.

[11] The oak however bears more things besides[12] its fruit than any other tree; as the small gall[13] and its

[6] *cf.* 3. 3. 8; 3. 5. 5.
[7] *cf.* 3. 16. 1. *i.e.* the kermes gall (whence Eng. 'crimson').
[8] βότρυον UMVAld., supported by G. and Plin. 16. 120; but some editors read βρύον on the strength of 3. 11. 4. and *C.P.* 2. 11. 4.
[9] ἀλλά τοι conj. W.; ἀλλὰ καὶ Ald.
[10] *cf.* 3. 3. 8 n.
[11] Plin. 16. 28.
[12] παρὰ conj. W., *cf.* § 6; φέρει Ald.
[13] *cf.* 3. 5. 2.

τὴν πιττώδη μέλαιναν. ἔτι δὲ συκαμινῶδες ἄλλο τῇ μορφῇ πλὴν σκληρὸν καὶ δυσκάτακτον, σπάνιον δὲ τοῦτο· καὶ ἕτερον αἰδοιώδη σχέσιν ἔχον, τελειούμενον δ' ἔτι σκληρὸν κατὰ τὴν ἐπανάστασιν καὶ τετρυπημένον· προσεμφερὲς τρόπον τινὰ τοῦτ' ἐστὶ καὶ ταύρου κεφαλῇ, περικαταγνύμενον δὲ ἔνδοθεν ἔχει πυρῆνος ἐλάας ἰσοφυές. φύει δὲ καὶ τὸν ὑπ' ἐνίων καλούμενον πῖλον· τοῦτο δ' ἐστὶ σφαιρίον ἐριῶδες μαλακὸν περὶ πυρήνιον σκληρότερον πεφυκός, ᾧ χρῶνται πρὸς τοὺς λύχνους· καίεται γὰρ καλῶς, ὥσπερ καὶ ἡ μέλαινα κηκίς. φύει δὲ καὶ ἕτερον σφαιρίον κόμην ἔχον, τὰ μὲν ἄλλα ἀχρεῖον, κατὰ δὲ τὴν ἐαρινὴν ὥραν ἐπίβαπτον χυλῷ μελιτηρῷ καὶ κατὰ τὴν ἁφὴν καὶ κατὰ τὴν γεῦσιν.

5 Παραφύει δ' ἐνδοτέρω τῆς τῶν ῥαβδῶν μασχαλίδος ἕτερον σφαιρίον ἄμισχον ἢ καὶ κοιλόμισχον ἴδιον καὶ ποικίλον· τοὺς μὲν γὰρ ἐπανεστηκότας ὀμφαλοὺς ἐπιλεύκους ἢ ἐπεστιγμένους ἔχει μέλανας τὸ δ' ἀνὰ μέσον κοκκοβαφὲς καὶ λαμπρόν· ἀνοιγόμενον δ' ἐστὶ μέλαν καὶ ἐπίσαπρον. σπάνιον δὲ παραφύει καὶ λιθάριον κισσηροειδὲς ἐπὶ πλεῖον. ἔτι δ' ἄλλο τούτου σπανιαίτερον φυλλικὸν συμπεπιλημένον πρόμηκες σφαιρίον. ἐπὶ δὲ τοῦ φύλλου φύει κατὰ τὴν ῥάχιν σφαιρίον λευκὸν διαυγὲς ὑδατῶδες, ὅταν ἁπαλὸν ᾖ· τοῦτο δὲ καὶ

---

[1] πυρῆνος ἐλάας ἰσοφυὲς conj. W.; πυρῆνος ἐλαίᾳ εἰρουφυην UMV ; πυρῆνα ἐλαία εἰρουφύην Ald.

[2] περὶ πυρήνιον σκληρότερον I conj. ; περὶ πυρηνίου σκληρότητα U ; περὶ πυρηνίου σκληρότερον M ; περιπυρηνίου σκληρότερον VAld. W. prints the reading of U. For πῖλος see Index.

other black resinous gall. Again it has another growth, like a mulberry in shape, but hard and difficult to break; this however is not common. It has also another growth like the *penis* in shape, which, when it is further developed, makes a hard prominence and has a hole through it. This to a certain extent resembles also a bull's head, but, when split open, it contains inside a thing shaped like the stone of an olive.[1] The oak also produces what some call the 'ball'; this is a soft woolly spherical object enclosing a small stone which is harder,[2] and men use it for their lamps; for it burns well, as does the black gall. The oak also produces another hairy ball, which is generally useless, but in the spring season it is covered with a juice which is like honey both to touch and taste.

[3] Further the oak produces right inside the axil[4] of the branches another ball with no stalk or else[5] a hollow one; this is peculiar and of various colours: for the knobs which arise on it are whitish or black and spotted,[6] while the part between these is brilliant scarlet; but, when it is opened, it is black and rotten.[7] It also occasionally produces a small stone which more or less resembles pumice-stone; also, less commonly, there is a leaf-like ball, which is oblong and of close texture. Further the oak produces on the rib of the leaf a white transparent ball, which is watery, when it is young; and this sometimes con-

[3] Plin. 16. 29.

[4] ἐνδοτέρω . . . μασχαλίδος conj. R. Const.; ἐντεριώνης τῶν ῥοπῶν μασχαλίδας UAld. Plin., *l.c.*, *gignunt et alae ramorum eius pilulas.*

[5] ἢ ins. St.

[6] Plin., *l.c.*, *nigra varietate dispersa.*

[7] ἐπίσαπρον; Plin., *l.c.*, has *apertis amara inanitas est* whence ἐπίπικρον conj. Sch.

μύας ἐνίοτε ἐνδὸν ἴσχει. τελειούμενον δὲ σκληρύνεται κηκίδος μικρᾶς λείας τρόπον.

6 Ἡ μὲν οὖν δρῦς τοσαῦτα φέρει παρὰ τὸν καρπόν. οἱ γὰρ μύκητες ἀπὸ τῶν ῥιζῶν καὶ παρὰ τὰς ῥίζας φυόμενοι κοινοὶ καὶ ἑτέρων εἰσίν. ὡσαύτως δὲ καὶ ἡ ἰξία· καὶ γὰρ αὕτη φύεται καὶ ἐν ἄλλοις· ἀλλ' οὐδὲν ἧττον, ὥσπερ ἐλέχθη, πλειστοφόρον ἐστίν· εἰ δέ γε δὴ καθ' Ἡσίοδον φέρει μέλι καὶ μελίττας, ἔτι μᾶλλον· φαίνεται δ' οὖν καὶ ὁ μελιτώδης οὗτος χυλὸς ἐκ τοῦ ἀέρος ἐπὶ ταύτῃ μάλιστα προσίζειν. φασὶ δὲ καὶ ὅταν κατακαυθῇ γίνεσθαι λίτρον ἐξ αὐτῆς. ταῦτα μὲν οὖν ἴδια τῆς δρυός.

VIII. Πάντων δέ, ὥσπερ ἐλέχθη, τῶν δένδρων ὡς καθ' ἕκαστον γένος λαβεῖν διαφοραὶ πλείους εἰσίν· ἡ μὲν κοινὴ πᾶσιν, ᾗ διαιροῦσι τὸ θῆλυ καὶ τὸ ἄρρεν, ὧν τὸ μὲν καρποφόρον τὸ δὲ ἄκαρπον ἐπί τινων. ἐν οἷς δὲ ἄμφω καρποφόρα τὸ θῆλυ καλλικαρπότερον καὶ πολυκαρπότερον· πλὴν ὅσοι ταῦτα καλοῦσιν ἄρρενα, καλοῦσι γάρ τινες. παραπλησία δ' ἡ τοιαύτη διαφορὰ καὶ ὡς τὸ ἥμερον διῄρηται πρὸς τὸ ἄγριον. ἑτέρα δὲ κατ' εἶδος αὐτῶν τῶν ὁμογενῶν· ὑπὲρ ὧν λεκτέον ἅμα συνεμφαίνοντας καὶ τὰς ἰδίας μορφὰς τῶν μὴ φανερῶν καὶ γνωρίμων.

---

[1] Plin. 16. 31.
[2] Hes. *Op.* 233.
[3] Plin. 16. 16.
[4] λεκτέον add. Sch.

tains flies: but as it develops, it becomes hard, like a small smooth gall.

Such are the growths which the oak produces as well as its fruit. For as for the fungi[1] which grow from the roots or beside them, these occur also in other trees. So too with the oak-mistletoe; for this grows on other trees also. However, apart from that, the oak, as was said, produces more things than any other tree; and all the more so if, as Hesiod[2] says, it produces honey and even bees; however, the truth appears to be that this honey-like juice comes from the air and settles on this more than on other trees. They say also that, when the oak is burnt, nitre is produced from it. Such are the things peculiar to the oak.

Of 'male' and 'female' in trees: the oak as an example of this and other differences.

VIII. [3]Taking, as was said, all trees according to their kinds, we find a number of differences. Common to them all is that by which men distinguish the 'male' and the 'female,' the latter being fruit-bearing, the former barren in some kinds. In those kinds in which both forms are fruit-bearing the 'female' has fairer and more abundant fruit; however some call these the 'male' trees—for there are those who actually thus invert the names. This difference is of the same character as that which distinguishes the cultivated from the wild tree, while other differences distinguish different forms of the same kind; and these we must discuss,[4] at the same time indicating the peculiar forms, where these are not[5] obvious and easy to recognise.

[5] μὴ conj. St.; μήτε Ald.H.

2 Δρυὸς δὴ γένη—ταύτην γὰρ μάλιστα διαιροῦσι· καὶ ἔνιοί γε εὐθὺς τὴν μὲν ἥμερον καλοῦσι τὴν δ' ἀγρίαν οὐ τῇ γλυκύτητι τοῦ καρποῦ διαιροῦντες· ἐπεὶ γλυκύτατός γε ὁ τῆς φηγοῦ, ταύτην δ' ἀγρίαν ποιοῦσιν· ἀλλὰ τῷ μᾶλλον ἐν τοῖς ἐργασίμοις φύεσθαι καὶ τὸ ξύλον ἔχειν λειότερον, τὴν δὲ φηγὸν τραχὺ καὶ ἐν τοῖς ὀρεινοῖς—γένη μὲν οὖν οἱ μὲν τέτταρα ποιοῦσιν οἱ δὲ πέντε. διαλλάττουσι δ' ἔνια τοῖς ὀνόμασιν, οἷον τὴν τὰς γλυκείας φέρουσαν οἱ μὲν ἡμερίδα καλοῦντες οἱ δ' ἐτυμόδρυν. ὁμοίως δὲ καὶ ἐπ' ἄλλων. ὡς δ' οὖν οἱ περὶ τὴν Ἴδην διαιροῦσι, τάδ' ἐστὶ τὰ εἴδη· ἡμερὶς αἰγίλωψ πλατύφυλλος φηγὸς ἁλίφλοιος· οἱ δὲ εὐθύφλοιον καλοῦσιν. κάρπιμα μὲν πάντα· γλυκύτατα δὲ τὰ τῆς φηγοῦ, καθάπερ εἴρηται, καὶ δεύτερον τὰ τῆς ἡμερίδος, ἔπειτα τῆς πλατυφύλλου, καὶ τέταρτον ἡ ἁλίφλοιος, ἔσχατον δὲ
3 καὶ πικρότατον ἡ αἰγίλωψ. οὐχ ἅπασαι δὲ γλυκεῖαι ἐν τοῖς γένεσιν ἀλλ' ἐνίοτε καὶ πικραί, καθάπερ ἡ φηγός. διαφέρουσι δὲ καὶ τοῖς μεγέθεσι καὶ τοῖς σχήμασι καὶ τοῖς χρώμασι τῶν βαλανων. ἴδιον δὲ ἔχουσιν ἥ τε φηγὸς καὶ ἡ ἁλίφλοιος· ἀμφότεραι γὰρ παραλιθάζουσιν ἐν τοῖς ἄρρεσι καλουμένοις ἐξ ἄκρων τῶν βαλάνων ἑκατέρωθεν, αἱ μὲν πρὸς τῷ κελύφει αἱ δὲ πρὸς

[1] Plin. 16. 16 and 17.
[2] See Index, δρῦς and ἡμερίς. ἡμερίς, lit. 'cultivated oak.'
[3] Plin. 16. 20.

[1] Take then the various kinds of oak; for in this tree men recognise more differences than in any other. Some simply speak of a cultivated and a wild kind, not recognising any distinction made by the sweetness of the fruit; (for sweetest is that of the kind called Valonia oak, and this they make the wild kind), but distinguishing the cultivated kind by its growing more commonly on tilled land and having smoother timber, while the Valonia oak has rough wood and grows in mountain districts. Thus some make four kinds, others five. They also in some cases vary as to the names assigned; thus the kind which bears sweet fruit is called by some *hemeris*, by others 'true oak.' So too with other kinds. However, to take the classification given by the people of Mount Ida, these [2] are the kinds: *hemeris* (gall-oak), *aigilops* (Turkey-oak), 'broad-leaved' oak (scrub oak), Valonia oak, sea-bark oak, which some call 'straight-barked' oak. [3] All these bear fruit; but the fruits of Valonia oak are the sweetest, as has been said; second to these those of *hemeris* (gall-oak), third those of the 'broad-leaved' oak (scrub oak), fourth sea-bark oak, and last *aigilops* (Turkey-oak), whose fruits are very bitter. [4] However the fruit is not always sweet in the kinds specified as such [5]; sometimes it is bitter, that of the Valonia oak for instance. There are also differences in the size shape and colour of the acorns. Those of Valonia oak and sea-bark oak are peculiar; in both of these kinds on what are called the 'male' trees the acorns become stony at one end or the other; in one kind this hardening takes place in the end which is

[4] Plin. 16. 19-21.
[5] οὐχ . . . ἐνίοτε conj. W.; text defective in Ald.H.

αὐτῇ τῇ σαρκί. δι' ὃ καὶ ἀφαιρεθέντων ὅμοια γίνεται κοιλώματα τοῖς ἐπὶ τῶν ζώων.

4 Διαφέρουσι δὲ καὶ τοῖς φύλλοις καὶ τοῖς στελέχεσι καὶ τοῖς ξύλοις καὶ τῇ ὅλῃ μορφῇ. ἡ μὲν γὰρ ἡμερὶς οὐκ ὀρθοφυὴς οὐδὲ λεία οὐδὲ μακρά· περίκομος γὰρ ἡ φυτεία καὶ ἐπεστραμμένη καὶ πολυμάσχαλος, ὥστε ὀζώδη καὶ βραχεῖαν γίνεσθαι· τὸ δὲ ξύλον ἰσχυρὸν μὲν ἀσθενέστερον δὲ τῆς φηγοῦ· τοῦτο γὰρ ἰσχυρότατον καὶ ἀσαπέστατον. οὐκ ὀρθοφυὴς δὲ οὐδ' αὕτη ἀλλ' ἧττον ἔτι τῆς ἡμερίδος, τὸ δὲ στέλεχος παχύτατον, ὥστε καὶ τὴν ὅλην μορφὴν βραχεῖαν εἶναι· καὶ γὰρ ἡ φυτεία περίκομος καὶ ταύτῃ καὶ οὐκ εἰς ὀρθόν. ἡ δὲ αἰγίλωψ ὀρθοφυέστατον καὶ ὑψηλότατον καὶ λειότατον καὶ τὸ ξύλον εἰς μῆκος ἰσχυρότατον. οὐ φύεται δὲ ἐν τοῖς ἐργασίμοις ἢ σπανίως.

5 Ἡ δὲ πλατύφυλλος δεύτερον ὀρθοφυΐᾳ καὶ μήκει, πρὸς δὲ τὴν χρείαν τὴν οἰκοδομικὴν χείριστον μετὰ τὴν ἁλίφλοιον, φαῦλον δὲ καὶ εἰς τὸ καίειν καὶ ἀνθρακεύειν, ὥσπερ καὶ τὸ τῆς ἁλιφλοίου, καὶ θριπηδέστατον μετ' ἐκείνην· ἡ γὰρ ἁλίφλοιος παχὺ μὲν ἔχει τὸ στέλεχος χαῦνον δὲ καὶ κοῖλον ἐὰν ἔχῃ πάχος ὡς ἐπὶ τὸ πολύ, δι' ὃ καὶ ἀχρεῖον εἰς τὰς οἰκοδομάς· ἔτι δὲ σήπεται τάχιστα· καὶ γὰρ ἔνυγρόν ἐστι τὸ δένδρον· δι' ὃ καὶ κοίλη γίνεται. φασὶ δέ τινες οὐδ' ἐγκάρδιον εἶναι μόνῃ. λέγουσιν ὡς καὶ κεραυνοβλῆτες αὗται μόναι γίνονται καίπερ ὕψος οὐκ ἔχουσαι

---

[1] *i.e.* at the 'top' end; πρὸς: ? ἐν, πρὸς being repeated by mistake.

[2] ζώων MSS.; ᾠῶν conj. Palm.

[3] Plin. 16. 22.

attached to the cup, in the other in the flesh itself.[1] Wherefore, when the cups are taken off, we find a cavity like the visceral cavities in animals.[2]

3 There are also differences in leaves trunk timber and general appearance. *Hemeris* (gall-oak) is not straight-growing nor smooth nor tall, for its growth is very leafy[4] and twisted, with many side-branches, so that it makes a low much-branched tree: its timber is strong, but not so strong as that of the Valonia oak, for that is the strongest and the least liable to rot. This[5] kind too is not straight-growing, even less so than the *hemeris* (gall-oak), but the trunk is very thick, so that the whole appearance is stunted; for in growth this kind too is very leafy[4] and not erect. The *aigilops* (Turkey oak) is the straightest growing and also the tallest and smoothest, and its wood, cut lengthways, is the strongest. It does not grow on tilled land, or very rarely.

The 'broad-leaved' oak (scrub oak)[6] comes second as to straightness of growth and length of timber to be got from it, but for use in building it is the worst next after the sea-bark oak, and it is even poor wood for burning and making charcoal, as is also that of the sea-bark oak, and next after this kind it is the most worm-eaten. For the sea-bark oak has a thick trunk, but it is generally spongy and hollow when it is thick; wherefore it is useless for building. Moreover it rots very quickly, for the tree contains much moisture; and that is why it also becomes hollow; and some say that it is the only[7] oak which has no heart. And some of the Aeolians say that these are the only oaks which are struck by light-

[4] *i.e.* of bushy habit.
[5] αὕτη conj. Sch.; αὐτὴ UAld.
[6] Plin. 16. 23 and 24.
[7] μόνη conj. St.; μόνην Ald. H.

τῶν Αἰολέων τινές, οὐδὲ πρὸς τὰ ἱερὰ χρῶνται τοῖς ξύλοις. κατὰ μὲν οὖν τὰ ξύλα καὶ τὰς ὅλας μορφὰς ἐν τούτοις αἱ διαφοραί.

6 Κηκίδας δὲ πάντα φέρει τὰ γένη, μόνη δὲ εἰς τὰ δέρματα χρησίμην ἡ ἡμερίς. ἡ δὲ τῆς αἰγίλωπος καὶ τῆς πλατυφύλλου τῇ μὲν ὄψει παρομοία τῇ τῆς ἡμερίδος, πλὴν λειοτέρα, ἀχρεῖος δέ. φέρει καὶ τὴν ἑτέραν τὴν μέλαιναν ᾗ τὰ ἔρια βάπτουσιν. ὃ δὲ καλοῦσί τινες φάσκον ὅμοιον τοῖς ῥακίοις ἡ αἰγίλωψ μόνη φέρει πολιὸν καὶ τραχύ· καὶ γὰρ πηχυαῖον κατακρεμάννυται, καθάπερ τρύχος ὀθονίου μακρόν. φύεται δὲ τοῦτο ἐκ τοῦ φλοιοῦ καὶ οὐκ ἐκ τῆς κορύνης ὅθεν ἡ βάλανος, οὐδ' ἐξ ὀφθαλμοῦ ἀλλ' ἐκ τοῦ πλαγίου τῶν ἄνωθεν ὄζων. ἡ δ' ἁλίφλοιος ἐπίμελαν τοῦτο φύει καὶ βραχύ.

7 Οἱ μὲν οὖν ἐκ τῆς Ἴδης οὕτως διαιροῦσιν. οἱ δὲ περὶ Μακεδονίαν τέτταρα γένη ποιοῦσιν, ἐτυμόδρυν ᾗ τὰς γλυκείας, πλατύφυλλον ᾗ τὰς πικράς, φηγὸν ᾗ τὰς στρογγύλας, ἄσπριν· ταύτην δὲ οἱ μὲν ἄκαρπον ὅλως οἱ δὲ φαῦλον τὸν καρπόν, ὥστε μηδὲν ἐσθίειν ζῷον πλὴν ὑός, καὶ ταύτην ὅταν ἑτέραν μὴ ἔχῃ· καὶ τὰ πολλὰ λαμβάνεσθαι περικεφαλαίᾳ. μοχθηρὰ δὲ καὶ τὰ ξύλα· πελε-

---

[1] Plin. 16. 26.

[2] φάσκον . . . ῥακίοις conj. Sch. (ῥακίοις Salm.): φάσκος ὅμοιος τοῖς βραχείοις UP_2; φάσκον ὁμοίως τοῖς βραγχίοις Ald.H. Plin 16. 33, *cf.* 12. 108; Diosc. 1. 20; Hesych. *s.v.* φάσκος.

[3] τραχὺ conj. W.; βραχὺ UP. [4] κορύνης. *cf.* 3. 5. 1.

ning, although they are not lofty; nor do they use the wood for their sacrifices. Such then are the differences as to timber and general appearance.

[1] All the kinds produce galls, but only *hemeris* (gall-oak) produces one which is of use for tanning hides. That of *aigilops* (Turkey-oak) and that of the 'broad-leaved' oak (scrub oak) are in appearance like that of *hemeris* (gall-oak), but smoother and useless. This also produces the other gall, the black kind, with which they dye wool. The substance which some call tree-moss and which resembles rags[2] is borne only by the *aigilops* (Turkey-oak); it is grey and rough[3] and hangs down for a cubit's length, like a long shred of linen. This grows from the bark and not from the knob[4] whence the acorn starts; nor does it grow from an eye, but from the side of the upper boughs. The sea-bark oak also produces this, but it is blackish[5] and short.

Thus the people of Mount Ida distinguish. But the people of Macedonia make four kinds, 'true-oak,' or the oak which bears the sweet acorns, 'broad-leaved' oak (scrub oak), or that which bears the bitter ones, Valonia oak, or that which bears the round ones, and *aspris*[6] (Turkey-oak); [7] the last-named some say is altogether without fruit, some say it bears poor fruit, so that no animal eats it except the pig, and only he when he can get no others, and that after eating it the pig mostly gets an affection of the head.[8] The wood is also wretched; when hewn with the axe it is altogether

[5] ἐπίμελαν τοῦτο φύει conj. Scal.; ἐπιμ. τοῦτο φύσει U; ἐπὶ μελίαν τοῦτο φύει MVAld.

[6] See Index.

[7] Plin. 16. 24.

[8] περικεφαλαίᾳ: apparently the name of a disease.

κηθέντα μὲν ὅλως ἀχρεῖα· καταρήγνυται γὰρ καὶ διαπίπτει· ἀπελέκητα δὲ βελτίω, δι' ὃ καὶ οὕτω χρῶνται. μοχθηρὰ δὲ καὶ εἰς καῦσιν καὶ εἰς ἀνθρακείαν· ἀχρεῖος γὰρ ὅλως ὁ ἄνθραξ διὰ τὸ πηδᾶν καὶ σπινθηρίζειν πλὴν τοῖς χαλκεῦσι. τούτοις δὲ χρησιμώτερος τῶν ἄλλων· διὰ γὰρ τὸ ἀποσβέννυσθαι, ὅταν παύσηται φυσώμενος, ὀλίγος ἀναλίσκεται. [τὸ δὲ τῆς ἁλιφλοίου χρήσιμον εἰς τοὺς ἄξονας μόνον καὶ τὰ τοιαῦτα.] δρυὸς μὲν οὖν ταύτας ποιοῦσι τὰς ἰδέας.

IX. Τῶν δὲ ἄλλων ἐλάττους· καὶ σχεδὸν τά γε πλεῖστα διαιροῦσι ἄρρενι καὶ θήλει, καθάπερ εἴρηται, πλὴν ὀλίγων ὧν ἐστι καὶ ἡ πεύκη· πεύκης γὰρ τὸ μὲν ἥμερον ποιοῦσι τὸ δ' ἄγριον, τῆς δ' ἀγρίας δύο γένη· καλοῦσι δὲ τὴν μὲν Ἰδαίαν τὴν δὲ παραλίαν· τούτων δὲ ὀρθοτέρα καὶ μακροτέρα καὶ τὸ φύλλον ἔχουσα παχύτερον ἡ Ἰδαία, τὸ δὲ φύλλον λεπτότερον καὶ ἀμενηνότερον ἡ παραλία καὶ λειότερον τὸν φλοιὸν καὶ εἰς τὰ δέρματα χρήσιμον· ἡ δὲ ἑτέρα οὔ. καὶ τῶν στροβίλων ὁ μὲν τῆς παραλίας στρογγύλος τε καὶ διαχάσκων ταχέως, ὁ δὲ τῆς Ἰδαίας μακρότερος καὶ χλωρὸς καὶ ἧττον χάσκων ὡς ἂν ἀγριώτερος· τὸ δὲ ξύλον ἰσχυρότερον τὸ τῆς παραλίας· δεῖ γὰρ καὶ τὰς τοιαύτας διαφορὰς

[1] Plin. 16. 23.

[2] τὸ δὲ . . . τοιαῦτα : this sentence seems out of place, as ἁλίφλοιος was not one of the 'Macedonian' oaks mentioned above (Sch.).

useless, for it breaks in pieces and falls asunder; if it is not hewn with the axe it is better, wherefore they so use it. [1] It is even wretched for burning and for making charcoal; for the charcoal is entirely useless except to the smith, because it springs about and emits sparks. But for use in the smithy it is more serviceable than the other kinds, since, as it goes out when it ceases to be blown, little of it is consumed. [2] The wood of the sea-bark oak is only useful for wheel-axles and the like purposes. Such are the varieties of the oak[3] which men make out.

Of the differences in firs.

IX. [4] The differences between other trees are fewer; for the most part men distinguish them merely according as they are 'male' or 'female,' as has been said, except in a few cases including the fir; for in this tree they distinguish the wild and the cultivated[5] kinds, and make two wild kinds, calling one the 'fir of Ida' (Corsican pine[6]) the other the 'fir of the sea-shore' (Aleppo pine); of these the former is straighter and taller and has thicker leaves,[7] while in the latter the leaves are slenderer and weaker, and the bark is smoother and useful for tanning hides, which the other is not. Moreover the cone of the seaside kind is round and soon splits open, while that of the Idaean kind is longer and green and does not open so much, as being of wilder character. The timber of the seaside kind is stronger,—for one must note such differences also between trees of the

[3] T. describes πρῖνος σμῖλαξ, and φελλόδρυς in 3. 16, φελλός in 3. 17. 1.
[4] Plin. 16. 43.
[5] Stone pine. See Index.
[6] Plin. 16. 48.
[7] φύλλον W. conj.; ξύλον UMVP.

λαμβάνειν τῶν συγγενῶν· γνώριμοι γὰρ διὰ τὴν χρείαν.

2 Ὀρθότερον δὲ καὶ παχύτερον, ὥσπερ εἴπομεν, ἡ Ἰδαία, καὶ πρὸς τούτοις πιττωδέστερον ὅλως τὸ δένδρον, μελαντέρᾳ δὲ πίττῃ καὶ γλυκυτέρᾳ καὶ λεπτοτέρᾳ καὶ εὐωδεστέρᾳ, ὅταν ᾖ ὠμή· ἑψηθεῖσα δὲ χείρων ἐκβαίνει διὰ τὸ πολὺν ἔχειν τὸν ὀρρόν. ἐοίκασι δ' ἅπερ οὗτοι διαιροῦσιν ὀνόμασιν ἰδίοις οἱ ἄλλοι διαιρεῖν τῷ ἄρρενι καὶ θήλει. φασὶ δ' οἱ περὶ Μακεδονίαν καὶ ἄκαρπόν τι γένος ὅλως εἶναι πεύκης, καὶ τὸ μὲν ἄρρεν βραχύτερόν τε καὶ σκληροφυλλότερον, τὸ δὲ θῆλυ εὐμηκέστερον, καὶ τὰ φύλλα λιπαρὰ καὶ ἁπαλὰ καὶ κεκλιμένα μᾶλλον ἔχειν· ἔτι δὲ τὰ ξύλα τῆς μὲν ἄρρενος περίμητρα καὶ σκληρὰ καὶ ἐν ταῖς ἐργασίαις στρεφόμενα, τῆς δὲ θηλείας εὐεργὰ καὶ ἀστραβῆ καὶ μαλακώτερα.

3 Σχεδὸν δὲ κοινή τις ἡ διαφορὰ πάντων τῶν ἀρρένων καὶ θηλειῶν, ὡς οἱ ὑλοτόμοι φασίν. ἅπαν γὰρ τὸ ἄρρεν τῇ πελεκήσει καὶ βραχύτερον καὶ ἐπεστραμμένον μᾶλλον καὶ δυσεργότερον καὶ τῷ χρώματι μελάντερον, τὸ δὲ θῆλυ εὐμηκέστερον· ἐπεὶ καὶ τὴν αἰγίδα τὴν καλουμένην ἡ θήλεια τῆς πεύκης ἔχει· τοῦτο δ' ἐστὶ τὸ ἐγκάρδιον αὐτῆς·

---

[1] συγγενῶν conj. R. Const.; ἀγγείων UAld.; ἐγγείων MV mBas.

[2] γνώριμοι conj. R. Const.; γνώριμος UAld.H.; γνώριμα conj. W.

[3] ὀρθότερον conj. R. Const.; ὀξύτερον UMVAld.

[4] μελαντέρᾳ . . . εὐωδεστέρᾳ conj. W.; μελάντεραι δὲ πίττη καὶ γλυκύτεραι καὶ λεπτότεραι καὶ εὐωδέστεραι UMV; μελαντέρα

same kind,[1] since it is by their use that the different characters are recognised.[2]

The Idaean kind is, as we have said, of straighter[3] and stouter growth, and moreover the tree is altogether more full of pitch, and its pitch is blacker sweeter thinner and more fragrant[4] when it is fresh; though, when it is boiled, it turns out inferior,[5] because it contains so much watery matter. However it appears that the kinds which these people distinguish by special names are distinguished by others merely as 'male' and 'female.' The people of Macedonia say that there is also a kind of fir which bears no fruit whatever, in which the 'male'[6] (Aleppo pine) is shorter and has harder leaves, while the 'female' (Corsican pine) is taller and has glistening delicate leaves which are more pendent. Moreover the timber of the 'male' kind has much heart-wood,[7] is tough, and warps in joinery work, while that of the 'female' is easy to work, does not warp,[8] and is softer.

This distinction between 'male' and 'female' may, according to the woodmen, be said to be common to all trees. Any wood of a 'male' tree, when one comes to cut it with the axe, gives shorter lengths, is more twisted, harder to work, and darker in colour; while the 'female' gives better lengths. For it is the 'female' fir which contains what is called the *aigis*[9]; this is the heart of the tree; the

δὲ καὶ γλυκυτέρα καὶ λεπτοτέρα καὶ εὐωδεστέρα Ald. λεπτοτέρᾳ, ? less viscous.

[5] *cf.* 9. 2. 5; Plin. 16. 60. [6] Plin. 16. 47.

[7] περίμητρα conj. R. Const.: so Mold. explains; περιμήτρια UMV. *cf.* 3. 9. 6.

[8] ἀστραβῆ conj. R. Const.; εὐστραβῆ Ald.

[9] αἰγίδα: *cf.* 5. 1. 9; Plin. 16. 187.

αἴτιον δὲ ὅτι ἀπευκοτέρα καὶ ἧττον ἔνδᾳδος καὶ λειοτέρα καὶ εὐκτεανωτέρα. γίνεται δὲ ἐν τοῖς μέγεθος ἔχουσι τῶν δένδρων, ὅταν ἐκπεσόντα περισαπῇ τὰ λευκὰ τὰ κύκλῳ. τούτων γὰρ περιαιρεθέντων καὶ καταλειφθείσης τῆς μήτρας ἐκ ταύτης πελεκᾶται· ἔστι δὲ εὔχρουν σφόδρα καὶ λεπτόϊνον. ὃ δὲ οἱ περὶ τὴν Ἴδην δᾳδουργοὶ καλοῦσι συκῆν, τὸ ἐπιγιγνόμενον ἐν ταῖς πεύκαις, ἐρυθρότερον τὴν χροιὰν τῆς δᾳδός, ἐν τοῖς ἄρρεσίν ἐστι μᾶλλον· δυσῶδες δὲ τοῦτο καὶ οὐκ ὄζει δᾳδὸς οὐδὲ καίεται ἀλλ' ἀποπηδᾷ ἀπὸ τοῦ πυρός.

4 Πεύκης μὲν οὖν ταῦτα γένη ποιοῦσιν, ἥμερόν τε καὶ ἄγριον, καὶ τῆς ἀγρίας ἄρρενά τε καὶ θήλειαν καὶ τρίτην τὴν ἄκαρπον. οἱ δὲ περὶ τὴν Ἀρκαδίαν οὔτε τὴν ἄκαρπον λέγουσιν οὔτε τὴν ἥμερον πεύκην, ἀλλὰ πίτυν εἶναί φασι· καὶ γὰρ τὸ στέλεχος ἐμφερέστατον εἶναι τῇ πίτυϊ καὶ ἔχειν τήν τε λεπτότητα καὶ τὸ μέγεθος καὶ ἐν ταῖς ἐργασίαις ταὐτὸ τὸ ξύλον· τὸ γὰρ τῆς πεύκης καὶ παχύτερον καὶ λειότερον καὶ ὑψηλότερον εἶναι· καὶ τὰ φύλλα τὴν μὲν πεύκην ἔχειν πολλὰ καὶ λιπαρὰ καὶ βαθέα καὶ κεκλιμένα, τὴν δὲ πίτυν καὶ τὴν κωνοφόρον ταύτην ὀλίγα τε καὶ αὐχμωδέστερα καὶ πεφρικότα μᾶλλον· <ἄμφω δὲ τριχόφυλλα.> ἔτι δὲ τὴν πίτταν ἐμφερεστέραν τῆς

[1] *εὐκτεανωτέρα*: *εὐκτηδονωτέρα* conj. R. Const. *cf.* 5. 1. 9; but text is supported by Hesych. *s.v.* *ἰθυκτέανον*.

[2] I omit *καί* before *τὰ κύκλῳ*.

[3] Plin. 16. 44.

reason being that it is less resinous, less soaked with pitch, smoother, and of straighter grain.[1] This *aigis* is found in the larger trees, when, as they have fallen down, the white outside part[2] has decayed; when this has been stripped off and the core left, it is cut out of this with the axe; and it is of a good colour with fine fibre. However the substance which the torch-cutters of Mount Ida call the 'fig,'[3] which forms in the fir and is redder in colour than the resin, is found more in the 'male' trees; it has an evil smell, not like the smell of resin, nor will it burn, but it leaps away from the fire.

4 Such are the kinds of fir which they make out, the cultivated and the wild, the latter including the 'male' and the 'female' and also the kind which bears no fruit. However the Arcadians say that neither the sterile kind nor the cultivated is a fir, but a pine; for, they say, the trunk closely resembles the pine and has its slenderness, its stature, and the same kind[5] of wood for purposes of joinery, the trunk of the fir being thicker smoother and taller; moreover that the fir has many leaves, which are glossy massed together[6] and pendent, while in the pine and in the above-mentioned cone-bearing tree[7] the leaves are few and drier and stiffer; though in both the leaves are hair-like.[8] Also, they say, the pitch of this tree is more like that of the pine; for

4 ταῦτα γένη conj. R. Const. from G; ταῦτά γε UMVAld.; Plin. 16. 45-49.

5 ταὐτὸ conj. W.; αὐτὸ Ald.

6 βαθέα: δασέα conj. R. Const. *cf.* 3. 16. 2.

7 *i.e.* the cultivated πεύκη (so called). T. uses this periphrasis to avoid begging the question of the name.

8 ἄμφω δὲ τριχ. ins. here by Sch.; in MSS. and Ald. the words occur in § 5 after πιττωδέστερον.

πίτυος· καὶ γὰρ τὴν πίτυν ἔχειν ὀλίγην τε καὶ πικράν, ὥσπερ καὶ τὴν κωνοφόρον, τὴν δὲ πεύκην εὐώδη καὶ πολλήν. φύεται δ' ἐν μὲν τῇ Ἀρκαδίᾳ ἡ πίτυς ὀλίγη περὶ δὲ τὴν Ἠλείαν πολλή. οὗτοι μὲν οὖν ὅλῳ τῷ γένει διαμφισβητοῦσιν.

5 Ἡ δὲ πίτυς δοκεῖ τῆς πεύκης καὶ διαφέρειν τῷ λιπαρωτέρα τε εἶναι καὶ λεπτοφυλλοτέρα καὶ τὸ μέγεθος ἐλάττων καὶ ἧττον ὀρθοφυής· ἔτι δὲ τὸν κῶνον ἐλάττω φέρειν καὶ πεφρικότα μᾶλλον καὶ τὸ κάρυον πιττωδέστερον· καὶ τὰ ξύλα λευκότερα καὶ ὁμοιότερα τῇ ἐλάτῃ καὶ τὸ ὅλον ἄπευκα· διαφορὰν δ' ἔχει καὶ ταύτην μεγάλην πρὸς τὴν πεύκην· πεύκην μὲν γὰρ ἐπικαυθεισῶν τῶν ῥιζῶν οὐκ ἀναβλαστάνειν, τὴν πίτυν δέ φασί τινες ἀναβλαστάνειν, ὥσπερ καὶ ἐν Λέσβῳ ἐμπρησθέντος τοῦ Πυρραίων ὄρους τοῦ πιτυώδους. νόσημα δὲ ταῖς πεύκαις τοιοῦτόν τι λέγουσι συμβαίνειν οἱ περὶ τὴν Ἴδην ὥστ', ὅταν μὴ μόνον τὸ ἐγκάρδιον ἀλλὰ καὶ τὸ ἔξω τοῦ στελέχους ἔνδᾳδον γένηται, τηνικαῦτα ὥσπερ ἀποπνίγεσθαι. τοῦτο δὲ αὐτόματον συμβαίνει δι' εὐτροφίαν τοῦ δένδρου, ὡς ἄν τις εἰκάσειεν· ὅλον γὰρ γίνεται δᾴς· περὶ μὲν οὖν τὴν πεύκην ἴδιον τοῦτο πάθος.

6 Ἐλάτη δ' ἐστὶν ἡ μὲν ἄρρην ἡ δὲ θήλεια, διαφορὰς δ' ἔχουσα τοῖς φύλλοις· ὀξύτερα γὰρ καὶ κεντητικώτερα τὰ τοῦ ἄρρενος καὶ ἐπεστραμμένα μᾶλλον, δι' ὃ καὶ οὐλότερον τῇ ὄψει φαίνεται τὸ δένδρον ὅλον. καὶ τῷ ξύλῳ· λευκότερον γὰρ καὶ μαλακώτερον καὶ εὐεργέστερον τὸ τῆς θηλείας καὶ

---

[1] πικρὰν conj. R. Const. from G; μικρὰν VAld.

[2] καὶ ταύτην μεγάλην πρὸς conj. Sch.; καὶ τὴν μεγ. πρὸς UMV; μεγάλην πρὸς Ald.

in the pine too it is scanty and bitter,[1] as in this other cone-bearing tree, but in the fir it is fragrant and abundant. Now the pine is rare in Arcadia, but common in Elis. The Arcadians then dispute altogether the nomenclature.

The pine appears to differ also from the fir in being glossier and having finer leaves, while it is smaller in stature and does not grow so straight; also in bearing a smaller cone, which is stiffer and has a more pitchy kernel, while its wood is whiter, more like that of the silver-fir, and wholly free from pitch. And there is another great difference[2] between it and the fir; the fir, if it is burnt down to the roots, does not shoot up again, while the pine, according to some, will do so; for instance this happened in Lesbos,[3] when the pine-forest of Pyrrha[4] was burnt. The people of Ida say that the fir is liable to a kind of disease;—when not only the heart but the outer part of the trunk becomes glutted[5] with pitch, the tree then is as it were choked. This happens of its own accord through the excessive luxuriance of the tree, as one may conjecture; for it all turns into pitch-glutted wood. This then is an affection peculiar to the fir.

[6] The silver-fir is either 'male' or 'female,' and has differences in its leaves[7]; those of the 'male' are sharper more needle-like and more bent; wherefore the whole tree has a more compact appearance. There are also differences in the wood, that of the 'female' being whiter softer and easier to work,

3 ἐν Λέσβῳ conj. W. from G, and Plin. 16. 46; εἰς Λέσβον MSS.

4 On the W. of Lesbos, modern Caloni. *cf.* 2. 2. 6; Plin. *l.c.*

5 *cf.* 1. 6. 1; Plin. 16. 44.

6 Plin. 16. 48. 7 *cf.* 1. 8. 2.

τὸ ὅλον στέλεχος εὐμηκέστερον· τὸ δὲ τοῦ ἄρρενος ποικιλώτερον καὶ παχύτερον καὶ σκληρότερον καὶ περίμητρον μᾶλλον ὅλως δὲ φαυλότερον τὴν ὄψιν. ἐν δὲ τῷ κώνῳ τῷ μὲν τοῦ ἄρρενός ἐστι κάρυα ὀλίγα ἐπὶ τοῦ ἄκρου, τῷ δὲ τῆς θηλείας ὅλως οὐδέν' ὡς οἱ ἐκ Μακεδονίας ἔλεγον. ἔχει δὲ πτέρυγας τὸ φύλλον καὶ ἐπ' ἔλαττον, ὥστε τὴν ὅλην μορφὴν εἶναι θολοειδῆ καὶ παρόμοιον μάλιστα ταῖς Βοιωτίαις κυνέαις· πυκνὸν δὲ οὕτως ὥστε μήτε χιόνα διιέναι μήθ' ὑετόν. ὅλως δὲ καὶ τῇ ὄψει τὸ δένδρον καλόν· καὶ γὰρ ἡ βλάστησις ἰδία τις, ὥσπερ εἴρηται, παρὰ τὰς ἄλλας καὶ μόνη τάξιν ἔχουσα· τῷ δὲ μεγέθει μέγα καὶ πολὺ τῆς πεύκης εὐμηκέστερον.

Διαφέρει δὲ καὶ κατὰ τὸ ξύλον οὐ μικρόν· τὸ μὲν γὰρ τῆς ἐλάτης ἰνῶδες καὶ μαλακὸν καὶ κοῦφον, τὸ δὲ τῆς πεύκης δᾳδῶδες καὶ βαρὺ καὶ σαρκωδέστερον. ὄζους δὲ ἔχει πλείους μὲν ἡ πεύκη σκληροτέρους δ' ἡ ἐλάτη, σχεδὸν δὲ πάντων ὡς εἰπεῖν σκληροτέρους, τὸ δὲ ξύλον μαλακώτερον. ὅλως δὲ οἱ ὄζοι πυκνότατοι καὶ στερεώτατοι μόνον οὐ διαφανεῖς ἐλάτης καὶ πεύκης καὶ τῷ χρώματι δᾳδώδεις καὶ μάλιστα διάφοροι τοῦ ξύλου, μᾶλλον δὲ τῆς ἐλάτης. ἔχει δέ, ὥσπερ ἡ πεύκη τὴν αἰγίδα, καὶ ἡ ἐλάτη τὸ λευκὸν λοῦσσον

---

[1] παχύτερον conj. W.; πλατύτερον Ald.
[2] Plin. 16. 48 and 49. [3] For the tense see Intr. p. xx.
[4] φύλλον, *i.e.* the leafy shoot. Sch. considers φύλλον to be corrupt, and refers the following description to the cone; W. marks a lacuna after φύλλον. Pliny, *l.c.*, seems to have read φύλλον, but does not render καὶ ἐπ' ἔλαττον . . . κυνέαις. The words καὶ ἐπ' ἔλαττον can hardly be sound as they stand. For the description of the foliage *cf.* 1. 10. 5.

while the whole trunk is longer; that of the 'male' is less of a uniform colour thicker[1] and harder, has more heart-wood, and is altogether inferior in appearance. In the cone[2] of the 'male' are a few seeds at the apex, while that of the 'female,' according to what the Macedonians said,[3] contains none at all. The foliage[4] is feathered and the height disproportionate so that the general appearance of the tree is dome-like,[5] and closely resembles the Boeotian peasant's hat[6]; and it is so dense that neither snow nor rain penetrates it. And in general the tree has a handsome appearance; for its growth is somewhat peculiar, as has been said, compared with the others, it being the only one which is regular, and in stature it is large, much taller than the fir.

[7] There is also not a little difference in the wood: that of the silver-fir is fibrous[8] soft and light, that of the fir is resinous heavy and more fleshy. The fir has more knots,[9] but the silver-fir harder ones; indeed they may be said to be harder than those of any tree, though the wood otherwise is softer. And in general the knots of silver-fir and fir are of the closest and most solid[10] texture and almost[11] transparent: in colour they are like resin-glutted wood, and quite different from the rest of the wood; and this is especially so[12] in the silver-fir. And just as the fir has its *aigis*,[13] so the silver-fir has what is

[5] θολοειδῆ conj. Scal.; θηλοειδῆ U (erased); θηλοειδὲς MV; *ut concameratum imitetur* G; ? θολιοειδῆ; in Theocr. 15. 39. θολία seems to be a sun-hat.

[6] κυνέαις: *cf.* Hesych. *s.v.* κυνῆ Βοιωτία, apparently a hat worn in the fields.

[7] *cf.* 5. 1. 7. [8] *cf.* 5. 1. 5. [9] *cf.* 5. 1. 6.

[10] *cf.* 5. 1. 6, κερατώδεις. [11] οὐ ins. Sch.

[12] μᾶλλον δὲ conj. W.; μᾶλλον ἢ Ald. [13] *cf.* 3. 9. 3.

καλούμενον, οἷον ἀντίστροφον τῇ αἰγίδι, πλὴν τὸ μὲν λευκὸν ἡ δ' αἰγὶς εὔχρως διὰ τὸ ἔνδᾳδον. πυκνὸν δὲ καὶ λευκὸν γίνεται καὶ καλὸν ἐκ τῶν πρεσβυτέρων ἤδη δένδρων· ἀλλὰ σπάνιον τὸ χρηστόν, τὸ δὲ τυχὸν δαψιλές, ἐξ οὗ τά τε τῶν ζωγράφων πινάκια ποιοῦσι καὶ τὰ γραμματεῖα τὰ πολλά· τὰ δ' ἐσπουδασμένα ἐκ τοῦ βελτίονος.

8 Οἱ δὲ περὶ Ἀρκαδίαν ἀμφότερα καλοῦσιν αἰγίδα καὶ τὴν τῆς πεύκης καὶ τὴν τῆς ἐλάτης, καὶ εἶναι πλείω τὴν τῆς ἐλάτης ἀλλὰ καλλίω τὴν τῆς πεύκης· εἶναι γὰρ τῆς μὲν ἐλάτης πολλήν τε καὶ λείαν καὶ πυκνήν, τῆς δὲ πεύκης ὀλίγην, τὴν μέντοι οὖσαν οὐλοτέραν καὶ ἰσχυροτέραν καὶ τὸ ὅλον καλλίω. οὗτοι μὲν οὖν ἐοίκασι τοῖς ὀνόμασι διαφωνεῖν. ἡ δὲ ἐλάτη ταύτας ἔχει τὰς διαφορὰς πρὸς τὴν πεύκην καὶ ἔτι τὴν περὶ τὴν ἄμφαυξιν, ἣν πρότερον εἴπομεν.

X. Ὀξύη δ' οὐκ ἔχει διαφορὰς ἀλλ' ἐστὶ μονογενές· ὀρθοφυὲς δὲ καὶ λεῖον καὶ ἄνοζον καὶ πάχος καὶ ὕψος ἔχον σχεδὸν ἴσον τῇ ἐλάτῃ· καὶ τἆλλα δὲ παρόμοιον [τε] τὸ δένδρον· ξύλον δὲ εὔχρουν ἰσχυρὸν εὔινον καὶ φλοιὸν λεῖον καὶ παχύν, φύλλον δ' ἀσχιδὲς προμηκέστερον ἀπίου καὶ ἐπακάνθιζον ἐξ ἄκρου, ῥίζας οὔτε πολλὰς οὔτε κατὰ βάθους· ὁ δὲ καρπὸς λεῖος βαλανώδης ἐν ἐχίνῳ

[1] *cf.* Eur. *I.A.* 99; Hipp. 1254.
[2] τὰ δ' conj. Scal.; καὶ Ald.
[3] πεύκης conj. Scal. from G; ἐλάτης Ald.
[4] ἐλάτης conj. Scal. from G; πεύκης Ald.

called its white 'centre,' which answers, as it were, to the *aigis* of the fir, except that it is white, while the other is bright-coloured because it is glutted with pitch. It becomes close white and good in trees which are of some age, but it is seldom found in good condition, while the ordinary form of it is abundant and is used to make painters' boards and ordinary writing tablets,[1] superior ones being[2] made of the better form.

However the Arcadians call both substances *aigis*, alike that of the fir[3] and the corresponding part of the silver-fir,[4] and say that, though the silver-fir produces more, that of the fir is better; for that, though that of the silver-fir is abundant[5] smooth and close, that of the fir, though scanty, is compacter stronger and fairer in general. The Arcadians then appear to differ as to the names which they give. Such are the differences in the silver-fir as compared with the fir, and there is also that of having the *amphauxis*,[6] which we mentioned before.

Of beech, yew, hop-hornbeam, lime.

X. The beech presents no differences, there being but one kind. It is a straight-growing smooth and unbranched tree, and in thickness and height is about equal to the silver-fir, which it also resembles in other respects; the wood is of a fair colour strong and of good grain, the bark smooth and thick, the leaf undivided, longer than a pear-leaf, spinous at the tip,[7] the roots neither numerous nor running deep; the fruit is smooth like an acorn, enclosed in a shell,

[5] πολλὴν conj. Gesner; οὔλην UmBas.; ὅλην MVAld.
[6] *cf.* 3. 7. 1.
[7] *i.e.* mucronate. *cf.* 3. 11. 3.

πλὴν [οὐκ] ἀνακάνθῳ καὶ λείῳ, καὶ οὐχ ὡς ἡ διοσβάλανος ἀκανθώδει, προσεμφερὴς δὲ καὶ κατὰ γλυκύτητα καὶ κατὰ τὸν χυλὸν ἐκείνῳ. γίνεται δὲ καὶ ἐν τῷ ὄρει λευκή, ἣ καὶ χρήσιμον ἔχει τὸ ξύλον πρὸς πολλά· καὶ γὰρ πρὸς ἁμαξουργίαν καὶ πρὸς κλινοπηγίαν καὶ εἰς διφρουργίαν καὶ εἰς τραπεζίαν καὶ εἰς ναυπηγίαν· ἡ δ' ἐν τοῖς πεδίοις μέλαινα καὶ ἄχρηστος πρὸς ταῦτα· τὸν δὲ καρπὸν ἔχουσι παραπλήσιον.

2 Μονογενὴς δὲ καὶ ἡ μίλος, ὀρθοφυὴς δὲ καὶ εὐαυξὴς καὶ ὁμοία τῇ ἐλάτῃ, πλὴν οὐχ ὑψηλὸν οὕτω, πολυμάσχαλον δὲ μᾶλλον. ὅμοιον δὲ καὶ τὸ φύλλον ἔχει τῇ ἐλάτῃ, λιπαρώτερον δὲ καὶ μαλακώτερον. τὸ δὲ ξύλον ἡ μὲν ἐξ Ἀρκαδίας μέλαν καὶ φοινικοῦν, ἡ δ' ἐκ τῆς Ἴδης ξανθὸν σφόδρα καὶ ὅμοιον τῇ κέδρῳ, δι' ὃ καὶ τοὺς πωλοῦντάς φασιν ἐξαπατᾶν ὡς κέδρον πωλοῦντας· πᾶν γὰρ εἶναι καρδίαν, ὅταν ὁ φλοιὸς περιαιρεθῇ· ὅμοιον δὲ καὶ τὸν φλοιὸν ἔχειν καὶ τῇ τραχύτητι καὶ τῷ χρώματι τῇ κέδρῳ, ῥίζας δὲ μικρὰς καὶ λεπτὰς καὶ ἐπιπολαίους. σπάνιον δὲ τὸ δένδρον περὶ τὴν Ἴδην· περὶ δὲ Μακεδονίαν καὶ Ἀρκαδίαν πολύ· καὶ καρπὸν φέρει στρογγύλον μικρῷ μείζω κυάμου, τῷ χρώματι δ' ἐρυθρὸν καὶ μαλακόν· φασὶ δὲ τὰ μὲν λόφουρα ἐὰν φάγῃ τῶν φύλλων ἀποθνήσκειν, τὰ δὲ μηρυκάζοντα οὐδὲν πάσχειν. τὸν δὲ καρπὸν ἐσθίουσι καὶ τῶν ἀνθρώπων τινὲς καὶ ἔστιν ἡδὺς καὶ ἀσινής.

---

1 ἐχῖνος being otherwise used of a prickly case, such as that of the chestnut. πλὴν ἀνακ. καὶ λείῳ conj. W.; πλὴν οὐκ ἀνακάνθωι καὶ λείωι U; πλὴν οὐκ ἐν ἀκάνθῳ MVAld.

which is however without prickles[1] and smooth, not spinous,[2] like the chestnut, though in sweetness and flavour it resembles it. In mountain country it also grows white and has[3] timber which is useful for many purposes, for making carts beds chairs and tables, and for shipbuilding[4]; while the tree of the plains is black and useless for these purposes; but the fruit is much the same in both.

[5] The yew has also but one kind, is straight-growing, grows readily, and is like the silver-fir, except that it is not so tall and is more branched. Its leaf is also like that of the silver-fir, but glossier and less stiff. As to the wood, in the Arcadian yew it is black or red, in that of Ida bright yellow and like prickly cedar; wherefore they say that dealers practise deceit, selling it for that wood: for that it is all heart, when the bark is stripped off; its bark also resembles that of prickly cedar in roughness and colour, its roots are few slender and shallow. The tree is rare about Ida, but common in Macedonia and Arcadia; it bears a round fruit a little larger than a bean, which is red in colour and soft; and they say that, if beasts of burden[6] eat of the leaves they die, while ruminants take no hurt. Even men sometimes eat the fruit, which is sweet and harmless.

[2] ἀκανθώδει conj R. Const.; ἀκανθώδη Ald.H.
[3] λευκὴ ἢ καὶ conj. W.; λευκή τε καὶ Ald.H.
[4] *cf.* 5. 6. 4; 5. 7. 2 and 6.
[5] Plin. 16. 62. (description taken from this passage, but applied to *fraxinus*, apparently from confusion between μίλος and μελία).
[6] *cf.* 2. 7. 4 n.

3 Ἔστι δὲ καὶ ἡ ὄστρυς μονοειδής, ἣν καλοῦσί τινες ὀστρύαν, ὁμοφυὲς τῇ ὀξύᾳ τῇ τε φυτείᾳ καὶ τῷ φλοιῷ· φύλλα δὲ ἀπιοειδῆ τῷ σχήματι, πλὴν προμηκέστερα πολλῷ καὶ εἰς ὀξὺ συνηγμένα καὶ μείζω, πολύϊνα δέ, ἀπὸ τῆς μέσης εὐθείας καὶ μεγάλης τῶν ἄλλων πλευροειδῶς κατατεινουσῶν καὶ πάχος ἐχουσῶν· ἔτι δὲ ἐρρυτιδωμένα κατὰ τὰς ἶνας καὶ χαραγμὸν ἔχοντα κύκλῳ λεπτόν· τὸ δὲ ξύλον σκληρὸν καὶ ἄχρουν, ἔκλευκον· καρπὸν δὲ μικρὸν πρόμακρον ὅμοιον κριθῇ ξανθόν· ῥίζας δὲ ἔχει μετεώρους· ἔνυδρον δὲ καὶ φαραγγῶδες. λέγεται δὲ ὡς οὐκ ἐπιτήδειον εἰς οἰκίαν εἰσφέρειν· δυσθανατεῖν γάρ φασι καὶ δυστοκεῖν οὗ ἂν ᾖ.

4 Τῆς δὲ φιλύρας ἡ μὲν ἄρρην ἐστὶ ἡ δὲ θήλεια· διαφέρουσι δὲ τῇ μορφῇ τῇ ὅλῃ καὶ τῇ τοῦ ξύλου καὶ τῷ τὸ μὲν εἶναι κάρπιμον τὸ δ' ἄκαρπον. τὸ μὲν γὰρ τῆς ἄρρενος ξύλον σκληρὸν καὶ ξανθὸν καὶ ὀζωδέστερον καὶ πυκνότερόν ἐστι, ἔτι δ' εὐωδέστερον, τὸ δὲ τῆς θηλείας λευκότερον. καὶ ὁ φλοιὸς τῆς μὲν ἄρρενος παχύτερος καὶ περιαιρεθεὶς ἀκαμπὴς διὰ τὴν σκληρότητα, τῆς δὲ θηλείας λεπτότερος καὶ εὐκαμπής, ἐξ οὗ τὰς κίστας ποιοῦσιν· καὶ ἡ μὲν ἄκαρπος καὶ ἀνανθής, ἡ δὲ θήλεια καὶ ἄνθος ἔχει καὶ καρπόν· τὸ μὲν ἄνθος καλυκῶδες παρὰ τὸν τοῦ φύλλου μίσχον καὶ παρὰ

---

[1] *cf.* 1. 8. 2 (ὀστρυΐς), 3. 3. 1; *C.P.* 5. 12. 9 (ὀστρύη); Plin. 13. 117.

[2] μέσης . . . κατατεινουσῶν conj. Sch.; μέσης πλευροειδῶς τῶν ἄλλων εὐθειῶν καὶ μεγάλην κατατεινουσῶν Ald. *cf.* 1. 10. 2; 3. 17. 3.

The *ostrys* (hop-hornbeam),[1] which some call *ostrya*, has also but one kind: it is like the beech in growth and bark; its leaves are in shape like a pear's, except that they are much longer, come to a sharp point, are larger, and have many fibres, which branch out like ribs from a large straight one[2] in the middle, and are thick; also the leaves are wrinkled along the fibres and have a finely serrated edge; the wood is hard colourless and whitish; the fruit is small oblong and yellow like barley; it has shallow roots; it loves water and is found in ravines. It is said to be unlucky to bring it into the house, since, wherever it is, it is supposed to cause a painful death[3] or painful labour in giving birth.

[4] The lime has both 'male' and 'female' forms, which differ in their general appearance, in that of the wood, and in being respectively fruit-bearing and sterile. The wood of the 'male' tree is hard yellow more branched closer, and also more fragrant[5]; that of the 'female' is whiter. The bark of the 'male' is thicker, and, when stripped off, is unbending because of its hardness; that of the 'female' is thinner[6] and flexible; men make their writing-cases[7] out of it. The 'male' has neither fruit nor flower, but the 'female' has both flower and fruit; the flower is cup-shaped, and appears alongside of the stalk of the leaf, or alongside of next year's

[3] δυσθανατεῖν I conj.; δυσθάνατον P_2Ald.; δυσθανατᾶν conj. Sch., but δυσθανατᾶν has a desiderative sense.

[4] Plin. 16. 65.

[5] ἔτι δ' εὐωδ. inserted here by Sch.; *cf.* Plin., *l.c.* In Ald. the words, with the addition τὸ τῆς θηλείας, occur after ποιοῦσιν.

[6] λεπτότερος conj. Sch; λευκότερος Ald.

[7] *cf.* 3. 13. 1; Ar. *Vesp.* 529.

τὴν εἰς νέωτα κάχρυν ἐφ' ἑτέρου μίσχου, χλοερὸν δὲ ὅταν ᾖ καλυκῶδες, ἐκκαλυπτόμενον δὲ ἐπίξανθον· ἡ δὲ ἄνθησις ἅμα τοῖς ἡμέροις. ὁ δὲ καρπὸς στρογγύλος πρόμακρος ἡλίκος κύαμος ὅμοιος τῷ τοῦ κιττοῦ, γωνίας ἔχων ὁ ἁδρὸς πέντε οἷον ἰνῶν ἐξεχουσῶν καὶ εἰς ὀξὺ συναγομένων· ὁ δὲ μὴ ἁδρὸς ἀδιαρθρότερος· διακνιζόμενος δὲ ὁ ἁδρὸς ἔχει μίκρ' ἄττα καὶ λεπτὰ σπερμάτια ἡλίκα καὶ ὁ τῆς ἀδραφάξυος. τὸ δὲ φύλλον καὶ ὁ φλοιὸς ἡδέα καὶ γλυκέα· τὴν δὲ μορφὴν κιττῶδες τὸ φύλλον, πλὴν ἐκ προσαγωγῆς μᾶλλον ἡ περιφέρεια, κατὰ τὸ πρὸς τῷ μίσχῳ κυρτότατον, ἀλλὰ κατὰ μέσον εἰς ὀξύτερον τὴν συναγωγὴν ἔχον καὶ μακρότερον, ἔπουλον δὲ κύκλῳ καὶ κεχαραγμένον. μήτραν δ' ἔχει τὸ ξύλον μικρὰν καὶ οὐ πολὺ μαλακωτέραν τοῦ ἄλλου· μαλακὸν γὰρ καὶ τὸ ἄλλο ξύλον.

XI. Τῆς δὲ σφενδάμνου, καθάπερ εἴπομεν, δύο γένη ποιοῦσιν, οἱ δὲ τρία· ἓν μὲν δὴ τῷ κοινῷ προσαγορεύουσι σφένδαμνον, ἕτερον δὲ ζυγίαν, τρίτον δὲ κλινότροχον, ὡς οἱ περὶ Στάγειρα. διαφορὰ δ' ἐστὶ τῆς ζυγίας καὶ τῆς σφενδάμνου ὅτι ἡ μὲν σφένδαμνος λευκὸν ἔχει τὸ ξύλον καὶ εὐινότερον, ἡ δὲ ζυγία ξανθὸν καὶ οὖλον· τὸ δὲ φύλλον εὐμέγεθες ἄμφω, τῇ σχίσει ὅμοιον τῷ

---

[1] *cf.* 3. 5. 5. and 6.
[2] διακνιζόμενος: διασχιζόμενος, 'when split open,' conj. W.
[3] *cf.* 1. 12. 4; *C.P.* 6. 12. 7. [4] 3. 3. 1.
[5] προσαγορεύουσι conj. W. from G; προσαγορεύεται Ald.

winter-bud[1] on a separate stalk; it is green, when in the cup-like stage, but brownish as it opens; it appears at the same time as in the cultivated trees. The fruit is rounded oblong as large as a bean, resembling the fruit of the ivy; when mature, it has five angular projections, as it were, made by projecting fibres which meet in a point; the immature fruit is less articulated. When the mature fruit is pulled to pieces,[2] it shows some small fine seeds of the same size as those of orach. The leaf and the bark[3] are well flavoured and sweet; the leaf is like that of the ivy in shape, except that it rounds more gradually, being most curved at the part next the stalk, but in the middle contracting to a sharper and longer apex, and its edge is somewhat puckered and jagged. The timber contains little core, which is not much softer than the other part; for the rest of the wood is also soft.

Of maple and ash.

XI. Of the maple, as we have said,[4] some make[5] two kinds, some three; one they call by the general name 'maple,' another *zygia*, the third *klinotrokhos*[6]; this name, for instance, is used by the people of Stagira. The difference between *zygia* and maple proper is that the latter has white wood of finer fibre, while that of *zygia* is yellow and of compact texture. The leaf[7] in both trees is large, resembling that of the plane in the way in which it is

[6] κλινότροχον Ald.; κλινόστροχον U; ἰνότροχον conj. Salm. from Plin. 16. 66 and 67, *cursivenium* or *crassivenium*. Sch. thinks that the word conceals γλῖνος; *cf.* 3. 3. 1; 3. 11. 2.
[7] φύλλον conj. R. Const.; ξύλον UMVAld.H.G.

τῆς πλατάνου τετανὸν λεπτότερον δὲ καὶ ἀσαρκότερον καὶ μαλακώτερον καὶ προμηκέστερον· τὰ δὲ σχίσμαθ' ὅλα τ' εἰς ὀξὺ συνήκοντα καὶ οὐχ οὕτω μεσοσχιδῆ ἀλλ' ἀκροσχιδέστερα· οὐ πολυΐνα δὲ ὡς κατὰ μέγεθος· ἔχει δὲ καὶ φλοιὸν μικρῷ τραχύτερον τοῦ τῆς φιλύρας, ὑποπέλιον παχὺν καὶ πυκνότερον ἢ ὁ τῆς πίτυος καὶ ἀκαμπῆ· ῥίζαι δ' ὀλίγαι καὶ μετέωροι καὶ οὖλαι σχεδὸν αἱ πλεῖσται
2 καὶ αἱ τῆς ξανθῆς καὶ αἱ τῆς λευκῆς. γίνεται δὲ μάλιστα ἐν τοῖς ἐφύδροις, ὡς οἱ περὶ τὴν Ἴδην λέγουσι, καὶ ἔστι σπάνιον. περὶ ἄνθους δὲ οὐκ ᾔδεσαν· τὸν δὲ καρπὸν οὐ λίαν μὲν προμήκη, παρόμοιον δὲ τῷ παλιούρῳ πλὴν προμηκέστερον. οἱ δ' ἐν τῷ Ὀλύμπῳ τὴν μὲν ζυγίαν ὄρειον μᾶλλον, τὴν δὲ σφένδαμνον καὶ ἐν τοῖς πεδίοις φύεσθαι· εἶναι δὲ τὴν μὲν ἐν τῷ ὄρει φυομένην ξανθὴν καὶ εὔχρουν καὶ οὔλην καὶ στερεάν, ᾗ καὶ πρὸς τὰ πολυτελῆ τῶν ἔργων χρῶνται, τὴν δὲ πεδεινὴν λευκήν τε καὶ μανοτέραν καὶ ἧττον οὔλην· καλοῦσι δ' αὐτὴν ἔνιοι γλεῖνον, οὐ σφένδαμνον. . . . καὶ τῆς ἄρρενος οὐλότερα τὰ ξύλα συνεστραμμένα, καὶ ἐν τῷ πεδίῳ ταύτην φύεσθαι μᾶλλον καὶ βλαστάνειν πρωΐτερον.

3 Ἔστι δὲ καὶ μελιας γενη δυο. τούτων δ' ἡ μὲν ὑψηλὴ καὶ εὐμήκης ἐστὶ τὸ ξύλον ἔχουσα λευκὸν καὶ εὔινον καὶ μαλακώτερον καὶ ἀνοζό-

---

[1] τετανὸν: *cf.* 3. 12. 5; 3. 15. 6.
[2] σχίσμαθ' conj. R. Const. from G; σχίμαθ' Ald.Cam.; σχῆμαθ' Bas., which W. reads.
[3] ὅλα: ? ὅλως.
[4] *i.e.* do not run back so far.
[5] πολυΐνα conj. R. Const.; πολύ· ἵνα δὲ Ald.; πολύ· ἶνα δὲ M.

divided; it is smooth,[1] but more delicate, less fleshy, softer, longer in proportion to its breadth, and the divisions[2] all[3] tend to meet in a point, while they do not occur so much in the middle of the leaf,[4] but rather at the tip; and for their size the leaves have not many fibres.[5] The bark too is somewhat rougher than that of the lime, of blackish colour thick closer[6] than that of the Aleppo pine, and stiff; the roots are few shallow and compact for the most part, both those of the yellow and those of the white-wooded tree. This tree occurs chiefly in wet ground,[7] as the people of Mount Ida say, and is rare. About its flower they did[8] not know, but the fruit, they said, is not very oblong, but like that of Christ's thorn,[9] except that it is more oblong than that. But the people of Mount Olympus say that, while *zygia* is rather a mountain tree, the maple proper grows also in the plains; and that the form which grows in the mountains has yellow wood of a bright colour, which is of compact texture and hard, and is used even for expensive work, while that of the plains has white wood of looser make and less compact texture. And some call it *gleinos*[10] instead of maple. . . . .[11] The wood of the 'male' tree is of compacter texture and twisted; this tree, it is said, grows rather in the plain and puts forth its leaves earlier.

[12] There are also two kinds of ash. Of these one is lofty and of strong growth with white wood of good fibre, softer, with less knots, and of more compact

[6] πυκνότερον conj. Scal. from G; πυρώτερον UAld.

[7] ἐφύδροις: ὑφύδροις conj. Sch. *cf.* ὕφαμμος, ὑπόπετρος.

[8] *cf.* 3. 9. 6 n.; Intr. p. xx. [9] *cf.* 3. 18. 3.

[10] *cf.* 3. 3. 1; Plin. 16. 67.

[11] W. marks a lacuna: the description of the 'female' tree seems to be missing. [12] Plin. 16. 62–64.

τερον καὶ οὐλότερον· ἡ δὲ ταπεινοτέρα καὶ ἧττον εὐαυξὴς καὶ τραχυτέρα καὶ σκληροτέρα καὶ ξανθοτέρα. τὰ δὲ φύλλα τῷ μὲν σχήματι δαφνοειδῆ, πλατυφύλλου δάφνης, εἰς ὀξύτερον δὲ συνηγμένα, χαραγμὸν δέ τιν' ἔχοντα κύκλῳ καὶ ἐπακανθίζοντα· τὸ δὲ ὅλον, ὅπερ εἴποι τις ἂν φύλλον τῷ ἅμα φυλλορροεῖν, ἀφ' ἑνὸς μίσχου· καὶ περὶ μίαν οἷον ἶνα κατὰ γόνυ καὶ συζυγίαν τὰ φύλλα καθ' ἕκαστον πέφυκε, συχνῶν διεχουσῶν τῶν συζυγιῶν, ὁμοίως καὶ ἐπὶ τῆς οἴης. ἔστι δὲ τῶν μὲν βραχέα τὰ γόνατα καὶ αἱ συζυγίαι τὸ πλῆθος ἐλάττους, τῶν δὲ τῆς λευκῆς καὶ μακρὰ καὶ πλείους· καὶ τὰ καθ' ἕκαστον φύλλα μακρότερα καὶ στενότερα, τὴν δὲ χρόαν πρασώδη. φλοιὸν δὲ λεῖον ἔχει, καπυρὸν δὲ καὶ λεπτὸν καὶ τῇ

4 χρόᾳ πυρρόν. πυκνόρριζον δὲ καὶ παχύρριζον καὶ μετέωρον. καρπὸν δὲ οἱ μὲν περὶ τὴν Ἴδην οὐχ ὑπελάμβανον ἔχειν οὐδ' ἄνθος· ἔχει δ' ἐν λοβῷ λεπτῷ καρπὸν καρυηρὸν ὡς τῶν ἀμυγδαλῶν ὑπόπικρον τῇ γεύσει. φέρει δὲ καὶ ἕτερ' ἄττα οἷον βρύα, καθάπερ ἡ δάφνη, πλὴν στιφρότερα· καὶ ἕκαστον καθ' αὑτὸ σφαιροειδές, ὥσπερ τὰ τῶν πλατάνων· τούτων δὲ τὰ μὲν περὶ τὸν καρπόν, τὰ δ' ἀπηρτημένα πολύ, καὶ τὰ πλεῖστα οὕτω. φύεται δὲ ἡ μὲν λεία περὶ τὰ βαθυάγκη μάλιστα καὶ ἔφυδρα, ἡ δὲ τραχεῖα καὶ περὶ τὰ ξηρὰ καὶ πετρώδη. ἔνιοι δὲ καλοῦσι τὴν μὲν μελίαν

---

[1] οὐλότερον: ἀνουλότερον W. from Sch.'s conj.; ἄνουλος does not occur elsewhere, and T. uses μανός as the opposite of οὖλος.

[2] *i.e.* instead of considering the leaflet as the unit. For the description *cf.* 3. 12. 5; 3. 15. 4.

texture[1]; the other is shorter, less vigorous in growth, rougher harder and yellower. The leaves in shape are like those of the bay, that is, the broad-leaved bay, but they contract to a sharper point, and they have a sort of jagged outline with sharp points. The whole leaf (if one may consider this as[2] a 'leaf' because it is all shed at once) grows on a single stalk; on either side of a single fibre, as it were, the leaflets grow at a joint in pairs, which are numerous and distinct, just as in the sorb. In some leaves the joints are short[3] and the pairs fewer in number, but in those of the white kind the joint is long and the pairs more numerous, while the leaflets are longer narrower and leek-green in colour. Also this tree has a smooth bark, which is dry thin and red in colour. The roots are matted stout and shallow.[4] As to the fruit, the people of Ida supposed it to have none, and no flower either; however it has a nut-like fruit in a thin pod, like the fruit of the almond, and it is somewhat bitter in taste. And it also bears certain other things like winter-buds, as does the bay, but they are more solid,[5] and each separate one is globular, like those of the plane; some of these occur around the fruit, some, in fact the greater number,[6] are at a distance from it. The smooth kind[7] grows mostly in deep ravines and damp places, the rough kind occurs also in dry and rocky parts. Some, for instance the Macedonians, call the

[3] βραχέα conj. Scal. from G; τραχέα UAld.H.

[4] Bod. inserts οὐ before μετέωρον; *cf.* 3. 6. 5. (Idaean account.)

[5] στιφρότερα conj. Dalec.; στρυφνότερα MSS.

[6] πλεῖστα conj. R. Const.; πλεκτὰ UMVAld.

[7] *cf.* Plin., *l.c.*

τὴν δὲ βουμέλιον, ὥσπερ οἱ περὶ Μακεδονίαν.
5 μεῖζον δὲ καὶ μανότερον ἡ βουμέλιος, δι' ὃ καὶ ἧττον οὖλον. φύσει δὲ τὸ μὲν πεδεινὸν καὶ τραχύ, τὸ δ' ὀρεινὸν καὶ λεῖον· ἔστι δὲ ἡ μὲν ἐν τοῖς ὄρεσι φυομένη εὔχρους καὶ λεία καὶ στερεὰ καὶ γλίσχρα, ἡ δ' ἐν τῷ πεδίῳ ἄχρους καὶ μανὴ καὶ τραχεῖα. (τὸ δ' ὅλον ὡς εἰπεῖν τὰ δένδρα ὅσα καὶ ἐν τῷ πεδίῳ καὶ ἐν τῷ ὄρει φύεται, τὰ μὲν ὀρεινὰ εὔχροά τε καὶ στερεὰ καὶ λεῖα γίνεται, καθάπερ ὀξύη πτελέα τὰ ἄλλα· τὰ δὲ πεδεινὰ μανότερα καὶ ἀχρούστερα καὶ χείρω, πλὴν ἀπίου καὶ μηλέας καὶ ἀχράδος, ὡς οἱ περὶ τὸν Ὄλυμπόν φασι· ταῦτα δ' ἐν τῷ πεδίῳ κρείττω καὶ τῷ καρπῷ καὶ τοῖς ξύλοις· ἐν μὲν γὰρ τῷ ὄρει τραχεῖς καὶ ἀκανθώδεις καὶ ὀζώδεις εἰσίν, ἐν δὲ τῷ πεδίῳ λειότεροι καὶ μείζους καὶ τὸν καρπὸν ἔχουσι γλυκύτερον καὶ σαρκωδέστερον· μεγέθει δὲ αἰεὶ μείζω τὰ πεδεινά.)

XII. Κρανείας δὲ τὸ μὲν ἄρρεν τὸ δὲ θῆλυ, ἣν δὴ καὶ θηλυκρανείαν καλοῦσιν. ἔχουσι δὲ φύλλον μὲν ἀμυγδαλῇ ὅμοιον, πλὴν λιπωδέστερον καὶ παχύτερον, φλοιὸν δ' ἰνώδη λεπτόν· τὸ δὲ στέλεχος οὐ παχὺ λίαν, ἀλλὰ παραφύει ῥάβδους ὥσπερ ἄγνος· ἐλάττους δὲ ἡ θηλυκρανεία καὶ θαμνωδέστερόν ἐστιν. τοὺς δὲ ὄζους ὁμοίως ἔχουσιν ἄμφω τῇ ἄγνῳ καὶ κατὰ δύο καὶ κατ'

---

[1] *cf.* Plin., *l.c.*, and Index.

[2] μεῖζον δὲ καὶ μανότερον conj. W. from G; μ. δὲ καὶ μανότερα MVU (? μανότερον); μείζων δὲ καὶ μακροτέρα Ald.H.

one 'ash' (manna-ash), the other 'horse-ash[1]' (ash). The 'horse-ash' is a larger and more spreading[2] tree, wherefore it is of less compact appearance. It is naturally a tree of the plains and rough, while the other belongs to the mountains and is smooth[3]; the one which grows on the mountains is fair-coloured smooth hard and stunted, while that of the plains is colourless spreading and rough. (In general one may say of trees that grow in the plain and on the mountain respectively, that the latter are of fair colour hard and smooth,[4] as beech elm and the rest; while those of the plain are more spreading, of less good colour and inferior, except the pear apple[5] and wild pear, according to the people of Mount Olympus. These when they grow in the plain are better both in fruit and in wood; for on the mountain they are rough spinous and much branched, in the plain smoother larger and with sweeter and fleshier fruit. However the trees of the plain are always of larger size.)

Of cornelian cherry, cornel, 'cedars,' medlar, thorns, sorb.

XII. Of the cornelian cherry there is a 'male' and a 'female' kind (cornel), and the latter bears a corresponding name. Both have a leaf like that of the almond, but oilier and thicker; the bark is fibrous and thin, the stem is not very thick, but it puts out side-branches like the chaste-tree, those of the 'female' tree, which is more shrubby, being fewer. Both kinds have branches like those of the chaste-tree,

³ καὶ τραχὺ . . . λεῖον conj. Sch.; καὶ λεῖον . . . τραχὺ Ald.
⁴ λεῖα conj. Mold.; λευκὰ Ald.G.
⁵ μηλέας conj. Scal., *cf.* 3. 3. 2; μελίας UMAld.H.

ἀλλήλους· το δὲ ξύλον τὸ μὲν τῆς κρανείας
ἀκάρδιον καὶ στερεὸν ὅλον, ὅμοιον κέρατι τὴν
πυκνότητα καὶ τὴν ἰσχύν, τὸ δὲ τῆς θηλυκρανείας
ἐντεριώνην ἔχον καὶ μαλακώτερον καὶ κοιλαινό-
2 μενον· δι' ὃ καὶ ἀχρεῖον εἰς τὰ ἀκόντια. τὸ δ'
ὕψος τοῦ ἄρρενος δώδεκα μάλιστα πηχέων, ἡλίκη
τῶν σαρισσῶν ἡ μεγίστη· τὸ γὰρ ὅλον στέλεχος
ὕψος οὐκ ἴσχει. φασὶ δ' οἱ μὲν ἐν τῇ Ἴδῃ τῇ Τρῳάδι
τὸ μὲν ἄρρεν ἄκαρπον[1] εἶναι τὸ δὲ θῆλυ κάρπιμον.
πυρῆνα δ' ὁ καρπὸς εχει παραπλήσιον ἐλάᾳ, καὶ
ἐσθιόμενος γλυκὺς καὶ εὐώδης· ἄνθος δὲ ὅμοιον
τῷ τῆς ἐλάας, καὶ ἀπανθεῖ δὲ καὶ καρποφορεῖ
τὸν αὐτὸν τρόπον τῷ ἐξ ἑνὸς μίσχου πλείους
ἔχειν, σχεδὸν δὲ καὶ τοῖς χρόνοις παραπλησίως.
οἱ δ' ἐν Μακεδονίᾳ καρποφορεῖν μὲν ἄμφω φασὶν
τὸν δὲ τῆς θηλείας ἄβρωτον εἶναι· τὰς ῥίζας δ'
ὁμοίας ἔχει ταῖς ἄγνοις ἰσχυρὰς καὶ ἀνωλέθρους.
γίνεται δὲ καὶ περὶ τὰ ἔφυδρα καὶ οὐκ ἐν τοῖς
ξηροῖς μόνον· φύεται δὲ καὶ ἀπὸ σπέρματος καὶ
ἀπὸ παρασπάδος.

3 Κέδρον δὲ οἱ μέν φασιν εἶναι διττήν, τὴν μὲν Λυκίαν τὴν δὲ Φοινικῆν, οἱ δὲ μονοειδῆ, καθάπερ οἱ ἐν τῇ Ἴδῃ· παρόμοιον δὲ τῇ ἀρκεύθῳ, διαφέρει δὲ μάλιστα τῷ φύλλῳ· τὸ μὲν γὰρ τῆς κέδρου σκληρὸν καὶ ὀξὺ καὶ ἀκανθῶδες, τὸ δὲ τῆς ἀρκεύθου μαλακώτερον· δοκεῖ δὲ καὶ ὑψηλοφυέστερον εἶναι ἡ ἄρκευθος· οὐ μὴν ἀλλ' ἔνιοί γε οὐ διαιροῦσι

[1] The Idaeans are evidently responsible for this statement. T. himself (3. 4. 3) says the fruit is inedible.

[2] But (1. 11. 4) only certain varieties of the olive are said to have this character: the next statement seems also inconsistent with 3. 4. 3. Perhaps T. is still reproducing his Idaean authority.

arranged in pairs opposite one another. The wood of the 'male' tree has no heart, but is hard throughout, like horn in closeness and strength; whereas that of the 'female' tree has heart-wood and is softer and goes into holes; wherefore it is useless for javelins. The height of the 'male' tree is at most twelve cubits, the length of the longest Macedonian spear, the stem up to the point where it divides not being very tall. The people of Mount Ida in the Troad say that the 'male' tree is barren, but that the 'female' bears fruit. The fruit has a stone like an olive and is sweet to the taste and fragrant[1]; the flower is like that of the olive, and the tree produces its flowers and fruit in the same manner, inasmuch as it has several growing from one stalk,[2] and they are produced at almost the same time in both forms. However the people of Macedonia say that both trees bear fruit, though that of the 'female' is uneatable, and the roots are like those of the chaste-tree, strong and indestructible. This tree grows in wet ground and not only[3] in dry places; and it comes from seed, and also can be propagated from a piece torn off.

[4] The 'cedar,' some say, has two forms, the Lycian and the Phoenician[5]; but some, as the people of Mount Ida, say that there is only one form. It resembles the *arkeuthos* (Phoenician cedar), differing chiefly in the leaf, that of 'cedar' being hard sharp and spinous, while that of *arkeuthos* is softer: the latter tree also seems to be of taller growth. However some do not give them distinct names, but call

[3] μόνον ins. R. Const. from G.
[4] Plin. 13. 52. See Index κέδρος and ἄρκευθος.
[5] Φοινικῆν: Φοινικικὴν conj. W. *cf.* 9. 2. 3; Plin. *l.c.*

τοῖς ὀνόμασιν ἀλλ' ἄμφω καλοῦσι κέδρους, πλὴν παρασήμως τὴν κέδρον ὀξύκεδρον. ὀζώδη δ' ἄμφω καὶ πολυμάσχαλα καὶ ἐπεστραμμένα ἔχοντα τὰ ξύλα· μήτραν δ' ἡ μὲν ἄρκευθος ἔχει μικρὰν καὶ πυκνὴν καὶ ὅταν κοπῇ ταχὺ σηπομένην· ἡ δὲ κέδρος τὸ πλεῖστον ἐγκάρδιον καὶ ἀσαπές, ἐρυθροκάρδια δ' ἄμφω· καὶ ἡ μὲν τῆς
4 κέδρου εὐώδης ἡ δὲ τῆς ἑτέρας οὔ. καρπὸς δ' ὁ μὲν τῆς κέδρου ξανθὸς μύρτου μέγεθος ἔχων εὐώδης ἡδὺς ἐσθίεσθαι. ὁ δὲ τῆς ἀρκεύθου τὰ μὲν ἄλλα ὅμοιος, μέλας δὲ καὶ στρυφνὸς καὶ ὥσπερ ἄβρωτος· διαμένει δ' εἰς ἐνιαυτόν, εἶθ' ὅταν ἄλλος ἐπιφυῇ ὁ περυσινὸς ἀποπίπτει. ὡς δὲ οἱ ἐν Ἀρκαδίᾳ λέγουσι, τρεῖς ἅμα καρποὺς ἴσχει, τόν τε περυσινὸν οὔπω πέπονα καὶ τὸν προπερύσινον ἤδη πέπονα καὶ ἐδώδιμον καὶ τρίτον τὸν νέον ὑποφαίνει. ἔφη δὲ Σάτυρος καὶ κομίσαι τοὺς ὀρεοτύπους αὐτῷ ἀνανθεῖς ἄμφω. τὸν δὲ φλοιὸν ὅμοιον ἔχει κυπαρίττῳ τραχύτερον δέ· ῥίζας δὲ μανὰς ἀμφότεραι καὶ ἐπιπολαίους. φύονται περὶ τὰ πετρώδη καὶ χειμέρια καὶ τούτους τοὺς τόπους ζητοῦσι.

5 Μεσπίλης δ' ἐστὶ τρία γένη, ἀνθηδὼν σατάνειος ἀνθηδονοειδής, ὡς οἱ περὶ τὴν Ἴδην διαιροῦσι. φέρει δὲ ἡ μὲν σατάνειος τὸν καρπὸν μείζω καὶ λευκότερον καὶ χαυνότερον καὶ τοὺς πυρῆνας ἔχοντα μαλακωτέρους· αἱ δ' ἕτεραι

---

1 παρασήμως τὴν κέδρον U; π. τὸν κέδρον M; Ald. omits the article; παρασημασίᾳ κέδρου conj W.

2 μήτραν conj. Sch.; μᾶλλον UMVAld. Plin., 16. 198, supports μήτραν: he apparently read μήτραν δ' ἡ μὲν ἀ. ἔχει μᾶλλον

them both 'cedar,' distinguishing them however as 'the cedar'[1] and 'prickly cedar.' Both are branching trees with many joints and twisted wood. On the other hand *arkeuthos* has only a small amount of close core,[2] which, when the tree is cut, soon rots, while the trunk of 'cedar' consists mainly of heart and does not rot. The colour of the heart in each case is red: that of the 'cedar' is fragrant, but not that of the other. The fruit of 'cedar' is yellow, as large as the myrtle-berry, fragrant, and sweet to the taste. That of *arkeuthos* is like it in other respects, but black, of astringent taste and practically uneatable; it remains on the tree for a year, and then, when another grows, last year's fruit falls off. According to the Arcadians it has three fruits on the tree at once, last year's, which is not yet ripe, that of the year before last which is now ripe and eatable, and it also shews the new fruit. Satyrus[3] said that the wood-cutters gathered him specimens of both kinds which were flowerless. The bark is[4] like that of the cypress but rougher. Both[5] kinds have spreading shallow roots. These trees grow in rocky cold parts and seek out such districts.

[6] There are three kinds of *mespile*, *anthedon* (oriental thorn), *sataneios* (medlar) and *anthedonoeides* (hawthorn), as the people of mount Ida distinguish them. [7] The fruit of the medlar is larger paler more spongy and contains softer stones; in the other

πυκνήν; but the words *καὶ ὅταν . . . σηπομένην* (which P. does not render) seem inconsistent. ? ins. *οὐ* before *ταχὺ* Sch.

3 ? An enquirer sent out by the Lyceum: see Intr. p. xxi.

4 *ἔχει* conj. W.; *ἐδόκει* Ald.

5 *ἀμφότεραι* conj. W.; *ἀμφοτέρας* U; *ἀμφοτέρους* Ald.H.

6 Plin. 15. 84.

7 *cf.* *C.P.* 2. 8. 2; 6. 14. 4; 6. 16. 1.

ἐλάττω τέ τι καὶ εὐωδέστερον καὶ στρυφνότερον, ὥστε δύνασθαι πλείω χρόνον θησαυρίζεσθαι. πυκνότερον δὲ καὶ τὸ ξύλον τούτων καὶ ξανθότερον, τὰ δ' ἄλλα ὅμοιον. τὸ δ' ἄνθος πασῶν ὅμοιον ἀμυγδαλῇ, πλὴν οὐκ ἐρυθρὸν ὥσπερ ἐκεῖνο ἀλλ' ἐγχλωρότερον. . . . . . . μεγέθει μέγα τὸ δένδρον καὶ περίκομον. φύλλον δὲ τὸ μὲν ἐπὶ . . . . . . πολυσχιδὲς δὲ καὶ ἐν ἄκρῳ σελινοειδές, τὸ δ' ἐπὶ τῶν παλαιοτέρων πολυσχιδὲς σφόδρα καὶ ἐγγωνοειδὲς μείζοσι σχίσμασι, τετανὸν ἰνῶδες λεπτότερον σελίνου καὶ προμηκέστερον καὶ τὸ ὅλον καὶ τὰ σχίσματα, περικεχαραγμένον δὲ ὅλον· μίσχον δ' ἔχει λεπτὸν μακρόν· πρὸ τοῦ φυλλορροεῖν δ' ἐρυθραίνεται σφόδρα. πολύρριζον δὲ τὸ δένδρον καὶ βαθύρριζον· δι' ὃ καὶ χρόνιον καὶ δυσώλεθρον. καὶ τὸ ξύλον ἔχει πυκνὸν καὶ
6 στερεὸν καὶ ἀσαπές. φύεται δὲ καὶ ἀπὸ σπέρματος καὶ ἀπὸ παρασπάδος. νόσημα δὲ αὐτῶν ἐστιν ὥστε γηράσκοντα σκωληκόβρωτα γίνεσθαι· καὶ οἱ σκώληκες μεγάλοι καὶ ἴδιοι ἢ οἱ ἐκ τῶν δένδρων τῶν ἄλλων.

Τῶν δ' οἰῶν δύο γένη ποιοῦσι, τὸ μὲν δὴ καρποφόρον θῆλυ τὸ δὲ ἄρρεν ἄκαρπον· οὐ μὴν ἀλλὰ διαφέρουσι τοῖς καρποῖς, τῷ τὰς μὲν στρογγύλον τὰς δὲ προμήκη τὰς δ' ᾠοειδῆ φέρειν. διαφέρουσι δὲ καὶ τοῖς χυλοῖς· ὡς γὰρ ἐπὶ τὸ

---

[1] ἐλάττω τέ τι conj. W.; ἐλάττω εἰσὶ UAld.

[2] W. suggests that some words are missing here, as it does not appear to which kind of μεσπίλη the following description belongs; hence various difficulties. See Sch.

[3] Probably a lacuna in the text. W. thus supplies the sense: he suggests σικυοειδές for σελινοειδές.

kinds it is somewhat smaller,[1] more fragrant and of more astringent taste, so that it can be stored for a longer time. The wood also of these kinds is closer and yellower, though in other respects it does not differ. The flower in all the kinds is like the almond flower, except that it is not pink, as that is, but greenish . . . . .[2] In stature the tree is large and it has thick foliage. The leaf in the young tree is round[3] but much divided and like celery at the tip; but the leaf of older trees is very much divided and forms angles with larger divisions; it is smooth[4] fibrous thinner and more oblong than the celery leaf, both as a whole and in its divisions, and it has a jagged edge all round.[5] It has a long thin stalk, and the leaves turn bright red before they are shed. The tree has many roots, which run deep; wherefore it lives a long time and is hard to kill. The wood is close and hard and does not rot. The tree grows from seed and also from a piece torn off. It is subject to a disease which causes it to become worm-eaten[6] in its old age, and the worms are large and different[7] to those engendered by other trees.

[8] Of the sorb they make two kinds, the 'female' which bears fruit and the 'male' which is barren. There are moreover differences in the fruit of the 'female' kind; in some forms it is round, in others oblong and egg-shaped. There are also differences

[4] τετανὸν: *cf.* 3. 11. 1; 3. 15. 6.

[5] περικεχαραγμένον conj. Scal.; περικεθαρμένον U; περικεκαρμένον MVAld. *cf.* allusions to the leaf of μεσπίλη, 3. 13. 1; 3. 15. 6.

[6] *cf.* 4. 14. 10; Plin. 17. 221; Pall. 4. 10.

[7] ἴδιοι Ald. (for construction *cf.* Plat. *Gorg.* 481 c); ἰδίους UMV (the first ι corrected in U). W. adopts Sch.'s conj., ἡδίους, in allusion to the edible *cossus*: *cf.* Plin. *l.c.*

[8] Plin. 15. 85.

πᾶν εὐωδέστερα καὶ γλυκύτερα τὰ στρογγύλα,
τὰ δ' ᾠοειδῆ πολλάκις ἐστὶν ὀξέα καὶ ἧττον
7 εὐώδη. φύλλα δ' ἀμφοῖν κατὰ μίσχον μακρὸν
ἰνοειδῆ πεφύκασι στοιχηδὸν ἐκ τῶν πλαγίων
πτερυγοειδῶς, ὡς ἑνὸς ὄντος τοῦ ὅλου λοβοὺς δὲ
ἔχοντος ἐσχισμένους ἕως τῆς ἰνός· πλὴν διεστᾶσιν
ἀφ' ἑαυτῶν ὑπόσυχνον τὰ κατὰ μέρος· φυλλοβολεῖ δὲ οὐ κατὰ μέρος ἀλλὰ ὅλον ἅμα τὸ
πτερυγῶδες. εἰσὶ δὲ περὶ μὲν τὰ παλαιότερα
καὶ μακρότερα πλείους αἱ συζυγίαι, περὶ δὲ τὰ
νεώτερα καὶ βραχύτερα ἐλάττους, πάντων δὲ ἐπ'
ἄκρου τοῦ μίσχου φύλλον περιττόν, ὥστε καὶ
πάντ' εἶναι περιττά. τῷ δὲ σχήματι δαφνοειδῆ
τῆς λεπτοφύλλου, πλὴν χαραγμὸν ἔχοντα καὶ
βραχύτερα καὶ οὐκ εἰς ὀξὺ τὸ ἄκρον συνῆκον
ἀλλ' εἰς περιφερέστερον. ἄνθος δὲ ἔχει βοτρυῶδες ἀπὸ μιᾶς κορύνης ἐκ πολλῶν μικρῶν καὶ
8 λευκῶν συγκείμενον. καὶ ὁ καρπὸς ὅταν εὐκαρπῇ
βοτρυώδης· πολλὰ γὰρ ἀπὸ τῆς αὐτῆς κορύνης,
ὥστ' εἶναι καθάπερ κηρίον. σκωληκόβορος ἐπὶ
τοῦ δένδρου ὁ καρπὸς ἄπεπτος ὢν ἔτι γίνεται
μᾶλλον τῶν μεσπίλων καὶ ἀπίων καὶ ἀχράδων·
καίτοι πολὺ στρυφνότατος. γίνεται δὲ καὶ αὐτὸ
τὸ δένδρον σκωληκόβρωτον καὶ οὕτως αὐαίνεται
γηράσκον· καὶ ὁ σκώληξ ἴδιος ἐρυθρὸς δασύς.
καρποφορεῖ δ' ἐπιεικῶς νέα· τριετὴς γὰρ εὐθὺς
φύει. τοῦ μετοπώρου δ' ὅταν ἀποβάλῃ τὸ φύλλον,
εὐθὺς ἴσχει τὴν καχρυώδη κορύνην λιπαρὰν καὶ

---

[1] φύλλα . . . στοιχηδὸν conj. W.; φύλλον δ' ἀμφοῖν τὸ μὲν μίσχον μακρὸν ἰνοειδῆ· πεφ. [δὲ] στοιχηδὸν UMVAld.

[2] ἀφ' ἑαυτῶν (=ἀπ' ἀλλήλων) conj. Scal.; ἀπ' αὐτῶν U: so W., who however renders *inter se*.

in taste; the round fruits are generally more fragrant and sweeter, the oval ones are often sour and less fragrant. The leaves in both grow attached to a long fibrous stalk, and project on each side in a row[1] like the feathers of a bird's wing, the whole forming a single leaf but being divided into lobes with divisions which extend to the rib; but each pair are some distance apart,[2] and, when the leaves fall,[3] these divisions do not drop separately, but the whole wing-like structure drops at once. When the leaves are older and longer, the pairs are more numerous; in the younger and shorter leaves they are fewer; but in all at the end of the leaf-stalk there is an extra leaflet, so that the total number of leaflets is an odd number. In form the leaflets resemble[4] the leaves of the 'fine-leaved' bay, except that they are jagged and shorter and do not narrow to a sharp point but to a more rounded end. The flower[5] is clustering and made up of a number of small white blossoms from a single knob. The fruit too is clustering, when the tree fruits well; for a number of fruits are formed from the same knob, giving an appearance like a honeycomb. The fruit gets eaten by worms on the tree before it is ripe to a greater extent than that of medlar pear or wild pear, and yet it is much more astringent than any of these. The tree itself also gets worm-eaten, and so withers away as it ages; and the worm[6] which infests it is a peculiar one, red and hairy. This tree bears fruit when it is quite young, that is as soon as it is three years old. In autumn, when it has shed its leaves, it immediately produces its winter-bud-like knob,[7]

[3] Plin. 16. 92. [4] For construction *cf.* 3. 11. 3.
[5] *i.e.* inflorescence. [6] Plin. 17. 221. [7] *cf.* 3. 5. 5.

ἐπωδηκυῖαν ὡσὰν ἤδη βλαστικόν, καὶ διαμένει
9 τὸν χειμῶνα. ἀνάκανθον δέ ἐστι καὶ ἡ οἴη καὶ ἡ μεσπίλη· φλοιὸν δ' ἔχει λεῖον ὑπολίπαρον, ὅσαπερ μὴ γεράνδρυα, τὴν δὲ χρόαν ξανθὸν ἐπιλευκαίνοντα· τὰ δὲ γεράνδρυα τραχὺν καὶ μέλανα. τὸ δὲ δένδρον εὐμέγεθες ὀρθοφυὲς εὔρυθμον τῇ κόμῃ· σχεδὸν γὰρ ὡς ἐπὶ τὸ πολὺ στροβιλοειδὲς σχῆμα λαμβάνει κατὰ τὴν κόμην, ἐὰν μή τι ἐμποδίσῃ· τὸ δὲ ξύλον στερεὸν πυκνὸν ἰσχυρὸν εὔχρουν, ῥίζας δὲ οὐ πολλὰς μὲν οὐδὲ κατὰ βάθους, ἰσχυρὰς δὲ καὶ παχείας καὶ ἀνωλέθρους ἔχει. φύεται δὲ καὶ ἀπὸ ῥίζης καὶ ἀπὸ παρασπάδος καὶ ἀπὸ σπέρματος· τόπον δὲ ζητεῖ ψυχρὸν ἔνικμον, φιλόζωον δ' ἐν τούτῳ καὶ δυσώλεθρον· οὐ μὴν ἀλλὰ καὶ φύεται ἐν τοῖς ὄρεσιν.

XIII. Ἴδιον δὲ τῇ φύσει δένδρον ὁ κέρασός ἐστι· μεγέθει μὲν μέγα· καὶ γὰρ εἰς τέτταρας καὶ εἴκοσι πήχεις· ἔστι δ' ὀρθοφυὲς σφόδρα· πάχος δὲ ὥστε καὶ δίπηχυν τὴν περίμετρον ἀπὸ τῆς ῥίζης ἔχειν. φύλλον δ' ὅμοιον τῷ τῆς μεσπίλης σκληρὸν δὲ σφόδρα καὶ παχύτερον, ὥστε τῇ χροιᾷ πόρρωθεν φανερὸν εἶναι τὸ δένδρον. φλοιὸν δὲ τὴν λειότητα καὶ τὴν χρόαν καὶ τὸ πάχος ὅμοιον φιλύρᾳ, δι' ὃ καὶ τὰς κίστας ἐξ αὐτοῦ ποιοῦσιν ὥσπερ καὶ ἐκ τοῦ τῆς φιλύρας. περιπέφυκε δὲ οὗτος οὔτε ὀρθοφυὴς οὔτε κύκλῳ κατ' ἶσον ἀλλ' ἑλικηδὸν περιείληφε κάτωθεν ἄνω

---

1 ὅσαπερ μὴ conj. Bod.; ὥσπερ τὰ Ald.; ὥστε τὰ M.
2 κόμην Ald.H.; κορυφὴν conj. Sch.; *vertice* G.
3 Plin. 16. 125; *cf.* 16. 74; 17. 234.
4 παχύτερον: so quoted by Athen. 2. 34; πλατύτερον MSS

which is glistening and swollen as though the tree were just about to burst into leaf, and this persists through the winter. The sorb, like the medlar, is thornless; it has smooth rather shiny bark, (except when[1] the tree is old), which in colour is a whitish yellow; but in old trees it is rough and black. The tree is of a good size, of erect growth and with well balanced foliage; for in general it assumes a cone-like shape as to its foliage,[2] unless something interferes. The wood is hard close strong and of a good colour; the roots are not numerous and do not run deep, but they are strong and thick and indestructible. The tree grows from a root, from a piece torn off, or from seed, and seeks a cold moist position; in such a position it is tenacious of life and hard to kill: however it also grows on mountains.

Of bird-cherry, elder, willow.

XIII. [3] The *kerasos* (bird-cherry) is peculiar in character; it is of great stature, growing as much as twenty-four cubits high; and it is of very erect growth; as to thickness, it is as much as two cubits in circumference at the base. The leaves are like those of the medlar, but very tough and thicker,[4] so that the tree is conspicuous by its colour from a distance. The bark [5] in smoothness colour and thickness is like that of the lime; wherefore men make their writing-cases [6] from it, as from the bark of that tree. [7] This bark does not grow straight nor evenly all round the tree, but runs round it [8] in a spiral

[5] *cf.* 4. 15. 1; Hesych. *s.v.* κέρασος.
[6] *cf.* 3. 10. 4; Ar. *Vesp.* 529.
[7] περιπέφυκε . . . περιπεφυκός: text as restored by Sch. and others, following U as closely as possible.
[8] περιείληφε conj. R. Const.

προσάγων, ὥσπερ ἡ διαγραφὴ τῶν φύλλων· και
λοπιζόμενος οὗτος ἐκδέρεται, ἐκεῖνος δ' ἐπίτομος
2 γίνεται καὶ οὐ δύναται· μέρος δ' αὐτοῦ τι τὸν
αὐτὸν τρόπον ἀφαιρεῖται κατὰ πάχος σχιζόμενον
λεπτὸν ὡς ἂν φύλλον, τὸ δὲ λοιπὸν προσμένειν
τε δύναται καὶ σώζει τὸ δένδρον ὡσαύτως περιπεφυκός. περιαιρουμένου δὲ ὅταν λοπᾷ τοῦ
φλοιοῦ συνεκραίνει καὶ τότε τὴν ὑγρότητα· καὶ
ὅταν ὁ ἔξω χιτὼν περιαιρεθῇ, μόνον ὁ ὑπολιπὴς
ἐπιμελαίνεται ὥσπερ μυξώδει ὑγρασίᾳ, καὶ πάλιν
ὑποφύεται τῷ δευτέρῳ ἔτει χιτὼν ἄλλος ἀντ'
ἐκείνου πλὴν λεπτότερος. πέφυκε καὶ τὸ ξύλον
ὅμοιον ταῖς ἰσὶ τῷ φλοιῷ στρεπτῶς ἑλιττόμενον·
καὶ οἱ ῥάβδοι φύονται τὸν αὐτὸν τρόπον εὐθύς·
τοὺς ὄζους δ' αὐξανομένου συμβαίνει τοὺς μὲν
3 κάτω ἀεὶ ἀπόλλυσθαι τοὺς δ' ἄνω αὔξειν. τὸ δ'
ὅλον οὐ πολύοζον τὸ δένδρον ἀλλ' ἀνοζότερον
πολὺ τῆς αἰγείρου. πολύρριζον δὲ καὶ ἐπιπολαιόρριζον οὐκ ἄγαν δὲ παχύρριζον· ἡ δ'
ἐπιστροφὴ καὶ τῆς ῥίζης καὶ τοῦ φλοιοῦ τοῦ περὶ
αὐτὴν ἡ αὐτή. ἄνθος δὲ λευκὸν ἀπίῳ καὶ μεσπίλῃ
ὅμοιον, ἐκ μικρῶν ἀνθῶν συγκείμενον κηριῶδες.
ὁ δὲ καρπὸς ἐρυθρὸς ὅμοιος διοσπύρῳ τὸ σχῆμα,
τὸ δὲ μέγεθος ἡλίκον κύαμος, πλὴν τοῦ διοσπύρου
μὲν ὁ πυρὴν σκληρὸς τοῦ δὲ κεράσου μαλακός.
φύεται δ' ὅπου καὶ ἡ φίλυρα, τὸ δὲ ὅλον ὅπου
ποταμοὶ καὶ ἔφυδρα.

4 Φύεται δὲ καὶ ἡ ἀκτὴ μάλιστα παρ' ὕδωρ καὶ

---

[1] Which is an ellipse, the segment of a cylinder: so Sch. explains.

[2] ἐκεῖνος: *i.e.* lower down the trunk, where the spiral is less open. [3] ἐπίτομος: *cf.* 5. 1. 12.

(which becomes closer as it gets higher up the tree) like the outline of the leaves.[1] And this part of it can be stripped off by peeling, whereas with the other part[2] this is not possible and it has to be cut in short lengths.[3] In the same manner part is removed by being split off in flakes as thin as a leaf, while the rest can be left and protects the tree, growing about it as described. If the bark is stripped off when the tree is peeling, there is also at the time a discharge of the sap; further, when only the outside coat is stripped off, what remains turns black with a kind[4] of mucus-like moisture; and in the second year another coat grows to replace what is lost, but this is thinner. The wood in its fibres is like the bark, twisting spirally,[5] and the branches grow in the same manner from the first; and, as the tree grows, it comes to pass that the lower branches keep on perishing, while the upper ones increase. However the whole tree is not much branched, but has far fewer branches than the black poplar. Its roots are numerous and shallow and not very thick; and there is a similar twisting of the root and of the bark which surrounds it. [6]The flower is white, like that of the pear and medlar, composed of a number of small blossoms arranged like a honeycomb. The fruit is red, like that of *diospyros* in shape, and in size it is as large as a bean. However the stone of the *diospyros* fruit is hard, while that of the bird-cherry is soft. The tree grows where the lime grows, and in general where there are rivers and damp places.

[7]The elder also grows chiefly by water and in shady

[4] ὥσπερ conj. Sch.; περ MV; πως Ald.H.
[5] στρεπτῶς ἑλιττόμενον conj. Sch.; στρεπτῷ ἑλιττομένωι U; στρεπτῷ ἑλιττομένῳ Ald. [6] *cf.* 3. 12. 7. [7] Plin. 17. 151.

ἐν τοῖς σκιεροῖς, οὐ μὴν ἀλλὰ καὶ ἐν τοῖς μὴ τοιούτοις· θαμνῶδες δὲ ῥάβδοις ἐπετείοις αὐξανομέναις μέχρι τῆς φυλλορροίας εἰς μῆκος, εἶτα μετὰ ταῦτα εἰς πάχος· τὸ δὲ ὕψος τῶν ῥάβδων οὐ μέγα λίαν ἀλλὰ καὶ μάλιστα ὡς ἑξάπηχυ· τῶν δὲ στελεχῶν πάχος τῶν γερανδρύων ὅσον περικεφαλαίας, φλοιὸς δὲ λεῖος λεπτὸς καπυρός· τὸ δὲ ξύλον χαῦνον καὶ κοῦφον ξηρανθέν, ἐντεριώνην δὲ ἔχον μαλακήν, ὥστε δι' ὅλου καὶ κοιλαίνεσθαι τὰς ῥάβδους, ἐξ ὧν καὶ τὰς βακτηρίας ποιοῦσι τὰς κούφας. ξηρανθὲν δὲ ἰσχυρὸν καὶ ἀγήρων ἐὰν βρέχηται, κἂν ᾖ λελοπισμένον· λοπίζεται δὲ αὐτόματον ξηραινόμενον. ῥίζας δὲ ἔχει μετεώρους οὐ πολλὰς δὲ οὐδὲ μεγάλας.

5 φύλλον δὲ τὸ μὲν καθ' ἕκαστον μαλακόν, πρόμηκες ὡς τὸ τῆς πλατυφύλλου δάφνης, μεῖζον δὲ καὶ πλατύτερον καὶ περιφερέστερον ἐκ μέσου καὶ κάτωθεν, τὸ δ' ἄκρον εἰς ὀξὺ μᾶλλον συνῆκον κύκλῳ δ' ἔχον χαραγμόν· τὸ δὲ ὅλον, περὶ ἕνα μίσχον παχὺν καὶ ἰνώδη ὡσὰν κλωνίον τὰ μὲν ἔνθεν τὰ δὲ ἔνθεν κατὰ γόνυ καὶ συζυγίαν πεφύκασι τῶν φύλλων διέχοντα ἀπ' ἀλλήλων, ἓν δὲ ἐξ ἄκρου τοῦ μίσχου. ὑπέρυθρα δὲ τὰ φύλλα ἐπιεικῶς καὶ χαῦνα καὶ σαρκώδη· φυλλορροεῖ δὲ τοῦτο ὅλον, διόπερ φύλλον ἄν τις εἴποι τὸ ὅλον. ἔχουσι δὲ καὶ οἱ κλῶνες οἱ νέοι γωνοειδῆ τινα.

6 τὸ δ' ἄνθος λευκὸν ἐκ μικρῶν λευκῶν πολλῶν ἐπὶ τῇ τοῦ μίσχου σχίσει κηριῶδες· εὐωδίαν

---

[1] περικεφαλαίας, some part of a ship's prow: so Pollux.
[2] καπυρός conj. Sch.; καὶ πυρσός U (?); καὶ πυρρός V; καὶ πουρός M.
[3] Sc. pith.

places, but likewise in places which are not of this character. It is shrubby, with annual branches which go on growing in length till the fall of the leaf, after which they increase in thickness. The branches do not grow to a very great height, about six cubits at most. The thickness of the stem of old trees is about that of the 'helmet'[1] of a ship; the bark is smooth thin and brittle[2]; the wood is porous and light when dried, and has a soft heart-wood,[3] so that the boughs are hollow right through, and men make of them their light walking-sticks. When dried it is strong and durable if it is soaked, even if it is stripped of the bark; and it strips itself of its own accord as it dries. The roots are shallow and neither numerous nor large. The single leaflet is soft and oblong, like the leaf of the 'broad-leaved' bay, but larger broader and rounder at the middle and base, though the tip narrows more to a point and is jagged[4] all round. The whole leaf is composed of leaflets growing about a single thick fibrous stalk, as it were, to which they are attached at either side in pairs at each joint; and they are separate from one another, while one is attached to the tip of the stalk. The leaves are somewhat reddish porous and fleshy: the whole is shed in one piece; wherefore one may consider the whole structure as a 'leaf.'[5] The young twigs too have certain crooks[6] in them. The flower[7] is white, made up of a number of small white blossoms attached to the point where the stalk divides, in form like a honeycomb, and it has the heavy

[4] χαραγμόν conj. R. Const. from G; παραγμόν UMV; σπαραγμόν Ald. [5] *cf.* 3. 11. 3 n.

[6] γωνοειδῆ U; ? γωνιοειδῆ; G seems to have read γονατοειδῆ; Sch. considers the text defective or mutilated.

[7] *cf.* 3. 12. 7 n.

δὲ ἔχει λειριώδη ἐπιβαρεῖαν. ἔχει δὲ καὶ τὸν καρπὸν ὁμοίως πρὸς ἑνὶ μίσχῳ παχεῖ βοτρυώδη δέ· γίνεται δὲ καταπεπαινόμενος μέλας, ὠμὸς δὲ ὢν ὀμφακώδης· μεγέθει δὲ μικρῷ μείζων ὀρόβου· τὴν ὑγρασίαν δὲ οἰνώδη τῇ ὄψει· καὶ τὰς χεῖρας τελειούμενοι βάπτονται καὶ τὰς κεφαλάς· ἔχει δὲ καὶ τὰ ἐντὸς σησαμοειδῆ τὴν ὄψιν.

Πάρυδρον δὲ καὶ ἡ ἰτέα καὶ πολυειδές· ἡ μὲν μέλαινα καλουμένη τῷ τὸν φλοιὸν ἔχειν μέλανα καὶ φοινικοῦν, ἡ δὲ λευκὴ τῷ λευκόν. καλλίους δὲ ἔχει τὰς ῥάβδους καὶ χρησιμωτέρας εἰς τὸ πλέκειν ἡ μέλαινα, ἡ δὲ λευκὴ καπυρωτέρας. ἔστι δὲ καὶ τῆς μελαίνης καὶ τῆς λευκῆς ἔνιον γένος μικρὸν καὶ οὐκ ἔχον αὔξησιν εἰς ὕψος· ὥσπερ καὶ ἐπ' ἄλλων τοῦτο δένδρων, οἷον κέδρου φοίνικος. καλοῦσι δ' οἱ περὶ Ἀρκαδίαν οὐκ ἰτέαν ἀλλὰ ἑλίκην τὸ δένδρον· οἴονται δέ, ὥσπερ ἐλέχθη· καὶ καρπὸν ἔχειν αὐτὴν γόνιμον.

XIV. Ἔστι δὲ τῆς πτελέας δύο γένη, καὶ τὸ μὲν ὀρειπτελέα καλεῖται τὸ δὲ πτελέα· διαφέρει δὲ τῷ θαμνωδέστερον εἶναι τὴν πτελέαν εὐαυξέστερον δὲ τὴν ὀρειπτελέαν. φύλλον δὲ ἀσχιδὲς περικεχαραγμένον ἡσυχῇ, προμηκέστερον δὲ τοῦ τῆς ἀπίου·

[1] καταπεπαινόμενος conj. W.; καὶ πεπ. VAld.

[2] καὶ . . . βάπτονται I conj., following Scal., W., etc., but keeping closer to U: certain restoration perhaps impossible; καὶ τὰς χεῖρας τελείους ἀναβλάστει δὲ καὶ τὰς κεφαλάς U; χεῖρας δὲ τελείυυς· ἀναβλασεῖ MV; om. G.

[3] Plin. 16. 174 and 175.

fragrance of lilies. The fruit is in like manner attached to a single thick stalk, but in a cluster: as it becomes quite ripe,[1] it turns black, but when unripe it is like unripe grapes; in size the berry is a little larger than the seed of a vetch; the juice is like wine in appearance, and in it men bathe[2] their hands and heads when they are being initiated into the mysteries. The seeds inside the berry are like sesame.

[3] The willow also grows by the water, and there are many kinds. There is that which is called the black willow[4] because its bark is black and red, and that which is called the white[4] from the colour of its bark. The black kind has boughs which are fairer and more serviceable for basket-work, while those of the white are more brittle.[5] There is a form both of the black and of the white which is small and does not grow to a height,—just as there are dwarf forms of other trees, such as prickly cedar and palm. The people of Arcadia call the tree[6] not 'willow' but *helike*: they believe, as was said,[7] that it bears fruitful seed.

Of elm, poplars, alder, [semyda, *bladder-senna*].

XIV. [8] Of the elm there are two kinds, of which one is called the 'mountain elm,' the other simply the 'elm': the difference is that the latter is shrubbier, while the mountain elm grows more vigorously. The leaf is undivided and slightly jagged, longer than that of the pear, but rough

[4] See Index.
[5] καπυρωτέρας conj. Sch.; καὶ πυρωτέρας U; καὶ πυροτέρας MVAld. *cf.* 3. 13. 4.
[6] Sc. ἰτέα generally. [7] 3. 1. 2. [8] Plin. 16. 72.

τραχυ δὲ και οὐ λεῖον. μέγα δὲ τὸ δένδρον καὶ τῷ ὕψει καὶ τῷ μεγέθει. πολὺ δ' οὐκ ἔστι περὶ τὴν Ἴδην ἀλλὰ σπάνιον· τόπον δὲ ἔφυδρον φιλεῖ. τὸ δὲ ξύλον ξανθὸν καὶ ἰσχυρὸν καὶ εὔινον καὶ γλίσχρον· ἅπαν γὰρ καρδία· χρῶνται δ' αὐτῷ καὶ πρὸς θυρώματα πολυτελῆ, καὶ χλωρὸν μὲν εὔτομον ξηρὸν δὲ δύστομον. ἄκαρπον δὲ νομίζουσιν, ἀλλ' ἐν ταῖς κωρυκίσι τὸ κόμμι καὶ θηρί' ἄττα κωνωποειδῆ φέρει. τὰς δὲ κάχρυς ἰδίας ἴσχει τοῦ μετοπώρου πολλὰς καὶ μικρὰς καὶ μελαίνας, ἐν δὲ ταῖς ἄλλαις ὥραις οὐκ ἐπέσκεπται.

2 Ἡ δὲ λεύκη καὶ ἡ αἴγειρος μονοειδής, ὀρθοφυῆ δὲ ἄμφω, πλὴν μακρότερον πολὺ καὶ μανότερον καὶ λειότερον ἡ αἴγειρος, τὸ δὲ σχῆμα τῶν φύλλων παρόμοιον. ὅμοιον δὲ καὶ τὸ ξύλον τεμνόμενον τῇ λευκότητι. καρπὸν δ' οὐδέτερον τούτων οὐδὲ ἄνθος ἔχειν δοκεῖ.

Ἡ κερκὶς δὲ παρόμοιον τῇ λεύκῃ καὶ τῷ μεγέθει καὶ τῷ τοὺς κλάδους ἐπιλεύκους ἔχειν· τὸ δὲ φύλλον κιττῶδες μὲν ἀγώνιον δὲ ἐκ τοῦ ἄλλου, τὴν δὲ μίαν προμήκη καὶ εἰς ὀξὺ συνήκουσαν· τῷ δὲ χρώματι σχεδὸν ὅμοιον τὸ ὕπτιον καὶ τὸ πρανές· μίσχῳ δὲ προσηρτημένον μακρῷ καὶ λεπτῷ, δι' ὃ καὶ οὐκ ὀρθὸν ἀλλ' ἐγκεκλιμένον. φλοιὸν δὲ τραχύτερον τῆς λεύκης καὶ μᾶλλον ὑπόλεπρον, ὥσπερ ὁ τῆς ἀχράδος. ἄκαρπον δέ.

3 Μονογενὲς δὲ καὶ ἡ κλήθρα· φύσει δὲ καὶ

---

[1] γλίσχρον conj. St.; αἰσχρόν Ald.H. *cf.* 5. 3. 4.
[2] *cf.* 5. 5. 2.
[3] *cf.* τὸ θυλακῶδες τοῦτο, 3. 7. 3; 2. 8. 3 n.; 9. 1. 2.

rather than smooth. The tree is large, being both tall and wide-spreading. It is not common about Ida, but rare, and likes wet ground. The wood is yellow strong fibrous and tough[1]; for it is all heart. Men use it for expensive doors[2]: it is easy to cut when it is green, but difficult when it is dry. The tree is thought to bear no fruit, but in the 'wallets'[3] it produces its gum and certain creatures like gnats; and it has in autumn its peculiar 'winter-buds'[4] which are numerous small and black, but these have not been observed at other seasons.

The abele and the black poplar have each but a single kind: both are of erect growth, but the black poplar is much taller and of more open growth, and is smoother, while the shape of its leaves is similar to those of the other. The wood also of both, when cut, is much the same in whiteness. Neither of these trees appears to have fruit or flower.[5]

The aspen is a tree resembling the abele both in size and in having whitish branches, but the leaf is ivy-like: while however it is otherwise without angles, its one angular[6] projection is long and narrows to a sharp point: in colour the upper and under sides are much alike. The leaf is attached to a long thin stalk: wherefore the leaf is not set straight, but has a droop.[7] The bark of the abele is rougher and more scaly, like that of the wild pear, and it bears no fruit.

The alder also has but one form: in growth it is

[4] κάχρυς, here probably a gall, mistaken for winter-bud.
[5] *cf.*, however, 3. 3. 4; 4. 10. 2, where T. seems to follow a different authority.
[6] Supply γωνίαν from ἀγώνιον·
[7] ἐγκεκλιμένον: sc. is not in line with the stalk.

ὀρθοφυές, ξύλον δ' ἔχον μαλακὸν καὶ ἐντεριώνην μαλακήν, ὥστε δι' ὅλου κοιλαίνεσθαι τὰς λεπτὰς ῥάβδους. φύλλον δ' ὅμοιον ἀπίῳ, πλὴν μεῖζον καὶ ἰνωδέστερον. τραχύφλοιον δὲ καὶ ὁ φλοιὸς ἔσωθεν ἐρυθρός, δι' ὃ καὶ βάπτει τὰ δέρματα. ῥίζας δὲ ἐπιπολαίους . . . ἡλίκον δάφνης. φύεται δὲ ἐν τοῖς ἐφύδροις ἀλλόθι δ' οὐδαμοῦ.

4 [Σημύδα δὲ τὸ μὲν φύλλον ἔχει ὅμοιον τῇ Περσικῇ καλουμένῃ καρύᾳ πλὴν μικρῷ στενότερον, τὸν φλοιὸν δὲ ποικίλον, ξύλον δὲ ἐλαφρόν· χρήσιμον δὲ εἰς βακτηρίας μόνον εἰς ἄλλο δὲ οὐδέν.

Ἡ δὲ κολυτέα ἔχει τὸ μὲν φύλλον ἐγγὺς τοῦ τῆς ἰτέας, πολύοζον δὲ καὶ πολύφυλλον καὶ τὸ δένδρον ὅλως μέγα· τὸν δὲ καρπὸν ἔλλοβον, καθάπερ τὰ χεδροπά· λοβοῖς γὰρ πλατέσι καὶ οὐ στενοῖς τὸ σπερμάτιον τὸ ἐνὸν μικρὸν καὶ οὐ μέγα· σκληρὸν δὲ μετρίως οὐκ ἄγαν· οὐδὲ πολύκαρπον ὡς κατὰ μέγεθος. σπάνιον δὲ τὸ ἐν λοβοῖς ἔχειν τὸν καρπόν· ὀλίγα γὰρ τοιαῦτα τῶν δένδρων.]

XV. Ἡ δὲ Ἡρακλεωτικὴ καρύα—φύσει γὰρ καὶ τοῦτ' ἄγριον τῷ τε μηδὲν ἢ μὴ πολὺ χείρω γίνεσθαι <ἢ> τῶν ἡμέρων τὸν καρπόν, καὶ τῷ δύνασθαι χειμῶνας ὑποφέρειν καὶ τῷ πολὺ φύεσθαι κατὰ τὰ ὄρη καὶ πολύκαρπον ἐν τοῖς ὀρείοις· ἔτι δὲ τῷ μηδὲ στελεχῶδες ἀλλὰ θαμ-

---

[1] Part of the description of the flower, and perhaps of the fruit, seems to be missing. Sch.

[2] *cf.* 4. 8. 1; but in 1. 4. 3 the alder is classed with 'amphibious' trees, and in 3. 3. 1 with 'trees of the plain.'

[3] *Betulam*, G from Plin. 16. 74.

also erect, and it has soft wood and a soft heart-wood, so that the slender boughs are hollow throughout. The leaf is like that of the pear, but larger and more fibrous. It has rough bark, which on the inner side is red: wherefore it is used for dyeing hides. It has shallow roots . . . [1] the flower is as large as that of the bay. It grows in wet places [2] and nowhere else.

The *semyda* [3] has a leaf like that of the tree called the 'Persian nut' (walnut), but it is rather narrower: the bark is variegated and the wood light: it is only of use for making walking-sticks and for no other purpose.

The bladder-senna [4] has a leaf near that of the willow, but is many-branched and has much foliage; and the tree altogether is a large one. The fruit is in a pod, as in leguminous plants: the pods in fact are broad rather than narrow, and the seed in them is comparatively small, and is moderately hard, but not so very hard. For its size the tree does not bear much fruit. It is uncommon to have the fruit in a pod; in fact there are few such trees.

Of filbert, terebinth, box, krataigos.

XV. The filbert is also naturally a wild tree, in that its fruit is little, if at all, inferior to that of the tree in cultivation, that it can stand winter, that it grows commonly on the mountains, and that it bears abundance of fruit in mountain regions [5]; also because it does not make a trunk, but is shrubby with

[4] Sch. remarks that the description of κολυτέα is out of place: *cf.* 3. 17. 2. W. thinks the whole section spurious. The antitheses in the latter part suggest a different context, in which κολυτέα was described by comparison with some other tree. [5] ὀρείοις conj. W.; φοραῖς Ald.

νῶδες εἶναι ῥάβδοις ἄνευ μασχαλῶν καὶ ἀνόζοις μακραῖς δὲ καὶ παχείαις ἐνίαις·—οὐ μὴν ἀλλὰ καὶ ἐξημεροῦται. διαφορὰν δὲ ἔχει τῷ τὸν καρπὸν ἀποδιδόναι βελτίω καὶ μεῖζον τὸ φύλλον· κεχαραγμένον δ' ἀμφοῖν· ὁμοιότατον τὸ τῆς κλήθρας, πλὴν πλατύτερον καὶ αὐτὸ τὸ δένδρον μεῖζον. καρπιμώτερον δ' αἰεὶ γίνεται κατα-
2 κοπτόμενον τὰς ῥάβδους. γένη δὲ δύο ἀμφοῖν· αἱ μὲν γὰρ στρογγύλον αἱ δὲ πρόμακρον φέρουσι τὸ κάρυον· ἐκλευκότερον δὲ τὸ τῶν ἡμέρων. καὶ καλλικαρπεῖ μάλιστά γ' ἐν τοῖς ἐφύδροις. ἐξημεροῦται δὲ τὰ ἄγρια μεταφυτευόμενα. φλοιὸν δ' ἔχει λεῖον ἐπιπόλαιον λεπτὸν λιπαρὸν ἰδίως στιγμὰς λευκὰς ἔχοντα ἐν αὐτῷ· τὸ δὲ ξύλον σφόδρα γλίσχρον, ὥστε καὶ τὰ λεπτὰ πάνυ ῥαβδία περιλοπίσαντες κανέα ποιοῦσι, καὶ τὰ παχέα δὲ καταξύσαντες. ἔχει δὲ καὶ ἐντεριώνην λεπτὴν ξανθήν, ᾗ κοιλαίνεται. ἴδιον δ' αὐτῶν τὸ περὶ τὸν ἴουλον, ὥσπερ εἴπομεν.

3 Τῆς δὲ τερμίνθου τὸ μὲν ἄρρεν τὸ δὲ θῆλυ. τὸ μὲν οὖν ἄρρεν ἄκαρπον, δι' ὃ καὶ καλοῦσιν ἄρρεν· τῶν δὲ θηλειῶν ἡ μὲν ἐρυθρὸν εὐθὺς φέρει τὸν καρπὸν ἡλίκον φακὸν ἄπεπτον, ἡ δὲ χλοερὸν ἐνέγκασα μετὰ ταῦτα ἐρυθραίνει, καὶ ἅμα τῇ ἀμπέλῳ πεπαίνουσα τὸ ἔσχατον ποιεῖ μέλανα, μέγεθος ἡλίκον κύαμον, ῥητινώδη δὲ καὶ θυωδέστερον. ἔστι δὲ τὸ δένδρον περὶ μὲν τὴν Ἴδην καὶ Μακεδονίαν βραχὺ θαμνῶδες ἐστραμμένον, περὶ δὲ Δαμασκὸν τῆς Συρίας μέγα καὶ πολὺ καὶ καλόν· ὄρος γάρ τί φασιν εἶναι πάμμεστον

---

[1] *cf.* *C.P.* 2. 12. 6. [2] *cf.* *Geop.* 10. 68.
[3] λεῖον conj. W.; πλέον UMVAld.

unbranched stems without knots; though some of these are long and stout. Nevertheless it also submits to cultivation. The cultivated form differs in producing better fruit and larger leaves; in both forms the leaf has a jagged edge: the leaf of the alder most closely resembles it, but is broader, and the tree itself is bigger. [1] The filbert is always more fruitful if it has its slender boughs cut off. [2] There are two kinds of each sort; some have a round, others an oblong nut; that of the cultivated tree is paler, and it fruits best in damp places. The wild tree becomes cultivated by being transplanted. Its bark is smooth,[3] consisting of one layer, thin glossy and with peculiar white blotches on it. The wood is extremely tough, so that men make baskets even of the quite thin twigs, having stripped them of their bark, and of the stout ones when they have whittled them. Also it has a small amount of yellow heart-wood, which makes[4] the branches hollow. Peculiar to these trees is the matter of the catkin, as we mentioned.[5]

[6] The terebinth has a 'male' and a 'female' form. The 'male' is barren, which is why it is called 'male'; the fruit of one of the 'female' forms is red from the first and as large as an unripe[7] lentil; the other produces a green fruit which subsequently turns red, and, ripening at the same time as the grapes, becomes eventually black and is as large as a bean, but resinous and somewhat aromatic. About Ida and in Macedonia the tree is low shrubby and twisted, but in the Syrian Damascus, where it abounds, it is tall and handsome; indeed they say

[4] ἧ Ald.H.; ἣ W. with U. *cf.* 3. 13. 4.
[5] 3. 7. 3. [6] Plin. 13. 54.
[7] καὶ before ἄπεπτον om. St.

4 τερμίνθων, ἄλλο δ' οὐδὲν πεφυκέναι. ξύλον δὲ ἔχει γλίσχρον καὶ ῥίζας ἰσχυρὰς κατὰ βάθους, καὶ τὸ ὅλον ἀνώλεθρον· ἄνθος δὲ ὅμοιον τῷ τῆς ἐλάας, τῷ χρώματι δὲ ἐρυθρόν. φύλλον, περὶ ἕνα μίσχον πλείω δαφνοειδῆ κατὰ συζυγίαν, ὥσπερ καὶ τὸ τῆς οἴης· καὶ τὸ ἐξ ἄκρου περιττόν· πλὴν ἐγγωνιώτερον τῆς οἴης καὶ δαφνοειδέστερον δὲ κύκλῳ καὶ λιπαρὸν ἅπαν ἅμα τῷ καρπῷ. φέρει δὲ καὶ κωρυκώδη τινὰ κοῖλα, καθάπερ ἡ πτελέα, ἐν οἷς θηρίδια ἐγγίγνεται κωνωποειδῆ· ἐγγίγνεται δέ τι καὶ ῥητινῶδες ἐν τούτοις καὶ γλίσχρον· οὐ μὴν ἐνθεῦτέν γε ἡ ῥητίνη συλλέγεται ἀλλ' ἀπὸ τοῦ ξύλου. ὁ δὲ καρπὸς οὐκ ἀφίησι ῥητίνης πλῆθος, ἀλλὰ προσέχεται μὲν ταῖς χερσί, κἂν μὴ πλυθῇ μετὰ τὴν συλλογὴν συνέχεται· πλυνόμενος δὲ ὁ μὲν λευκὸς καὶ ἄπεπτος ἐπιπλεῖ, ὁ δὲ μέλας ὑφίσταται.

5 Ἡ δὲ πύξος μεγέθει μὲν οὐ μεγάλη, τὸ δὲ φύλλον ὅμοιον ἔχει μυρρίνῳ. φύεται δ' ἐν τοῖς ψυχροῖς τόποις καὶ τραχέσι· καὶ γὰρ τὰ Κύτωρα τοιοῦτον, οὗ ἡ πλείστη γίνεται· ψυχρὸς δὲ καὶ ὁ Ὄλυμπος ὁ Μακεδονικός· καὶ γὰρ ἐνταῦθα γίνεται πλὴν οὐ μεγάλη· μεγίστη δὲ καὶ καλλίστη ἐν Κύρνῳ· καὶ γὰρ εὐμήκεις καὶ πάχος ἔχουσαι πολὺ παρὰ τὰς ἄλλας. δι' ὃ καὶ τὸ μέλι οὐχ ἡδὺ ὄζον τῆς πύξου.

---

[1] *πλείω*: sc. *φύλλα*, in loose apposition to *φύλλον*. Apparently the leaf is said to resemble that of *οἴη* in its composite structure, but that of the bay in shape: *cf.* 3. 12. 7.

[2] *ἅπαν ἅμα* conj. W.; *ἅμα ἅπαν* UAld.

[3] *cf.* 2. 8. 3; 3. 7. 3; 3. 14. 1. *κωρυκώδη* conj. R. Const.; *κορυώδη* Ald.; *κωρυώδη* H.; *καρυώδη* mBas.

that there is a certain hill which is covered with terebinths, though nothing else grows on it. It has tough wood and strong roots which run deep, and the tree as a whole is impossible to destroy. The flower is like that of the olive, but red in colour. The leaf is made up of a number of leaflets,[1] like bay leaves, attached in pairs to a single leaf-stalk. So far it resembles the leaf of the sorb; there is also the extra leaflet at the tip: but the leaf is more angular than that of the sorb, and the edge resembles more the leaf of the bay; the leaf is glossy all over,[2] as is the fruit. It bears also some hollow bag-like[3] growths, like the elm, in which are found little creatures like gnats; and resinous sticky matter is found also in these bags; but the resin is gathered from the wood and not from these. The fruit does not discharge much resin, but it clings to the hands, and, if it is not washed after gathering, it all sticks together; if it is washed, the part which is white and unripe floats,[4] but the black part sinks.

The box is not a large tree, and it has a leaf like that of the myrtle. It grows in cold rough places; for of this character is Cytora,[5] where it is most abundant. The Macedonian Olympus is also a cold region;[6] for there too it grows, though not to a great size. It is largest and fairest in Corsica,[7] where the tree grows taller and stouter than anywhere else; wherefore the honey there is not sweet, as it smells of the box.

[4] ἐπιπλεῖ conj. R. Const. from G; ἐπὶ πλεῖον Ald.; ἐπὶ πλεῖ (erased) U.
[5] *cf.* *Cytore buxifer*, Catull. 4. 13; Plin. 16. 70.
[6] *cf.* 5. 7. 7.
[7] Κύρνῳ conj. R. Const. from Plin. *l.c.*; Κυρήνωι U; Κυρήνῃ Ald.

6 Πλήθει δὲ πολὺ κράταιγός ἐστιν, οἱ δὲ κραταιγόνα καλοῦσιν· ἔχει δὲ τὸ μὲν φύλλον ὅμοιον μεσπίλῃ τετανόν, πλὴν μεῖζον ἐκείνου καὶ πλατύτερον ἢ προμηκέστερον, τὸν δὲ χαραγμὸν οὐκ ἔχον ὥσπερ ἐκεῖνο. γίνεται δὲ τὸ δένδρον οὔτε μέγα λίαν οὔτε παχύ· τὸ δὲ ξύλον ποικίλον ἰσχυρὸν ξανθόν· ἔχει δὲ φλοιὸν λεῖον ὅμοιον μεσπίλῃ· μονόρριζον δ' εἰς βάθος ὡς ἐπὶ τὸ πολύ. καρπὸν δ' ἔχει στρογγύλον ἡλίκον ὁ κότινος· πεπαινόμενος δὲ ξανθύνεται καὶ ἐπιμελαίνεται· κατὰ δὲ τὴν γεῦσιν καὶ τὸν χυλὸν μεσπιλῶδες· διόπερ οἷον ἀγρία μεσπίλη δόξειεν ἂν εἶναι. μονοειδὲς δὲ καὶ οὐκ ἔχον διαφοράς.

XVI. Ὁ δὲ πρῖνος φύλλον μὲν ἔχει δρυῶδες, ἔλαττον δὲ καὶ ἐπακανθίζον, τὸν δὲ φλοιὸν λειότερον δρυός. αὐτὸ δὲ τὸ δένδρον μέγα, καθάπερ ἡ δρῦς, ἐὰν ἔχῃ τόπον καὶ ἔδαφος· ξύλον δὲ πυκνὸν καὶ ἰσχυρόν· βαθύρριζον δὲ ἐπιεικῶς καὶ πολύρριζον. καρπὸν δὲ ἔχει βαλανώδη· μικρὰ δὲ ἡ βάλανος· περικαταλαμβάνει δὲ ὁ νέος τὸν ἔνον· ὀψὲ γὰρ πεπαίνει, δι' ὃ καὶ διφορεῖν τινές φασι. φέρει δὲ παρὰ τὴν βάλανον καὶ κόκκον τινὰ φοινικοῦν· ἴσχει δὲ καὶ ἰξίαν καὶ ὑφέαρ· ὥστε ἐνίοτε συμβαίνει τέτταρας ἅμα καρποὺς ἔχειν αὐτόν, δύο μὲν τοὺς ἑαυτοῦ δύο δ' ἄλλους τόν τε τῆς ἰξίας καὶ τὸν τοῦ ὑφέαρος. καὶ τὴν

[1] Quoted by Athen. 2. 34; *cf.* Plin. 16. 120; 26. 99; 27. 62 and 63.

[2] τετανόν: *cf.* 3. 11. 1; 3. 12. 5. Athen., *l.c.*, has τεταμένον.

[3] ἐκεῖνο Athen. *l.c.*; κἀκεῖνο Ald.

[4] ξανθὸν before ἰσχυρόν Athen. *l.c.*

[1]The *krataigos* is a very common tree; some call it *krataigon*. It has a smooth[2] leaf like that of the medlar, but longer, and its breadth is greater than its length, while the edge is not jagged like that[3] of the medlar. The tree does not grow very tall or thick; its wood is mottled strong and brown[4]; it has a smooth bark like that of the medlar; it has generally a single root, which runs deep. The fruit is round and as large as that of the wild olive[5]; as it ripens it turns brown and black; in taste and flavour it is like that of the medlar; wherefore this might seem to be a sort of wild form of that tree.[6] There is only one form of it and it shews no variation.

Of certain other oaks, arbutus, andrachne, wig-tree.

XVI. The kermes-oak[7] has a leaf like that of the oak, but smaller and spinous,[8] while its bark is smoother than that of the oak. The tree itself is large, like the oak, if it has space and root-room; the wood is close and strong; it roots fairly deep and it has many roots. The fruit is like an acorn, but the kermes-oak's acorn is small; the new one overtakes that of last year, for it ripens late.[9] Wherefore some say that it bears twice. Besides the acorn it bears a kind of scarlet berry[10]; it also has oak-mistletoe[11] and mistletoe; so that sometimes it happens that it has four fruits on it at once, two which are its own and two others, namely those of the oak-mistletoe[11] and

[5] κότινος Athen. *l.c.*; κόψιμος UMVAld.
[6] μεσπίλη added from Athen. *l.c.*
[7] *cf.* 3. 7. 3. [8] *cf.* 3. 16. 2. [9] *cf.* 3. 4. 1, 4 and 6.
[10] Plin. 16. 32; Simon. *ap.* Plut. *Theseus* 17.
[11] *cf.* *C.P.* 2. 17. 1.

μὲν ἰξίαν φέρει ἐκ τῶν πρὸς βορρᾶν, τὸ δὲ ὕφεαρ ἐκ τῶν πρὸς μεσημβρίαν.

2 Οἱ δὲ περὶ Ἀρκαδίαν δένδρον τι σμίλακα καλοῦσιν, ὅ ἐστιν ὅμοιον τῷ πρίνῳ, τὰ δὲ φύλλα οὐκ ἀκανθώδη ἔχει ἀλλ' ἁπαλώτερα καὶ βαθύτερα καὶ διαφορὰς ἔχοντα πλείους· οὐδὲ τὸ ξύλον ὥσπερ ἐκεῖνο στερεὸν καὶ πυκνόν, ἀλλὰ καὶ μαλακὸν ἐν ταῖς ἐργασίαις.

3 Ὃ δὲ καλοῦσιν οἱ Ἀρκάδες φελλόδρυν τοιάνδε ἔχει τὴν φύσιν· ὡς μὲν ἁπλῶς εἰπεῖν ἀνὰ μέσον πρίνου καὶ δρυός ἐστιν· καὶ ἔνιοί γε ὑπολαμβάνουσιν εἶναι θῆλυν πρῖνον· δι' ὃ καὶ ὅπου μὴ φύεται πρῖνος τούτῳ χρῶνται πρὸς τὰς ἁμάξας καὶ τὰ τοιαῦτα, καθάπερ οἱ περὶ Λακεδαίμονα καὶ Ἠλείαν. καλοῦσι δὲ οἵ γε Δωριεῖς καὶ ἀρίαν τὸ δένδρον· ἔστι δὲ μαλακώτερον μὲν καὶ μανότερον τοῦ πρίνου, σκληρότερον δὲ καὶ πυκνότερον τῆς δρυός· καὶ τὸ χρῶμα φλοϊσθέντος τοῦ ξύλου λευκότερον μὲν τοῦ πρίνου, οἰνωπότερον δὲ τῆς δρυός· τὰ δὲ φύλλα προσέοικε μὲν ἀμφοῖν, ἔχει δὲ μείζω μὲν ἢ ὡς πρῖνος ἐλάττω δὲ ἢ ὡς δρῦς· καὶ τὸν καρπὸν τοῦ μὲν πρίνου κατὰ μέγεθος ἐλάττω ταῖς ἐλαχίσταις δὲ βαλάνοις ἴσον, καὶ γλυκύτερον μὲν τοῦ πρίνου πικρότερον δὲ τῆς δρυός. καλοῦσι δέ τινες τὸν μὲν τοῦ πρίνου καὶ τὸν ταύτης καρπὸν ἄκυλον, τὸν δὲ τῆς δρυὸς βάλανον. μήτραν δὲ ἔχει φανερωτέραν ἢ ὁ πρῖνος· καὶ ἡ μὲν φελλόδρυς τοιαύτην τινὰ ἔχει φύσιν.

---

[1] Plin. 16. 19. See Index.
[2] βαθύτερα MSS.; εὐθύτερα conj. Dalec.
[3] Plin. *l.c.* See Index.

of the mistletoe. It produces the oak-mistletoe on the north side and the mistletoe on the south.

The Arcadians have a tree which they call *smilax* [1] (holm-oak), which resembles the kermes-oak, but has not spinous leaves, its leaves being softer and longer [2] and differing in several other ways. Nor is the wood hard and close like that of the kermes-oak, but quite soft to work.

The tree which the Arcadians call 'cork-oak' [3] (holm-oak) has this character:—to put it generally, it is between the kermes-oak and the oak; and some suppose it to be the 'female' kermes-oak; wherefore, where the kermes-oak does not grow, they use this tree for their carts and such-like purposes; for instance it is so used by the peoples of Lacedaemon and Elis. The Dorians also call the tree *aria*.[4] Its wood is softer and less compact than that of the kermes-oak, but harder and closer than that of the oak. When it is barked,[5] the colour of the wood is paler than that of the kermes-oak, but redder than that of the oak. The leaves resemble those of both trees, but they are somewhat large, if we consider the tree as a kermes-oak, and somewhat small if we regard it as an oak. The fruit is smaller in size than that of the kermes-oak, and equal to the smallest acorns; it is sweeter than that of the kermes-oak, bitterer than that of the oak. Some call the fruit of the kermes-oak and of the *aria* 'mast,' [6] keeping the name 'acorn' for the fruit of the oak. It has a core which is more obvious than in kermes-oak. Such is the character of the 'cork-oak.'

[4] Already described; *cf.* 3. 4. 2; 3. 17. 1.
[5] *cf.* Paus. *Arcadia*, 8. 12.
[6] ἄκυλον: *cf.* Hom. *Od.* 10. 242.

4 Ἡ δὲ κόμαρος, ἡ τὸ μεμαίκυλον φέρουσα τὸ ἐδώδιμον, ἐστὶ μὲν οὐκ ἄγαν μέγα, τὸν δὲ φλοιὸν ἔχει λεπτὸν μὲν παρόμοιον μυρίκῃ, τὸ δὲ φύλλον μεταξὺ πρίνου καὶ δάφνης. ἀνθεῖ δὲ τοῦ Πυανεψιῶνος· τὰ δὲ ἄνθη πέφυκεν ἀπὸ μιᾶς κρεμάστρας ἐπ' ἄκρων βοτρυδόν· τὴν δὲ μορφὴν ἕκαστόν ἐστιν ὅμοιον μύρτῳ προμήκει καὶ τῷ μεγέθει δὲ σχεδὸν τηλικοῦτον· ἄφυλλον δὲ καὶ κοῖλον ὥσπερ ᾠὸν ἐκκεκολαμμένον τὸ στόμα δὲ ἀνεῳγμένον· ὅταν δ' ἀπανθήσῃ, καὶ ἡ πρόσφυσις τετρύπηται, τὸ δ' ἀπανθῆσαν λεπτὸν καὶ ὥσπερ σφόνδυλος περὶ ἄτρακτον ἢ κάρνειος Δωρικός· ὁ δὲ καρπὸς ἐνιαυτῷ πεπαίνεται, ὥσθ' ἅμα συμβαίνει τοῦτόν τ' ἔχειν καὶ τὸν ἕτερον ἀνθεῖν.

5 Παρόμοιον δὲ τὸ φύλλον καὶ ἡ ἀνδράχλη ἔχει τῷ κομάρῳ, μέγεθος οὐκ ἄγαν μέγα· τὸν δὲ φλοιὸν λεῖον ἔχει καὶ περιρρηγνύμενον· καρπὸν δ' ἔχει ὅμοιον τῇ κομάρῳ.

6 Ὅμοιον δ' ἐστὶ τούτοις τὸ φύλλον καὶ τὸ τῆς κοκκυγέας· τὸ δὲ δένδρον μικρόν. ἴδιον δὲ ἔχει τὸ ἐκπαπποῦσθαι τὸν καρπόν· τοῦτο γὰρ οὐδ' ἐφ' ἑνὸς ἀκηκόαμεν ἄλλου δένδρου. ταῦτα μὲν οὖν κοινότερα πλείοσι χώραις καὶ τόποις.

---

1 Plin. 15. 98 and 99; Diosc. 1. 122. 2 October.

3 ἐκκεκολαμμένον MV, *cf.* Arist. *H.A.* 6. 3; ἐγκεκολαμμένον UAld. 4 *cf.* 1. 13. 3.

5 κάρνειος, an unknown word, probably corrupt; κίονος Δωρικοῦ conj. Sch., 'drum of a Doric column.' *cf.* Athen. 5. 39.

[1] The arbutus, which produces the edible fruit called *memaikylon*, is not a very large tree; its bark is thin and like that of the tamarisk, the leaf is between that of the kermes-oak and that of the bay. It blooms in the month Pyanepsion [2]; the flowers grow in clusters at the end of the boughs from a single attachment; in shape each of them is like an oblong myrtle flower and it is of about the same size; it has no petals, but forms a cup like an empty eggshell,[3] and the mouth is open: when the flower drops off, there is a hole [4] also through the part by which it is attached, and the fallen flower is delicate and like a whorl on a spindle or a Doric *karneios*.[5] The fruit takes a year to ripen, so that it comes to pass that this and the new flower are on the tree together.

[6] The andrachne has a leaf like that of the arbutus and is not a very large tree; the bark is smooth [7] and cracked,[8] the fruit is like that of the arbutus.

The leaf of the wig-tree [9] is also like that of the last named tree, but it is a small tree. Peculiar to it is the fact that the fruit passes into down [10]: we have not heard of such a thing in any other tree. These trees are found in a good many positions and regions.

6 Plin. 13. 120.

7 λεῖον conj. Sch.; λευκὸν UAld. In Pletho's excerpt the passage has λεῖον, and Plin., *l.c.*, evidently read λεῖον.

8 περιρρηγνύμενον. Plin., *l.c.*, seems to have read περιπηγνύμενον. *cf.* 1. 5. 2; 9. 4. 3.

9 Plin. 13. 121. κοκκυγέας conj. Sch. after Plin. *l.c.*, *cf.* Hesych. *s.v.* κεκκοκυγωμένην; κοκκομηλέας U; κοκκυμηλέας P_2Ald.

10 ἐκπαπποῦσθαι: *fructum amittere lanugine* Plin. *l.c.* *cf.* 6. 8. 4.

XVII. Ἔνια δὲ ἰδιώτερα, καθάπερ καὶ ὁ φελλός· γίνεται μὲν ἐν Τυρρηνίᾳ, τὸ δὲ δένδρον ἐστὶ στελεχῶδες μὲν καὶ ὀλιγόκλαδον, εὐμηκες δ' ἐπιεικῶς καὶ εὐαυξές· ξύλον ἰσχυρόν· τὸν δὲ φλοιὸν παχὺν σφόδρα καὶ καταρρηγνύμενον, ὥσπερ ὁ τῆς πίτυος, πλὴν κατὰ μείζω. τὸ δὲ φύλλον ὅμοιον ταῖς μελίαις παχὺ προμηκέστερον· οὐκ ἀείφυλλον ἀλλὰ φυλλοβολοῦν. καρπὸν δὲ [αἰεὶ] φέρει βαλανηρὸν ὅμοιον τῇ ἀρίᾳ. περιαιροῦσι δὲ τὸν φλοιὸν καί φασι δεῖν πάντα ἀφαιρεῖν, εἰ δὲ μὴ χεῖρον γίνεται τὸ δένδρον· ἐξαναπληροῦται δὲ πάλιν σχεδὸν ἐν τρισὶν ἔτεσιν.

2 Ἴδιον δὲ καὶ ἡ κολουτέα περὶ Λιπάραν· δένδρον μὲν εὐμέγεθες, τὸν δὲ καρπὸν φέρει ἐν λοβοῖς ἡλίκον φακόν, ὅς πιαίνει τὰ πρόβατα θαυμαστῶς. φύεται δὲ ἀπὸ σπέρματος καὶ ἐκ τῆς τῶν προβάτων κόπρου κάλλιστα. ὥρα δὲ τῆς φυτείας ἅμα Ἀρκτούρῳ δυομένῳ· δεῖ δὲ φυτεύειν προβρέχοντας ὅταν ἤδη διαφύηται ἐν τῷ ὕδατι. φύλλον δ' ἔχει παρόμοιον τήλει. βλαστάνει δὲ τὸ πρῶτον μονοφυὲς ἐπὶ ἔτη μάλιστα τρία ἐν οἷς καὶ τὰς βακτηρίας τέμνουσι· δοκοῦσι γὰρ εἶναι καλαί· καὶ ἐάν τις κολούσῃ ἀποθνήσκει· καὶ γὰρ ἀπαράβλαστόν ἐστιν· εἶτα σχίζεται καὶ ἀποδενδροῦται τῷ τετάρτῳ ἔτει.

---

[1] Plin. 16. 34.
[2] Τυρρηνίᾳ conj. R. Const.; πυρρηνίαι UMV; πυρρηνίᾳ Ald.
[3] αἰεὶ must be corrupt: probably repeated from ἀείφυλλον.
[4] βαλανηρὸν conj. Sch.; βαλανήφορον UMVAld.
[5] ἀρίᾳ conj. R. Const. from G; ἀγρίᾳ P_2MVAld.; ἀγρίαι U.

Of cork-oak, kolutea, koloitia, *and of certain other trees peculiar to particular localities.*

XVII. 1 Some however are more local, such as the cork-oak: this occurs in Tyrrhenia[2]; it is a tree with a distinct trunk and few branches, and is fairly tall and of vigorous growth. The wood is strong, the bark very thick and cracked, like that of the Aleppo pine, save that the cracks are larger. The leaf is like that of the manna-ash, thick and somewhat oblong. The tree is not evergreen but deciduous. It has always[3] an acorn-like[4] fruit like that of the *aria*[5] (holm-oak). They strip off the bark,[6] and they say that it should all be removed,[7] otherwise the tree deteriorates: it is renewed again in about three years.

The *kolutea*[8] too is a local tree, occurring in the Lipari islands. It is a tree of good size, and bears its fruit, which is as large as a lentil, in pods; this fattens sheep wonderfully. It grows from seed, and also grows very well from sheep-droppings. The time for sowing it is the setting of Arcturus; and one should first soak the seed and sow it when it is already sprouting in the water. It has a leaf like *telis*[9] (fenugreek). At first it grows for about three years with a single stem, and in this period men cut their walking-sticks from it; for it seems that it makes excellent ones. And, if the top is cut off during this period, it dies, for it makes no side-shoots. After this period it divides, and in the fourth year develops into a tree.

6 *cf.* 1. 5. 2; 4. 15. 1; Plin. 17. 234.
7 ἀφαιρεῖν conj. Coraës; διαιρεῖν P_2Ald.
8 *cf.* 1. 11. 2; 3. 17. 3.
9 τήλει conj. R. Const. from G, *faeno graeco*; τίλει UMV; τύλῃ Ald.

3 Ἡ δὲ περὶ τὴν Ἴδην, ἣν καλοῦσι κολοιτίαν, ἕτερον εἶδός ἐστιν, θαμνοειδὲς δὲ καὶ ὀζῶδες καὶ πολυμάσχαλον, σπάνιον δέ, οὐ πολύ· ἔχει δὲ φύλλον δαφνοειδὲς πλατυφύλλου δάφνης, πλὴν στρογγυλώτερον καὶ μεῖζον ὥσθ' ὅμοιον φαίνεσθαι τῷ τῆς πτελέας, προμηκέστερον δέ, τὴν χρόαν ἐπὶ θάτερα χλοερὸν ὄπισθεν δὲ ἐπιλευκαῖνον, καὶ πολύϊνον ἐκ τῶν ὄπισθεν ταῖς λεπταῖς ἰσὶ ἔκ τε τῆς ῥάχεως καὶ μεταξὺ τῶν πλευροειδῶν ἀπὸ τῆς μέσης κατατεινουσῶν· φλοιὸν δ' οὐ λεῖον ἀλλ' οἷον τὸν τῆς ἀμπέλου· τὸ δὲ ξύλον σκληρὸν καὶ πυκνόν· ῥίζας δὲ ἐπιπολαίους καὶ λεπτὰς καὶ μανὰς οὐλὰς δ' ἐνίοτε, καὶ ξανθὰς σφόδρα. καρπὸν δὲ οὐκ ἔχειν φασὶν οὐδὲ ἄνθος· τὴν δὲ κορυνώδη κάχρυν καὶ τοὺς ὀφθαλμοὺς τοὺς παρὰ τὰ φύλλα λείους σφόδρα καὶ λιπαροὺς καὶ λευκοὺς τῷ σχήματι δὲ καχρυώδεις· ἀποκοπὲν δὲ καὶ ἐπικαυθὲν παραφύεται καὶ ἀναβλαστάνει.

4 Ἴδια δὲ καὶ τάδε τὰ περὶ τὴν Ἴδην ἐστίν, οἷον ἥ τε Ἀλεξάνδρεια καλουμένη δάφνη καὶ συκῆ τις καὶ ἄμπελος. τῆς μὲν οὖν δάφνης ἐν τούτῳ τὸ ἴδιον, ὅτι ἐπιφυλλόκαρπόν ἐστιν, ὥσπερ καὶ ἡ κεντρομυρρίνη· ἀμφότεραι γὰρ τὸν καρπὸν ἔχουσιν ἐκ τῆς ῥάχεως τοῦ φύλλου.

5 Ἡ δὲ συκῆ θαμνῶδες μὲν καὶ οὐχ ὑψηλόν, πάχος δ' ἔχον ὥστε καὶ πηχυαῖον εἶναι τὴν περίμετρον· τὸ δὲ ξύλον ἐπεστραμμένον γλίσχρον· κάτωθεν μὲν λεῖον καὶ ἄνοζον ἄνωθεν δὲ περί-

---

[1] κολοιτίαν (? κολοιτέαν) U. *cf.* 1. 11. 2; 3. 17. 2. Whichever spelling is correct should probably be adopted in all three places. [2] *cf.* 3. 11. 3.

The tree found about Mount Ida, called *koloitia*,[1] is a distinct kind and is shrubby and branching with many boughs; but it is rather rare. It has a leaf like that of the 'broad-leaved' bay,[2] but rounder and larger, so that it looks like that of the elm, but it is more oblong: the colour on both sides is green, but the base is whitish; in this part it is very fibrous, because of its fine fibres which spring partly from the midrib,[3] partly between the ribs[4] (so to call them) which run out from the midrib. The bark is not smooth but like that of the vine; the wood is hard and close, the roots are shallow slender and spreading, (though sometimes they are compact), and they are very yellow. They say that this shrub has no fruit nor flower, but has its knobby winter-bud and its 'eyes'; these grow alongside of the leaves, and are very smooth glossy and white, and in shape are like a winter-bud. When the tree is cut or burnt down, it grows from the side and springs up again.

There are also three trees peculiar to Mount Ida, the tree called Alexandrian laurel, a sort of fig, and a 'vine' (currant grape). The peculiarity of the laurel is that it bears fruit on its leaves, like the 'prickly myrtle' (butcher's broom): both have their fruit on the midrib of the leaf.

The 'fig'[5] is shrubby and not tall, but so thick that the stem is a cubit in circumference. The wood is twisted and tough; below it is smooth and unbranched, above it has thick foliage: the colour both

[3] ἔκ τε τῆς ῥαχέως καὶ conj. W.; καὶ ταῖς ῥίζαις καὶ Ald. *cf.* 3. 10. 3, and ἐκ τῆς ῥαχέως below, 3. 17. 4.

[4] πλευροειδῶν: πλευροειδῶς conj. St.

[5] See Index. Plin. 15. 68; *cf.* Athen. 3. 11.

κομον· χρῶμα δὲ καὶ φύλλου καὶ φλοιοῦ πελιόν, τὸ δὲ σχῆμα τῶν φύλλων ὅμοιον τῷ τῆς φιλύρας καὶ μαλακὸν καὶ πλατὺ καὶ τὸ μέγεθος παραπλήσιον· ἄνθος μεσπιλῶδες καὶ ἀνθεῖ ἅμα τῇ μεσπίλῃ. ὁ δὲ καρπός, ὃν καλοῦσι σῦκον, ἐρυθρὸς ἡλίκος ἐλάας πλὴν στρογγυλώτερος, ἐσθιόμενος δὲ μεσπιλώδης· ῥίζας δὲ ἔχει παχείας ὡσὰν συκῆς ἡμέρου καὶ γλίσχρας. ἀσαπὲς δέ ἐστι τὸ δένδρον καὶ καρδίαν ἔχει στερεὰν οὐκ ἐντεριώνην.

6 Ἡ δὲ ἄμπελος φύεται μὲν τῆς Ἴδης περὶ τὰς Φαλάκρας καλουμένας· ἔστι δὲ θαμνῶδες ῥαβδίοις μικροῖς· τείνονται δὲ οἱ κλῶνες ὡς πυγωνιαῖοι, πρὸς οἷς ῥᾶγές εἰσιν ἐκ πλαγίου μέλαιναι τὸ μέγεθος ἡλίκος κύαμος γλυκεῖαι· ἔχουσι δὲ ἐντὸς γιγαρτῶδές τι μαλακόν· φύλλον στρογγύλον ἀσχιδὲς μικρόν.

XVIII. Ἔχει δὲ καὶ τἆλλα σχεδὸν ὄρη φύσεις τινὰς ἰδίας τὰ μὲν δένδρων τὰ δὲ θάμνων τὰ δ' ἄλλων ὑλημάτων. ἀλλὰ γὰρ περὶ μὲν τῆς ἰδιότητος εἴρηται πλεονάκις ὅτι γίνεται καθ' ἑκάστους τόπους. ἡ δὲ ἐν αὐτοῖς τοῖς ὁμογενέσιν διαφορά, καθάπερ ἡ τῶν δένδρων καὶ τῶν θάμνων, ὁμοίως ἐστὶ καὶ τῶν ἄλλων, ὥσπερ εἴρηται, τῶν πλείστων, ὥσπερ καὶ ῥάμνου καὶ παλιούρου καὶ οἴσου [καὶ οἴτου] καὶ ῥοῦ καὶ κιττοῦ καὶ βάτου καὶ ἑτέρων πολλῶν.

---

[1] Lit. grape-stone.
[2] I omit ἡ before διαφορά with Sch.

of leaf and bark is a dull green, the shape of the leaf is like that of the lime; it is soft and broad, and in size it also corresponds; the flower is like that of the medlar, and the tree blooms at the same time as that tree. The fruit, which they call a 'fig,' is red, and as large as an olive, but it is rounder and is like the medlar in taste; the roots are thick like those of the cultivated fig, and tough. The tree does not rot, and it has a solid heart, instead of ordinary heart-wood.

The 'vine' (currant grape) grows about the place called Phalakrai in the district of Ida; it is shrubby with small twigs; the branches are about a cubit long, and attached to them at the side are black berries, which are the size of a bean and sweet; inside they have a sort of soft stone[1]; the leaf is round undivided and small

Of the differences in various shrubs—buckthorn, withy, Christ's thorn, bramble, sumach, ivy, smilax, [spindle-tree].

XVIII. Most other mountains too have certain peculiar products, whether trees shrubs or other woody plants. However we have several times remarked as to such peculiarities that they occur in all regions. Moreover the variation[2] between things of the same kind which we find in trees obtains also among shrubs and most other things, as has been said: for instance, we find it in buckthorn Christ's thorn withy[3] sumach ivy bramble and many others.

[3] [καὶ οἴτου] bracketed by W.; καὶ ἴσου Ald.; καὶ ἴσου καὶ οἴτου MVP; καὶ οἴσου καὶ υἶτοι U. Only οἶσος is mentioned in the following descriptions.

Ῥάμνος τε γάρ ἐστιν ἡ μὲν μέλαινα ἡ δὲ λευκή, καὶ ὁ καρπὸς διάφορος, ἀκανθοφόροι δὲ ἄμφω.

Τοῦ τε οἴσου τὸ μὲν λευκὸν τὸ δὲ μέλαν· καὶ τὸ ἄνθος ἑκατέρου καὶ ὁ καρπὸς κατὰ λόγον ὁ μὲν λευκὸς ὁ δὲ μέλας· ἔνιοι δὲ καὶ ὥσπερ ἀνὰ μέσον, ὧν καὶ τὸ ἄνθος ἐπιπορφυρίζει καὶ οὔτε οἰνωπὸν οὔτε ἔκλευκόν ἐστιν ὥσπερ τῶν ἑτέρων. ἔχει δὲ καὶ τὰ φύλλα λεπτότερα καὶ λειότερα καὶ τὰς ῥάβδους τὸ λευκόν.

Ὅ τε παλίουρος ἔχει διαφορὰς . . . ἅπαντα δὲ ταῦτα καρποφόρα. καὶ ὅ γε παλίουρος ἐν λοβῷ τινι τὸν καρπὸν ἔχει καθαπερεὶ φύλλῳ, ἐν ᾧ τρία ἢ τέτταρα γίνεται. χρῶνται δ' αὐτῷ πρὸς τὰς βῆχας οἱ ἰατροὶ κόπτοντες· ἔχει γάρ τινα γλισχρότητα καὶ λίπος, ὥσπερ τὸ τοῦ λίνου σπέρμα. φύεται δὲ καὶ ἐπὶ τοῖς ἐφύδροις καὶ ἐν τοῖς ξηροῖς, ὥσπερ ὁ βάτος. [οὐχ ἧττον δέ ἐστι τὸ δένδρον πάρυδρον.] φυλλοβόλον δὲ καὶ οὐχ ὥσπερ ἡ ῥάμνος ἀείφυλλον.

Ἔτι δὲ καὶ τοῦ βάτου πλείω γένη, μεγίστην δὲ ἔχοντες διαφορὰν ὅτι ὁ μὲν ὀρθοφυὴς καὶ ὕψος ἔχων, ὁ δ' ἐπὶ τῆς γῆς καὶ εὐθὺς κάτω νεύων καὶ ὅταν συνάπτῃ τῇ γῇ ῥιζούμενος πάλιν, ὃν δὴ καλοῦσί τινες χαμαίβατον. τὸ δὲ κυνόσβατον τὸν καρπὸν ὑπέρυθρον ἔχει καὶ παραπλήσιον τῷ τῆς ῥόας· ἔστι δὲ θάμνου καὶ δένδρου μεταξὺ καὶ παρόμοιον ταῖς ῥόαις, τὸ δὲ φύλλον ἀκανθῶδες.

---

[1] *cf.* 1. 9. 4; 3. 18. 12; *C.P.* 1. 10. 7.

[2] Some words are missing, which described various forms of παλίουρος, alluded to in πάντα ταῦτα (Sch.). *cf.* 4. 3. 3, where an African παλίουρος is described.

[3] καθαπερεὶ φύλλῳ conj. W., *cf.* 3. 11. 2; καθάπερ τὸ φύλλον UMV.

[1] Thus of buckthorn there is the black and the white form, and there is difference in the fruit, though both bear thorns.

Of the withy there is a black and a white form; the flower and fruit of each respectively correspond in colour to the name; but some specimens are, as it were, intermediate, the flower being purplish, and neither wine-coloured nor whitish as in the others. The leaves in the white kind are also slenderer and smoother, as also are the branches.

There is variation also in the Christ's thorn . . .[2] all these forms are fruit-bearing. Christ's thorn has its fruit in a sort of pod, resembling a leaf,[3] which contains three or four seeds. Doctors bruise[4] them and use them against coughs; for they have a certain viscous and oily character, like linseed. The shrub grows in wet and dry places alike, like the bramble.[5] But it is deciduous, and not evergreen like buckthorn.

Of the bramble again there are several kinds, shewing very great variation; one is erect and tall, another runs along the ground and from the first bends downwards, and, when it touches the earth, it roots again; this some call the 'ground bramble.' The 'dog's bramble' (wild rose) has a reddish fruit, like that of the pomegranate[6]; and, like the pomegranate, it is intermediate between a shrub and a tree; but the leaf is spinous.[7]

[4] κόπτοντες: for the tense *cf.* 3. 17. 2, προβρέχοντας.

[5] οὐχ . . . πάρυδρον probably a gloss, W.

[6] ῥόαις UMV (?) Ald.; ῥοδαῖς conj. Sch. from Plin. 16. 180. Athen. (2. 82) cites the passage with παραπ. τῇ ῥοίᾳ. The Schol. on Theocr. 5. 92 seems to have traces of *both* readings.

[7] ἀκανθῶδες conj. Sch. from Schol. on Theocr. (see last note), which quotes the passage with ἀκανθῶδες; ἀγνῶδες UAld.; so also Athen. *l.c.* Plin. (24. 121) seems to have read ἰχνῶδες (*vestigio hominis simile*).

5 Τῆς δὲ ῥοῦ τὸ μὲν ἄρρεν τὸ δὲ θῆλυ καλοῦσι τῷ τὸ μὲν ἄκαρπον εἶναι τὸ δὲ κάρπιμον. οὐκ ἔχει δὲ οὐδὲ τὰς ῥάβδους ὑψηλὰς οὐδὲ παχείας, φύλλον δ' ὅμοιον πτελέᾳ πλὴν μικρὸν προμηκέστερον καὶ ἐπίδασυ. τῶν δὲ κλωνίων τῶν νέων ἐξ ἴσου τὰ φύλλα εἰς δύο, κατ' ἄλληλα δὲ ἐκ τῶν πλαγίων ὥστε στοιχεῖν. βάπτουσι δὲ τούτῳ καὶ οἱ σκυτοδέψαι τὰ δέρματα τὰ λευκά. ἄνθος λευκὸν βοτρυῶδες, τῷ σχήματι δὲ τὸ ὁλοσχερὲς ὄστλιγγας ἔχον ὥσπερ καὶ ὁ βότρυς· ἀπανθήσαντος δὲ ὁ καρπὸς ἅμα τῇ σταφυλῇ ἐρυθραίνεται, καὶ γίνονται οἷον φακοὶ λεπτοὶ συγκείμενοι· βοτρυῶδες δὲ τὸ σχῆμα καὶ τούτων. ἔχει δὲ τὸ φαρμακῶδες τοῦτο ὃ καλεῖται ῥοῦς ἐν αὐτῷ ὀστῶδες, ὃ καὶ τῆς ῥοῦ διηττημένης ἔχει πολλάκις· ῥίζα δ' ἐπιπόλαιος καὶ μονοφυὴς ὥστε ἀνακάμπτεσθαι ῥᾳδίως ὁλόρριζα· τὸ δὲ ξύλον ἐντεριώνην ἔχει, εὔφθαρτον δὲ καὶ κοπτόμενον. ἐν πᾶσι δὲ γίγνεται τοῖς τόποις, εὐθενεῖ δὲ μάλιστα ἐν τοῖς ἀργιλώδεσι.

6 Πολυειδὴς δὲ ὁ κιττός· καὶ γὰρ ἐπίγειος, ὁ δὲ εἰς ὕψος αἰρόμενος· καὶ τῶν ἐν ὕψει πλείω γένη. τρία δ' οὖν φαίνεται τὰ μέγιστα ὅ τε λευκὸς καὶ ὁ μέλας καὶ τρίτον ἡ ἕλιξ. εἴδη δὲ καὶ ἑκάστου τούτων πλείω. λευκὸς γὰρ ὁ μὲν τῷ καρπῷ μόνον, ὁ δὲ καὶ τοῖς φύλλοις ἐστί. πάλιν δὲ τῶν λευκοκάρπων μόνον ὁ μὲν ἁδρὸν καὶ πυκνὸν καὶ συνεστηκότα τὸν καρπὸν ἔχει καθαπερεὶ σφαῖραν,

---

1 Plin. 13. 55; 24. 91.
2 στοιχεῖν: *cf.* 3. 5. 3; Plin. 13. 55.
3 βοτρυῶδες conj. W.; βοτρυηδόν U; βοτρυδόν Ald.
4 ὁ ῥοῦς masc. *cf.* Diosc. 1. 108.

[1] Of the sumach they recognise a 'male' and a 'female' form, the former being barren, the latter fruit-bearing. The branches are not lofty nor stout, the leaf is like that of the elm, but small more oblong and hairy. On the young shoots the leaves grow in pairs at equal distances apart, corresponding to each other on the two sides, so that they are in regular rows.[2] Tanners use this tree for dyeing white leather. The flower is white and grows in clusters; the general form of it, with branchlets, is like that of the grape-bunch; when the flowering is over, the fruit reddens like the grape, and the appearance of it is like small lentils set close together; the form of these too is clustering.[3] The fruit contains the drug called by the same name,[4] which is a bony substance; it is often still found even when the fruit has been put through a sieve. The root is shallow and single, so that these trees are easily bent right over,[5] root and all. The wood has heart-wood, and it readily perishes and gets worm-eaten.[6] The tree occurs in all regions, but flourishes most in clayey soils.

[7] The ivy also has many forms; one kind grows on the ground, another grows tall, and of the tall-growing ivies there are several kinds. However the three most important seem to be the white the black and the *helix*. And of each of these there are several forms. Of the 'white' one is white only in its fruit, another in its leaves also. Again to take only white-fruited sorts, one of these has its fruit well formed close and compact like a ball; and this

[5] *i.e.* nearly uprooted by wind.
[6] κοπτόμενον: *cf.* 8. 11. 2, 3 and 5.
[7] Plin. 16. 144–147.

ὃν δὴ καλοῦσί τινες κορυμβίαν, οἱ δ' Ἀθήνῃσιν Ἀχαρνικόν. ὁ δὲ ἐλάττων διακεχυμένος ὥσπερ καὶ ὁ μέλας· ἔχει δὲ καὶ ὁ μέλας διαφορὰς ἀλλ' οὐχ ὁμοίως φανεράς.

7 Ἡ δὲ ἕλιξ ἐν μεγίσταις διαφοραῖς· καὶ γὰρ τοῖς φύλλοις πλεῖστον διαφέρει τῇ τε μικρότητι καὶ τῷ γωνοειδῆ καὶ εὐρυθμότερα εἶναι· τὰ δὲ τοῦ κιττοῦ περιφερέστερα καὶ ἁπλᾶ· καὶ τῷ μήκει τῶν κλημάτων καὶ ἔτι τῷ ἄκαρπος εἶναι. διατείνονται γάρ τινες τῷ μὴ ἀποκιττοῦσθαι τῇ φύσει τὴν ἕλικα ἀλλὰ τὴν ἐκ τοῦ κιττοῦ τελειουμένην. (εἰ δὲ πᾶσα ἀποκιττοῦται, καθάπερ τινές φασιν, ἡλικίας ἂν εἴη καὶ διαθέσεως οὐκ εἴδους διαφορά, καθάπερ καὶ τῆς ἀπίου πρὸς τὴν ἀχράδα.) πλὴν τό γε φύλλον καὶ ταύτης πολὺ διαφέρει πρὸς τὸν κιττόν. σπάνιον δὲ τοῦτο καὶ ἐν ὀλίγοις ἐστὶν ὥστε παλαιούμενον μεταβάλλειν,
8 ὥσπερ ἐπὶ τῆς λεύκης καὶ τοῦ κρότωνος. εἴδη δ' ἐστὶ πλείω τῆς ἕλικος, ὡς μὲν τὰ προφανέστατα καὶ μέγιστα λαβεῖν τρία, ἥ τε χλοερὰ καὶ ποιώδης ἥπερ καὶ πλείστη, καὶ ἑτέρα ἡ λευκή, καὶ τρίτη ἡ ποικίλη, ἣν δὴ καλοῦσί τινες Θρᾳκίαν.

[1] *cf.* Theocr. 11. 46. [2] Plin. 16. 145 foll.
[3] *i e.* is the most 'distinct' of the ivies.
[4] *cf.* 1. 10. 1; Diosc. 2. 179.
[5] *i.e.* as an explanation of the barrenness of *helix*.
[6] *i.e.* and so becomes fertile.
[7] διατείνονται: *cf.* *C.P.* 4. 6. 1. διατ. τῷ . . . apparently = "insist on the view that," . . . but the dative is strange. The sentence, which is highly elliptical, is freely emended by most editors.

kind some call *korymbias*, but the Athenians call it the 'Acharnian' ivy. Another kind is smaller and loose in growth like the black ivy.[1] There are also variations in the black kind, but they are not so well marked.

[2] The *helix* presents the greatest differences[3]; the principal difference is in the leaves,[4] which are small angular and of more graceful proportions, while those of the ivy proper are rounder and simple; there is also difference in the length of the twigs, and further in the fact that this tree is barren. For,[5] as to the view that the *helix* by natural development turns into the ivy,[6] some insist[7] that this is not so, the only true ivy according to these being that which was ivy from the first[8]; (whereas if, as some say, the *helix* invariably[9] turns into ivy, the difference would be merely one of age and condition, and not of kind, like the difference between the cultivated and the wild pear). However the leaf even of the full-grown *helix* is very different from that of the ivy, and it happens but rarely and in a few specimens that in this plant a change in the leaf occurs as it grows older, as it does in the abele and the castor-oil plant.[10] [11] There are several forms of the *helix*, of which the three most conspicuous and important are the green 'herbaceous' kind (which is the commonest), the white, and the variegated, which some call the 'Thracian' *helix*. Each of these appears to

[8] *i.e.* and *helix* being a distinct plant which is always barren.

[9] πᾶσα conj. Sch.; πᾶς Ald.

[10] Sc. as well as in *ivy*; *cf.* 1. 10. 1, where this change is said to be characteristic of these three trees. (The rendering attempted of this obscure section is mainly from W.'s note.)

[11] Plin. 16. 148 foll.

ἑκάστη δὲ τούτων δοκεῖ διαφέρειν· καὶ γὰρ τῆς χλοώδους ἡ μὲν λεπτοτέρα καὶ ταξιφυλλοτέρα καὶ ἔτι πυκνοφυλλοτέρα, ἡ δ' ἧττον πάντα ταῦτ' ἔχουσα. καὶ τῆς ποικίλης ἡ μὲν μεῖζον ἡ δ' ἔλαττον τὸ φύλλον, καὶ τὴν ποικιλίαν διαφέρουσα. ὡσαύτως δὲ καὶ τὰ τῆς λευκῆς τῷ μεγέθει καὶ τῇ χροιᾷ διαφέρουσιν. εὐαυξεστάτη δὲ ἡ ποιώδης καὶ ἐπὶ πλεῖστον προϊοῦσα. φανερὰν δ' εἶναί φασιν τὴν ἀποκιττουμένην οὐ μόνον τοῖς φύλλοις ὅτι μείζω καὶ πλατύτερα ἔχει ἀλλὰ καὶ τοῖς βλαστοῖς· εὐθὺς γὰρ ὀρθοὺς ἔχει, καὶ οὐχ ὥσπερ ἡ ἑτέρα κατακεκαμμένη, καὶ διὰ τὴν λεπτότητα καὶ διὰ τὸ μῆκος· τῆς δὲ κιττώδους καὶ βραχύτεροι καὶ παχύτεροι. καὶ ὁ κιττὸς ὅταν ἄρχηται σπερμοῦσθαι μετέωρον ἔχει καὶ ὀρθὸν τὸν βλαστόν.

9 Πολύρριζος μὲν οὖν ἅπας κιττὸς καὶ πυκνόρριζος συνεστραμμένος ταῖς ῥίζαις καὶ ξυλώδεσι καὶ παχείαις καὶ οὐκ ἄγαν βαθύρριζος, μάλιστα δ' ὁ μέλας, καὶ τοῦ λευκοῦ ὁ τραχύτατος καὶ ἀγριώτατος· δι' ὃ καὶ χαλεπὸς παραφύεσθαι πᾶσι τοῖς δένδροις· ἀπόλλυσι γὰρ πάντα καὶ ἀφαναίνει παραιρούμενος τὴν τροφήν. λαμβάνει δὲ μάλιστα πάχος οὗτος καὶ ἀποδενδροῦται καὶ γίνεται αὐτὸ καθ' αὑτὸ κιττοῦ δένδρον. ὡς δ' ἐπὶ τὸ πλεῖον εἶναι πρὸς ἑτέρῳ φιλεῖ καὶ ζητεῖ καὶ ὥσπερ

10 ἐπαλλόκαυλόν ἐστιν. ἔχει δ' εὐθὺς καὶ τῆς

---

[1] ταξιφυλλοτέρα conj. W. from Plin. 16. 149, *folia in ordinem digesta*; μακροφυλλοτέρα MSS. *cf.* 1. 10. 8.

[2] κατακεκαμμένη conj. W.; κατακεκαυμένη UAld.; κατακεκαμμένους conj. Sch.

[3] κιττώδους MSS.; ποώδους conj. St.

[4] *cf. C.P.* 1. 16. 4.

present variations; of the green one form is slenderer and has more regular[1] and also closer leaves, the other has all these characteristics in a less degree. Of the variegated kind again one sort has a larger, one a smaller leaf, and the variegation is variable. In like manner the various forms of the white *helix* differ in size and colour. The 'herbaceous' kind is the most vigorous and covers most space. They say that the form which is supposed to turn into ivy is clearly marked not only by its leaves, because they are larger and broader, but also by its shoots; for these are straight from the first, and this form does not bend over[2] like the other; also because the shoots are slenderer and larger, while those of the ivy-like[3] form are shorter and stouter. [4]The ivy too, when it begins to seed, has its shoots upward-growing and erect.

All ivies have numerous close roots, which are tangled together woody and stout, and do not run very deep; but this is specially true of the black kind and of the roughest and wildest forms of the white. Wherefore it is mischievous to plant this against any tree; for it destroys and starves any tree by withdrawing the moisture. This form also more than the others grows stout and becomes tree-like, and in fact becomes itself an independent ivy tree, though in general it likes and seeks to be[5] against another tree, and is, as it were, parasitic.[6] [7]Moreover from the first it has also this natural

[5] *εἶναι* conj. W.; *αἰεὶ* UM; *ἀεὶ* Ald.

[6] *i.e.* depends on another tree; not, of course, in the strict botanical sense. *cf.* 3. 18. 11. *ἐπαλλόκαυλον* conj. Scal.; *ἐπαυλόκαλον* MVAld.U (with *υ* corrected). *cf.* *περιαλλόκαυλος*, 7. 8. 1; *C.P.* 2. 18. 2.

[7] Plin. 16. 152.

φύσεώς τι τοιοῦτον· ἐκ γὰρ τῶν βλαστῶν ἀφίησιν ἀεὶ ῥίζας ἀνὰ μέσον τῶν φύλλων, αἷσπερ ἐνδύεται τοῖς δένδροις καὶ τοῖς τειχίοις οἷον ἐξεπίτηδες πεποιημέναις ὑπὸ τῆς φύσεως· δι' ὃ καὶ ἐξαιρούμενος τὴν ὑγρότητα καὶ ἕλκων ἀφαναίνει, καὶ ἐὰν ἀποκοπῇ κάτωθεν δύναται διαμένειν καὶ ζῆν. ἔχει δὲ καὶ ἑτέραν διαφορὰν κατὰ τὸν καρπὸν οὐ μικράν· ὁ μὲν γὰρ ἐπίγλυκύς ἐστιν ὁ δὲ σφόδρα πικρὸς καὶ τοῦ λευκοῦ καὶ τοῦ μέλανος· σημεῖον δ' ὅτι τὸν μὲν ἐσθίουσιν οἱ ὄρνιθες τὸν δ' οὔ. τὰ μὲν οὖν περὶ τὸν κιττὸν οὕτως ἔχει.

11 Ἡ δὲ σμῖλάξ ἐστι μὲν ἐπαλλόκαυλον, ὁ δὲ καυλὸς ἀκανθώδης καὶ ὥσπερ ὀρθάκανθος, τὸ δὲ φύλλον κιττῶδες μικρὸν ἀγώνιον, κατὰ τὴν μίσχου πρόσφυσιν τυληρόν. ἴδιον δ' ὅτι τήν τε διὰ μέσου ταύτην ὥσπερ ῥάχιν λεπτὴν ἔχει καὶ τὰς στημονίους διαλήψεις οὐκ ἀπὸ ταύτης, ὥσπερ τὰ τῶν ἄλλων, ἀλλὰ περὶ αὐτὴν περιφερεῖς ἠγμένας ἀπὸ τῆς προσφύσεως τοῦ μίσχου τῷ φύλλῳ. παρὰ δὲ τοῦ καυλοῦ τὰ γόνατα καὶ παρὰ τὰς διαλείψεις τὰς φυλλικὰς ἐκ τῶν αὐτῶν μίσχων τοῖς φύλλοις παραπέφυκεν ἴουλος λεπτὸς καὶ ἑλικτός· ἄνθος δὲ λευκὸν καὶ εὐῶδες λείρινον·

---

[1] σμῖλαξ: ? μῖλαξ W. *cf.* 1. 10. 5; Plin. 16. 153–155.
[2] ἐπαλλόκαυλον conj. Sch.; ἐπαυλόκαυλον V. *cf.* 3. 18. 10.
[3] καυλὸς conj. R. Const.; καρπὸς UMVAld.
[4] τυληρόν conj. W.; νοτηρόν Ald.U (corrected).
[5] ταύτην: *cf.* τὸ θυλακῶδες τοῦτο, 3. 7. 3. Is the pronoun

characteristic, that it regularly puts forth roots from the shoots between the leaves, by means of which it gets a hold of trees and walls, as if these roots were made by nature on purpose. Wherefore also by withdrawing and drinking up the moisture it starves its host, while, if it is cut off below, it is able to survive and live. There are also other not inconsiderable differences in the fruit; both in the white and in the black kind it is in some cases rather sweet, in others extremely bitter; in proof whereof birds eat one but not the other. Such are the facts about ivy.

The smilax[1] is parasitic,[2] but its stem[3] is thorny and has, as it were, straight thorns; the leaf is ivy-like small and without angles, and makes a callus[4] at the junction with the stalk. A peculiarity of it is its conspicuous[5] slender midrib, so to call it, which divides it in two; also the fact that the thread-like branchings[6] do not start from this, as in other leaves, but are carried in circles round it, starting from the junction of the leaflet with the leaf. And at the joints of the stem[7] and the spaces between the leaves there grows from the same stalk as the leaves a fine spiral tendril.[8] The flower is white and fragrant like a lily[9] The fruit

deictic, referring to an actual specimen shewn in lecture? *cf.* also 4. 7. 1.

[6] διαλήψεις Ald.; διαλείψεις UMV. A mistake probably due to διαλείψεις below, where it is right. διάληψις is the Aristotelian word for a 'division.'

[7] τοῦ καυλοῦ τὰ γόνατα conj. Sch.; τὸν καυλὸν τὸν ἄτονον Ald.

[8] This must be the meaning of ἴουλος here, qualified by ἑλικτός; but elsewhere it = catkin. *cf.* 3. 5. 5.

[9] λείρινον conj. R. Const. from Plin. *l.c.* *olente lilium*; ἠρινόν UAld.

τὸν δὲ καρπὸν ἔχει προσεμφερῆ τῷ στρύχνῳ καὶ
τῷ μηλώθρῳ καὶ μάλιστα τῇ καλουμένῃ σταφυλῇ
12 ἀγρίᾳ· κατακρέμαστοι δ' οἱ βότρυες κιττοῦ τρό-
πον· παρεγγίζει δ' ὁ παραθριγκισμὸς πρὸς τὴν
σταφυλήν· ἀπὸ γὰρ ἑνὸς σημείου οἱ μίσχοι οἱ
ῥαγικοί. ὁ δὲ καρπὸς ἐρυθρός, ἔχων πυρῆνας τὸ
μὲν ἐπὶ πᾶν δύο, ἐν τοῖς μείζοσι τρεῖς ἐν δὲ τοῖς
μικροῖς ἕνα· σκληρὸς δ' ὁ πυρὴν εὖ μάλα καὶ τῷ
χρώματι μέλας ἔξωθεν. ἴδιον δὲ τὸ τῶν βοτρύων,
ὅτι ἐκ πλαγίων τε τὸν καυλὸν παραθριγκίζουσιν,
καὶ κατ' ἄκρον ὁ μέγιστος βότρυς τοῦ καυλοῦ,
ὥσπερ ἐπὶ τῆς ῥάμνου καὶ τοῦ βάτου. τοῦτο δὲ
δῆλον ὡς καὶ ἀκρόκαρπον καὶ πλαγιόκαρπον.
13 [Τὸ δ' εὐώνυμος καλούμενον δένδρον φύεται μὲν
ἄλλοθί τε καὶ τῆς Λέσβου ἐν τῷ ὄρει τῷ Ὀρδύν-
νῳ καλουμένῳ· ἔστι δὲ ἡλίκον ῥόα καὶ τὸ φύλλον
ἔχει ῥοῶδες, μεῖζον δὲ ἢ χαμαιδάφνης, καὶ μαλακὸν δὲ ὥσπερ ἡ ῥόα. ἡ δὲ βλάστησις ἄρχεται
μὲν αὐτῷ περὶ τὸν Ποσειδεῶνα· ἀνθεῖ δὲ τοῦ
ἦρος· τὸ δὲ ἄνθος ὅμοιον τὴν χρόαν τῷ λευκῷ
ἴῳ· ὄζει δὲ δεινὸν ὥσπερ φόνου. ὁ δὲ καρπὸς
ἐμφερὴς τὴν μορφὴν μετὰ τοῦ κελύφους τῷ τοῦ
σησάμου λοβῷ· ἔνδοθεν δὲ στερεὸν πλὴν διῃρη-
μένον κατὰ τὴν τετραστοιχίαν. τοῦτο ἐσθιό-

---

[1] Presumably σ. ὁ ἐδώδιμος· See Index.

[2] παρεγγίζει δ' ὁ παραθριγκισμὸς I conj., cf. παραθριγκίζουσι below; παρωγγύζει δὲ παραθρινακίζει δὲ ὡς U; παραγγίζει δὲ παραθρηνακίζει δὲ ὡς MV; παραθριγκίζει δὲ ὡς conj. W.

is like the *strykhnos*[1] and the *melothron* (bryony), and most of all like the berry which is called the 'wild grape' (bryony). The clusters hang down as in the ivy, but the regular setting[2] of the berries resembles the grape-cluster more closely; for the stalks which bear the berries start from a single point. The fruit is red, having generally two stones, the larger ones three and the smaller one; the stone is very hard and in colour black outside. A peculiarity of the clusters is that they make a row[3] along the sides of the stalk, and the longest cluster is at the end of the stalk, as in the buckthorn and the bramble. It is clear that the fruit is produced both at the end and at the sides.

[4] The tree called the spindle-tree[5] grows, among other places, in Lesbos, on the mountain called Ordynnos.[6] It is as large as the pomegranate and has a leaf like that of that tree, but larger than that of the periwinkle,[7] and soft, like the pomegranate leaf. It begins to shoot about the month Poseideon,[8] and flowers in the spring; the flower in colour is like the gilliflower, but it has a horrible smell, like shed blood.[9] The fruit, with its case, is like the pod of sesame[10]; inside it is hard, but it splits easily according to its four divisions. This tree, if eaten

[3] *παραθριγκίζουσιν* conj. Sch.; *παραθρυγκίζουσαν* U (corrected); *παραθρυγγίζουσι* M.

[4] This section down to the word *ἀνόχῳ* is clearly out of place: *εὐώνυμος* was not one of the plants proposed for discussion 3. 18. 1. It should come somewhere among the descriptions of trees characteristic of special localities.

[5] Plin. 13. 118. [6] *cf.* Plin. 5. 140.

[7] This irrelevant comparison probably indicates confusion in the text, as is shewn also by Pletho's excerpt of part of this section: see Sch.

[8] January. [9] *φόνον*: *cf.* 6. 4. 6. [10] *cf.* 8. 5. 2.

μενον ὑπὸ τῶν προβάτων ἀποκτιννύει, καὶ τὸ φύλλον καὶ ὁ καρπός, καὶ μάλιστα τὰς αἶγας ἐὰν μὴ καθάρσεως τύχῃ. καθαίρεται δὲ ἀνόχῳ.] περὶ μὲν οὖν δένδρων καὶ θάμνων εἴρηται· ἐν δὲ τοῖς ἑξῆς περὶ τῶν λειπομένων λεκτέον.

by sheep, is fatal[1] to them, both the leaf and the fruit, and it is especially fatal to goats unless they are purged by it; and the purging is effected by diarrhoea.[2] So we have spoken of trees and shrubs; in what follows we must speak of the plants which remain.

[1] In Pletho's excerpt (see above) this is said of periwinkle.
[2] *i.e.* and not by vomiting.

BOOK IV

Δ

I. Αἱ μὲν οὖν διαφοραὶ τῶν ὁμογενῶν τεθεώρηνται πρότερον. ἅπαντα δ᾽ ἐν τοῖς οἰκείοις τόποις καλλίω γίνεται καὶ μᾶλλον εὐσθενεῖ· καὶ γὰρ τοῖς ἀγρίοις εἰσὶν ἑκάστοις οἰκεῖοι, καθάπερ τοῖς ἡμέροις· τὰ μὲν γὰρ φιλεῖ τοὺς ἐφύδρους καὶ ἑλώδεις, οἷον αἴγειρος λεύκη ἰτέα καὶ ὅλως τὰ παρὰ τοὺς ποταμοὺς φυόμενα, τὰ δὲ τοὺς εὐσκεπεῖς καὶ εὐηλίους, τὰ δὲ μᾶλλον τοὺς παλισκίους. πεύκη μὲν γὰρ ἐν τοῖς προσείλοις καλλίστη καὶ μεγίστη, ἐν δὲ τοῖς παλισκίοις ὅλως οὐ φύεται· ἐλάτη δὲ ἀνάπαλιν ἐν τοῖς παλισκίοις καλλίστη τοῖς δ᾽ εὐείλοις οὐχ ὁμοίως.

2 Ἐν Ἀρκαδίᾳ γοῦν περὶ τὴν Κράνην καλουμένην τόπος ἐστί τις κοῖλος καὶ ἄπνους, εἰς ὃν οὐδέποθ᾽ ὅλως ἥλιον ἐμβάλλειν φασίν· ἐν τούτῳ δὲ πολὺ διαφέρουσιν αἱ ἐλάται καὶ τῷ μήκει καὶ τῷ πάχει, οὐ μὴν ὁμοίως γε πυκναὶ οὐδ᾽ ὡραῖαι ἀλλ᾽ ἥκιστα, καθάπερ καὶ αἱ πεῦκαι αἱ ἐν τοῖς παλισκίοις· δι᾽ ὃ καὶ πρὸς τὰ πολυτελῆ τῶν ἔργων, οἷον θυρώματα καὶ εἴ τι ἄλλο σπουδαῖον, οὐ χρῶνται τούτοις ἀλλὰ πρὸς τὰς ναυπηγίας μᾶλλον καὶ τὰς οἰκοδομάς· καὶ γὰρ δοκοὶ κάλλι-

BOOK IV

Of the Trees and Plants special to particular Districts and Positions.

Of the importance of position and climate.

I. The differences between trees of the same kind have already been considered. Now all grow fairer and are more vigorous in their proper positions; for wild, no less than cultivated trees, have each their own positions: some love wet and marshy ground, as black poplar abele willow, and in general those that grow by rivers; some love exposed[1] and sunny positions; some prefer a shady place. The fir is fairest and tallest in a sunny position, and does not grow at all in a shady one; the silver-fir on the contrary is fairest in a shady place, and not so vigorous in a sunny one.

Thus there is in Arcadia near the place called Krane a low-lying district sheltered from wind, into which they say that the sun never strikes; and in this district the silver-firs excel greatly in height and stoutness, though they have not such close grain nor such comely wood, but quite the reverse,—like the fir when it grows in a shady place. Wherefore men do not use these for expensive work, such as doors or other choice articles, but rather for ship-building and house-building. For excellent

[1] εὐσκεπεῖς should mean 'sheltered,' but cannot in this context, nor in *C.P.* 1. 13. 11 and 12: the word seems to have been confused with εὔσκοπος.

σται καὶ ταυεῖαι καὶ κέραιαι αἱ ἐκ τούτων, ἔτι δ' ἱστοὶ τῷ μήκει διαφέροντες ἀλλ' οὐχ ὁμοίως ἰσχυροί· καὶ ἐκ τῶν προσείλων ἅμα τῇ βραχύτητι πυκνότεροί τε ἐκείνων καὶ ἰσχυρότεροι γίνονται.

3 Χαίρει δὲ σφόδρα καὶ ἡ μίλος τοῖς παλισκίοις καὶ ἡ πάδος καὶ ἡ θραύπαλος. περὶ δὲ τὰς κορυφὰς τῶν ὀρέων καὶ τοὺς ψυχροὺς τόπους θυία μὲν φύεται καὶ εἰς ὕψος, ἐλάτη δὲ καὶ ἄρκευθος φύεται μὲν οὐκ εἰς ὕψος δέ, καθάπερ καὶ περὶ τὴν ἄκραν Κυλλήνην· φύεται δὲ καὶ ἡ κήλαστρος ἐπὶ τῶν ἄκρων καὶ χειμεριωτάτων. ταῦτα μὲν οὖν ἄν τις θείη φιλόψυχρα· τὰ δ' ἄλλα πάντα ὡς εἰπεῖν [οὐ] μᾶλλον χαίρει τοῖς προσείλοις. οὐ μὴν ἀλλὰ καὶ τοῦτο συμβαίνει κατὰ τὴν χώραν τὴν οἰκείαν ἑκάστῳ τῶν δένδρων. ἐν Κρήτῃ γοῦν φασιν ἐν τοῖς Ἰδαίοις ὄρεσι καὶ ἐν τοῖς Λευκοῖς καλουμένοις ἐπὶ τῶν ἄκρων ὅθεν οὐδέποτ' ἐπιλείπει χιὼν κυπάριττον εἶναι· πλείστη γὰρ αὕτη τῆς ὕλης καὶ ὅλως ἐν τῇ νήσῳ καὶ ἐν τοῖς ὄρεσιν.

4 Ἔστι δέ, ὥσπερ καὶ πρότερον εἴρηται, καὶ τῶν ἀγρίων καὶ τῶν ἡμέρων τὰ μὲν ὀρεινὰ τὰ δὲ πεδεινὰ μᾶλλον. ἀναλογίᾳ δὲ καὶ ἐν αὐτοῖς τοῖς ὄρεσι τὰ μὲν ἐν τοῖς ὑποκάτω τὰ δὲ περὶ τὰς κορυφάς, ὥστε καὶ καλλίω γίνεται καὶ εὐσθενῆ πανταχοῦ δὲ καὶ πάσης τῆς ὕλης πρὸς βορρᾶν τὰ ξύλα πυκνότερα καὶ οὐλότερα καὶ ἁπλῶς καλλίω· καὶ ὅλως δὲ πλείω ἐν τοῖς προσβορείοις φύεται. αὐξάνεται δὲ καὶ ἐπιδίδωσι τὰ πυκνὰ

---

[1] I omit αἱ before κέραιαι with P.

[2] ἅμα I conj.; ἀλλὰ Ald.; om. W. after Sch.; ἀλλ' ἅμα conj. St.

rafters beams and yard-arms[1] are made from these, and also masts of great length which are not however equally strong; while masts made of trees grown in a sunny place are necessarily[2] short but of closer grain and stronger than the others.

Yew *pados* and joint-fir rejoice exceedingly in shade. On mountain tops and in cold positions odorous cedar grows even to a height, while silver-fir and Phoenician cedar grow, but not to a height,—for instance on the top of Mount Cyllene; and holly also grows in high and very wintry positions. These trees then we may reckon as cold-loving; all others, one may say in general, prefer a sunny position. However this too depends partly on the soil appropriate to each tree; thus they say that in Crete on the mountains of Ida and on those called the White Mountains the cypress is found on the peaks whence the snow never disappears; for this is the principal tree both in the island generally and in the mountains.

Again, as has been said[3] already, both of wild and of cultivated trees some belong more to the mountains, some to the plains. And on the mountains themselves in proportion to the height some grow fairer[4] and more vigorous in the lower regions, some about the peaks. However it is true of all trees anywhere that with a north aspect the wood is closer and more compact[5] and better generally; and, generally speaking, more trees grow in positions facing the north. Again trees which are close

[3] 3. 2. 4.

[4] Something seems to have dropped out before ὥστε.

[5] οὐλότερα conj. W. from mutilated word in U; καλλιώτερα MV; καλλίω Ald.

μὲν ὄντα μᾶλλον εἰς μῆκος, δι' ὃ καὶ ἄνοζα καὶ εὐθέα καὶ ὀρθοφυῆ γίνεται, καὶ κωπεῶνες ἐκ τούτων κάλλιστοι· <τὰ δὲ μανὰ> μᾶλλον εἰς βάθος καὶ πάχος, δι' ὃ καὶ σκολιώτερα καὶ ὀζωδέστερα καὶ τὸ ὅλον στερεώτερα καὶ πυκνότερα φύεται.

5 Σχεδὸν δὲ τὰς αὐτὰς ἔχει διαφορὰς τούτοις καὶ ἐν τοῖς παλισκίοις καὶ ἐν τοῖς εὐείλοις καὶ ἐν τοῖς ἀπνόοις καὶ εὐπνόοις· ὀζωδέστερα γὰρ καὶ βραχύτερα καὶ ἧττον εὐθέα τὰ ἐν τοῖς εὐείλοις ἢ τοῖς προσηνέμοις. ὅτι δὲ ἕκαστον ζητεῖ καὶ χώραν οἰκείαν καὶ κρᾶσιν ἀέρος φανερὸν τῷ τὰ μὲν φέρειν ἐνίους τόπους τὰ δὲ μὴ φέρειν μήτε αὐτὰ γιγνόμενα μήτε φυτευόμενα ῥᾳδίως, ἐὰν δὲ καὶ ἀντιλάβηται μὴ καρποφορεῖν, ὥσπερ ἐπὶ τοῦ φοίνικος ἐλέχθη καὶ τῆς Αἰγυπτίας συκαμίνου καὶ ἄλλων· εἰσὶ γὰρ πλείω καὶ ἐν πλείοσι χώραις τὰ μὲν ὅλως οὐ φυόμενα τὰ δὲ φυόμενα μὲν ἀναυξῆ δὲ καὶ ἄκαρπα καὶ τὸ ὅλον φαῦλα. περὶ ὧν ἴσως λεκτέον ἐφ' ὅσον ἔχομεν ἱστορίας.

II. Ἐν Αἰγύπτῳ γάρ ἐστιν ἴδια δένδρα πλείω, ἥ τε συκάμινος καὶ ἡ περσέα καλουμένη καὶ ἡ βάλανος καὶ ἡ ἄκανθα καὶ ἕτερ' ἄττα.

Ἔστι δὲ ἡ μὲν συκάμινος παραπλησία πως τῇ ἐνταῦθα συκαμίνῳ· καὶ γὰρ τὸ φύλλον παρόμοιον

---

[1] κωπεῶνες: *cf.* 5. 1. 7.
[2] τὰ δὲ μανὰ add. W.
[3] *cf.* 5. 1. 8.
[4] 2. 2. 10.
[5] ὅλως . . . μὲν conj. W.; ὅλως οὐ φυτευόμενα U; ὅλως φυτευόμενα MVPAld.

together grow and increase more in height, and so become unbranched straight and erect, and the best oar-spars[1] are made from these, while those that grow far apart[2] are of greater bulk and denser habit[3]; wherefore they grow less straight and with more branches, and in general have harder wood and a closer grain.

Such trees exhibit nearly the same differences, whether the position be shady or sunny, windless or windy; for trees growing in a sunny or windy position are more branched shorter and less straight. Further that each tree seeks an appropriate position and climate is plain from the fact that some districts bear some trees but not others; (the latter do not grow there of their own accord, nor can they easily be made to grow), and that, even if they obtain a hold, they do not bear fruit—as was said[4] of the date-palm the sycamore and others; for there are many trees which in many places either do not grow at all, or,[5] if they do, do not thrive nor bear fruit, but are in general of inferior quality. And perhaps we should discuss this matter, so far as our enquiries go.

Of the trees special to Egypt, and of the carob.

II. [6] Thus in Egypt there are a number of trees which are peculiar[7] to that country, the sycamore the tree called *persea* the *balanos* the acacia and some others.

Now the sycamore to a certain extent resembles the tree which bears that name[8] in our country; its

[6] Plin. 13. 56 and 57.
[7] ἴδια conj. R. Const.; ἔνια Ald.
[8] *i.e.* mulberry. See Index.

ἔχει καὶ τὸ μέγεθος καὶ τὴν ὅλην πρόσοψιν, τὸν δὲ καρπὸν ἰδίως φέρει παρὰ τὰ ἄλλα, καθάπερ ἐλέχθη καὶ ἐν τοῖς ἐξ ἀρχῆς· οὐ γὰρ ἀπὸ τῶν βλαστῶν οὐδ᾽ ἀπὸ τῶν ἀκρεμόνων ἀλλ᾽ ἐκ τοῦ στελέχους, μέγεθος μὲν ἡλίκον σῦκον καὶ τῇ ὄψει δὲ παραπλήσιον, τῷ χυλῷ δὲ καὶ τῇ γλυκύτητι τοῖς ὀλύνθοις, πλὴν γλυκύτερον πολὺ καὶ κεγχραμίδας ὅλως οὐκ ἔχοντα, πλήθει δὲ πολύν. καὶ πέττειν οὐ δύναται μὴ ἐπικνισθέντα· ἀλλ᾽ ἔχοντες ὄνυχας σιδηροῦς ἐπικνίζουσιν· ἃ δ᾽ ἂν ἐπικνισθῇ τεταρταῖα πέττεται· τούτων δ᾽ ἀφαιρεθέντων πάλιν ἄλλα φύεται καὶ ἄλλα καὶ ἐκ τοῦ αὐτοῦ τόπου μηδὲν παραλλάττοντα· καὶ τοῦθ᾽ οἱ μὲν τρὶς οἱ δὲ πλεονάκις φασὶ γίνεσθαι.

2 πολύοπον δὲ τὸ δένδρον σφόδρα ἐστὶ καὶ τὸ ξύλον αὐτοῦ εἰς πολλὰ χρήσιμον. ἴδιον δὲ ἔχειν δοκεῖ παρὰ τἆλλα· τμηθὲν γὰρ εὐθὺς χλωρόν ἐστι· αὐαίνεται δὲ ἐμβύθιον· εἰς βόθρον δὲ ἐμβάλλουσι καὶ εἰς τὰς λίμνας εὐθὺς καὶ ταριχεύουσι· βρεχόμενον δ᾽ ἐν τῷ βυθῷ ξηραίνεται· καὶ ὅταν τελέως ξηρὸν γένηται, τότε ἀναφέρεται καὶ ἐπινεῖ καὶ δοκεῖ τότε καλῶς τεταριχεῦσθαι· γίνεται γὰρ κοῦφον καὶ μανόν. ἡ μὲν οὖν συκάμινος ἔχει ταύτας τὰς ἰδιότητας.

3 Ἔοικε δέ τις παραπλησία ἡ φύσις εἶναι καὶ τῆς ἐν Κρήτῃ καλουμένης Κυπρίας συκῆς· καὶ γὰρ ἐκείνη φέρει τὸν καρπὸν ἐκ τοῦ στελέχους καὶ ἐκ τῶν παχυτάτων ἀκρεμόνων, πλὴν ὅτι βλαστόν τινα ἀφίησι μικρὸν ἄφυλλον ὥσπερ ῥιζίον, πρὸς ᾧ γε ὁ καρπός. τὸ δὲ στέλεχος μέγα

---

[1] 1. 1. 7; *cf.* 1. 14 2.

[2] *cf.* *C.P.* 1. 17. 9; Diosc. 1. 127; Athen. 2. 36. This

leaf is similar, its size, and its general appearance; but it bears its fruit in a quite peculiar manner, as was said at the very outset[1]; it is borne not on the shoots or branches, but on the stem; in size it is as large as a fig, which it resembles also in appearance, but in flavour and sweetness it is like the 'immature figs,' except that it is much sweeter and contains absolutely no seeds, and it is produced in large numbers. It cannot ripen unless it is scraped; but they scrape it with iron 'claws'[2]; the fruits thus scraped ripen in four days. If these are removed, others and others again grow from exactly the same point, and this some say occurs three times over, others say it can happen more times than that. Again the tree is very full of sap, and its wood is useful for many purposes. There is another peculiar property which it appears to possess; when it is cut, it is at first green, but it dries in deep water[3]; they put it at once in a hole or in pools and so season it; and it becomes dry by being soaked in the deep water, and when it is completely dry, it is fetched up and floats and is then thought to be duly seasoned; for it is now light and porous. Such are the peculiarities of the sycamore.

Somewhat similar appears to be the character of the tree which in Crete is called the 'Cyprian fig'[4] (sycamore). For this also bears its fruit on the stem and on the thickest branches; but in this case there is a small leafless shoot, like a root, to which the fruit is attached. The stem is large and like the

scraping was the prophet Amos' occupation: *cf.* Amos 7. 14. comm.

[3] ἐμβύθιον conj. W.; εἰς βύθον UMVPAld. ? ἐν βύθῳ ὄν.

[4] See Index. *cf.* Athen. 3. 11; Plin. 13. 58; Diosc. 1. 127. 3.

καὶ παρόμοιον τῇ λεύκῃ, φύλλον δὲ τῇ πτελέᾳ. πεπαίνει δὲ τέτταρας καρπούς, ὅσαιπερ αὐτοῦ καὶ αἱ βλαστήσεις· οὐδένα δὲ πεπαίνει μὴ ἐπιτμηθέντος τοῦ ἐρινοῦ καὶ ἐκρυεντος τοῦ ὀποῦ. ἡ δὲ γλυκύτης προσεμφερὴς τῷ σύκῳ καὶ τὰ ἔσωθεν τοῖς ἐρινοῖς· μέγεθος ἡλίκον κοκκύμηλον.

4 (Ταύτῃ δὲ παραπλησία καὶ ἣν οἱ Ἴωνες κερωνίαν καλοῦσιν· ἐκ τοῦ στελέχους γὰρ καὶ αὕτη φέρει τὸν πλεῖστον καρπόν, ἀπὸ δὲ τῶν ἀκρεμόνων, ὥσπερ εἴπομεν, ὀλίγον. ὁ δὲ καρπὸς ἔλλοβος, ὃν καλοῦσί τινες Αἰγύπτιον σῦκον διημαρτηκότες· οὐ γίνεται γὰρ ὅλως περὶ Αἴγυπτον ἀλλ' ἐν Συρίᾳ καὶ ἐν Ἰωνίᾳ δὲ καὶ περὶ Κνίδον καὶ Ῥόδον. ἀείφυλλον δὲ καὶ ἄνθος ἔκλευκον ἔχον καί τι βαρύτητος, μὴ μετεωρίζον δὲ σφόδρα καὶ ὅλως ἐκ τῶν κάτω παραβλαστητικὸν ἄνωθεν δὲ ὑποξηραινόμενον. ἔχει δὲ ἅμα καὶ τὸν ἔνον καὶ τὸν νέον καρπόν· ἀφαιρουμένου γὰρ θατέρου μετὰ Κύνα καὶ ὁ ἕτερος εὐθὺς φανερὸς κυούμενος· κύεται γὰρ ὥσπερ βότρυς ὁμοσχήμων· εἶτ' αὐξηθεὶς ἀνθεῖ περὶ Ἀρκτοῦρον καὶ ἰσημερίαν· ἀπὸ τούτου δὴ διαμένει τὸν χειμῶνα μέχρι Κυνός. ἡ μὲν οὖν ὁμοιότης ὅτι στελεχόκαρπα καὶ ταῦτα· διαφοραὶ δὲ αἱ εἰρημέναι πρὸς τὴν συκάμινον.)

5 Ἐν Αἰγύπτῳ δ' ἐστὶν ἕτερον ἡ περσέα καλούμενον, τῇ μὲν προσόψει μέγα καὶ καλόν, παραπλήσιον δὲ μάλιστα τῇ ἀπίῳ καὶ φύλλοις καὶ ἄνθεσι καὶ ἀκρεμόσι καὶ τῷ ὅλῳ σχήματι· πλὴν

---

[1] ὅσαιπερ conj. R. Const., etc., *cf.* Athen. *l.c.*; ὅσα ὑπὲρ αὐτοῦ U (corrected); ὅσα ὑπὲρ αὐτὸν M; ὅσα ὑπὲρ αὐτοῦ Ald.
[2] Plin. 13. 59. [3] 1. 14. 2.

abele, but the leaf is like that of the elm. It ripens its fruit four times a year, having also[1] four periods of growth; but it ripens no fruit unless the 'fig' is split and the juice let out. The sweet taste resembles that of the fig, and the inside of the fruit is like that of wild figs: it is as large as a plum.

[2] (Like this too is the tree which the Ionians call carob; for this too bears most of its fruit on the stem, though it bears a little also on the branches, as we said.[3] The fruit is in a pod; some call it the 'Egyptian fig'—erroneously; for it does not occur at all in Egypt, but in Syria and Ionia and also in Cnidos and Rhodes. It is evergreen and has a whitish flower and is somewhat acrid; it does not attain to a great height, and it sends out side-shoots entirely from its lower parts, while it withers above. It has on it at the same time both last year's fruit and the new fruit; for if the one is removed after the rising of the dog-star, immediately the other is seen swelling up; for there swells[4] up as it were another similar cluster. This then increases and flowers about the rising of Arcturus and the equinox; and thenceforward it[5] persists through the winter to the rising of the dog-star. The likeness then consists in the fact that these trees too bear fruit on their stems, and the differences between them and the sycamore are as has been said.)

[6] In Egypt there is another tree called the *persea*, which in appearance is large and fair, and it most resembles the pear in leaves flowers branches and general form, but it is evergreen, while the other is

[4] κυεται conj. W. from G; κύει MSS.
[5] *i.e.* the cluster, now in the fruit stage.
[6] Plin. 13. 60 and 61.

τὸ μὲν ἀείφυλλον τὸ δὲ φυλλοβόλον. καρπον δὲ φέρει πολὺν καὶ πᾶσαν ὥραν· περικαταλαμβάνει γὰρ ὁ νέος ἀεὶ τὸν ἔνον· πέττει δὲ ὑπὸ τοὺς ἐτησίας· τὸν δ' ἄλλον ὠμότερον ἀφαιροῦσι καὶ ἀποτιθέασιν. ἔστι δὲ τὸ μέγεθος ἡλίκον ἄπιος, τῷ σχήματι δὲ πρόμακρος ἀμυγδαλώδης, χρῶμα δὲ αὐτοῦ ποιῶδες. ἔχει δὲ ἐντὸς κάρυον, ὥσπερ τὸ κοκκύμηλον, πλὴν ἔλαττον πολὺ καὶ μαλακώτερον· τὴν δὲ σάρκα γλυκεῖαν σφόδρα καὶ ἡδεῖαν καὶ εὔπεπτον· οὐδὲν γὰρ ἐνοχλεῖ πολὺ προσενεγκαμένων. εὔριζον δὲ τὸ δένδρον καὶ μήκει καὶ πάχει καὶ πλήθει πολύ· ἔχει δὲ καὶ ξύλον ἰσχυρὸν καὶ καλὸν τῇ ὄψει μέλαν, ὥσπερ ὁ λωτός, ἐξ οὗ καὶ τὰ ἀγάλματα καὶ τὰ κλινία καὶ τραπέζια καὶ τἆλλα τὰ τοιαῦτα ποιοῦσιν.

6 Ἡ δὲ βάλανος ἔχει μὲν τὴν προσηγορίαν ἀπὸ τοῦ καρποῦ· φύλλον δ' αὐτῇ παραπλήσιον τῷ τῆς μυρρίνης πλὴν προμηκέστερον. ἔστι δὲ τὸ δένδρον εὐπαχὲς μὲν καὶ εὐμέγεθες, οὐκ εὐφυὲς δὲ ἀλλὰ παρεστραμμένον. τοῦ καρποῦ δὲ τοῖς κελύφεσι χρῶνται οἱ μυρεψοὶ κόπτοντες· εὐῶδες γὰρ ἔχει τὸν δὲ καρπὸν αὐτὸν ἀχρεῖον. ἔστι δὲ καὶ τῷ μεγέθει καὶ τῇ ὄψει παραπλήσιος τῷ τῆς καππάριος· ξύλον δὲ ἰσχυρὸν καὶ εἰς ἄλλα τε χρήσιμον καὶ εἰς τὰς ναυπηγίας.

7 Τὸ δὲ καλούμενον κουκιόφορόν ἐστιν ὅμοιον τῷ φοίνικι· τὴν δὲ ὁμοιότητα κατὰ τὸ στέλεχος ἔχει καὶ τὰ φύλλα· διαφέρει δὲ ὅτι ὁ μὲν φοῖνιξ μονοφυὲς καὶ ἁπλοῦν ἐστι, τοῦτο δὲ προσαυξηθὲν σχίζεται καὶ γίνεται δίκρουν, εἶτα πάλιν ἑκάτερον

[1] ἀποτιθέασιν conj. R. Const. from G (*recondunt*); τιθέασι UMVAld.

deciduous. It bears abundant fruit and at every season, for the new fruit always overtakes that of last year. It ripens its fruit at the season of the etesian winds: the other fruit they gather somewhat unripe and store[1] it. In size it is as large as a pear, but in shape it is oblong, almond-shaped, and its colour is grass-green. It has inside a stone like the plum, but much smaller and softer; the flesh is sweet and luscious and easily digested; for it does no hurt if one eats it in quantity. The tree has good roots as to length thickness and number. Moreover its wood is strong and fair in appearance, black like the nettle-tree: out of it men make their images beds tables and other such things.

[2] The *balanos* gets its name from its fruit[3]; its leaf is like that of the myrtle[4] but it is longer. The tree is of a good stoutness[5] and stature, but not of a good shape, being crooked. The perfumers use the husks of the fruit, which they bruise; for this is fragrant, though the fruit itself is useless. In size and appearance it is like the fruit of the caper; the wood is strong and useful for shipbuilding and other purposes.

[6] The tree called the doum-palm is like the date-palm; the resemblance is in the stem and the leaves, but it differs in that the date-palm is a tree with a single undivided stem, while the other, as it increases, splits and becomes forked,[7] and then each of the two

[2] Plin. 13. 61.
[3] *i.e.* it is like an acorn (*βάλανος*).
[4] *μυρρίνης* MVPAld.; *μυρίκης* U.
[5] *εὐπαχὲς* conj. Sch.; *εὐπαθὲς* U; *ἀπαθὲς* Ald.H.
[6] Plin. 13. 62.
[7] *cf.* 2. 6. 9, where the same tree is evidently indicated. *δίκρουν* conj. Salm., Scal., etc.; *ἄκρον* UAld.H.

τουτων ὁμοιως· ἔτι δὲ τὰς ῥάβδους βραχείας ἔχει σφόδρα καὶ οὐ πολλας. χρῶνται δὲ τῷ φύλλῳ, καθάπερ τῷ φοίνικι, πρὸς τὰ πλέγματα. καρπὸν δὲ ἴδιον ἔχει πολὺ διαφέροντα καὶ μεγέθει καὶ σχήματι καὶ χυλῷ· μέγεθος μὲν γὰρ ἔχει σχεδὸν χειροπληθές· στρογγύλον δὲ καὶ οὐ προμήκη· χρῶμα ἐπίξανθον· χυλὸν δὲ γλυκὺν καὶ εὔστομον· οὐκ ἀθρόον δέ, ὥσπερ ὁ φοῖνιξ, ἀλλὰ κεχωρισμένον καθ' ἕνα· πυρῆνα δὲ μέγαν καὶ σφόδρα σκληρόν, ἐξ οὗ τοὺς κρίκους τορνεύουσι τοὺς εἰς τοὺς στρωματεῖς τοὺς διαποικίλους· διαφέρει δὲ πολὺ τὸ ξύλον τοῦ φοίνικος· τὸ μὲν γὰρ μανὸν καὶ ἰνῶδες καὶ χαῦνον, τὸ δὲ πυκνὸν καὶ βαρὺ καὶ σαρκῶδες καὶ διατμηθὲν οὖλον σφόδρα καὶ σκληρόν ἐστιν. καὶ οἵ γε δὴ Πέρσαι πάνυ ἐτίμων αὐτὸ καὶ ἐκ τούτου τῶν κλινῶν ἐποιοῦντο τοὺς πόδας.

8 Ἡ δὲ ἄκανθα καλεῖται μὲν διὰ τὸ ἀκανθῶδες ὅλον τὸ δένδρον εἶναι πλὴν τοῦ στελέχους· καὶ γὰρ ἐπὶ τῶν ἀκρεμόνων καὶ ἐπὶ τῶν βλαστῶν καὶ ἐπὶ τῶν φύλλων ἔχει. μεγέθει δὲ μέγα, καὶ γὰρ δωδεκάπηχυς ἐξ αὐτῆς ἐρέψιμος ὕλη τέμνεται. διττὸν δὲ τὸ γένος αὐτῆς, ἡ μὲν γάρ ἐστι λευκὴ ἡ δὲ μέλαινα· καὶ ἡ μὲν λευκὴ ἀσθενής τε καὶ εὔσηπτος· ἡ δὲ μέλαινα ἰσχυροτέρα τε καὶ ἄσηπτος, δι' ὃ καὶ ἐν ταῖς ναυπηγίαις χρῶνται πρὸς τὰ ἐγκοίλια αὐτῇ. τὸ δένδρον δὲ οὐκ ἄγαν ὀρθοφυές. ὁ δὲ καρπὸς ἔλλοβος, καθάπερ τῶν χεδροπῶν, ᾧ χρῶνται οἱ ἐγχώριοι πρὸς τὰ δέρματα ἀντὶ κηκίδος. τὸ δ' ἄνθος καὶ τῇ ὄψει καλόν, ὥστε καὶ στεφάνους ποιεῖν ἐξ αὐτοῦ, καὶ φαρμα-

branches forks again: moreover the twigs are very short and not numerous. They use the leaf, like the palm-leaf, for plaiting. It has a peculiar fruit, very different from that of the date-palm in size form and taste; for in size it is nearly big enough to fill the hand, but it is round rather than long; the colour is yellowish, the flavour sweet and palatable. It does not grow bunched together, like the fruit of the date-palm, but each fruit grows separately; it has a large and very hard stone, out of which they turn the rings for embroidered bed-hangings.[1] The wood is very different to that of the date-palm; whereas the latter is of loose texture fibrous and porous,[2] that of the doum-palm is close heavy and flesby, and when split is exceedingly compact and hard. The Persians[3] used to esteem it highly and made the feet of their couches out of it.

[4]The *akantha* (acacia) is so called because the whole tree is spinous (*akanthodes*) except the stem; for it has spines on the branches shoots and leaves. It is of large stature, since lengths of timber for roofing of twelve cubits are cut from it. There are two kinds, the white and the black; the white is weak and easily decays, the black is stronger and less liable to decay; wherefore they use it in shipbuilding for the ribs.[5] The tree is not very erect in growth. The fruit is in a pod, like that of leguminous plants, and the natives use it for tanning hides instead of gall. [6]The flower is very beautiful in appearance, so that they make garlands of it, and it has medicinal

[1] Plin. *l.c.*, *velares annulos*; *cf.* Athen. 12. 71, *ad fin.*
[2] χαῦνον conj. Sch.; χλωρὸν Ald.
[3] *i.e.* during their occupation of Egypt.
[4] Plin 13. 63; Athen. 15. 25.
[5] *cf.* Hdt. 2. 96. [6] *cf.* Athen. *l.c.*

κῶδες, δι' ὃ καὶ συλλέγουσιν οἱ ἰατροί. γίνεται δὲ ἐκ ταύτης καὶ τὸ κόμμι· καὶ ῥέει καὶ πληγείσης καὶ αὐτόματον ἄνευ σχάσεως. ὅταν δὲ κοπῇ, μετὰ τρίτον ἔτος εὐθὺς ἀναβεβλάστηκε· πολὺ δὲ τὸ δένδρον ἐστί, καὶ δρυμὸς μέγας περὶ τὸν Θηβαϊκὸν νόμον, οὗπερ καὶ ἡ δρῦς καὶ ἡ περσέα πλείστη καὶ ἡ ἐλάα.

9 Καὶ γὰρ ἡ ἐλάα περὶ τοῦτον τὸν τόπον ἐστί, τῷ ποταμῷ μὲν οὐκ ἀρδευομένη, πλείω γὰρ ἢ τριακόσια στάδια ἀπέχει, ναματιαίοις δ' ὕδασιν· εἰσὶ γὰρ κρῆναι πολλαί. τὸ δ' ἔλαιον οὐδὲν χεῖρον τοῦ ἐνθάδε, πλὴν κακωδέστερον διὰ τὸ σπανίοις τοῖς ἁλσὶ χρῆσθαι· φύσει δὲ τὸ ξύλον τοῦ δένδρου καὶ σκληρὸν καὶ παραπλήσιον τεμνόμενον τὴν χρόαν τῷ λωτίνῳ.

10 Ἄλλο δέ τι δένδρον ἡ κοκκυμηλέα, μέγα μὲν τῷ μεγέθει καὶ τὴν φύσιν τοῦ καρποῦ ὅμοιον τοῖς μεσπίλοις, καὶ τὸ μέγεθος παραπλήσιον πλὴν ἔχοντα πυρῆνα στρογγύλον· ἄρχεται δὲ ἀνθεῖν μηνὸς Πυανεψιῶνος, τὸν δὲ καρπὸν πεπαίνει περὶ ἡλίου τροπὰς χειμερινάς· ἀείφυλλον δ' ἐστίν. οἱ δὲ περὶ τὴν Θηβαΐδα κατοικοῦντες διὰ τὴν ἀφθονίαν τοῦ δένδρου ξηραίνουσι τὸν καρπὸν καὶ τὸν πυρηνα ἐξαιροῦντες κόπτουσι καὶ ποιοῦσι παλάθας.

11 Ὕλημα δὲ ἴδιόν τι φύεται περὶ Μέμφιν, οὐ κατὰ φύλλα καὶ βλαστοὺς καὶ τὴν ὅλην μορφὴν

---

[1] *cf.* Hdt. *l.c.*
[2] σχάσεως conj. R. Const.; σχίσεως Ald.
[3] πλείστη conj. R. Const.; πλεκτὴ UMVAld.
[4] *cf.* *C.P.* 6. 8. 7, where this olive is said to produce *no* oil.
[5] *cf.* Strabo, 17. 1. 35.

properties, wherefore physicians gather it. [1] Gum is also produced from it, which flows both when the tree is wounded and also of its own accord without any incision [2] being made. When the tree is cut down, after the third year it immediately shoots up again; it is a common tree, and there is a great wood of it in the Thebaid, where grow the oak, the *persea* in great abundance,[3] and the olive.

[4] For the olive also grows in that district, though it is not watered by the river, being more than 300 furlongs distant from it, but by brooks; for there are many springs. The oil produced is not inferior to that of our country, except that it has a less pleasing smell,[5] because it has not a sufficient natural supply of salt.[6] The wood of the tree is hard in character, and, when split, is like in colour [7] to that of the nettle-tree.

[8] There is another tree, the (Egyptian) plum (sebesten), which is of great stature, and the character of its fruit [9] is like the medlar (which it resembles in size), except that it has a round stone. It begins to flower in the month Pyanepsion,[10] and ripens its fruit about the winter solstice, and it is evergreen.[11] The inhabitants of the Thebaid, because of the abundance of the tree, dry the fruit; they take out the stones, bruise it, and make cakes of it.

There is a peculiar bush [12] which grows about Memphis, whose peculiarity does not lie in its leaves

[6] σπανίοις . . . φύσει conj. W.; σπανίως τοῖς ἁλσὶ χρ. τῇ φύσει Ald.; so U, but omitting τῇ.

[7] *i.e.* black. *cf.* 4. 3. 1. [8] Plin. 13. 64 and 65.

[9] τοῦ καρποῦ add. Scal. from G and Plin. *l.c.* [10] October.

[11] ἀείφυλλον conj. Scal. from G and Plin. *l.c.*; φύλλον UMVAld.

[12] *Mimosa asperata*; see Index, App. (2). ὕλημα conj. Scal. from G (*materia*); οἴδημα MAld. U (corrected).

ἔχον τὸ ἴδιον ἀλλ' εἰς τὸ συμβαῖνον περὶ αὐτὸ πάθος·[1] ἡ μὲν γὰρ πρόσοψις ἀκανθώδης ἐστὶν αὐτοῦ, καὶ τὸ φύλλον παρόμοιον ταῖς πτερίσιν· ὅταν δέ τις ἅψηται τῶν κλωνίων, ὥσπερ ἀφαναινόμενα τὰ φύλλα συμπίπτειν φασὶν εἶτα μετά τινα χρόνον ἀναβιώσκεσθαι πάλιν καὶ θάλλειν.[2] καὶ τὰ μὲν ἴδια τῆς χώρας, ὅσα γ' ἂν δένδρα τις ἢ θάμνους εἴποι, τά γ' ἐπιφανέστατα ταῦτ' ἐστί. περὶ γὰρ τῶν ἐν τῷ ποταμῷ καὶ τοῖς ἕλεσιν ὕστερον ἐροῦμεν, ὅταν καὶ περὶ τῶν ἄλλων ἐνύδρων.

12 ["Απαντα δὲ ἐν τῇ χώρᾳ τὰ δένδρα τὰ τοιαῦτα μεγάλα καὶ τοῖς μήκεσι καὶ τοῖς πάχεσιν· ἐν γοῦν Μέμφιδι τηλικοῦτο δένδρον εἶναι λέγεται τὸ πάχος, ὃ τρεῖς ἄνδρες οὐ δύνανται περιλαμβάνειν. ἔστι δὲ καὶ τμηθὲν τὸ ξύλον καλόν· πυκνόν τε γὰρ σφόδρα καὶ τῷ χρώματι λωτοειδές.]

III. Ἐν Λιβύῃ δὲ ὁ λωτὸς πλεῖστος καὶ κάλλιστος καὶ ὁ παλίουρος καὶ ἔν τισι μέρεσι τῇ τε Νασαμωνικῇ καὶ παρ' Ἄμμωνι καὶ ἄλλοις ὁ φοῖνιξ· ἐν δὲ τῇ Κυρηναίᾳ κυπάρισσος καὶ ἐλάαι τε κάλλισται καὶ ἔλαιον πλεῖστον. ἰδιώτατον δὲ πάντων τὸ σίλφιον· ἔτι κρόκον πολὺν ἡ χώρα φέρει καὶ εὔοσμον. ἔστι δὲ τοῦ λωτοῦ τὸ μὲν ὅλον δένδρον ἴδιον εὐμέγεθες ἡλίκον ἄπιος ἢ μικρὸν ἔλαττον· φύλλον δὲ ἐντομὰς ἔχον καὶ πρινῶδες· τὸ μὲν ξύλον μέλαν· γένη δὲ αὐτοῦ πλείω διαφορὰς ἔχοντα τοῖς καρποῖς· ὁ δὲ καρπὸς

[1] πάθος: *cf.* 1. 1. 1 n.
[2] *cf.* Schol. ad Nic. *Ther.* 683 of a sensitive plant called σκορπίουρος or ἰσχύουσα. ἀφαναινόμενα conj. Scal.; ἀφαυλινόμενα UMVP_2Ald.

shoots and general form, but in the strange property[1] which belongs to it. Its appearance is spinous and the leaf is like ferns, but, when one touches the twigs, they say that the leaves as it were wither up[2] and collapse and then after a time come to life again and flourish. Such are the most conspicuous things peculiar to the country, to speak only of trees or shrubs. For we will speak later of the things which grow in the river and the marshes, when we come to speak of the other water plants.

[3] All the trees of this kind in that country are large, both in height and stoutness; thus at Memphis there is said to be a tree of such girth that three men cannot embrace it. The wood too, when split, is good, being of extremely close grain and in colour like the nettle-tree.

Of the trees and shrubs special to Libya.

III. [4] In Libya the *lotos* is most abundant and fairest; so also is the Christ's thorn, and in some parts, such as the Nasamonian district and near the temple of Zeus Ammon, the date-palm. In the Cyrenaica the cypress grows and the olives are fairest and the oil most abundant. Most special of all to this district is the silphium, and the land also bears abundant fragrant saffron-crocus. As to the *lotos*—the whole tree is peculiar, of good stature, as tall as a pear-tree, or nearly so; the leaf is divided and like that of the kermes-oak, and the wood is black. There are several sorts, which differ in their fruits; the fruit

[3] This section is evidently out of place; its probable place is at the end of § 10, so that the description will belong to the 'Egyptian plum.'

[4] See Index. Plin. 13. 104–106.

ἡλίκος κύαμος, πεπαινεται δέ, ὥσπερ οἱ βότρυες, μεταβάλλων τὰς χροιάς· φύεται δὲ, καθάπερ τὰ μύρτα, παρ' ἄλληλα πυκνὸς ἐπὶ τῶν βλαστῶν· ἐσθιόμενος δ' ὁ ἐν τοῖς Λωτοφάγοις καλουμένοις γλυκὺς καὶ ἡδὺς καὶ ἀσινὴς καὶ ἔτι πρὸς τὴν κοιλίαν ἀγαθός· ἡδίων δ' ὁ ἀπύρηνος, ἔστι γὰρ καὶ τοιοῦτόν τι γένος· ποιοῦσι δὲ καὶ οἶνον ἐξ αὐτοῦ.

2 Πολὺ δὲ τὸ δένδρον καὶ πολύκαρπον· τό γ' οὖν Ὀφέλλου στρατόπεδον, ἡνίκα ἐβάδιζεν εἰς Καρχηδόνα, καὶ τούτῳ φασὶ τραφῆναι πλείους ἡμέρας ἐπιλιπόντων τῶν ἐπιτηδείων. ἔστι μὲν οὖν καὶ ἐν τῇ νήσῳ τῇ Λωτοφαγιτίδι καλουμένῃ πολύς· αὕτη δ' ἐπίκειται καὶ ἀπέχει μικρόν· οὐ μὴν οὐθέν γε μέρος ἀλλὰ πολλῷ πλεῖον ἐν τῇ ἠπείρῳ· πλεῖστον γὰρ ὅλως ἐν τῇ Λιβύῃ, καθάπερ εἴρηται, τοῦτο καὶ ὁ παλίουρός ἐστιν· ἐν γὰρ Εὐεσπερίσι τούτοις καυσίμοις χρῶνται. διαφέρει δὲ οὗτος ὁ λωτὸς τοῦ παρὰ τοῖς Λωτοφάγοις.

3 Ὁ δὲ παλίουρος θαμνωδέστερος τοῦ λωτοῦ· φύλλον δὲ παρόμοιον ἔχει τῷ ἐνταῦθα, τὸν δὲ καρπὸν διάφορον· οὐ γὰρ πλατὺν ἀλλὰ στρογγύλον καὶ ἐρυθρόν, μέγεθος δὲ ἡλίκον τῆς κέδρου ἢ μικρῷ μεῖζον· πυρῆνα δὲ ἔχει οὐ συνεσθιόμενον καθάπερ ταῖς ῥοαῖς· ἡδὺν δὲ τὸν καρπόν· καὶ ἐάν τις οἶνον ἐπιχέῃ καὶ αὐτὸν ἡδίω γίνεσθαί φασι καὶ τὸν οἶνον ἡδίω ποιεῖν.

---

1 *cf.* Hdt. 4. 177; Athen. 14. 651; Scyl. *Peripl.* Lotophagi.

2 A ruler of Cyrene, who invaded Carthaginian territory in conjunction with Agathocles, B.C. 308.

3 τῇ Λωτοφαγιτίδι conj. W.; τῇ Λωτοφαγίᾳ Φάριδι UMAld.

4 μέρος: μείων conj. Sch. (*non minor* G).

is as large as a bean, and in ripening like grapes it changes its colour: it grows, like myrtle-berries, close together on the shoots; to eat, that which grows among the people called the Lotus-eaters[1] is sweet pleasant and harmless, and even good for the stomach; but that which has no stone is pleasanter (for there is also such a sort), and they also make wine from it.

The tree is abundant and produces much fruit; thus the army of Ophellas,[2] when it was marching on Carthage, was fed, they say, on this alone for several days, when the provisions ran short. It is abundant also in the island called the island of the Lotus-eaters;[3] this lies off the mainland at no great distance: it grows however in no less quantity,[4] but even more abundantly[5] on the mainland; for, as has been said,[6] this tree is common in Libya generally as well as the Christ's thorn; for in the islands called Euesperides[7] they use these trees as fuel. However this *lotos*[8] differs from that found in the land of the Lotus-eaters.

[9] The (Egyptian) 'Christ's thorn' is more shrubby than the *lotos*; it has a leaf like the tree of the same name of our country, but the fruit is different; for it is not flat, but round and red, and in size as large as the fruit of the prickly cedar or a little larger; it has a stone which is not eaten with the fruit, as in the case of the pomegranate, but the fruit is sweet, and, if one pours wine over it, they say that it becomes sweeter and that it makes the wine sweeter.

[5] πλεῖον U; ? πλείων with MV.
[6] 4. 3. 1. [7] *cf.* Hdt. 4. 191.
[8] *cf.* Hdt. 2. 96.
[9] See Index. Plin. 13. 111.

4 Ἔνιοι δὲ τὸ τοῦ λωτοῦ δένδρον θαμνῶδες εἶναι καὶ πολύκλαδον, τῷ στελέχει δὲ εὐπαχές· τὸν δὲ καρπὸν μέγα τὸ κάρυον ἔχειν· τὸ δ' ἐκτὸς οὐ σαρκῶδες ἀλλὰ δερματωδέστερον· ἐσθιόμενον δὲ οὐχ οὕτω γλυκὺν ὡς εὔστομον· καὶ τὸν οἶνον ὃν ἐξ αὐτοῦ ποιοῦσιν οὐ διαμένειν ἀλλ' ἢ δύο ἢ τρεῖς ἡμέρας εἶτ' ὀξύνειν. ἡδίω μὲν οὖν τὸν καρπὸν τὸν ἐν τοῖς Λωτοφάγοις, ξύλον δὲ κάλλιον τὸ ἐν Κυρηναίᾳ· θερμοτέραν δὲ εἶναι τὴν χώραν τὴν τῶν Λωτοφάγων· τοῦ ξύλου δὲ τὴν ῥίζαν εἶναι μελαντέραν μὲν πολὺ πυκνὴν δὲ ἧττον καὶ εἰς ἐλάττω χρησίμην· εἰς γὰρ τὰ ἐγχειρίδια καὶ τὰ ἐπικολλήματα χρῆσθαι, τῷ ξύλῳ δὲ εἴς τε τοὺς αὐλοὺς καὶ εἰς ἄλλα πλείω.

5 Ἐν δὲ τῇ μὴ ὑομένῃ τῆς Λιβύης ἄλλα τε πλείω φύεσθαι καὶ φοίνικας μεγάλους καὶ καλούς· οὐ μὴν ἀλλ' ὅπου μὲν φοῖνιξ ἁλμυρίδα τε εἶναι καὶ ἔφυδρον τὸν τόπον, οὐκ ἐν πολλῷ δὲ βάθει ἀλλὰ μάλιστα ἐπ' ὀργυίαις τρισίν. τὸ δ' ὕδωρ ἔνθα μὲν γλυκὺ σφόδρα ἔνθα δὲ ἁλυκὸν πλησίον ὄντων ἀλλήλοις· ὅπου δὲ τὰ ἄλλα φύεται ξηρὸν καὶ ἄνυδρον· ἐνιαχοῦ δὲ καὶ τὰ φρέατα εἶναι ἑκατὸν ὀργυιῶν, ὥστε ὑποζυγίοις ἀπὸ τροχηλιᾶς ἀνιμᾶν· δι' ὃ καὶ θαυμαστὸν πῶς ποτε ὠρύχθη τηλικαῦτα βάθη· τὸ δ' οὖν τῶν ὑδάτων τῶν ὑπὸ τοὺς φοίνικας καὶ ἐν Ἄμμωνος εἶναι διαφορὰν ἔχον τὴν εἰρημένην. φύεσθαι δὲ ἐν τῇ μὴ ὑομένῃ τὸ θύμον πολὺ καὶ ἄλλα ἴδιά τε καὶ πλείω γίνεσθαι

---

1 Sch. after Scal. places this section before § 3, making the account of this tree consecutive. 2 Plin. 13. 17. 104–106.

3 εὔπαχες conj R. Const.; εὐσταχές U; εὔσταχες MP$_2$Ald.

4 *cf.* Hdt. 2. 96.

[1] Some say that the *lotos* [2] is shrubby and much branched, though it has a stout [3] stem; and that the stone in the fruit is large, while the outside is not flesby but somewhat leathery; and that to eat it is not so much sweet as palatable; and that the wine which they make out of it does not keep more than two or three days, after which it gets sour; and so that the fruit [4] found in the Lotus-eaters' country is sweeter, while the wood in the Cyrenaica is better; and that the country of the Lotus-eaters is hotter; and that the root is much blacker than the wood, but of less close grain, and of use for fewer purposes; for they use it only for dagger-handles and tessellated work,[5] while the wood is used for pipes and many other things.

In the part of Libya where no rain falls they say that, besides many other trees, there grow tall and fine date-palms; however they add that, where the date-palm is found, the soil [6] is salt and contains water, and that at no great depth, not more than three fathoms. They say also that the water is in some places quite sweet, but in others quite close by it is brackish; that where however other things grow, the soil is dry and waterless; and that in places even the wells are a hundred fathoms deep, so that they draw water by means of a windlass worked by beasts. Wherefore it is wonderful how at any time digging to such depths was carried out. Such, they say, is the special character of the water supply which feeds the date-palms in the district also of the temple of Zeus Ammon. Further it is said that in the land where no rain falls thyme [7] is

[5] ἐπικολλήματα: lit. 'pieces glued on'; *cf.* Plin. *l.c.*
[6] *cf.* Hdt. 3. 183.
[7] θύμον nıBas. H.; θάμνον UMVAld. *cf.* 6. 2. 3.

ἐνταῦθα, καὶ πτῶκα καὶ δορκάδα καὶ στρουθὸν
6 καὶ ἕτερα τῶν θηρίων. ἀλλὰ ταῦτα μὲν ἄδηλον
εἰ ἐκτοπίζει που πιόμενα· (διὰ γὰρ τὸ τάχος
δύναται μακράν τε καὶ ταχὺ παραγενέσθαι),
ἄλλως τε κεὶ δι' ἡμερῶν τινων πίνουσι, καθάπερ
καὶ τὰ ἥμερα παρὰ τρίτην ἢ τετάρτην ποτίζεται
ταῦτα· τὸ δὲ τῶν ἄλλων ζώων, οἷον ὄφεων
σαυρῶν καὶ τῶν τοιούτων, φανερὸν ὅτι ἄποτα.
τοὺς δὲ Λίβυας λέγειν ὅτι τὸν ὄνον ἐσθίει ταῦτα
ὃς καὶ παρ' ἡμῖν γίνεται, πολύπουν τε καὶ μέλαν
συσπειρώμενον εἰς ἑαυτό· τοῦτον δὲ πολύν τε
γίνεσθαι σφόδρα καὶ ὑγρὸν τὴν φύσιν εἶναι.
7 Δρόσον δὲ ἀεὶ πίπτειν ἐν τῇ μὴ ὑομένῃ πολλήν,
ὥστε δῆλον ὅτι τὸν μὲν φοίνικα καὶ εἴ τι ἄλλο
φύεται ἐν ἀνύδροις τό τε ἐκ τῆς γῆς ἀνιὸν ἐκτρέφει
καὶ πρὸς τούτῳ ἡ δρόσος. ἱκανὴ γὰρ ὡς κατὰ
μεγέθη καὶ τὴν φύσιν αὐτῶν ξηρὰν οὖσαν καὶ ἐκ
τοιούτων συνεστηκυῖαν. καὶ δένδρα μὲν ταῦτα
πλεῖστα καὶ ἰδιώτατα. περὶ δὲ τοῦ σιλφίου
λεκτέον ὕστερον ποῖόν τι τὴν φύσιν.

IV. Ἐν δὲ τῇ Ἀσίᾳ παρ' ἑκάστοις ἴδι' ἄττα
τυγχάνει· τὰ μὲν γὰρ φέρουσιν αἱ χῶραι τὰ δ'

---

1 *Lepus Aegyptiacus.* *cf.* Arist. *H.A.* 8. 28.
2 ὡς κατὰ conj. Scal. from G; ὥστε τὰ Ald.H.

abundant, and that there are various other peculiar plants there, and that there are found the hare[1] gazelle ostrich and other animals. However it is uncertain whether these do not migrate in order to find drink somewhere, (for by reason of their fleetness they are able to appear at a distant place in a short space of time), especially if they can go for several days without drinking, even as these animals, when domesticated, are only given drink every third or fourth day. While as to other animals, such as snakes lizards and the like, it is plain that they go without drink. And we are told that according to the Libyans, these animals eat the wood-louse, which is of the same kind that is found also in our country, being black, with many feet, and rolling itself into a ball; this, they say, is extremely common and is juicy by nature.

They say also that dew always falls abundantly in the land in which no rain falls, so that it is plain that the date-pahn, as well as anything else which grows in waterless places, is kept alive by the moisture which rises from the ground, and also by the dew. For the latter is sufficient, considering[2] the size of such trees and their natural character, which is dry and formed of dry components. And trees of that character are most abundant in, and most specially belong to such country. The character of the silphium we must discuss later.

Of the trees and herbs special to Asia.

IV. In different parts of Asia also there are special trees, for the soil of the various regions produces some but not others. [3]Thus they say that

[3] Plin. 16. 144.

οὐ φύουσιν· οἷον κιττὸν καὶ ἐλάαν οὔ φασιν εἶναι τῆς Ἀσίας ἐν τοῖς ἄνω τῆς Συρίας ἀπὸ θαλάττης πένθ᾽ ἡμερῶν· ἀλλ᾽ ἐν Ἰνδοῖς φανῆναι κιττὸν ἐν τῷ ὄρει τῷ Μηρῷ καλουμένῳ, ὅθεν δὴ καὶ τὸν Διόνυσον εἶναι μυθολογοῦσι. δι᾽ ὃ καὶ Ἀλέξανδρος ἀπ᾽ ἐξοδίας λέγεται ἀπιὼν ἐστεφανωμένος κιττῷ εἶναι καὶ αὐτὸς καὶ ἡ στρατιά· τῶν δὲ ἄλλων ἐν Μηδίᾳ μόνον· περικλείειν γὰρ αὕτη δοκεῖ καὶ συνάπτειν πως τῷ Πόντῳ. καίτοι γε διεφιλοτιμήθη Ἅρπαλος ἐν τοῖς παραδείσοις τοῖς περὶ Βαβυλῶνα φυτεύων πολλάκις καὶ πραγματευόμενος, ἀλλ᾽ οὐδὲν ἐποίει πλέον· οὐ γὰρ ἐδύνατο ζῆν ὥσπερ τἆλλα τὰ ἐκ τῆς Ἑλλάδος. τοῦτο μὲν οὖν οὐ δέχεται ἡ χώρα διὰ τὴν τοῦ ἀέρος κρᾶσιν· ἀναγκαίως δὲ δέχεται καὶ πύξον καὶ φίλυραν· καὶ γὰρ περὶ ταῦτα πονοῦσιν οἱ ἐν τοῖς παραδείσοις. ἕτερα δὲ ἴδια φέρει καὶ δένδρα
2 καὶ ὑλήματα· καὶ ἔοικεν ὅλως ὁ τόπος ὁ πρὸς ἀνατολὰς καὶ μεσημβρίαν ὥσπερ καὶ ζῷα καὶ φυτὰ φέρειν ἴδια παρὰ τοὺς ἄλλους· οἷον ἥ τε Μηδία χώρα καὶ Περσὶς ἄλλα τε ἔχει πλείω καὶ τὸ μῆλον τὸ Μηδικὸν ἢ τὸ Περσικὸν καλούμενον. ἔχει δὲ τὸ δένδρον τοῦτο φύλλον μὲν ὅμοιον καὶ σχεδὸν ἴσον τῷ τῆς ἀνδράχλης, ἀκάνθας δὲ οἵας ἄπιος ἢ ὀξυάκανθος, λείας δὲ καὶ ὀξείας σφόδρα καὶ ἰσχυράς· τὸ δὲ μῆλον οὐκ ἐσθίεται μέν,

---

[1] ἐλάαν conj. Spr.; ἐλάτην MSS. *cf.* Hdt. 1. 193; Xen. *Anab.* 4. 4. 13; Arr. *Ind.* 40.

[2] κιττὸν conj. W., *cf.* Arr. *Anab.* 5. 1. 6; καὶ τὴν UMV; καὶ τῷ Ald.H.

[3] λέγεται add. W.

[4] ἐξοδίας UMVP; Ἰνδίας W. with Ald.

[5] κιττῷ εἶναι conj. W.; εἶτα μεῖναι U; εἶτα μὴ εἶναι MVPAld.

ivy and olive[1] do not grow in Asia in the parts of Syria which are five days' journey from the sea; but that in India ivy[2] appears on the mountain called Meros, whence, according to the tale, Dionysus came. Wherefore it is said[3] that Alexander, when he came back from an expedition,[4] was crowned with ivy,[5] himself and his army. But elsewhere in Asia it is said to grow only in Media, for that country seems in a way to surround and join on to the Euxine Sea.[6] However,[7] when Harpalus took great pains over and over again to plant it in the gardens of Babylon, and made a special point of it, he failed: since it could not live like the other things introduced from Hellas. The country then does not[8] admit this plant on account of the climate, and it grudgingly admits the box and the lime; for even these give much trouble to those engaged in the gardens. It also produces some peculiar trees and shrubs. And in general the lands of the East and South appear to have peculiar plants, as they have peculiar animals; for instance, Media and Persia have, among many others, that which is called the 'Median' or 'Persian apple' (citron).[9] This tree[10] has a leaf like to and almost identical with that of the andrachne, but it has thorns like those of the pear[11] or white-thorn, which however are smooth and very sharp and strong. The 'apple' is not

[6] *i.e.* and so Greek plants may be expected to grow there. But the text is probably defective; *cf.* the citation of this passage, Plut. *Quaest. Conv.* 3. 2. 1.

[7] καίτοι γε. This sentence does not connect properly with the preceding.

[8] οὐ add. Sch.

[9] Plin. 12. 15 and 16; cited also Athen. 3. 26.

[10] *cf.* Verg *G.* 2. 131–135.

[11] ἄπιος: ? here=ἀχράς R. Const. *cf. C.P.* 1. 15. 2.

εὔοσμον δὲ πάνυ καὶ τὸ φύλλον τοῦ δένδρου· κἂν εἰς ἱμάτια τεθῇ τὸ μῆλον ἄκοπα διατηρεῖ. χρήσιμον δ' ἐπειδὰν τύχῃ <τις> πεπωκὼς φάρμακον <θανάσιμον· δοθὲν γὰρ ἐν οἴνῳ διακόπτει τὴν κοιλίαν καὶ ἐξάγει τὸ φάρμακον·> καὶ πρὸς στόματος εὐωδίαν· ἐὰν γάρ τις ἑψήσῃ ἐν ζωμῷ, ἢ ἐν ἄλλῳ τινὶ τὸ ἔσωθεν τοῦ μήλου ἐκπιέσῃ εἰς τὸ στόμα καὶ καταροφήσῃ, ποιεῖ τὴν ὀσμὴν ἡδεῖαν.

3 σπείρεται δὲ τοῦ ἦρος εἰς πρασιὰς ἐξαιρεθὲν τὸ σπέρμα διειργασμένας ἐπιμελῶς, εἶτα ἀρδεύεται διὰ τετάρτης ἢ πέμπτης ἡμέρας· ὅταν δὲ ἁδρὸν ᾖ, διαφυτεύεται πάλιν τοῦ ἔαρος εἰς χωρίον μαλακὸν καὶ ἔφυδρον καὶ οὐ λίαν λεπτόν· φιλεῖ γὰρ τὰ τοιαῦτα. φέρει δὲ τὰ μῆλα πᾶσαν ὥραν· τὰ μὲν γὰρ ἀφῄρηται τὰ δὲ ἀνθεῖ τὰ δὲ ἐκπέττει. τῶν δὲ ἀνθῶν ὅσα, ὥσπερ εἴπομεν, ἔχει καθάπερ ἠλακάτην ἐκ μέσου τιν' ἐξέχουσαν, ταῦτά ἐστι γόνιμα, ὅσα δὲ μὴ ἄγονα. σπείρεται δὲ καὶ εἰς ὄστρακα διατετρημένα, καθάπερ καὶ οἱ φοίνικες. τοῦτο μὲν οὖν, ὥσπερ εἴρηται, περὶ τὴν Περσίδα καὶ τὴν Μηδίαν ἐστίν.

4 Ἡ δὲ Ἰνδικὴ χώρα τήν τε καλουμένην ἔχει συκῆν, ἣ καθίησιν ἐκ τῶν κλάδων τὰς ῥίζας ἀν' ἕκαστον ἔτος, ὥσπερ εἴρηται πρότερον· ἀφίησι δὲ οὐκ ἐκ τῶν νέων ἀλλ' ἐκ τῶν ἔνων καὶ ἔτι παλαιοτέρων· αὗται δὲ συνάπτουσαι τῇ γῇ ποιοῦσιν ὥσπερ δρύφακτον κύκλῳ περὶ τὸ δένδρον, ὥστε γίνεσθαι καθάπερ σκηνήν, οὗ δὴ καὶ

---

[1] τις add. W. from Athen. *l.c.*; θανάσιμον . . . φάρμακον add. Sch. from Athen. *l.c.* [2] Plin. 11. 278; 12. 16.

[3] ἁδρὸν ᾖ W. from Athen. *l.c.*, whence διαφυτεύεται W. etc. for διαφυτεύηται Ald.H. ἁδρόν τι UMVAld.

eaten, but it is very fragrant, as also is the leaf of the tree. And if the 'apple' is placed among clothes, it keeps them from being moth-eaten. It is also useful when one[1] has drunk deadly poison; for being given in wine it upsets the stomach and brings up the poison; also for producing sweetness of breath;[2] for, if one boils the inner part of the 'apple' in a sauce, or squeezes it into the mouth in some other medium, and then inhales it, it makes the breath sweet. The seed is taken from the fruit and sown in spring in carefully tilled beds, and is then watered every fourth or fifth day. And, when it is growing vigorously,[3] it is transplanted, also in spring, to a soft well-watered place, where the soil is not too fine; for such places it loves. And it bears its 'apples' at all seasons; for when some have been gathered, the flower of others is on the tree and it is ripening others. Of the flowers, as we have said,[4] those which have, as it were, a distaff[5] projecting in the middle are fertile, while those that have it not are infertile. It is also sown, like date-palms, in pots[6] with a hole in them. This tree, as has been said, grows in Persia and Media.

[7] The Indian land has its so-called 'fig-tree' (banyan), which drops its roots from its branches every year, as has been said above[8]; and it drops them, not from the new branches, but from those of last year or even from older ones; these take hold of the earth and make, as it were, a fence about the tree, so that it becomes like a tent, in

[4] 1. 13. 4. [5] *i.e.* the pistil.

[6] Plin. 12. 16, *fictilibus in vasis, dato per cavernas radicibus spiramento*: the object, as Plin. explains, was to export it for medical use.

[7] Plin. 12. 22 and 23. [8] 1. 7. 3.

εἰώθασι διατρίβειν. εἰσὶ δὲ αἱ ῥίζαι φυόμεναι διάδηλοι πρὸς τοὺς βλαστούς· λευκότεραι γὰρ καὶ δασεῖαι καὶ σκολιαὶ καὶ ἄφυλλοι. ἔχει δὲ καὶ τὴν ἄνω κόμην πολλήν, καὶ τὸ ὅλον δένδρον εὔκυκλον καὶ τῷ μεγέθει μέγα σφόδρα· καὶ γὰρ ἐπὶ δύο στάδια ποιεῖν φασι τὴν σκιάν· καὶ τὸ πάχος τοῦ στελέχους ἔνια πλειόνων ἢ ἑξήκοντα βημάτων, τὰ δὲ πολλὰ τετταράκοντα. τὸ δέ γε φύλλον οὐκ ἔλαττον ἔχει πέλτης, καρπὸν δὲ σφόδρα μικρὸν ἡλίκον ἐρέβινθον ὅμοιον δὲ σύκῳ· δι' ὃ καὶ ἐκάλουν αὐτὸ οἱ Ἕλληνες συκῆν· ὀλίγον δὲ θαυμαστῶς τὸν καρπὸν οὐχ ὅτι κατὰ τὸ τοῦ δένδρου μέγεθος ἀλλὰ καὶ τὸ ὅλον. φύεται δὲ καὶ τὸ δένδρον περὶ τὸν Ἀκεσίνην ποταμόν.

5 Ἔστι δὲ καὶ ἕτερον δένδρον καὶ τῷ μεγέθει μέγα καὶ ἡδύκαρπον θαυμαστῶς καὶ μεγαλόκαρπον· καὶ χρῶνται τροφῇ τῶν Ἰνδῶν οἱ σοφοὶ καὶ μὴ ἀμπεχόμενοι.

Ἕτερον δὲ οὗ τὸ φύλλον τὴν μὲν μορφὴν πρόμηκες τοῖς τῶν στρουθῶν πτεροῖς ὅμοιον, ἃ παρατίθενται παρὰ τὰ κράνη, μῆκος δὲ ὡς διπηχυαῖον.

Ἄλλο τέ ἐστιν οὗ ὁ καρπὸς μακρὸς καὶ οὐκ εὐθὺς ἀλλὰ σκολιὸς ἐσθιόμενος δὲ γλυκύς. οὗτος ἐν τῇ κοιλίᾳ δηγμὸν ἐμποιεῖ καὶ δυσεντερίαν, δι' ὃ Ἀλέξανδρος ἀπεκήρυξε μὴ ἐσθίειν. ἔστι δὲ καὶ ἕτερον οὗ ὁ καρπὸς ὅμοιος τοῖς κρανέοις.

---

[1] οὗ conj. W.; αἷς UMVAld.

[2] ἄφυλλοι conj. Dalec.; δίφυλλοι UVAld.; so also MH., omitting καί.

[3] ἑξήκοντα . . . τετταράκοντα MSS.; ἑξ . . τεττάρων conj. Salm. *cf.* Plin. *l.c.*; Strabo 15. 1. 21.

which[1] men sometimes even live. The roots as they grow are easily distinguished from the branches, being whiter hairy crooked and leafless.[2] The foliage above is also abundant, and the whole tree is round and exceedingly large. They say that it extends its shade for as much as two furlongs; and the thickness of the stem is in some instances more than sixty[3] paces, while many specimens are as much as forty[3] paces through. The leaf is quite as large as a shield,[4] but the fruit is very small,[5] only as large as a chick-pea, and it resembles a fig. And this is why the Greeks[6] named this tree a 'fig-tree.' The fruit is curiously scanty, not only relatively to the size of the tree, but absolutely. The tree also grows near the river Akesines.[7]

There is also another tree[8] which is very large and has wonderfully sweet and large fruit; it is used for food by the sages of India who wear no clothes.

There is another tree[9] whose leaf is oblong in shape, like the feathers of the ostrich; this they fasten on to their helmets, and it is about two cubits long.

There is also another[10] whose fruit is long and not straight, but crooked, and it is sweet to the taste. This causes griping in the stomach and dysentery; wherefore Alexander ordered that it should not be eaten. There is also another[11] whose fruit is like the fruit of the cornelian cherry.

4 πέλτη: a small round shield. 5 *cf.* *C.P.* 2. 10. 2.
6 *i.e.* in Alexander's expedition. 7 Chenab.
8 Jack-fruit. See Index App. (3). Plin. 12. 24.
9 Banana. See Index App. (4).
10 Mango. See Index App. (5). Plin. 12. 24.
11 Jujube. See Index App. (6).

Καὶ ἕτερα δὲ πλείω καὶ διαφέροντα τῶν ἐν τοῖς Ἕλλησιν ἀλλ' ἀνώνυμα. θαυμαστὸν δ' οὐδὲν τῆς ἰδιότητος· σχεδὸν γάρ, ὥς γε δή τινές φασιν, οὐθὲν ὅλως τῶν δένδρων οὐδὲ τῶν ὑλημάτων οὐδὲ τῶν ποιωδῶν ὅμοιόν ἐστι τοῖς ἐν τῇ Ἑλλάδι πλὴν ὀλίγων.

6 Ἴδιον δὲ καὶ ἡ ἐβένη τῆς χώρας ταύτης· ταύτης δὲ δύο γένη, τὸ μὲν εὔξυλον καὶ καλὸν τὸ δὲ φαῦλον. σπάνιον δὲ τὸ καλὸν θάτερον δὲ πολύ. τὴν δὲ χρόαν οὐ θησαυριζομένη λαμβάνει τὴν εὔχρουν ἀλλ' εὐθὺς τῇ φύσει. ἔστι δὲ τὸ δένδρον θαμνῶδες, ὥσπερ ὁ κύτισος.

7 Φασὶ δ' εἶναι καὶ τέρμινθον, οἱ δ' ὅμοιον τερμίνθῳ, ὃ τὸ μὲν φύλλον καὶ τοὺς κλῶνας καὶ τἆλλα πάντα ὅμοια ἔχει τῇ τερμίνθῳ τὸν δὲ καρπὸν διάφορον· ὅμοιον γὰρ ταῖς ἀμυγδαλαῖς. εἶναι γὰρ καὶ ἐν Βάκτροις τὴν τέρμινθον ταύτην καὶ κάρυα φέρειν ἡλίκα ἀμύγδαλα διὰ τὸ μὴ μεγάλα· καὶ τῇ ὄψει δὲ παρόμοια, πλὴν τὸ κέλυφος οὐ τραχύ, τῇ δ' εὐστομίᾳ καὶ ἡδονῇ κρείττω τῶν ἀμυγδάλων. δι' ὃ καὶ χρῆσθαι τοὺς ἐκεῖ μᾶλλον.

8 Ἐξ ὧν δὲ τὰ ἱμάτια ποιοῦσι τὸ μὲν φύλλον ὅμοιον ἔχει τῇ συκαμίνῳ, τὸ δὲ ὅλον φυτὸν τοῖς κυνορόδοις ὅμοιον. φυτεύουσι δὲ ἐν τοῖς πεδίοις αὐτὸ κατ' ὄρχους, δι' ὃ καὶ πόρρωθεν ἀφορῶσι ἄμπελοι φαίνονται. ἔχει δὲ καὶ φοίνικας ἔνια

---

[1] Plin. 12. 25.

[2] See Index. Plin. 12. 17-19.

[3] Pistachio-nut. See Index App. (7). Plin. 12. 25. Nic. *Ther.* 894.

There are also many more[1] which are different to those found among the Hellenes, but they have no names. There is nothing surprising in the fact that these trees have so special a character; indeed, as some say, there is hardly a single tree or shrub or herbaceous plant, except quite a few, like those in Hellas.

The ebony[2] is also peculiar to this country; of this there are two kinds, one with good handsome wood, the other inferior. The better sort is rare, but the inferior one is common. It does not acquire its good colour by being kept, but it is natural to it from the first. The tree is bushy, like laburnum.

Some say that a 'terebinth'[3] grows there also, others that it is a tree like the terebinth; this in leaf twigs and all other respects resembles that tree, but the fruit is different, being like almonds. In fact they say that this sort of terebinth grows also in Bactria and bears nuts only as big as almonds, inasmuch as they are not large for the size of the tree[4]; and they closely resemble almonds in appearance, except that the shell is not rough; and in palatableness and sweetness they are superior to almonds; wherefore the people of the country use them in preference to almonds.

[5]The trees from which they make their clothes have a leaf like the mulberry, but the whole tree resembles the wild rose. They plant them in the plains in rows, wherefore, when seen from a distance, they look like vines. Some parts also have many

[4] διὰ . . . μέγαλα: Sch. omits these words, and W. considers them corrupt; but G seems to have had them in his text. The translation is tentative.

[5] Cotton-plant. *cf.* 4. 7. 7 and 8. Plin. 12. 25.

μέρη πολλούς. καὶ ταῦτα μὲν ἐν δένδρου φύσει.

9 Φέρει δὲ καὶ σπέρματα ἴδια, τὰ μὲν τοῖς χεδροποῖς ὅμοια τὰ δὲ τοῖς πυροῖς καὶ ταῖς κριθαῖς. ἐρέβινθος μὲν γὰρ καὶ φακὸς καὶ τἆλλα τὰ παρ' ἡμῖν οὐκ ἔστιν· ἕτερα δ' ἐστὶν ὥστε παραπλήσια ποιεῖν τὰ ἑψήματα καὶ μὴ διαγιγνώσκειν, ὥς φασιν, ἂν μή τις ἀκούσῃ. κριθαὶ δὲ καὶ πυροὶ καὶ ἄλλο τι γένος ἀγρίων κριθῶν, ἐξ ὧν καὶ ἄρτοι ἡδεῖς καὶ χόνδρος καλός. ταύτας οἱ ἵπποι ἐσθίοντες τὸ πρῶτον διεφθείροντο, κατὰ μικρὸν δὲ οὖν ἐθισθέντες ἐν ἀχύροις οὐδὲν ἔπασχον.

10 Μάλιστα δὲ σπείρουσι τὸ καλούμενον ὄρυζον, ἐξ οὗ τὸ ἕψημα. τοῦτο δὲ ὅμοιον τῇ ζειᾷ καὶ περιπτισθὲν οἷον χόνδρος εὔπεπτον δέ, τὴν ὄψιν πεφυκὸς ὅμοιον ταῖς αἴραις καὶ τὸν πολὺν χρόνον ἐν ὕδατι, ἀποχεῖται δὲ οὐκ εἰς στάχυν ἀλλ' οἷον φόβην, ὥσπερ ὁ κέγχρος καὶ ὁ ἔλυμος. ἄλλο δὲ ὃ ἐκάλουν οἱ Ἕλληνες φακόν· τοῦτο δὲ ὅμοιον μὲν τῇ ὄψει καὶ τὸ βούκερας, θερίζεται δὲ περὶ Πλειάδος δύσιν.

11 Διαφέρει δὲ καὶ αὕτη ἡ χώρα τῷ τὴν μὲν φέρειν ἔνια τὴν δὲ μὴ φέρειν· ἡ γὰρ ὀρεινὴ καὶ ἄμπελον ἔχει καὶ ἐλάαν καὶ τὰ ἄλλα ἀκρόδρυα· πλὴν ἄκαρπον τὴν ἐλάαν, καὶ σχεδὸν καὶ τὴν φύσιν ὥσπερ μεταξὺ κοτίνου καὶ ἐλάας ἐστὶ καὶ

---

1 *cf.* 8. 4. 2. whence it appears that the original text here contained a fuller account. Plin. 18. 71.

2 *Sorghum halepense.* 3 Sc. of Alexander.

4 The verb seems to have dropped out (W.).

date-palms. So much for what come under the heading of 'trees.'

These lands bear also peculiar grains, some like those of leguminous plants, some like wheat and barley. For the chick-pea lentil and other such plants found in our country do not occur; but there are others, so that they make similar mashes, and one cannot, they say, tell the difference, unless one has been told. They have however barley wheat[1] and another kind of wild barley,[2] which makes sweet bread and good porridge. When the horses[3] ate this, at first it proved fatal to them, but by degrees they became accustomed to it mixed with bran and took no hurt.

But above all they sow the cereal called rice, of which they make their mash. This is like rice-wheat, and when bruised makes a sort of porridge, which is easily digested; in its appearance as it grows it is like darnel, and for most of its time of growth it is[4] in water; however it shoots[5] up not into an ear, but as it were into a plume,[6] like the millet and Italian millet. There was another plant[7] which the Hellenes[8] called lentil; this is like in appearance to 'ox-horn' (fenugreek), but it is reaped about the setting of the Pleiad.

Moreover this country shews differences in that part of it bears certain things which another part does not; thus the mountain country has the vine and olive and the other fruit-trees; but the olive is barren,[9] and in its character it is as it were almost between a wild and a cultivated olive, and so it

[5] ἀποχεῖται: *cf.* 8. 8. 1. [6] *cf.* 8. 3. 4.
[7] *Phaseolus Mungo*; see Index App. (8).
[8] *i.e.* of Alexander's expedition. [9] Plin. 12. 14.

τῇ ὅλῃ μορφῇ· καὶ τὸ φύλλον τοῦ μὲν πλατύτερον τοῦ δὲ στενότερον. ταῦτα μὲν οὖν κατὰ τὴν Ἰνδικήν.

12 Ἐν δὲ τῇ Ἀρίᾳ χώρᾳ καλουμένῃ ἄκανθά ἐστιν, ἐφ' ἧς γίνεται δάκρυον ὅμοιον τῇ σμύρνῃ καὶ τῇ ὄψει καὶ τῇ ὀσμῇ· τοῦτο δὲ ὅταν ἐπιλάμψῃ ὁ ἥλιος καταρρεῖ. πολλὰ δὲ καὶ ἄλλα παρὰ τὰ ἐνταῦθα καὶ ἐν τῇ χώρᾳ καὶ ἐν τοῖς ποταμοῖς γίνεται. ἐν ἑτέροις δὲ τόποις ἐστὶν ἄκανθα λευκὴ τρίοζος, ἐξ ἧς καὶ σκυτάλια καὶ βακτηρίας ποιοῦσιν· ὀπώδης δὲ καὶ μανή· ταύτην δὲ καλοῦσιν Ἡρακλέους.

Ἄλλο δὲ ὕλημα μέγεθος μεν ἡλίκον ῥάφανος, τὸ δὲ φύλλον ὅμοιον δάφνῃ καὶ τῷ μεγέθει καὶ τῇ μορφῇ. τοῦτο δ' εἴ τι φάγοι ἐναποθνήσκει. δι' ὃ καὶ ὅπου ἵπποι τούτους ἐφύλαττον διὰ χειρῶν.

13 Ἐν δὲ τῇ Γεδρωσίᾳ χώρᾳ πεφυκέναι φασὶν ἓν μὲν ὅμοιον τῇ δάφνῃ φύλλον ἔχον, οὗ τὰ ὑποζύγια καὶ ὁτιοῦν εἰ φάγοι μικρὸν ἐπισχόντα διεφθείροντο παραπλησίως διατιθέμενα καὶ σπώμενα ὁμοίως τοῖς ἐπιλήπτοις.

Ἕτερον δὲ ἄκανθάν τινα εἶναι· ταύτην δὲ φύλλον μὲν οὐδὲν ἔχειν πεφυκέναι δ' ἐκ μιᾶς ῥίζης· ἐφ' ἑκάστῳ δὲ τῶν ὄζων ἄκανθαν ἔχειν ὀξεῖαν σφόδρα, καὶ τούτων δὲ καταγνυμένων ἢ προστριβομένων ὀπὸν ἐκρεῖν πολύν, ὃς ἀποτυφλοῖ

---

[1] καὶ σχεδὸν . . . μορφῇ conj. W.; σχεδὸν δὲ καὶ τὴν φύσιν ὥσπερ μετ. κοτ. καὶ ἐλ. ἐστι δὲ τῇ ὅλῃ μορφῇ καὶ τὸ φ. Ald.; so also U, omitting the first *καί*.

[2] *Balsamodendron Mukul*; see Index App. (9). Plin. 12 33.

is also in its general appearance,[1] and the leaf is broader than that of the one and narrower than that of the other. So much for the Indian land.

In the country called Aria there is a 'thorn'[2] on which is found a gum resembling myrrh[3] in appearance and smell, and this drops when the sun shines on it. There are also many other plants besides those of our land, both in the country and in its rivers. In other parts there is a white 'thorn'[4] which branches in three, of which they make batons and sticks; its wood is sappy and of loose texture, and they call it the thorn 'of Herakles.'

There is another shrub[5] as large as a cabbage, whose leaf is like that of the bay in size and shape. And if any animal should eat this, it is certain to die of it. Wherefore, wherever there were horses,[6] they kept them under control.

In Gedrosia they say that there grows one tree[7] with a leaf like that of the bay, of which if the beasts or anything else ate, they very shortly died with the same convulsive symptoms as in epilepsy.

And they say that another tree[8] there is a sort of 'thorn' (spurge), and that this has no leaf and grows from a single root; and on each of its branches it has a very sharp spine, and if these are broken or bruised a quantity of juice flows out, which blinds animals or

[3] σμύρνῃ conj. Sch. from 9. 1. 2; Plin. *l.c.*; τῇ ἰλλυρίᾳ Ald. H.
[4] See Index.
[5] *Asafoetida*; see Index App. (10). Plin. 12. 33.
[6] *i.e.* in Alexander's expedition. Probably a verb, such as ὠσφραίνοντο, has dropped out after ἵπποι (Sch.). *Odore equos invitans* Plin. *l.c.*
[7] *Nerium odorum*; see Index App. (11). *cf.* 4. 4. 13; Strabo 15. 2. 7; Plin. *l.c.*
[8] Plin. *l.c.*; Arrian, *Anab.* 6. 22. 7.

τἆλλα ζῷα πάντα καὶ πρὸς τοὺς ἀνθρώπους εἴ τις προσραίνειεν αὐτοῖς. ἐν δὲ τόποις τισὶ πεφυκέναι τινὰ βοτάνην, ὑφ᾽ ᾗ συνεσπειρωμένους ὄφεις εἶναι μικροὺς σφόδρα· τούτοις δ᾽ εἴ τις ἐμβὰς πληγείη θνήσκειν. ἀποπνίγεσθαι δὲ καὶ ἀπὸ τῶν φοινικων τῶν ὠμῶν εἴ τις φάγοι, καὶ τοῦτο ὕστερον κατανοηθῆναι. τοιαῦται μὲν οὖν δυνάμεις καὶ ζώων καὶ φυτῶν ἴσως καὶ παρ᾽ ἄλλοις εἰσί.

14 Περιττότερα δὲ τῶν φυομενων καὶ πλεῖστον ἐξηλλαγμένα πρὸς τὰ ἄλλα τὰ εὔοσμα τὰ περὶ Ἀραβίαν καὶ Συρίαν καὶ Ἰνδούς· οἷον ὅ τε λιβανωτὸς καὶ ἡ σμύρνα καὶ ἡ κασία καὶ τὸ ὀποβάλσαμον καὶ τὸ κινάμωμον καὶ ὅσα ἄλλα τοιαῦτα· περὶ ὧν ἐν ἄλλοις εἴρηται διὰ πλειόνων. ἐν μὲν οὖν τοῖς πρὸς ἕω τε καὶ μεσημβρίαν καὶ ταῦτ᾽ ἴδια καὶ ἕτερα δὲ τούτων πλείω ἐστίν.

V. Ἐν δὲ τοῖς πρὸς ἄρκτον οὐχ ὁμοίως· οὐθὲν γὰρ ὅτι ἄξιον λόγου λέγεται παρὰ τὰ κοινὰ τῶν δένδρων ἃ καὶ φιλόψυχρά τε τυγχάνει καὶ ἔστι καὶ παρ᾽ ἡμῖν, οἷον πεύκη δρῦς ἐλάτη πύξος διοσβάλανος φίλυρα καὶ τὰ ἄλλα δὲ τὰ τοιαῦτα· σχεδὸν γὰρ οὐδὲν ἕτερον παρὰ ταῦτά ἐστιν, ἀλλὰ τῶν ἄλλων ὑλημάτων ἔνια ἃ τοὺς ψυχροὺς μᾶλλον ζητεῖ τόπους, καθάπερ κενταύριον ἀψίνθιον, ἔτι δὲ τὰ φαρμακώδη ταῖς ῥίζαις καὶ τοῖς ὀποῖς, οἷον ἐλλέβορος ἐλατήριον σκαμμωνία, σχεδὸν πάντα τὰ ῥιζοτομούμενα.

2 Τὰ μὲν γὰρ ἐν τῷ Πόντῳ καὶ τῇ Θράκῃ γίνεται,

---

[1] τὰ ἄλλα δὲ: ? om. τὰ; δὲ om. Sch.

even a man, if any drops of it should fall on him. Also they say that in some parts grows a herb under which very small snakes lie coiled up, and that, if anyone treads on these and is bitten, he dies. They also say that, if anyone should eat of unripe dates, he chokes to death, and that this fact was not discovered at first. Now it may be that animals and plants have such properties elsewhere also.

Among the plants that grow in Arabia Syria and India the aromatic plants are somewhat exceptional and distinct from the plants of other lands; for instance, frankincense myrrh cassia balsam of Mecca cinnamon and all other such plants, about which we have spoken at greater length elsewhere. So in the parts towards the east and south there are these special plants and many others besides.

Of the plants special to northern regions.

V. In the northern regions it is not so, for nothing worthy of record is mentioned except the ordinary trees which love the cold and are found also in our country, as fir oak silver-fir box chestnut lime, as well as other similar trees. There is hardly any other[1] besides these; but of shrubs there are some which for choice[2] seek cold regions, as centaury and wormwood, and further those that have medicinal properties in their roots and juices, such as hellebore squirting cucumber scammony, and nearly all those whose roots are gathered.[3]

Some of these grow in Pontus and Thrace, some

[2] I have moved μᾶλλον, which in the MSS. comes before τῶν ἄλλων.
[3] *i.e.* which have medicinal uses.

τὰ δὲ περὶ τὴν Οἴτην καὶ τὸν Παρνασὸν καὶ τὸ Πήλιον καὶ τὴν Ὄσσαν καὶ τὸ Τελέθριον· καὶ ἐν τούτοις δέ τινές φασι πλεῖστον· πολλὰ δὲ καὶ ἐν τῇ Ἀρκαδίᾳ καὶ ἐν τῇ Λακωνικῇ· φαρμακώδεις γὰρ καὶ αὗται. τῶν δὲ εὐωδῶν οὐδὲν ἐν ταύταις, πλὴν ἶρις ἐν τῇ Ἰλλυρίδι καὶ περὶ τὸν Ἀδρίαν· ταύτῃ γὰρ χρηστὴ καὶ πολὺ διαφέρουσα τῶν ἄλλων· ἀλλ' ἐν τοῖς ἀλεεινοῖς καὶ τοῖς πρὸς μεσημβρίαν ὥσπερ ἀντικείμενα τὰ εὐώδη. ἔχουσι δὲ καὶ κυπάριττον οἱ ἀλεεινοὶ μᾶλλον, ὥσπερ Κρήτη Λυκία Ῥόδος, κέδρον δὲ καὶ τὰ Θρᾴκια ὄρη καὶ τὰ Φρύγια.

8 Τῶν δὲ ἡμερουμένων ἥκιστά φασιν ἐν τοῖς ψυχροῖς ὑπομένειν δάφνην καὶ μυρρίνην, καὶ τούτων δὲ ἧττον ἔτι τὴν μυρρίνην· σημεῖον δὲ λέγουσιν ὅτι ἐν τῷ Ὀλύμπῳ δάφνη μὲν πολλή, μύρρινος δὲ ὅλως οὐκ ἔστιν. ἐν δὲ τῷ Πόντῳ περὶ Παντικάπαιον οὐδ' ἕτερον καίπερ σπουδαζόντων καὶ πάντα μηχανωμένων πρὸς τὰς ἱεροσύνας· συκαῖ δὲ πολλαὶ καὶ εὐμεγέθεις καὶ ῥοιαὶ δὲ περισκεπαζόμεναι· ἄπιοι δὲ καὶ μηλέαι πλεῖσται καὶ παντοδαπώταται καὶ χρησταί· αὗται δ' ἐαριναὶ πλὴν εἰ ἄρα ὄψιαι· τῆς δὲ ἀγρίας ὕλης ἐστὶ δρῦς πτελέα μελία καὶ ὅσα τοιαῦτα· πεύκη δὲ καὶ ἐλάτη καὶ πίτυς οὐκ ἔστιν οὐδὲ ὅλως οὐδὲν ἔνδᾳδον· ὑγρὰ δὲ αὕτη καὶ χείρων πολὺ τῆς Σινωπικῆς, ὥστ' οὐδὲ πολὺ χρῶνται αὐτῇ πλὴν πρὸς τὰ ὑπαίθρια. ταῦτα

1 Τελέθριον conj. Sch. (in Euboea), *cf.* 9. 15. 4; Πελέθριον UMVP; Παρθένιον Ald.G.

2 Whose rhizome was used for perfumes; *cf.* 1. 7. 2; *de odor.* 22. 23. 28. 32; Dykes, *The Genus Iris*, p. 237, gives an interesting account of the modern uses of 'orris-root.'

about Oeta Parnassus Pelion Ossa and Telethrion,[1] and in these parts some say that there is great abundance; so also is there in Arcadia and Laconia, for these districts too produce medicinal plants. But of the aromatic plants none grows in these lands, except the iris[2] in Illyria on the shores of the Adriatic; for here it is excellent and far superior to that which grows elsewhere; but in hot places and those which face the south the fragrant plants grow, as if by contrast to the medicinal plants. And the warm places have also the cypress in greater abundance; for instance, Crete Lycia Rhodes, while the prickly cedar grows in the Thracian and the Phrygian mountains.

Of cultivated plants they say that those least able to thrive in cold regions are the bay and myrtle, especially the myrtle, and they give for proof[3] that on Mount Olympus the bay is abundant, but the myrtle does not occur at all. In Pontus about Panticapaeum neither grows, though they are anxious to grow them and take special pains[4] to do so for religious purposes. But there are many well grown fig-trees and pomegranates, which are given shelter; pears and apples are abundant in a great variety of forms and are excellent. These are spring-fruiting trees, except that they may fruit later here than elsewhere. Of wild trees there are oak elm manna-ash and the like (while there is no fir silver-fir nor Aleppo pine, nor indeed any resinous tree). But the wood of such trees[5] in this country is damp and much inferior to that of Sinope, so that they do not much use it except for outdoor purposes. These

[3] Plin. 16. 137.
[4] Plin., *l.c.*, says that Mithridates made this attempt.
[5] *i.e.* oak, etc.

μὲν οὖν περὶ τὸν Πόντον ἢ ἔν τισί γε τόποις αὐτοῦ.

4 Ἐν δὲ τῇ Προποντίδι γίνεται καὶ μύρρινος καὶ δάφνη πολλαχοῦ ἐν τοῖς ὄρεσιν. ἴσως δ' ἔνια καὶ τῶν τόπων ἴδια θετέον· ἕκαστοι γὰρ ἔχουσι τὰ διαφέροντα, ὥσπερ εἴρηται, κατὰ τὰς ὕλας οὐ μόνον τῷ βελτίω καὶ χείρω τὴν αὐτὴν ἔχειν ἀλλὰ καὶ τῷ φέρειν ἢ μὴ φέρειν· οἷον ὁ μὲν Τμῶλος ἔχει καὶ ὁ Μύσιος Ὄλυμπος πολὺ τὸ κάρυον καὶ τὴν διοσβάλανον, ἔτι δὲ ἄμπελον καὶ μηλέαν καὶ ῥόαν· ἡ δὲ Ἴδη τὰ μὲν οὐκ ἔχει τούτων τὰ δὲ σπάνια· περὶ δὲ Μακεδονίαν καὶ τὸν Πιερικὸν Ὄλυμπον τὰ μὲν ἔστι τὰ δ' οὐκ ἔστι τούτων· ἐν δὲ τῇ Εὐβοίᾳ καὶ περὶ τὴν Μαγνησίαν τὰ μὲν Εὐβοϊκὰ πολλὰ τῶν δὲ ἄλλων οὐθέν· οὐδὲ δὴ περὶ τὸ Πέλιον οὐδὲ τὰ ἄλλα τὰ ἐνταῦθα ὄρη.

5 Βραχὺς δ' ἐστὶ τόπος ὃς ἔχει καὶ ὅλως τὴν ναυπηγήσιμον ὕλην· τῆς μὲν γὰρ Εὐρώπης δοκεῖ τὰ περὶ τὴν Μακεδονίαν καὶ ὅσα τῆς Θρᾴκης καὶ περὶ Ἰταλίαν· τῆς δὲ Ἀσίας τά τε ἐν Κιλικίᾳ καὶ τὰ ἐν Σινώπῃ καὶ Ἀμίσῳ, ἔτι δὲ ὁ Μύσιος Ὄλυμπος καὶ ἡ Ἴδη πλὴν οὐ πολλήν· ἡ γὰρ Συρία κέδρον ἔχει καὶ ταύτῃ χρῶνται πρὸς τὰς τριήρεις.

6 Ἀλλὰ καὶ τὰ φίλυδρα καὶ τὰ παραποτάμια ταῦθ' ὁμοίως· ἐν μὲν γὰρ τῷ Ἀδρίᾳ πλάτανον οὔ φασιν εἶναι πλὴν περὶ τὸ Διομήδους ἱερόν· σπανίαν δὲ καὶ ἐν Ἰταλίᾳ πάσῃ· καίτοι πολλοὶ καὶ μεγάλοι ποταμοὶ παρ' ἀμφοῖν· ἀλλ' οὐκ

[1] See Index.

[2] καὶ ὅσα: text probably defective, but sense clear. ? καὶ ὅσα τῆς Θ. ἔχει καὶ τὰ περὶ Ἰ·

are the trees of Pontus, or at least of certain districts of that country.

In the land of Propontis myrtle and bay are found in many places on the mountains. Perhaps however some trees should be put down as special to particular places. For each district, as has been said, has different trees, differing not only in that the same trees occur but of variable quality, but also as to producing or not producing some particular tree. For instance, Tmolus and the Mysian Olympus have the hazel and chestnut[1] in abundance, and also the vine apple and pomegranate; while Mount Ida has some of these not at all and others only in small quantity; and in Macedonia and on the Pierian Olympus some of these occur, but not others; and in Euboea and Magnesia the sweet chestnut[1] is common, but none of the others is found; nor yet on Pelion or the other mountains of that region.

Again it is only a narrow extent of country which produces wood fit for shipbuilding at all, namely in Europe the Macedonian region, and certain parts[2] of Thrace and Italy; in Asia Cilicia Sinope and Amisus, and also the Mysian Olympus, and Mount Ida; but in these parts it is not abundant. For Syria has Syrian cedar, and they use this for their galleys.

The like is true of trees which love water and the riverside; in the Adriatic region they say that the plane is not found, except near the Shrine of Diomedes,[3] and that it is scarce throughout Italy[4]; yet there are many large rivers in both countries, in spite of which the localities do not seem to

[3] On one of the islands of Diomedes, off the coast of Apulia; now called Isole di Tremiti. *cf.* Plin. 12. 6.

[4] *cf.* 2. 8. 1 n.

ἔοικε φέρειν ὁ τόπος· ἐν Ῥηγίῳ γοῦν ἃς Διονύσιος πρεσβύτερος ὁ τύραννος ἐφύτευσεν ἐν τῷ παραδείσῳ, αἵ εἰσι νῦν ἐν τῷ γυμνασίῳ, φιλοτιμηθεῖσαι[1] οὐ δεδύνηνται λαβεῖν μέγεθος.

7 Ἔνιοι δὲ πλείστην ἔχουσι πλάτανον, οἱ δὲ πτελέαν καὶ ἰτέαν, οἱ δὲ μυρίκην, ὥσπερ ὁ Αἷμος. ὥστε τὰ μὲν τοιαῦτα, καθάπερ ἐλέχθη, τῶν τόπων ἴδια θετέον ὁμοίως ἔν τε τοῖς ἀγρίοις καὶ τοῖς ἡμέροις. οὐ μὴν ἀλλὰ τάχ' ἂν εἴη καὶ τούτων ἐπί τινων ὥστε διακοσμηθέντων δύνασθαι τὴν χώραν φέρειν, ὃ καὶ νῦν ξυμβαῖνον ὁρῶμεν καὶ ἐπὶ ζώων ἐνίων καὶ φυτῶν.

VI. Μεγίστην δὲ διαφορὰν αὐτῆς τῆς φύσεως τῶν δένδρων καὶ ἁπλῶς τῶν ὑλημάτων ὑποληπτέον ἣν καὶ πρότερον εἴπομεν, ὅτι τὰ μὲν ἔγγαια τὰ δ' ἔνυδρα τυγχάνει, καθάπερ τῶν ζώων, καὶ τῶν φυτῶν· οὐ μόνον ἐν τοῖς ἕλεσι καὶ ταῖς λίμναις καὶ τοῖς ποταμοῖς γὰρ ἀλλὰ καὶ ἐν τῇ θαλάττῃ[2] φύεται καὶ ὑλήματα ἔνια ἔν τε τῇ ἔξω καὶ δένδρα· ἐν μὲν γὰρ τῇ περὶ ἡμᾶς μικρὰ πάντα τὰ φυόμενα, καὶ οὐδὲν ὑπερέχον ὡς εἰπεῖν τῆς θαλάττης· ἐν ἐκείνῃ δὲ καὶ τὰ τοιαῦτα καὶ ὑπερέχοντα, καὶ ἕτερα δὲ μείζω δένδρα.

2 Τὰ μὲν οὖν περὶ ἡμᾶς ἐστι τάδε· φανερώτατα μὲν καὶ κοινότατα πᾶσιν τό τε φῦκος καὶ τὸ βρύον καὶ ὅσα ἄλλα τοιαῦτα· φανερώτατα δὲ καὶ

---

[1] φιλοτιμηθεῖσαι conj. St.; φιλοτιμηθεὶς MSS.; Plin. 12. 7.
[2] θαλάττης conj. Scal from G; ἐλάτης Ald. H.

produce this tree. At any rate those which King Dionysius the Elder planted at Rhegium in the park, and which are now in the grounds of the wrestling school and are thought much of,[1] have not been able to attain any size.

Some of these regions however have the plane in abundance, and others the elm and willow, others the tamarisk, such as the district of Mount Haemus. Wherefore such trees we must, as was said, take to be peculiar to their districts, whether they are wild or cultivated. However it might well be that the country should be able to produce some of these trees, if they were carefully cultivated: this we do in fact find to be the case with some plants, as with some animals.

Of the aquatic plants of the Mediterranean.

VI. However the greatest difference in the natural character itself of trees and of tree-like plants generally we must take to be that mentioned already, namely, that of plants, as of animals, some belong to the earth, some to water. Not only in swamps, lakes and rivers, but even in the sea there are some tree-like growths, and in the ocean there are even trees. In our own sea all the things that grow are small, and hardly any of them rise above the surface[2]; but in the ocean we find the same kinds rising above the surface, and also other larger trees.

Those found in our own waters are as follows: most conspicuous of those which are of general occurrence are seaweed[3] oyster-green and the like; most obvious of those peculiar to certain parts are the

[1] Plin. 13. 135.

ἰδιώτατα κατὰ τοὺς τόπους ἐλάτη συκῆ δρῦς ἄμπελος φοῖνιξ. τούτων δὲ τὰ μὲν πρόσγεια τὰ δὲ πόντια τὰ δ' ἀμφοτέρων τῶν τόπων κοινά. καὶ τὰ μὲν πολυειδῆ, καθάπερ τὸ φῦκος, τὰ δὲ μίαν ἰδέαν ἔχοντα. τοῦ γὰρ φύκους τὸ μέν ἐστι πλατύφυλλον ταινιοειδὲς χρῶμα ποῶδες ἔχον, ὃ δὴ καὶ πράσον καλοῦσί τινες, οἱ δὲ ζωστῆρα· ῥίζαν δὲ ἔχει δασεῖαν ἔξωθεν ἔνδοθεν δὲ λεπυριώδη, μακρὰν δὲ ἐπιεικῶς καὶ εὐπαχῆ παρομοίαν τοῖς κρομυογητείοις.

3 Τὸ δὲ τριχόφυλλον, ὥσπερ τὸ μάραθον, οὐ ποῶδες ἀλλ' ἔξωχρον οὐδὲ ἔχον καυλὸν ἀλλ' ὀρθόν πως ἐν αὑτῷ· φύεται δὲ τοῦτο ἐπὶ τῶν ὀστράκων καὶ τῶν λίθων, οὐχ ὥσπερ θάτερον πρὸς τῇ γῇ· πρόσγεια δ' ἄμφω· καὶ τὸ μὲν τριχόφυλλον πρὸς αὐτῇ τῇ γῇ, πολλάκις δὲ ὥσπερ ἐπικλύζεται μόνον ὑπὸ τῆς θαλάττης, θάτερον δὲ ἀνωτέρω.

4 Γίνεται δὲ ἐν μὲν τῇ ἔξω τῇ περὶ Ἡρακλέους στήλας θαυμαστόν τι τὸ μέγεθος· ὥς φασι, καὶ τὸ πλάτος μεῖζον ὡς παλαιστιαῖον. φέρεται δὲ τοῦτο εἰς τὴν ἔσω θάλατταν ἅμα τῷ ῥῷ τῷ ἔξωθεν καὶ καλοῦσιν αὐτὸ πράσον· ἐν ταύτῃ δ' ἔν τισι τόποις ὥστ' ἐπάνω τοῦ ὀμφαλοῦ. λέγεται δὲ ἐπέτειον εἶναι καὶ φύεσθαι μὲν τοῦ ἦρος λήγοντος, ἀκμάζειν δὲ τοῦ θέρους, τοῦ μετοπώρου δὲ φθίνειν, κατὰ δὲ τὸν χειμῶνα ἀπόλλυσθαι καὶ ἐκπίπτειν. ἅπαντα δὲ καὶ τἆλλα τὰ φυόμενα χείρω καὶ ἀμαυρότερα γίνεσθαι τοῦ χειμῶνος.

---

[1] See Index: συκῆ, δρῦς, etc.

[2] ταινιοειδὲς conj. Dalec.; τετανοειδὲς UP$_2$Ald.H.; τὰ τενοειδὲς MV.

[3] cf. Diosc. 4. 99; Plin. 136.

sea-plants called 'fir' 'fig' 'oak' 'vine' 'palm.'[1] Of these some are found close to land, others in the deep sea, others equally in both positions. And some have many forms, as seaweed, some but one. Thus of seaweed there is the broad-leaved kind, riband-like[2] and green in colour, which some call 'green-weed' and others 'girdle-weed.' This has a root which on the outside is shaggy, but the inner part is made of several coats, and it is fairly long and stout, like *kromyogeteion* (a kind of onion).

3 Another kind has hair-like leaves like fennel, and is not green but pale yellow; nor has it a stalk, but it is, as it were, erect in itself; this grows on oyster-shells and stones, not, like the other, attached to the bottom; but both are plants of the shore, and the hair-leaved kind grows close to land, and sometimes is merely washed over by the sea[4]; while the other is found further out.

Again in the ocean about the pillars of Heracles there is a kind[5] of marvellous size, they say, which is larger, about a palmsbreadth.[6] This is carried into the inner sea along with the current from the outer sea, and they call it 'sea-leek' (riband-weed); and in this sea in some parts it grows higher than a man's waist. It is said to be annual and to come up at the end of spring, and to be at its best in summer, and to wither in autumn, while in winter it perishes and is thrown up on shore. Also, they say, all the other plants of the sea become weaker and feebler in winter. These then are, one may say, the

[4] *i.e.* grows above low water mark.
[5] See Index: φῦκος (2).
[6] *i.e.* the 'leaf': the comparison is doubtless with τὸ πλατὺ, § 2; ὡς UMVAld.; ἢ W. after Sch.'s conj.

ταῦτα μὲν οὖν οἷον πρόσγεια περί γε τὴν θάλατταν. τὸ δὲ πόντιον φῦκος ὃ οἱ σπογγιεῖς ἀνακολυμβῶσι πελάγιον.

5 Καὶ ἐν Κρήτῃ δὲ φύεται πρὸς τῇ γῇ ἐπὶ τῶν πετρῶν πλεῖστον καὶ κάλλιστον ᾧ βάπτουσιν οὐ μόνον τὰς ταινίας ἀλλὰ καὶ ἔρια καὶ ἱμάτια· καὶ ἕως ἂν ᾖ πρόσφατος ἡ βαφή, πολὺ καλλίων ἡ χρόα τῆς πορφύρας· γίνεται δ' ἐν τῇ προσβόρρῳ καὶ πλεῖον καὶ κάλλιον, ὥσπερ αἱ σπογγιαὶ καὶ ἄλλα τοιαῦτα.

6 Ἄλλο δ' ἐστὶν ὅμοιον τῇ ἀγρώστει· καὶ γὰρ τὸ φύλλον παραπλήσιον ἔχει καὶ τὴν ῥίζαν γονατώδη καὶ μακρὰν καὶ πεφυκυῖαν πλαγίαν, ὥσπερ ἡ τῆς ἀγρώστιδος· ἔχει δὲ καὶ καυλὸν καλαμώδη, καθάπερ ἡ ἄγρωστις· μεγέθει δὲ ἔλαττον πολὺ τοῦ φύκους.

Ἄλλο δὲ τὸ βρύον, ὃ φύλλον μὲν ἔχει ποῶδες τῇ χρόᾳ, πλατὺ δὲ καὶ οὐκ ἀνόμοιον ταῖς θριδακίναις, πλὴν ῥυτιδωδέστερον καὶ ὥσπερ συνεσπασμένον. καυλὸν δὲ οὐκ ἔχει, ἀλλ' ἀπὸ μιᾶς ἀρχῆς πλείω τὰ τοιαῦτα καὶ πάλιν ἀπ' ἄλλης· φύεται δὲ ἐπὶ τῶν λίθων τὰ τοιαῦτα πρὸς τῇ γῇ καὶ τῶν ὀστράκων. καὶ τὰ μὲν ἐλάττω σχεδὸν ταῦτ' ἐστίν.

7 Ἡ δὲ δρῦς καὶ ἡ ἐλάτη παράγειοι μὲν ἄμφω· φύονται δ' ἐπὶ λίθοις καὶ ὀστράκοις ῥίζας μὲν οὐκ ἔχουσαι, προσπεφυκυῖαι δὲ ὥσπερ αἱ λεπάδες. ἀμφότεραι μὲν οἷον σαρκόφυλλα· προμηκέστερον δὲ τὸ φύλλον πολὺ καὶ παχύτερον τῆς ἐλάτης

---

1 Plin. 13. 136, *cf.* 32. 22; Diosc. 4. 99.
2 litmus; see Index, φῦκος (5).
3 Plin. *l.c.*; grass-wrack, see Index, φῦκος (6).

sea-plants which are found near the shore. But the 'seaweed of ocean,' which is dived for by the sponge-fishers, belongs to the open sea.

[1] In Crete there is an abundant and luxuriant growth[2] on the rocks close to land, with which they dye not only their ribbons, but also wool and clothes. And, as long as the dye is fresh, the colour is far more beautiful than the purple dye; it occurs on the north coast in greater abundance and fairer, as do the sponges and other such things.

[3] There is another kind like dog's-tooth grass; the leaf is very like, the root is jointed and long, and grows out sideways, like that of that plant; it has also a reedy stalk like the same plant, and in size it is much smaller than ordinary seaweed.

[4] Another kind is the oyster-green, which has a leaf green in colour, but broad and not unlike lettuce leaves; but it is more wrinkled[5] and as it were crumpled. It has no stalk, but from a single starting-point grow many of the kind, and again from another starting-point. These things grow on stones close to land and on oyster-shells. These are about all the smaller kinds.

[6] The 'sea-oak' and 'sea-fir' both belong to the shore; they grow on stones and oyster-shells, having no roots, but being attached to them like limpets.[7] Both have more or less fleshy leaves; but the leaf of the 'fir' grows much longer and stouter, and is[8]

[4] Plin. 13. 137; 27. 56; *βρύον* conj. Scal. from G and Plin. *l.c.*; *βότρυον* UAld.H.

[5] *ῥυτιδωδέστερον* conj. Scal. from G and Plin. *l.c.*; *χρυσιωδέστερον* Ald.; *ῥυσιωδέστερον* mBas.

[6] Plin. *l.c.* [7] *λεπάδες* Ald.; *λοπάδες* W. with UMV.

[8] *προμηκέστερον . . . πέφυκε καὶ* conj. W.; *προμ. δὲ τὸ φύλλον παχὺ καὶ παχύτερον τῆς ἐλάτης· πολὺ δὲ καὶ* Ald.

πέφυκε καὶ οὐκ ἀνόμοιον τοῖς τῶν ὀσπρίων λοβοῖς, κοῖλον δ' ἔνδοθεν καὶ οὐδὲν ἔχον ἐν αὐτοῖς· τὸ δὲ τῆς δρυὸς λεπτὸν καὶ μυρικωδέστερον· χρῶμα δ' ἐπιπόρφυρον ἀμφοῖν. ἡ δὲ ὅλη μορφὴ τῆς μὲν ἐλάτης ὀρθὴ καὶ αὐτῆς καὶ τῶν ἀκρεμόνων, της δὲ δρυὸς σκολιωτέρα καὶ μᾶλλον ἔχουσα πλάτος·

8 γίνεται δὲ ἄμφω καὶ πολύκαυλα καὶ <μονόκαυλα,> μονοκαυλότερον δὲ ἡ ἐλάτη· τὰς δὲ ἀκρεμονικὰς ἀποφύσεις ἡ μὲν ἐλάτη μακρὰς ἔχει καὶ εὐθείας καὶ μανάς, ἡ δὲ δρῦς βραχυτέρας και σκολιωτέρας καὶ πυκνοτέρας. τὸ δ' ὅλον μέγεθος ἀμφοτέρων ὡς πυγωνιαῖον ἢ μικρὸν ὑπεραῖρον, μεῖζον δὲ ὡς ἁπλῶς εἰπεῖν τὸ τῆς ἐλάτης. χρήσιμον δὲ ἡ δρῦς εἰς βαφὴν ἐρίων ταῖς γυναιξίν. ἐπὶ μὲν τῶν ἀκρεμόνων προσηρτημένα τῶν ὀστρακοδέρμων ζώων ἔνια· καὶ κάτω δὲ πρὸς αὐτῷ τῷ καυλῷ περιπεφυκότων τινῶν γ' ὅλῳ, ἐν τούτοις δεδυκότες ὀνίννοι τε καὶ ἄλλ' ἄττα καὶ τὸ ὅμοιον πολύποδι.

9 Ταῦτα μὲν οὖν πρόσγεια καὶ ῥάδια θεωρηθῆναι· φασὶ δέ τινες καὶ ἄλλην δρῦν εἶναι ποντίαν ἣ καὶ καρπὸν φέρει, καὶ ἡ βάλανος αὐτῆς χρησίμη· τοὺς δὲ σκινθοὺς καὶ κολυμβητὰς λέγειν ὅτι καὶ ἕτεραι μεγάλαι τινὲς τοῖς μεγέθεσιν εἴησαν.

Ἡ δὲ ἄμπελος ἀμφοτέρωσε γίνεται· καὶ γὰρ πρὸς τῇ γῇ καὶ ποντία· μείζω δ' ἔχει καὶ τὰ φύλλα καὶ τὰ κλήματα καὶ τὸν καρπὸν ἡ ποντία.

Ἡ δὲ συκῆ ἄφυλλος μὲν τῷ δὲ μεγέθει οὐ μεγάλη, χρῶμα δὲ τοῦ φλοιοῦ φοινικοῦν.

---

[1] αὐτοῖς Ald.H.; αὐτῷ conj. W.
[2] I have inserted μονόκαυλα.

not unlike the pods of pulses, but is hollow inside and contains nothing in the 'pods.'[1] That of the 'oak' is slender and more like the tamarisk; the colour of both is purplish. The whole shape of the 'fir' is erect, both as to the stem and the branches, but that of the 'oak' is less straight and the plant is broader. Both are found both with many stems and with one,[2] but the 'fir' is more apt to have a single stem. The branchlike outgrowths in the 'fir' are long straight and spreading, while in the 'oak' they are shorter less straight and closer. The whole size of either is about a cubit or rather more, but in general that of the 'fir' is the longer. The 'oak' is useful to women for dyeing wool. To the branches are attached certain creatures with shells, and below they are also found attached to the stem itself, which in some cases they completely cover;[3] and among these are found millepedes and other such creatures, including the one which resembles a cuttlefish.

These plants occur close to land and are easy to observe; but some report[4] that there is another 'sea oak' which even bears fruit and has a useful 'acorn,' and that the sponge fishers[5] and divers told them that there were other large kinds.

[6] The 'sea-vine' grows under both conditions, both close to land and in the deep sea; but the deep sea form has larger leaves branches and fruit.

[7] The 'sea-fig' is leafless and not of large size, and the colour of the bark is red.

[3] τινῶν γ' ὅλῳ conj. W.; τινῶν ὅλων Ald.; τινῶν γε ὅλων U; text uncertain: the next clause has no connecting particle.

[4] Plin. 13. 137.

[5] σκίνθους, a *vox nihili*: perhaps conceals a proper name, *e.g.* Σικελικοὺς; σπογγεῖς conj. St.

[6] Plin. 13. 138. [7] Plin. *l.c.*

10 Ὁ δὲ φοῖνιξ ἐστι μὲν πόντιον βραχυστέλεχες δὲ σφόδρα, καὶ σχεδὸν εὐθεῖαι αἱ ἐκφύσεις τῶν ῥάβδων· καὶ κάτωθεν οὐ κύκλῳ αὗται, καθάπερ τῶν ῥάβδων αἱ ἀκρεμόνες, ἀλλ' ὡσὰν ἐν πλάτει κατὰ μίαν συνεχεῖς, ὀλιγαχοῦ δὲ καὶ ἀπαλλάττουσαι. τῶν δὲ ῥάβδων ἢ τῶν ἀποφύσεων τούτων ὁμοία τρόπον τινὰ ἡ φύσις τοῖς τῶν ἀκανθῶν φύλλοις τῶν ἀκανικῶν, οἷον σόγκοις καὶ τοῖς τοιούτοις, πλὴν ὀρθαὶ καὶ οὐχ, ὥσπερ ἐκεῖνα, περικεκλασμεναι καὶ τὸ φύλλον ἔχουσαι διαβεβρωμένον ὑπὸ τῆς ἅλμης· ἐπεὶ τό γε δι' ὅλου ἥκειν τὸν μέσον γε καυλὸν καὶ ἡ ἄλλη ὄψις παραπλησία. τὸ δὲ χρῶμα καὶ τούτων καὶ τῶν καυλῶν καὶ ὅλου τοῦ φυτοῦ ἐξέρυθρόν τε σφόδρα καὶ φοινικοῦν.

Καὶ τὰ μὲν ἐν τῇδε τῇ θαλάττῃ τοσαῦτά ἐστιν. ἡ γὰρ σπογγιὰ καὶ αἱ ἀπλυσίαι καλούμεναι καὶ εἴ τι τοιοῦτον ἑτέραν ἔχει φύσιν.

VII. Ἐν δὲ τῇ ἔξω τῇ περὶ Ἡρακλέους στήλας τό τε πράσον, ὥσπερ εἴρηται, φύεται καὶ τὰ ἀπολιθούμενα ταῦτα, οἷον θῦμα καὶ τὰ δαφνοειδῆ καὶ τὰ ἄλλα. τῆς δὲ ἐρυθρᾶς καλουμένης ἐν τῇ Ἀραβίᾳ μικρὸν ἐπάνω Κόπτου ἐν μὲν τῇ γῇ

---

[1] κάτωθεν . . , ἀπαλλάττουσαι probably beyond certain restoration: I have added καὶ before κάτωθεν (from G), altered κυκλωθὲν to κύκλῳ, put a stop before καὶ κάτωθεν and restored ἀπαλλάττουσαι (Ald.H.). [2] *cf.* 6. 4. 8; 7. 8. 3.

[3] περικεκλασμένα, *i.e.* towards the ground. *cf.* Diosc. 3. 68 and 69, where Plin. (27. 13) renders (φύλλα) ὑποπερικλᾶται *ad terram infracta.*

The 'sea-palm' is a deep-sea plant, but with a very short stem, and the branches which spring from it are almost straight; and these under water are not set all round the stem, like the twigs which grow from the branches, but extend, as it were, quite flat in one direction, and are uniform; though occasionally they are irregular.[1] The character of these branches or outgrowths to some extent resembles the leaves of thistle-like spinous plants, such as the sow-thistles[2] and the like, except that they are straight and not bent over[3] like these, and have their leaves eaten away by the brine; in the fact that the central stalk[4] at least runs through the whole, they resemble these, and so does the general appearance. The colour both of the branches and of the stalks and of the plant as a whole is a deep red or scarlet.

Such are the plants found in this sea. For sponges and what are called *aplysiai*[5] and such-like growths are of a different character.

Of the aquatic plants of the 'outer sea' (i.e. *Atlantic, Persian Gulf, etc.*).

VII. In the outer sea near the pillars of Heracles grows the 'sea-leek,' as has been said[6]; also the well known[7] plants which turn to stone, as *thyma*, the plants like the bay and others. And in the sea called the Red Sea[8] a little above Coptos[9]

[4] *i.e.* midrib.

[5] Some kind of sponge. ἀπλυσίαι conj. R. Const.; πλύσιαι UAld.; πλυσίαι M; πλουσίαι V.

[6] 4. 6. 4.

[7] ταῦτα: *cf.* 3. 7. 3; 3. 18. 11.

[8] Plin. 13. 139.

[9] Κόπτου conj. Scal.; κόπου MV; κόλπου UAld.; *Capto* G and Plin. *l.c.*

δένδρον οὐδὲν φύεται πλὴν τῆς ἀκάνθης τῆς διψάδος καλουμένης· σπανία δὲ καὶ αὕτη διὰ τὰ καύματα καὶ τὴν ἀνυδρίαν· οὐχ ὕει γὰρ ἀλλ' ἢ δι' ἐτῶν τεττάρων ἢ πέντε καὶ τότε λάβρως καὶ ἐπ' ὀλίγον χρόνον.

Ἐν δὲ τῇ θαλάττῃ φύεται, καλοῦσι δ' αὐτὰ δάφνην καὶ ἐλάαν. ἔστι δὲ ἡ μὲν δάφνη ὁμοία τῇ ἀρίᾳ ἡ δὲ ἐλάα <τῇ ἐλάᾳ> τῷ φύλλῳ· καρπὸν δὲ ἔχει ἡ ἐλάα παραπλήσιον ταῖς ἐλάαις· ἀφίησι δὲ καὶ δάκρυον, ἐξ οὗ οἱ ἰατροὶ φάρμακον ἔναιμον συντιθέασιν ὃ γίνεται σφόδρα ἀγαθόν. ὅταν δὲ ὕδατα πλείω γένηται, μύκητες φύονται πρὸς τῇ θαλάττῃ κατά τινα τόπον, οὗτοι δὲ ἀπολιθοῦνται ὑπὸ τοῦ ἡλίου. ἡ δὲ θάλαττα θηριώδης· πλείστους δὲ ἔχει τοὺς καρχαρίας, ὥστε μὴ εἶναι κολυμβῆσαι.

Ἐν δὲ τῷ κόλπῳ τῷ καλουμένῳ Ἡρώων, ἐφ' ὃν καταβαίνουσιν οἱ ἐξ Αἰγύπτου, φύεται μὲν δάφνη τε καὶ ἐλάα καὶ θύμον, οὐ μὴν χλωρά γε ἀλλὰ λιθοειδῆ τὰ ὑπερέχοντα τῆς θαλάττης, ὅμοια δὲ καὶ τοῖς φύλλοις καὶ τοῖς βλαστοῖς τοῖς χλωροῖς. ἐν δὲ τῷ θύμῳ καὶ τὸ τοῦ ἄνθους χρῶμα διάδηλον ὡσὰν μήπω τελέως ἐξηνθηκός. μήκη δὲ τῶν δενδρυφίων ὅσον εἰς τρεῖς πήχεις.

Οἱ δέ, ὅτε ἀνάπλους ἦν τῶν ἐξ Ἰνδῶν ἀποσταλέντων ὑπὸ Ἀλεξάνδρου, τὰ ἐν τῇ θαλάττῃ φυόμενα, μέχρι οὗ μὲν ἂν ᾖ ἐν τῷ ὑγρῷ, χρῶμά φασιν ἔχειν ὅμοιον τοῖς φυκίοις, ὁπόταν δ' ἐξ-

---

[1] *cf* Strabo 16. 1. 147. [2] See Index

[3] The name of a tree seems to have dropped out: I have inserted τῇ ἐλάᾳ: *cf.* ταῖς ἐλάαις below. Bretzl suggests ἰδέᾳ for ἀρίᾳ.

in Arabia there grows on the land no tree except that called the 'thirsty' acacia, and even this is scarce by reason of the heat and the lack of water; for it never rains except at intervals of four or five years, and then the rain comes down heavily and is soon over.

[1] But there are plants in the sea, which they call 'bay' and 'olive' (white mangrove[2]). In foliage the 'bay' is like the *aria* (holm-oak), the 'olive' like the real olive.[3] The latter has a fruit like olives, and it also discharges a gum,[4] from which the physicians[4] compound a drug[5] for stanching blood, which is extremely effective. And when there is more rain than usual, mushrooms grow in a certain place close to the sea, which are turned to stone by the sun. The sea is full of beasts, and produces sharks[6] in great numbers, so that diving is impossible.

In the gulf called 'the Gulf of the Heroes,'[7] to which the Egyptians go down, there grow a 'bay,' an 'olive,' and a 'thyme'; these however are not green, but like stones so far as they project above the sea, but in leaves and shoots they are like their green namesakes. In the 'thyme' the colour of the flower is also conspicuous, looking as though the flower had not yet completely developed. These treelike growths are about three cubits in height.

[8] Now some, referring to the occasion when there was an expedition of those returning from India sent out by Alexander, report that the plants which grow in the sea, so long as they are kept damp, have a colour

[4] *cf.* Diosc. 1. 105 and 106.
[5] *cf.* Athen. 4. 83; Plin. 12. 77.
[6] Plin. 13. 139. [7] *cf.* 9. 4. 4. [8] Plin. 13. 140.

συνεχθέντα τεθῇ πρὸς τὸν ἥλιον, ἐν ὀλίγῳ χρόνῳ ἐξομοιοῦσθαι τῷ ἁλί. φύεσθαι δὲ καὶ σχοίνους λιθίνους παρ' αὐτὴν τὴν θάλατταν, οὓς οὐδεὶς ἂν διαγνοίη τῇ ὄψει πρὸς τοὺς ἀληθινούς. θαυμασιώτερον δέ τι τούτου λέγουσι· φύεσθαι γὰρ δενδρύφι' ἄττα τὸ μὲν χρῶμα ἔχοντα ὅμοιον κέρατι βοὸς τοῖς δὲ ὄζοις τραχέα καὶ ἀπ' ἄκρου πυρρά· ταῦτα δὲ θραύεσθαι μὲν εἰ συγκλῴη τις· ἐκ δὲ τούτων πυρὶ ἐμβαλλόμενα, καθάπερ τὸν σίδηρον, διάπυρα γινόμενα πάλιν ὅταν ἀποψύχοιτο καθίστασθαι καὶ τὴν αὐτὴν χρόαν λαμβάνειν.

4 Ἐν δὲ ταῖς νήσοις ταῖς ὑπὸ τῆς πλημμυρίδος καταλαμβανομέναις δένδρα μεγάλα πεφυκέναι ἡλίκαι πλάτανοι καὶ αἴγειροι αἱ μέγισται· συμβαίνειν δέ, ὅθ' ἡ πλημμυρὶς ἐπέλθοι, τὰ μὲν ἄλλα κατακρύπτεσθαι ὅλα, τῶν δὲ μεγίστων ὑπερέχειν τοὺς κλάδους, ἐξ ὧν τὰ πρυμνησία ἀνάπτειν, εἶθ' ὅτε πάλιν ἄμπωτις γίνοιτο ἐκ τῶν ῥιζῶν. ἔχειν δὲ τὸ δένδρον φύλλον μὲν ὅμοιον τῇ δάφνῃ, ἄνθος δὲ τοῖς ἴοις καὶ τῷ χρώματι καὶ τῇ ὀσμῇ, καρπὸν δὲ ἡλίκον ἐλάα καὶ τοῦτον εὐώδη σφόδρα· καὶ τὰ μὲν φύλλα οὐκ ἀποβάλλειν, τὸ δὲ ἄνθος καὶ τὸν καρπὸν ἅμα τῷ φθινοπώρῳ γίνεσθαι, τοῦ δὲ ἔαρος ἀπορρεῖν.

5 Ἄλλα δ' ἐν αὐτῇ τῇ θαλάττῃ πεφυκέναι, ἀείφυλλα μὲν τὸν δὲ καρπὸν ὅμοιον ἔχειν τοῖς θέρμοις.

Περὶ δὲ τὴν Περσίδα τὴν κατὰ τὴν Καρμανίαν, καθ' ὃ ἡ πλημμυρὶς γίνεται, δένδρα ἐστὶν εὐμεγέθη ὅμοια τῇ ἀνδράχλῃ καὶ τῇ μορφῇ καὶ τοῖς φύλλοις· καρπὸν δὲ ἔχει πολὺν ὅμοιον τῷ χρώματι ταῖς

like sea-weeds, but that when they are taken out and put in the sun, they shortly become like salt. They also say that rushes of stone grow close to the sea, which none could distinguish at sight from real rushes. They also report a more marvellous thing than this; they say that there are certain tree-like growths which in colour resemble an ox-horn, but whose branches are rough, and red at the tip; these break if they are doubled up, and some of them, if they are cast on a fire, become red-hot like iron, but recover when they cool and assume their original colour.

[1] On the islands which get covered by the tide they say that great trees[2] grow, as big as planes or the tallest poplars, and that it came to pass that, when the tide[3] came up, while the other things were entirely buried, the branches of the biggest trees projected and they fastened the stern cables to them, and then, when the tide ebbed again, fastened them to the roots. And that the tree has a leaf like that of the bay, and a flower like gilliflowers in colour and smell, and a fruit the size of that of the olive, which is also very fragrant. And that it does not shed its leaves, and that the flower and the fruit form together in autumn and are shed in spring.

[4] Also they say there are plants which actually grow in the sea, which are evergreen and have a fruit like lupins.

[5] In Persia in the Carmanian district, where the tide is felt, there are trees[6] of fair size like the andrachne in shape and in leaves; and they bear much fruit like

[1] Plin. 13. 141.
[2] Mangroves. See Index App. (12).
[3] *cf.* Arr. *Anab.* 6. 22. 6.
[4] Plin. *l.c.* Index App. (13).
[5] Plin. 12. 37.
[6] White mangroves. Index App. (14).

ἀμυγδάλαις ἔξωθεν, τὸ δ' ἐντὸς συνελίττεται καθάπερ συνηρτημένον πᾶσιν. ὑποβέβρωται δὲ ταῦτα τὰ δένδρα πάντα κατὰ μέσον ὑπὸ τῆς θαλάττης καὶ ἕστηκεν ὑπὸ τῶν ῥιζῶν, ὥσπερ πολύπους. ὅταν γὰρ ἡ ἄμπωτις γένηται θεωρεῖν
6 ἔστιν. ὕδωρ δὲ ὅλως οὐκ ἔστιν ἐν τῷ τόπῳ· καταλείπονται δέ τινες διώρυχες δι' ὧν διαπλέουσιν· αὗται δ' εἰσὶ θαλάττης· ᾧ καὶ δῆλον οἴονταί τινες ὅτι τρέφονται ταύτῃ καὶ οὐ τῷ ὕδατι, πλὴν εἴ τι ταῖς ῥίζαις ἐκ τῆς γῆς ἕλκουσιν. εὔλογον δὲ καὶ τοῦθ' ἁλμυρὸν εἶναι· καὶ γὰρ οὐδὲ κατὰ βάθους αἱ ῥίζαι. τὸ δὲ ὅλον ἓν τὸ γένος εἶναι τῶν τ' ἐν τῇ θαλάττῃ φυομένων καὶ τῶν ἐν τῇ γῇ ὑπὸ τῆς πλημμυρίδος καταλαμβανομένων· καὶ τὰ μὲν ἐν τῇ θαλάττῃ μικρὰ καὶ φυκώδη φαινόμενα, τὰ δ' ἐν τῇ γῇ μεγάλα καὶ χλωρὰ καὶ ἄνθος εὔοδμον ἔχοντα, καρπὸν δὲ οἷον θέρμος.

7 Ἐν Τύλῳ δὲ τῇ νήσῳ, κεῖται δ' αὕτη ἐν τῷ Ἀραβίῳ κόλπῳ, τὰ μὲν πρὸς ἕω τοσοῦτο πλῆθος εἶναί φασι δένδρων ὅτ' ἐκβαίνει ἡ πλημμυρὶς ὥστ' ἀπωχυρῶσθαι. πάντα δὲ ταῦτα μεγέθη μὲν ἔχειν ἡλίκα συκῆ, τὸ δὲ ἄνθος ὑπερβάλλον τῇ εὐωδίᾳ, καρπὸν δὲ ἄβρωτον ὅμοιον τῇ ὄψει τῷ θέρμῳ. φέρειν δὲ τὴν νῆσον καὶ τὰ δένδρα τὰ ἐριοφόρα πολλά. ταυτα δὲ φύλλον μὲν ἔχειν παρόμοιον τῇ ἀμπέλῳ πλὴν μικρόν, καρπὸν δὲ οὐδένα φέρειν· ἐν ᾧ δὲ τὸ ἔριον ἡλίκον μῆλον ἐαρινὸν συμμεμυκός· ὅταν δὲ ὡραῖον ῇ, ἐκπετάν-

---

[1] Plin. *l.c.* *Sicco litore radicibus nudis polyporum modo complexae steriles arenas aspectantur*: he appears to have had a fuller text.

in colour to almonds on the outside, but the inside is coiled up as though the kernels were all united. 1 These trees are all eaten away up to the middle by the sea and are held up by their roots, so that they look like a cuttle-fish. For one may see this at ebb-tide. And there is no rain at all in the district, but certain channels are left, along which they sail, and which are part of the sea. Which, some think, makes it plain that the trees derive nourishment from the sea and not from fresh water, except what they draw up with their roots from the land. And it is reasonable to suppose that this too is brackish; for the roots do not run to any depth. In general they say that the trees which grow in the sea and those which grow on the land and are overtaken by the tide are of the same kind, and that those which grow in the sea are small and look like seaweed, while those that grow [2] on land are large and green and have a fragrant flower and a fruit like a lupin.

In the island of Tylos,[3] which is situated in the Arabian gulf,[4] they say that on the east side there is such a number of trees when the tide goes out that they make a regular fence. All these are in size as large as a fig-tree, the flower is exceedingly fragrant, and the fruit, which is not edible, is like in appearance to the lupin. They say that the island also produces the 'wool-bearing' tree (cotton-plant) in abundance. This has a leaf like that of the vine, but small, and bears no fruit; but the vessel in which the 'wool' is contained is as large as a spring apple,

[2] *φυκώδη φαινόμενα τὰ δ' ἐν* conj. W.; *φυκ. φυ. δ' ἐν* MVAld.; U has *φερόμενα* (?).

[3] *cf.* 5. 4. 6; Plin. 12. 38 and 39; modern name Bahrein.

[4] *i.e.* Persian Gulf.

νυσθαι καὶ ἐξείρειν τὸ ἔριον, ἐξ οὗ τὰς σινδόνας ὑφαίνουσι, τὰς μὲν εὐτελεῖς τὰς δὲ πολυτελεστάτας.

Γίνεται δὲ τοῦτο καὶ ἐν Ἰνδοῖς, ὥσπερ ἐλέχθη, καὶ ἐν Ἀραβίᾳ. εἶναι δὲ ἄλλα δένδρα τὸ ἄνθος ἔχοντα ὅμοιον τῷ λευκοΐῳ, πλὴν ἄοδμον καὶ τῷ μεγέθει τετραπλάσιον τῶν ἴων. καὶ ἕτερον δέ τι δένδρον πολύφυλλον ὥσπερ τὸ ῥόδον· τοῦτο δὲ τὴν μὲν νύκτα συμμύειν ἅμα δὲ τῷ ἡλίῳ ἀνιόντι διοίγνυσθαι, μεσημβρίας δὲ τελέως διεπτύχθαι, πάλιν δὲ τῆς δείλης συνάγεσθαι κατὰ μικρὸν καὶ τὴν νύκτα συμμύειν· λέγειν δὲ καὶ τοὺς ἐγχωρίους ὅτι καθεύδει. γίνεσθαι δὲ καὶ φοίνικας ἐν τῇ νήσῳ καὶ ἀμπέλους καὶ τἆλλα ἀκρόδρυα καὶ συκᾶς οὐ φυλλορροούσας. ὕδωρ δὲ οὐράνιον γίνεσθαι μέν, οὐ μὴν χρῆσθαί γε πρὸς τοὺς καρπούς· ἀλλ᾽ εἶναι κρήνας ἐν τῇ νήσῳ πολλάς, ἀφ᾽ ὧν πάντα βρέχειν, ὃ καὶ συμφέρειν μᾶλλον τῷ σίτῳ καὶ τοῖς δένδρεσιν. δι᾽ ὃ καὶ ὅταν ὕσῃ τοῦτο ἐπαφιέναι καθαπερεὶ καταπλύνοντας ἐκεῖνο. καὶ τὰ μὲν ἐν τῇ ἔξω θαλάττῃ δένδρα τά γε νῦν τεθεωρημένα σχεδὸν τοσαῦτά ἐστιν.

VIII. Ὑπὲρ δὲ τῶν ἐν τοῖς ποταμοῖς καὶ τοῖς ἕλεσι καὶ ταῖς λίμναις μετὰ ταῦτα λεκτέον. τρία δέ ἐστιν εἴδη τῶν ἐν τούτοις, τὰ μὲν δένδρα τὰ δ᾽

---

1 *ἐξείρειν* conj. W.; *ἐξειαίρειν* P_2; *ἐξαίρειν* Ald. 2 4. 5. 8.
3 Tamarind. See Index App. (15). Plin. 12. 40.
4 *πλὴν ἄοδμον* conj. H. Steph.; *πλείονα ὄδμον* UMAld.
5 *τῷ μεγέθει καὶ* I conj.; *καὶ τῷ μεγέθει* UMVP; *καὶ* om. Ald.
6 Tamarind also. See Index App. (16). 7 *i.e.* leaflets.
8 *Ficus laccifera*. See Index App. (17). *οὐ φυλλορροούσας* conj. W., *cf.* G and Plin. *l.c.*; *αἳ φυλλορροοῦσιν* Ald.H.

and closed, but when it is ripe, it unfolds and puts forth[1] the 'wool,' of which they weave their fabrics, some of which are cheap and some very expensive.

This tree is also found, as was said,[2] in India as well as in Arabia. They say that there are other trees[3] with a flower like the gilliflower, but scentless[4] and in size[5] four times as large as that flower. And that there is another tree[6] with many leaves[7] like the rose, and that this closes at night, but opens at sunrise, and by noon is completely unfolded; and at evening again it closes by degrees and remains shut at night, and the natives say that it goes to sleep. Also that there are date-palms on the island and vines and other fruit-trees, including evergreen[8] figs. Also that there is water from heaven, but that they do not use it for the fruits, but that there are many springs on the island, from which they water everything, and that this is more beneficial[9] to the corn and the trees. Wherefore, even when it rains, they let this water over the fields,[10] as though they were washing away the rain water. Such are the trees as so far observed which grow in the outer sea.

Of the plants of rivers, marshes, and lakes, especially in Egypt.

VIII. Next we must speak of plants which live in rivers marshes and lakes. Of these there are three classes, trees, plants of 'herbaceous'[11] character, and

[9] ὃ καὶ συμφέρειν conj. Sch.; ἃ καὶ συμφέρει Ald.; U has συμφέρειν.

[10] *cf.* *C.P.* 2. 5. 5, where Androsthenes, one of Alexander's admirals, is given as the authority for this statement.

[11] The term τὰ ποιώδη seems to be given here a narrower connotation than usual, in order that τὰ λοχμώδη may be distinguished.

ὥσπερ ποιώδη τὰ δὲ λοχμώδη. λέγω δὲ ποιώδη μὲν οἷον τὸ σέλινον τὸ ἕλειον καὶ ὅσα ἄλλα τοιαῦτα· λοχμώδη δὲ κάλαμον κύπειρον φλεὼ σχοῖνον βούτομον, ἅπερ σχεδὸν κοινὰ πάντων τῶν ποταμῶν καὶ τῶν τοιούτων τόπων.

Ἐνιαχοῦ δὲ καὶ βάτοι καὶ παλίουροι καὶ τὰ ἄλλα δένδρα, καθάπερ ἰτέα λεύκη πλάτανος. τὰ μὲν οὖν μέχρι τοῦ κατακρύπτεσθαι, τὰ δὲ ὥστε μικρὸν ὑπερέχειν, τῶν δὲ αἱ μὲν ῥίζαι καὶ μικρὸν τοῦ στελέχους ἐν τῷ ὑγρῷ, τὸ δὲ ἄλλο σῶμα πᾶν ἔξω. τοῦτο γὰρ καὶ ἰτέᾳ καὶ κλήθρᾳ καὶ πλατάνῳ καὶ φιλύρᾳ καὶ πᾶσι τοῖς φιλύδροις συμβαίνει.

2 Σχεδὸν δὲ καὶ ταῦτα κοινὰ πάντων τῶν ποταμῶν ἐστιν· ἐπεὶ καὶ ἐν τῷ Νείλῳ πέφυκεν· οὐ μὴν πολλή γε ἡ πλάτανος, ἀλλὰ σπανιωτέρα ἔτι ταύτης ἡ λεύκη, πλείστη δὲ μελία καὶ βουμέλιος. τῶν γοῦν ἐν Αἰγύπτῳ φυομένων τὸ μὲν ὅλον πολὺ πλῆθός ἐστιν πρὸς τὸ ἀριθμήσασθαι καθ' ἕκαστον· οὐ μὴν ἀλλ' ὥς γε ἁπλῶς εἰπεῖν ἅπαντα ἐδώδιμα καὶ χυλοὺς ἔχοντα γλυκεῖς. διαφέρειν δὲ δοκεῖ τῇ γλυκύτητι καὶ τῷ τρόφιμα μάλιστα εἶναι τρία ταῦτα, ὅ τε πάπυρος καὶ τὸ καλούμενον σάρι καὶ τρίτον ὃ μνάσιον καλοῦσι.

3 Φύεται δὲ ὁ πάπυρος οὐκ ἐν βάθει τοῦ ὕδατος ἀλλ' ὅσον ἐν δύο πήχεσιν, ἐνιαχοῦ δὲ καὶ ἐν ἐλάττονι. πάχος μὲν οὖν τῆς ῥίζης ἡλίκον καρπὸς χειρὸς ἀνδρὸς εὐρώστου, μῆκος δὲ ὑπὲρ τετράπηχυ· φύεται δὲ ὑπὲρ τῆς γῆς αὐτῆς, πλαγίας ῥίζας εἰς τὸν πηλὸν καθιεῖσα λεπτὰς καὶ πυκνάς, ἄνω δὲ τοὺς παπύρους καλουμένους τριγώνους,

[1] τῶν γοῦν κ.τ.λ.: text probably defective; what follows appears to relate to τὰ ποιώδη.

plants growing in clumps. By 'herbaceous' I mean here such plants as the marsh celery and the like; by 'plants growing in clumps' I mean reeds galingale *phleo* rush sedge—which are common to almost all rivers and such situations.

And in some such places are found brambles Christ's thorn and other trees, such as willow abele plane. Some of these are water plants to the extent of being submerged, while some project a little from the water; of some again the roots and a small part of the stem are under water, but the rest of the body is altogether above it. This is the case with willow alder plane lime, and all water-loving trees.

These too are common to almost all rivers, for they grow even in the Nile. However the plane is not abundant by rivers, while the abele is even more scarce, and the manna-ash and ash are commonest. At any rate of those[1] that grow in Egypt the list is too long to enumerate separately; however, to speak generally, they are all edible and have sweet flavours. But they differ in sweetness, and we may distinguish also three as the most useful for food, namely the papyrus, the plant called *sari*, and the plant which they call *mnasion*.

[2]The papyrus does not grow in deep water, but only in a depth of about two cubits, and sometimes shallower. The thickness of the root is that of the wrist of a stalwart man, and the length above four cubits[3]; it grows above the ground itself, throwing down slender matted roots into the mud, and producing above the stalks which give it its name 'papyrus'; these are three-cornered and about ten

² Plin. 13. 71–73.

³ τετράπηχυ: δέκα πήχεις MSS. See next note.

μέγεθος ὡς δέκα πήχεις, κόμην ἔχοντας ἀχρεῖον ἀσθενῆ καρπὸν δὲ ὅλως οὐδένα· τούτους δ' ἀναδί-
4 δωσι κατὰ πολλὰ μέρη. χρῶνται δὲ ταῖς μὲν ῥίζαις ἀντὶ ξύλων οὐ μόνον τῷ κάειν ἀλλὰ καὶ τῷ σκεύη ἄλλα ποιεῖν ἐξ αὐτῶν παντοδαπά· πολὺ γὰρ ἔχει τὸ ξύλον καὶ καλόν. αὐτὸς δὲ ὁ πάπυρος πρὸς πλεῖστα χρήσιμος· καὶ γὰρ πλοῖα ποιοῦσιν ἐξ αὐτοῦ, καὶ ἐκ τῆς βίβλου ἱστία τε πλέκουσι καὶ ψιάθους καὶ ἐσθῆτά τινα καὶ στρωμνὰς καὶ σχοινία τε καὶ ἕτερα πλείω. καὶ ἐμφανέστατα δὴ τοῖς ἔξω τὰ βιβλία· μάλιστα δὲ καὶ πλείστη βοήθεια πρὸς τὴν τροφὴν ἀπ' αὐτοῦ γίνεται. μασῶνται γὰρ ἅπαντες οἱ ἐν τῇ χώρᾳ τὸν πάπυρον καὶ ὠμὸν καὶ ἑφθὸν καὶ ὀπτόν· καὶ τὸν μὲν χυλὸν καταπίνουσι, τὸ δὲ μάσημα ἐκβάλλουσιν. ὁ μὲν οὖν πάπυρος τοιοῦτός τε καὶ ταύτας παρέχεται τὰς χρείας. γίνεται δὲ καὶ ἐν Συρίᾳ περὶ τὴν λίμνην ἐν ᾗ καὶ ὁ κάλαμος ὁ εὐώδης· ὅθεν καὶ Ἀντίγονος εἰς τὰς ναῦς ἐποιεῖτο τὰ σχοινία.

5 Τὸ δὲ σάρι φύεται μὲν ἐν τῷ ὕδατι περὶ τὰ ἕλη καὶ τὰ πεδία, ἐπειδὰν ὁ ποταμὸς ἀπέλθῃ, ῥίζαν δὲ ἔχει σκληρὰν καὶ συνεστραμμένην, καὶ ἐξ αὐτῆς φύεται τὰ σαρία καλούμενα· ταῦτα δὲ μῆκος μὲν ὡς δύο πήχεις, πάχος δὲ ἡλίκον ὁ δάκτυλος ὁ μέγας τῆς χειρός· τρίγωνον δὲ καὶ τοῦτο, καθάπερ ὁ πάπυρος, καὶ κόμην ἔχον παραπλήσιον. μασώμενοι δὲ ἐκβάλλουσι καὶ τοῦτο τὸ μάσημα, τῇ ῥίζῃ δὲ οἱ σιδηρουργοὶ χρῶνται· τὸν γὰρ ἄνθρακα ποιεῖ χρηστὸν διὰ τὸ σκληρὸν εἶναι τὸ ξύλον.

6 Τὸ δὲ μνάσιον ποιῶδές ἐστιν, ὥστ' οὐδεμίαν παρέχεται χρείαν πλὴν τὴν εἰς τροφήν.

cubits[1] long, having a plume which is useless and weak, and no fruit whatever; and these stalks the plant sends up at many points. They use the roots instead of wood, not only for burning, but also for making a great variety of articles; for the wood is abundant and good. The 'papyrus' itself[2] is useful for many purposes; for they make boats from it, and from the rind they weave sails mats a kind of raiment coverlets ropes and many other things. Most familiar to foreigners are the papyrus-rolls made of it; but above all the plant also is of very great use in the way of food.[3] For all the natives chew the papyrus both raw boiled and roasted; they swallow the juice and spit out the quid. Such is the papyrus and such its uses. It grows also in Syria about the lake in which grows also sweet-flag; and Antigonus made of it the cables for his ships.

[4]The *sari* grows in the water in marshes and plains, when the river has left them; it has a hard twisted root, and from it grow what they call the *saria*[5]; these are about two cubits long and as thick as a man's thumb; this stalk too is three-cornered, like the papyrus, and has similar foliage. This also they chew, spitting out the quid; and smiths use the root, for it makes excellent charcoal, because the wood is hard.

Mnasion is herbaceous, so that it has no use except for food.

[1] δέκα πήχεις: τετραπήχεις MSS. The two numbers seem to have changed places (Bartels ap. Sch.). *cf.* Plin. *l.c.*

[2] *i.e.* the stalk.

[3] *cf.* Diod. 1. 80. [4] Plin. 13. 128.

[5] *i.e.* stalks, like those of the papyrus.

Καὶ τὰ μὲν γλυκύτητι διαφέροντα ταῦτά ἐστι. φύεται δὲ καὶ ἕτερον ἐν τοῖς ἕλεσι καὶ ταῖς λίμναις ὃ οὐ συνάπτει τῇ γῇ, τὴν μὲν φύσιν ὅμοιον τοῖς κρίνοις, πολυφυλλότερον δὲ καὶ παρ' ἄλληλα τὰ φύλλα καθάπερ ἐν διστοιχίᾳ· χρῶμα δὲ χλωρὸν ἔχει σφόδρα. χρῶνται δὲ οἱ ἰατροὶ πρός τε τὰ γυναικεῖα αὐτῷ καὶ πρὸς τὰ κατάγματα.

[Ταῦτα δὲ γίνεται ἐν τῷ ποταμῷ εἰ μὴ ὁ ῥοῦς ἐξέφερεν· συμβαίνει δὲ ὥστε καὶ ἀποφέρεσθαι· ἕτερα δ' ἀπ' αὐτῶν πλείω.]

Ὁ δὲ κύαμος φύεται μὲν ἐν τοῖς ἕλεσι καὶ λίμναις, καυλὸς δὲ αὐτοῦ μῆκος μὲν ὁ μακρότατος εἰς τέτταρας πήχεις, πάχος δὲ δακτυλιαῖος, ὅμοιος δὲ καλάμῳ μαλακῷ ἀγονάτῳ. διαφύσεις δὲ ἔνδοθεν ἔχει δι' ὅλου διειλημμένας ὁμοίας τοῖς κηρίοις· ἐπὶ τούτῳ δὲ ἡ κωδύα, παρομοία σφηκίῳ περιφερεῖ, καὶ ἐν ἑκάστῳ τῶν κυττάρων κύαμος μικρὸν ὑπεραίρων αὐτῆς, πλῆθος δὲ οἱ πλεῖστοι τριάκοντα. τὸ δὲ ἄνθος διπλάσιον ἢ μήκωνος, χρῶμα δὲ ὅμοιον ῥόδῳ κατακορές· ἐπάνω δὲ τοῦ ὕδατος ἡ κωδύα. παραφύεται δὲ φύλλα μεγάλα παρ' ἕκαστον τῶν κυάμων, ὧν ἴσα τὰ μεγέθη πετάσῳ Θετταλικῇ τὸν αὐτὸν ἔχοντα καυλὸν τῷ τῶν κυάμων. συντρίψαντι δ' ἕκαστον τῶν κυάμων φανερόν ἐστι τὸ πικρὸν συνεστραμμένον, ἐξ

---

[1] *Ottelia alismoeides.* See Index App. (18).

[2] ταῦτα . . . πλείω conj. W. after Sch.; I have also transposed the two sentences, after Sch. The whole passage in [ ] (which is omitted by G) is apparently either an interpolation or defective. σημαίνει δὲ ὥσπερ καὶ ἀποφέρεσθαι· ἕτερα δὲ ἀπ' αὐτῶν τὰ πλεῖα· ταῦτα δὲ γίνεται ἐν τῷ ποταμῷ· εἰ μὴ ὁ ῥοῦς ἐξέφερεν Ald.; so also U, but αὐτῶν πλείω.

Such are the plants which excel in sweetness of taste. There is also another plant[1] which grows in the marshes and lakes, but which does not take hold of the ground; in character it is like a lily, but it is more leafy, and has its leaves opposite to one another, as it were in a double row; the colour is a deep green. Physicians use it for the complaints of women and for fractures.

Now these plants grow in the river, unless the stream has thrown them up on land; it sometimes happens that they are borne down the stream, and that then other plants grow from them.[2]

[3] But the 'Egyptian bean' grows in the marshes and lakes; the length of its stalk at longest is four cubits, it is as thick as a man's finger, and resembles a pliant[4] reed without joints. Inside it has tubes which run distinct from one another right through, like a honey-comb: on this is set the 'head,' which is like a round wasps' nest, and in each of the cells is a 'bean,' which slightly projects from it; at most there are thirty of these. The flower is twice as large as a poppy's, and the colour is like a rose, of a deep shade; the 'head' is above the water. Large leaves grow at the side of each plant, equal[5] in size to a Thessalian hat[6]; these have a stalk exactly like that[7] of the plant. If one of the 'beans' is crushed, you find the bitter substance coiled up, of which the

[3] Plin. 18. 121 and 122.

[4] *μαλακῷ* Ald.H.G Plin. *l.c.* Athen. 3. 2 cites the passage with *μακρῷ*.

[5] *ἴσα* conj. W.; *καὶ* Ald.

[6] *πετάσῳ* conj. Sch. from Diosc. 2. 106; *πίλῳ* Ald.H.; *οἱ πέτασοι* are mentioned below (§ 9) without explanation. The comparison is omitted by G and Plin. *l.c.*

[7] *i.e.* that which carries the *κωδύα*·

8 οὗ γίνεται ὁ πῖλος. τὰ μὲν οὖν περὶ τὸν καρπὸν τοιαῦτα. ἡ δὲ ῥίζα παχυτέρα τοῦ καλάμου τοῦ παχυτάτου καὶ διαφύσεις ὁμοίως ἔχουσα τῷ καυλῷ. ἐσθίουσι δ' αὐτὴν καὶ ὠμὴν καὶ ἑφθὴν καὶ ὀπτήν, καὶ οἱ περὶ τὰ ἕλη τούτῳ σίτῳ χρῶνται. φύεται μὲν οὖν ὁ πολὺς αὐτόματος· οὐ μὴν ἀλλὰ καὶ καταβάλλουσιν ἐν πηλῷ ἀχυρώσαντες εὖ μάλα πρὸς τὸ κατενεχθῆναί τε καὶ μεῖναι καὶ μὴ διαφθαρῆναι· καὶ οὕτω κατασκευάζουσι τοὺς κυαμῶνας· ἂν δ' ἅπαξ ἀντιλάβηται, μένει διὰ τέλους. ἰσχυρὰ γὰρ ἡ ῥίζα καὶ οὐ πόρρω τῆς τῶν καλάμων πλὴν ἐπακανθίζουσα· δι' ὃ καὶ ὁ κροκόδειλος φεύγει μὴ προσκόψῃ τῷ ὀφθαλμῷ τῷ μὴ ὀξὺ καθορᾶν· γίνεται δὲ οὗτος καὶ ἐν Συρίᾳ καὶ κατὰ Κιλικίαν, ἀλλ' οὐκ ἐκπέττουσιν αἱ χῶραι· καὶ περὶ Τορώνην τῆς Χαλκιδικῆς ἐν λίμνῃ τινὶ μετρίᾳ τῷ μεγέθει· καὶ αὐτοῦ πέττεται τελέως καὶ τελεοκαρπεῖ.

9 Ὁ δὲ λωτὸς καλούμενος φύεται μὲν ὁ πλεῖστος ἐν τοῖς πεδίοις, ὅταν ἡ χώρα κατακλυσθῇ. τούτου δὲ ἡ μὲν τοῦ καυλοῦ φύσις ὁμοία τῇ τοῦ κυάμου, καὶ οἱ πέτασοι δὲ ὡσαύτως, πλὴν ἐλάττους καὶ λεπτότεροι. ἐπιφύεται δὲ ὁμοίως ὁ καρπὸς τῷ τοῦ κυάμου. τὸ ἄνθος αὐτοῦ λευκὸν ἐμφερὲς τῇ στενότητι τῶν φύλλων τοῖς τοῦ κρίνου, πολλὰ δὲ καὶ πυκνὰ ἐπ' ἀλλήλοις φύεται. ταῦτα δὲ ὅταν μὲν ὁ ἥλιος δύῃ συμμύει καὶ συγκαλύπτει τὴν κωδύαν, ἅμα δὲ τῇ ἀνατολῇ διοί-

[1] ὁ πῖλος UMV; ἡ πῖλος Ald.H.; ?=*germen* Sch.
[2] *cf.* Diosc. 2. 107.
[3] καὶ καταβ. conj. W.; καταβ. Ald.; καταβ. δ' UMV.
[4] Plin. 13. 107 and 108.

pilos[1] is made. So much for the fruit. The root is thicker than the thickest reed, and is made up of distinct tubes, like the stalk. [2]They eat it both raw boiled and roasted, and the people of the marshes make this their food. It mostly grows of its own accord; however they also sow[3] it in the mud, having first well mixed the seed with chaff, so that it may be carried down and remain in the ground without being rotted; and so they prepare the 'bean' fields, and if the plant once takes hold it is permanent. For the root is strong and not unlike that of reeds, except that it is prickly on the surface. Wherefore the crocodile avoids it, lest he may strike his eye on it, since he has not sharp sight. This plant also grows in Syria and in parts of Cilicia, but these countries cannot ripen it; also about Torone in Chalcidice in a certain lake of small size; and this lake ripens it perfectly and matures its fruit.

[4]The plant called the *lotos* (Nile water-lily) grows chiefly in the plains when the land is inundated. The character of the stalk of this plant is like that of the 'Egyptian bean,' and so are the 'hat-like' leaves,[5] except that they are smaller and slenderer. And the fruit[6] grows on the stalk in the same way as that of the 'bean.' The flower is white, resembling in the narrowness of its petals those of the lily,[7] but there are many petals growing close one upon another. When the sun sets, these close[8] and cover up the 'head,' but with sunrise they open and

[5] *cf.* 4. 8. 7.

[6] καρπὸς conj. W.; λωτὸς MSS. Possibly the fruit was specially called λωτός.

[7] *cf.* Hdt. 2. 92; Diosc. 4. 113.

[8] δύῃ, συμμύει conj. St.; συμμύει MV; συμμύῃ U; συμμύῃ (omitting καὶ) Ald. H.

γεται καὶ ὑπὲρ τοῦ ὕδατος γίνεται. τοῦτο δὲ
ποιεῖ μέχρι ἂν ἡ κωδύα ἐκτελεωθῇ καὶ τὰ ἄνθη
10 περιρρυῇ. τῆς δὲ κωδύας τὸ μέγεθος ἡλίκον
μήκωνος τῆς μεγίστης, καὶ διέζωσται ταῖς κατα-
τομαῖς τὸν αὐτὸν τρόπον τῇ μήκωνι· πλὴν πυκνό-
τερος ἐν ταύταις ὁ καρπός. ἔστι δὲ παρόμοιος
τῷ κέγχρῳ. ἐν δὲ τῷ Εὐφράτῃ τὴν κωδύαν φασὶ
καὶ τὰ ἄνθη δύνειν καὶ ὑποκαταβαίνειν τῆς ὀψίας
μέχρι μεσῶν νυκτῶν καὶ τῷ βάθει πόρρω· οὐδὲ
γὰρ καθιέντα τὴν χεῖρα λαβεῖν εἶναι. μετὰ δὲ
ταῦτα ὅταν ὄρθρος ᾖ πάλιν ἐπανιέναι καὶ πρὸς
ἡμέραν ἔτι μᾶλλον, ἅμα τῷ ἡλίῳ φανερὸν <ὂν>
ὑπὲρ τοῦ ὕδατος καὶ ἀνοίγειν τὸ ἄνθος, ἀνοιχθέν-
τος δὲ ἔτι ἀναβαίνειν· συχνὸν δὲ τὸ ὑπεραῖρον
11 εἶναι τὸ ὕδωρ. τὰς δὲ κωδύας ταύτας οἱ Αἰγύ-
πτιοι συνθέντες εἰς τὸ αὐτὸ σήπουσιν· ἐπὰν δὲ
σαπῇ τὸ κέλυφος, ἐν τῷ ποταμῷ κλύζοντες ἐξαι-
ροῦσι τὸν καρπόν, ξηράναντες δὲ καὶ πτίσαντες
ἄρτους ποιοῦσι καὶ τούτῳ χρῶνται σιτίῳ. ἡ δὲ
ῥίζα τοῦ λωτοῦ καλεῖται μὲν κόρσιον, ἐστὶ δὲ
στρογγύλη, τὸ μέγεθος ἡλίκον μῆλον Κυδώνιον·
φλοιὸς δὲ περίκειται περὶ αὐτὴν μέλας ἐμφερὴς
τῷ κασταναϊκῷ καρύῳ· τὸ δὲ ἐντὸς λευκόν, ἑψό-
μενον δὲ καὶ ὀπτώμενον γίνεται λεκιθῶδες, ἡδὺ δὲ
ἐν τῇ προσφορᾷ· ἐσθίεται δὲ καὶ ὠμή, ἀρίστη
δὲ ἐν [τῷ] ὕδατι ἑφθὴ καὶ ὀπτή. καὶ τὰ μὲν
ἐν τοῖς ὕδασιν σχεδὸν ταῦτά ἐστιν.
12 Ἐν δὲ τοῖς ἀμμώδεσι χωρίοις, ἅ ἐστιν οὐ πόρρω

---

[1] *cf.* Diosc. *l.c.* [2] *cf.* *C.P.* 2. 19. 1; Plin. 13. 109.
[3] ὀψίας conj. W. from Plin. *l.c.*; ? ὀψίας ὥρας.
[4] <ὂν> add. W.
[5] κέλυφος *i.e.* fruit: καρπόν *i.e.* seeds.

appear above the water. This the plant does until the 'head' is matured and the flowers have fallen off. [1] The size of the 'head' is that of the largest poppy, and it has grooves all round it in the same way as the poppy, but the fruit is set closer in these. This is like millet. [2] In the Euphrates they say that the 'head' and the flowers sink and go under water in the evening [3] till midnight, and sink to a considerable depth; for one can not even reach them by plunging one's hand in; and that after this, when dawn comes round, they rise and go on rising towards day-break, being [4] visible above the water when the sun appears; and that then the plant opens its flower, and, after it is open, it still rises; and that it is a considerable part which projects above the water. These 'heads' the Egyptians heap together and leave to decay, and when the 'pod' [5] has decayed, they wash the 'head' in the river and take out the 'fruit,' [5] and, having dried and pounded [6] it, they make loaves of it, which they use for food. The root of the *lotos* is called *korsion*,[7] and it is round and about the size of a quince; it is enclosed in a black 'bark,' like the shell of a chestnut. The inside is white; but when it is boiled or roasted, it becomes of the colour of the yolk of an egg and is sweet to taste. The root is also eaten raw, though it is best when boiled in water or roasted.[8] Such are the plants found in water.

In sandy places which are not [9] far from the river

[6] πτίσαντες: *cf*. Hdt. 2. 92. [7] *cf*. Strabo 17. 2. 4.

[8] ἐσθίεται . . . ὀπτή conj. Sch. from Plin. *l.c.* and G; ἐσθ. δὲ καὶ ὠμόν· ἀρίστη δὲ ἐν τοῖς ὕδασιν αὐτὴ ὠμή Ald.; ἀρίστη δὲ καὶ τοῖς ὕδασιν αὐτὴν UMV, then ομή U, ὠμῆ V, ὠμή M; ἀρίστη δὲ ἐν τῷ ὕδατι ἑφθὴ ἢ καὶ ὀπτή H.

[9] οὐ was apparently not in Pliny's text; (21. 88.)

τοῦ ποταμοῦ, φύεται κατὰ γῆς ὃ καλεῖται μαλιναθάλλη, στρογγύλον τῷ σχήματι μέγεθος δὲ ἡλίκον μέσπιλον ἀπύρηνον δὲ ἄφλοιον· φύλλα δὲ ἀφίησιν ἀπ' αὐτοῦ ὅμοια κυπείρῳ· ταῦτα συνάγοντες οἱ κατὰ τὴν χώραν ἕψουσιν ἐν βρυτῷ τῷ ἀπὸ τῶν κριθῶν καὶ γίνεται γλυκέα σφόδρα· χρῶνται δὲ πάντες ὥσπερ τραγήμασι.

13 Τοῖς δὲ βουσὶ καὶ τοῖς προβάτοις ἅπαντα μὲν τὰ φυόμενα ἐδώδιμά ἐστιν, ἓν δέ τι γένος ἐν ταῖς λίμναις καὶ τοῖς ἕλεσι φύεται διαφέρον, ὃ καὶ χλωρὸν νέμονται καὶ ξηραίνοντες παρέχουσι κατὰ χειμῶνα τοῖς βουσὶν ὅταν ἐργάσωνται· καὶ τὰ σώματα ἔχουσιν εὖ σίτου ἄλλο λαμβάνοντες οὐθέν.

14 Ἔστι δὲ καὶ ἄλλο παραφυόμενον αὐτόματον ἐν τῷ σίτῳ· τοῦτο δέ, ὅταν ὁ σῖτος ᾖ καθαρός, ὑποπτίσαντες καταβάλλουσι τοῦ χειμῶνος ὑγρὰν εἰς γῆν· βλαστήσαντος δὲ τεμόντες καὶ ξηράναντες παρέχουσι καὶ τοῦτο βουσὶ καὶ ἵπποις καὶ τοῖς ὑποζυγίοις σὺν τῷ καρπῷ τῷ ἐπιγινομένῳ· ὁ δὲ καρπὸς μέγεθος μὲν ἡλίκον σήσαμον, στρογγύλος δὲ καὶ τῷ χρώματι χλωρός, ἀγαθὸς δὲ διαφερόντως. ἐν Αἰγύπτῳ μὲν οὖν τὰ περιττὰ σχεδὸν ταῦτα ἄν τις λάβοι.

IX. Ἕκαστοι δὲ τῶν ποταμῶν ἐοίκασιν ἴδιόν τι φέρειν, ὥσπερ καὶ τῶν χερσαίων. ἐπεὶ οὐδὲ ὁ τρίβολος ἐν ἅπασιν οὐδὲ πανταχοῦ φύεται, ἀλλ' ἐν τοῖς ἑλώδεσι τῶν ποταμῶν· ἐν μεγίστῳ δὲ βάθει πενταπήχει ἢ μικρῷ μείζονι, καθάπερ

---

1 Plin. *l.c. anthalium*, whence Salm. conj. ἀνθάλλιον.
2 *Saccharum biflorum*. See Index App. (19).
3 εὖ σίτου ἄλλο conj. W.; εὐσιτοῦντα Ald.

there grows under ground the thing called *malinathalle*[1]; this is round in shape and as large as a medlar, but has no stone and no bark. It sends out leaves like those of galingale. These the people of the country collect and boil in beer made from barley, and they become extremely sweet, and all men use them as sweetmeats.

All the things that grow in such places may be eaten by oxen and sheep, but there is one kind of plant[2] which grows in the lakes and marshes which is specially good for food: they graze their cattle on it when it is green, and also dry it and give it in the winter to the oxen after their work; and these keep in good condition when they have no other[3] kind of food.

There is also another plant[4] which comes up of its own accord among the corn; this, when the harvest is cleared, they crush slightly[5] and lay during the winter on[6] moist ground; when it shoots, they cut and dry it and give this also to the cattle and horses and beasts of burden with the fruit which forms on it. The fruit in size is as large as sesame, but round and green in colour, and exceedingly good. Such one might take to be specially remarkable plants of Egypt.

IX. Every river seems to hear some peculiar plant, just as does each part of the dry land. [7] For not even the water-chestnut grows in all rivers nor everywhere, but only in marshy rivers, and only in those whose depth is not more or not much more than five cubits,

[4] *Corchorus trilocularis.* See Index App. (20).
[5] G seems to have read ὑποπτίσαντες (*leviter pinsentes*); ὑποπτήσαντες W. with Ald.H.
[6] εἰς conj. W.; τὴν Ald.
[7] Plin. 21. 98; Diosc. 4. 15.

περὶ τὸν Στρυμόνα· σχεδὸν δὲ ἐν τοσούτῳ καὶ ὁ κάλαμος καὶ τὰ ἄλλα. ὑπερέχει δὲ οὐθὲν αὐτοῦ πλὴν αὐτὰ τὰ φύλλα ὥσπερ ἐπινέοντα καὶ κρύπτοντα τὸν τρίβολον, ὁ δὲ τρίβολος αὐτὸς ἐν τῷ ὕδατι νεύων εἰς βυθόν. τὸ δὲ φύλλον ἐστὶ πλατὺ προσεμφερὲς τῷ τῆς πτελέας, μίσχον δὲ
2 ἔχει σφόδρα μακρόν· ὁ δὲ καυλὸς ἐξ ἄκρου παχύτατος, ὅθεν τὰ φύλλα καὶ ὁ καρπός, τὰ δὲ κάτω λεπτότερος ἀεὶ μέχρι τῆς ῥίζης· ἔχει δὲ ἀποπεφυκότα ἀπ' αὐτοῦ τριχώδη τὰ μὲν πλεῖστα παράλληλα τὰ δὲ καὶ παραλλάττοντα, κάτωθεν ἀπὸ τῆς ῥίζης μεγάλα τὰ δὲ ἄνω ἀεὶ ἐλάττω προϊοῦσιν, ὥστε τὰ τελευταῖα μικρὰ πάμπαν εἶναι καὶ τὴν διαφορὰν μεγάλην τὴν ἀπὸ τῆς ῥίζης πρὸς τὸν καρπόν. ἔχει δὲ ἐκ τοῦ ἑνὸς καυλοῦ καὶ παραβλαστήματα πλείω· καὶ γὰρ τρία καὶ τέτταρα, μέγιστον δ' αἰεὶ τὸ πλησιαίτερον τῆς ῥίζης, εἶτα τὸ μετὰ τοῦτο καὶ τὰ ἄλλα κατὰ λόγον. τὸ δὲ παραβλάστημά ἐστιν ὥσπερ καυλὸς ἄλλος λεπτότερος μὲν τοῦ πρώτου, τὰ δὲ φύλλα καὶ τὸν καρπὸν ἔχων ὁμοίως. ὁ δὲ καρπὸς μέλας καὶ σκληρὸς σφόδρα. ῥίζαν δὲ ἡλίκην καὶ ποίαν ἔχει σκεπτέον. ἡ μὲν οὖν φύσις τοιαύτη. φύεται μὲν ἀπὸ τοῦ καρποῦ τοῦ πίπτοντος καὶ ἀφίησι βλαστὸν τοῦ ἦρος·
3 φασὶ δὲ οἱ μὲν εἶναι ἐπέτειον οἱ δὲ διαμένειν τὴν μὲν ῥίζαν εἰς χρόνον, ἐξ ἧς καὶ τὴν βλάστησιν εἶναι τοῦ καυλοῦ. τοῦτο μὲν οὖν σκεπτέον. ἴδιον δὲ παρὰ τἆλλα τὸ τῶν παραφυομένων ἐκ τοῦ καυλοῦ τριχωδῶν· οὔτε γὰρ φύλλα ταῦτα οὔτε καυλός· ἐπεὶ τό γε τῆς παραβλαστήσεως κοινὸν καλάμου καὶ ἄλλων.

as the Strymon. (In rivers of such a depth grow also reeds and other plants.) No part of it projects from the water except just the leaves; these float as it were and conceal the 'chestnut,' which is itself under water and bends down towards the bottom. The leaf is broad, like that of the elm, and has a very long stalk. The stem is thickest at the top, whence spring the leaves and the fruit; below it gets thinner down to the root. It has springing from it hair-like growths, most of which are parallel to each other, but some are irregular; below, starting from the root, they are large, but, as one gets higher up the plant, they become smaller, so that those at the top are quite small and there is a great contrast between the root and the top where the fruit grows. The plant also has on the same stalk several side-growths; of these there are three or four, and the largest is always that which is nearer to the root, the next largest is the one next above it, and so on in proportion: this sidegrowth is like another stalk, but slenderer than the original one, though like that it has leaves and fruit. The fruit is black and extremely hard. The size and character of the root are matter for further enquiry. Such is the character of this plant. It grows from the fruit which falls, and begins to grow in spring. Some say that it is annual, others that the root persists for a time, and that from it grows the new stalk. This then is matter for enquiry. However quite peculiar to this plant is the hair-like character of the growths which spring from the stalk; for these are neither leaves nor stalk; though reeds and other things have also sidegrowths.

X. Τὰ μὲν οὖν ἴδια θεωρητέον ἰδίως δῆλον ὅτι, τὰ δὲ κοινὰ κοινῶς.[1] διαιρεῖν δὲ χρὴ καὶ ταῦτα κατὰ τοὺς τόπους, οἷον εἰ τὰ μὲν ἕλεια τὰ δὲ λιμναῖα τὰ δὲ ποτάμια μᾶλλον ἢ καὶ κοινὰ πάντων τῶν τόπων· διαιρεῖν δὲ καὶ ποῖα ταὐτὰ[2] ἐν τῷ ὑγρῷ καὶ τῷ ξηρῷ φύεται, καὶ ποῖα ἐν τῷ ὑγρῷ μόνον, ὡς ἁπλῶς εἰπεῖν πρὸς τὰ κοινότατα εἰρημένα πρότερον.[3]

Ἐν δ' οὖν τῇ λίμνῃ τῇ περὶ Ὀρχομενὸν τάδ' ἐστὶ τὰ φυόμενα δένδρα καὶ ὑλήματα, ἰτέα ἐλαίαγνος σίδη κάλαμος ὅ τε αὐλητικὸς καὶ ὁ ἕτερος κύπειρον φλεὼς τύφη, ἔτι γε μήνανθος ἴκμη καὶ τὸ καλούμενον ἴπνον. ὃ γὰρ προσαγορεύουσι λέμνα τούτου τὰ πλείω καθ' ὕδατός ἐστι.

2 Τούτων δὲ τὰ μὲν ἄλλα γνώριμα· ὁ δ' ἐλαίαγνος καὶ ἡ σίδη καὶ ἡ μήνανθος καὶ ἡ ἴκμη καὶ τὸ ἴπνον ἴσως μὲν φύεται καὶ ἑτέρωθι, προσαγορεύεται δὲ ἄλλοις ὀνόμασι· λεκτέον δὲ περὶ αὐτῶν. ἔστι δὲ ὁ μὲν ἐλαίαγνος φύσει μὲν θαμνῶδες καὶ παρόμοιον τοῖς ἄγνοις, φύλλον δὲ ἔχει τῷ μὲν σχήματι παραπλήσιον μαλακὸν δέ, ὥσπερ αἱ μηλέαι καὶ χνοῶδες. ἄνθος δὲ τῷ τῆς λεύκης ὅμοιον ἔλαττον· καρπὸν δὲ οὐδένα φέρει. φύεται δὲ ὁ πλεῖστος μὲν ἐπὶ τῶν πλοάδων νήσων· εἰσὶ γάρ τινες καὶ ἐνταῦθα πλοάδες, ὥσπερ ἐν Αἰγύπτῳ

---

[1] τὰ δὲ κοινὰ κοινῶς conj. Sch. from G; τὰ δὲ κοινῶς Ald. H.
[2] ταὐτὰ conj. Sch.: ταῦτα Ald.
[3] πρὸς τὰ κοιν. εἰρ. πρ. conj. W. supported by G; κοινότατα προσειρημένα πρότερον Ald. H.

Of the plants peculiar to the lake of Orchomenos (Lake Copaïs), especially its reeds; and of reeds in general.

X. Plants peculiar to particular places must be considered separately, while a general account may be given of those which are generally distributed.[1] But even the latter must be classified according to locality; thus some belong to marshes, others to lakes, others to rivers, or again others may be common to all kinds of locality: we must also distinguish which occur alike[2] in wet and in dry ground, and which only in wet ground, marking these off in a general way from those mentioned above as being most impartial.[3]

Now in the lake near Orchomenos grow the following trees and woody plants: willow goat-willow water-lily reeds (both that used for making pipes and the other kind) galingale *phleos* bulrush; and also 'moon-flower' duckweed and the plant called marestail: as for the plant called water-chickweed the greater part of it grows under water.[4]

Now of these most are familiar: the goat-willow water-lily 'moon-flower' duckweed and marestail probably grow also elsewhere, but are called by different names. Of these we must speak. The goat-willow is of shrubby habit and like the chaste-tree: its leaf resembles that leaf in shape, but it is soft like that of the apple,[5] and downy. The bloom[6] is like that of the abele, but smaller, and it bears no fruit. It grows chiefly on the floating islands; (for here too there are floating islands, as in the marshes

[4] τούτου τὰ πλείω καθ' ὑδ. conj. Sch.; τοῦτο πλείω τὸ καθ' ὑδ. UM; τοῦτο πλεῖον τὸ καθ' ὑδ. Ald.

[5] μηλέαι perhaps here = quince (μηλέα Κυδωνία).

[6] ἄνθος here = catkin.

περὶ τὰ ἕλη καὶ ἐν Θεσπρωτίδι καὶ ἐν ἄλλαις λίμναις· ἐλάττων δὲ καθ' ὕδατος· ὁ μὲν οὖν ἐλαίαγνος τοιοῦτον.

Ἡ δὲ σίδη την μὲν μορφήν ἐστιν ὁμοία τῇ μήκωνι· καὶ γὰρ τὸ ἄνω κυτινῶδες τοιοῦτον ἔχει, πλὴν μεῖζον ὡς κατὰ λόγον· μεγέθει δὲ ὅλος ὁ ὄγκος ἡλίκον μῆλον· ἔστι δὲ οὐ γυμνόν, ἀλλὰ ὑμένες περὶ αὐτὴν λευκοί, καὶ ἐπὶ τούτοις ἔξωθεν φύλλα ποώδη παραπλήσια τοῖς τῶν ῥόδων ὅταν ἐν κάλυξιν ὦσι, τέτταρα τὸν ἀριθμόν· ἀνοιχθεῖσα δὲ τοὺς κόκκους ἐρυθροὺς μὲν ἔχει τῷ σχήματι δὲ οὐχ ὁμοίους ταῖς ῥόαις ἀλλὰ περιφερεῖς μικροὺς δὲ καὶ οὐ πολλῷ μείζους κέγχρου· τὸν δὲ χυλὸν ὑδατώδη τινά, καθάπερ ὁ τῶν πυρῶν. ἁδρύνεται δὲ τοῦ θέρους, μίσχον δὲ ἔχει μακρόν. τὸ δὲ ἄνθος ὅμοιον ῥόδου κάλυκι, μεῖζον δὲ καὶ σχεδὸν διπλάσιον τῷ μεγέθει. τοῦτο μὲν οὖν καὶ τὸ φύλλον ἐπὶ τοῦ ὕδατος· μετὰ δὲ ταῦτα, ὅταν ἀπανθήσῃ καὶ συστῇ τὸ περικάρπιον, κατακλίνεσθαί φασιν εἰς τὸ ὕδωρ μᾶλλον, τέλος δὲ συνάπτειν τῇ γῇ καὶ τὸν καρπὸν ἐκχεῖν.

Καρποφορεῖν δὲ τῶν ἐν τῇ λίμνῃ τοῦτο καὶ τὸ βούτομον καὶ τὸν φλεών. εἶναι δὲ τοῦ βουτόμου μέλανα, τῷ δὲ μεγέθει παραπλήσιον τῷ τῆς σίδης. τοῦ δὲ φλεὼ τὴν καλουμένην ἀνθήλην,

---

1 ἐλάττων . . . ὕδατος: sense doubtful. G. seems to render a different reading.

2 *i.e.* the flower-head, which, as well as the plant, was called σίδη.

3 μήκωνι can hardly be right: suspected by H.

4 *cf.* Athen. 14. 64.

5 *i.e.* petals.

of Egypt, in Thesprotia, and in other lakes). When it grows under water, it is smaller.[1] Such is the goat-willow.

The water-lily[2] is in shape like the poppy.[3] For the top of it has this character, being shaped like the pomegranate flower,[4] but it is longer in proportion to the size of the plant. Its size in fact as a whole is that of an apple; but it is not bare, having round it white membranes,[5] and attached to these on the outside are grass-green 'leaves,'[6] like those of roses when they are still in bud, and of these there are four; when it is opened it shews its seeds, which are red; in shape however they are not like pomegranate[7] seeds, but round small and not much longer than millet seeds; the taste is insipid, like that of wheat-grains. It ripens in summer and has a long stalk. The flower is like a rose-bud, but larger, almost twice as large. Now this and the leaf float on the water; but later, when the bloom is over and the fruit-case[8] has formed, they say that it sinks deeper into the water, and finally reaches the bottom and sheds its fruit.

Of the plants of the lake they say that water-lily sedge and *phleos* bear fruit, and that that of the sedge is black, and in size like that of the water-lily. The fruit of *phleos* is what is called the 'plume,'[9]

[6] *i.e.* sepals.

[7] ῥόαις conj. Bod. from Nic. *Ther.* 887 and Schol.; ῥίζαις UMVAld H.

[8] περικάρπιον conj. W.; κατακάρπιον MSS. κατα- probably due to κατακλίνεσθαι.

[9] *cf.* Diosc. 3. 118. ἀνθήλην, sc. καρπὸν εἶναι. But Sch. suggests that further description of the fruit has dropped out, and that the clause ᾧ . . . κονίας does not refer to the fruit.

ᾧ χρῶνται πρὸς τὰς κονίας. τοῦτο δ' ἐστὶν οἷον πλακουντῶδές τι μαλακὸν ἐπίπυρρον. ἔτι δὲ καὶ τοῦ φλεὼ καὶ τοῦ βουτόμου τὸ μὲν θῆλυ ἄκαρπον, χρήσιμον δὲ πρὸς τὰ πλόκανα, τὸ δὲ ἄρρεν ἀχρεῖον.

Περὶ δὲ τῆς ἴκμης καὶ μηνάνθους καὶ τοῦ ἵπνου σκεπτέον.

5 Ἰδιώτατον δὲ τούτων ἐστὶν ἡ τύφη καὶ τῷ ἄφυλλον εἶναι καὶ τῷ μὴ πολύρριζον τοῖς ἄλλοις ὁμοίως· ἐπεὶ τἆλλα οὐχ ἧττον εἰς τὰ κάτω τὴν ὁρμὴν ἔχει καὶ τὴν δύναμιν· μάλιστα δὲ τὸ κύπειρον, ὥσπερ καὶ ἡ ἄγρωστις, δι' ὃ καὶ δυσώλεθρα καὶ ταῦτα καὶ ὅλως ἅπαν τὸ γένος τὸ τοιοῦτον. ἡ δὲ ῥίζα τοῦ κυπείρου πολύ τι τῶν ἄλλων παραλλάττει τῇ ἀνωμαλίᾳ, τῷ τὸ μὲν εἶναι παχύ τι καὶ σαρκῶδες αὐτῆς τὸ δὲ λεπτὸν καὶ ξυλῶδες· καὶ τῇ βλαστήσει καὶ τῇ γενέσει· φύεται γὰρ ἀπὸ τοῦ πρεμνώδους ἑτέρα λεπτὴ κατὰ πλάγιον, εἶτ' ἐν ταύτῃ συνίσταται πάλιν τὸ σαρκῶδες, ἐν ᾧ καὶ ὁ βλαστὸς ἀφ' οὗ ὁ καυλός· ἀφίησι δὲ καὶ εἰς βάθος τὸν αὐτὸν τρόπον ῥίζας, δι' ὃ καὶ πάντων μάλιστα δυσώλεθρον καὶ ἔργον ἐξελεῖν.

6 (Σχεδὸν δὲ παραπλησίως φύεται ἡ ἄγρωστις ἐκ τῶν γονάτων· αἱ γὰρ ῥίζαι γονατώδεις, ἐξ ἑκάστου δ' ἀφίησιν ἄνω βλαστὸν καὶ κάτωθεν ῥίζαν. ὡσαύτως δὲ καὶ ἡ ἄκανθα ἡ ἀκανώδης, ἀλλ' οὐ καλαμώδης οὐδὲ γονατώδης ἡ ῥίζα ταύ-

[1] κονίας: ? κονιάσεις (plastering), a conjecture mentioned by Sch.

and it is used as a soap-powder.[1] It is something like a cake, soft and reddish. Moreover the 'female' plant both of *phleos* and sedge is barren, but useful for basket-work,[2] while the 'male' is useless.

Duckweed 'moon-flower' and marestail require further investigation.

Most peculiar of these plants is the bulrush, both in being leafless and in not having so many roots as the others; for the others tend downwards quite as much as upwards, and shew their strength in that direction; and especially is this true of galingale, and also of dog's-tooth grass; wherefore these plants too and all others like them are hard to destroy. The root of galingale exceeds all the others in the diversity of characters which it shews, in that part of it is stout and flesby, part slender and woody. So also is this plant peculiar in its way of shooting and originating; for from the trunk-like stock[3] grows another slender root[4] sideways, and on this again forms the flesby part which contains the shoot from which the stalk springs.[5] In like manner it also sends out roots downwards; wherefore of all plants it is hardest to kill, and troublesome to get rid of.

(Dog's-tooth grass grows in almost the same way from the joints; for the roots are jointed, and from each joint it sends a shoot upwards and a root downwards. The growth of the spinous plant called corn-thistle[6] is similar, but it is not reedy and its

[2] *cf.* Hdt. 3. 98. [3] *i.e.* rhizome.

[4] *i.e.* stolon; *cf.* 1. 6. 8.

[5] ἀφ' οὗ ὁ καυλός transposed by W.; in Ald. these words come before ἐν ᾧ.

[6] ἡ ἀκανώδης I conj.; κεάνωνος UMV; κεάνωθος Ald.: ἡ κεάνωθος most edd.; G omits the word.

της. ταῦτα μὲν οὖν ἐπὶ πλεῖον διὰ τὴν ὁμοιότητα εἴρηται.)

Φύεται δ' ἐν ἀμφοῖν καὶ ἐν τῇ γῇ καὶ ἐν τῷ ὕδατι ἰτέα κάλαμος, πλὴν τοῦ αὐλητικοῦ, κύπειρον τύφη φλεὼς βούτομος· ἐν δὲ τῷ ὕδατι μόνον σίδη. περὶ γὰρ τῆς τύφης ἀμφισβητοῦσι. καλλίω δὲ καὶ μείζω τῶν ἐν ἀμφοῖν φυομένων αἰεὶ τὰ ἐν τῷ ὕδατι γίνεσθαί φασι. φύεσθαι δ' ἔνια τούτων καὶ ἐπὶ τῶν πλοάδων, οἷον τὸ κύπειρον καὶ τὸ βούτομον καὶ τὸν φλεών, ὥστε πάντα τὰ μέρη ταῦτα κατέχειν.

Ἐδώδιμα δ' ἐστὶ τῶν ἐν τῇ λίμνῃ τάδε· ἡ μὲν σίδη καὶ αὐτὴ καὶ τὰ φύλλα τοῖς προβάτοις, ὁ δὲ βλαστὸς τοῖς ὑσίν, ὁ δὲ καρπὸς τοῖς ἀνθρώποις. τοῦ δὲ φλεὼ καὶ τῆς τύφης καὶ τοῦ βουτόμου τὸ πρὸς ταῖς ῥίζαις ἁπαλόν, ὃ μάλιστα ἐσθίει τὰ παιδία. ῥίζα δ' ἐδώδιμος ἡ τοῦ φλεὼ μόνη τοῖς βοσκήμασιν. ὅταν δ' αὐχμὸς ᾖ καὶ μὴ γένηται τὸ κατὰ κεφαλὴν ὕδωρ, ἅπαντα αὐχμεῖ τὰ ἐν τῇ λίμνῃ, μάλιστα δὲ ὁ κάλαμος, ὑπὲρ οὗ καὶ λοιπὸν εἰπεῖν· ὑπὲρ γὰρ τῶν ἄλλων σχεδὸν εἴρηται.

XI. Τοῦ δὴ καλάμου δύο φασὶν εἶναι γένη, τόν τε αὐλητικὸν καὶ τὸν ἕτερον· ἓν γὰρ εἶναι τὸ γένος τοῦ ἑτέρου, διαφέρειν δὲ ἀλλήλων ἰσχύϊ <καὶ παχύτητι> καὶ λεπτότητι καὶ ἀσθενείᾳ· καλοῦσι δὲ τὸν μὲν ἰσχυρὸν καὶ παχὺν χαρακίαν τὸν δ' ἕτερον πλόκιμον· καὶ φύεσθαι τὸν μὲν

---

[1] *i.e.* we have gone beyond the list of typical plants of Orchomenus given 4. 10. 1, because we have found others of which much the same may be said.

[2] *cf.* 4. 10. 2.

[3] αὐτὴ: *cf.* 4. 10. 3 n.

root is not jointed. We have enlarged on these matters[1] because of the resemblance.)

The willow and the reed (not however the reed used for pipes) galingale bulrush *phleos* sedge grow both on land and in the water, water-lily only in the water. (As to bulrush indeed there is a difference of opinion.) However they say that those plants which grow in the water are always finer and larger than those that grow in both positions; also that some of these plants grow also on the floating islands,[2] for instance galingale sedge and *phleos*; thus all parts of the lake contain these plants.

Of the plants of the lake the parts good for food are as follows: of the water-lily both the flower[3] and the leaves are good for sheep, the young shoots for pigs, and the fruit for men. Of *phleos* galingale and sedge the part next the roots is tender, and is mostly eaten by children. The root of *phleos* is the only part which is edible by cattle. When there is a drought and there is no water from overhead,[4] all the plants of the lake are dried up, but especially the reed; of this it remains to speak, since we have said almost enough about the rest.

XI. [5] Of the reed there are said to be two kinds, the one used for making pipes and the other kind. For that of the latter there is only one kind, though individual plants differ in being strong and stout,[6] or on the other hand slender and weak. The strong stout one they call the 'stake-reed,' the other the 'weaving reed.' The latter they say grows on the

[4] κεφαλὴν UMVAld.; for the case *cf.* Xen. *Hell.* 7. 2. 8 and 11; κεφαλῆς conj. W.

[5] Plin. 16. 168 and 169.

[6] καὶ παχύτητι add. Dalec. from G.

πλόκιμον ἐπὶ τῶν πλοάδων τὸν δὲ χαρακίαν ἐπὶ τοῖς κώμυσι· κώμυθας δὲ καλοῦσι οὗ ἂν ᾖ συνηθροισμένος κάλαμος καὶ συμπεπλεγμένος ταῖς ῥίζαις· τοῦτο δὲ γίνεται καθ' οὓς ἂν τόπους τῆς λιμνης εὔγειον ᾖ χωρίον· γίνεσθαι δέ ποτε τὸν χαρακίαν καὶ οὗ ὁ αὐλητικός, μακρότερον μὲν τοῦ ἄλλου χαρακίου σκωληκόβρωτον δέ. τούτου μὲν οὖν ταύτας λέγουσι τὰς διαφοράς.

2 Περὶ δὲ τοῦ αὐλητικοῦ τὸ μὲν φύεσθαι δι' ἐννεατηρίδος, ὥσπερ τινές φασι, καὶ ταύτην εἶναι τὴν τάξιν οὐκ ἀληθές, ἀλλὰ τὸ μὲν ὅλον αὐξηθείσης γίνεται τῆς λίμνης· ὅτι δὲ τοῦτ' ἐδόκει συμβαίνειν ἐν τοῖς πρότερον χρόνοις μάλιστα δι' ἐννεατηρίδος, καὶ τὴν γένεσιν τοῦ καλάμου ταύτην ἐποίουν τὸ συμβεβηκὸς ὡς τάξιν λαμβάνον-
3 τες. γίνεται δὲ ὅταν ἐπομβρίας γενομένης ἐμμένῃ τὸ ὕδωρ δύ' ἔτη τοὐλάχιστον, ἂν δὲ πλείω καὶ καλλίων· τούτου δὲ μάλιστα μνημονεύουσι γεγονότος τῶν ὕστερον χρόνων ὅτε συνέβη τὰ περὶ Χαιρώνειαν· πρὸ τούτων γὰρ ἔφασαν ἔτη πλείω βαθυνθῆναι τὴν λίμνην· μετὰ δὲ ταῦτα ὕστερον, ὡς ὁ λοιμὸς ἐγένετο σφοδρός, πλησθῆναι μὲν αὐτήν, οὐ μείναντος δὲ τοῦ ὕδατος ἀλλ' ἐκλιπόντος χειμῶνος οὐ γενέσθαι τὸν κάλαμον· φασὶ γὰρ καὶ δοκεῖ βαθυνομένης τῆς λίμνης αὐξάνεσθαι τὸν κάλαμον εἰς μῆκος, μείναντα δὲ τὸν ἐπιόντα ἐνιαυτὸν ἁδρύνεσθαι· καὶ γίνεσθαι τὸν μὲν ἁδρυθέντα ζευγίτην, ᾧ δ' ἂν μὴ συμπαραμείνῃ τὸ

---

[1] κώμυσι: lit 'bundles.'
[2] δύ' ἔτη conj. W.; διετῆ UMVAld.
[3] B.C. 338.

floating islands, the stout form in the 'reed-beds'[1]; this name they give to the places where there is a thick mass of reed with its roots entangled together. This occurs in any part of the lake where there is rich soil. It is said that the 'stake-reed' is also sometimes found in the same places as the reed used for pipes, in which places it is longer than the 'stake-reed' found elsewhere, but gets worm-eaten. These then are the differences in reeds of which they tell.

As to the reed used for pipes, it is not true, as some say, that it only grows once in nine years and that this is its regular rule of growth; it grows in general whenever the lake is full: but, because in former days this was supposed to happen generally once in nine years, they made the growth of the reed to correspond, taking what was really an accident to be a regular principle. As a matter of fact it grows whenever after a rainy season the water remains in the lake for at least two years,[2] and it is finer if the water remains longer; this is specially remembered to have happened in recent times at the time of the battle of Chaeronea.[3] For before that period they told me that the lake was for several years deep[4]; and, at a time later than that, when there was a severe visitation of the plague, it filled up; but, as the water did not remain but failed in winter, the reed did not grow; for they say, apparently with good reason, that, when the lake is deep, the reed increases in height, and, persisting for the next year, matures its growth; and that the reed which thus matures is suitable for making a reed mouthpiece,[5] while that for which the water has not remained is

[4] ἔτη πλείω conj. Scal. from G; ἔτι πλείω UMV; ἔτι πλεῖον Ald.

[5] See n. on τὸ στόμα τῶν γλωττῶν, § 4.

ὕδωρ βομβυκίαν. τὴν μὲν οὖν γένεσιν εἶναι τοιαύτην.

4 Διαφέρειν δὲ τῶν ἄλλων καλάμων ὡς καθ' ὅλου λαβεῖν εὐτροφίᾳ τινὶ τῆς φύσεως· εὐπληθέστερον γὰρ εἶναι καὶ εὐσαρκότερον καὶ ὅλως δὲ θῆλυν τῇ προσόψει. καὶ γὰρ τὸ φύλλον πλατύτερον ἔχειν καὶ λευκότερον τὴν δὲ ἀνθήλην ἐλάττω τῶν ἄλλων, τινὰς δὲ ὅλως οὐκ ἔχειν, οὓς καὶ προσαγορεύουσιν εὐνουχίας· ἐξ ὧν ἄριστα μέν φασί τινες γίνεσθαι τὰ ζεύγη, κατορθοῦν δὲ ὀλίγα παρὰ τὴν ἐργασίαν.

Τὴν δὲ τομὴν ὡραίαν εἶναι πρὸ Ἀντιγενίδου μέν, ἡνίκ' ηὔλουν ἀπλάστως, ὑπ' Ἄρκτουρον Βοηδρομιῶνος μηνός· τὸν γὰρ οὕτω τμηθέντα συχνοῖς μὲν ἔτεσιν ὕστερον γίνεσθαι χρήσιμον καὶ προκαταυλήσεως δεῖσθαι πολλῆς, συμμύειν δὲ τὸ στόμα τῶν γλωττῶν, ὃ πρὸς τὴν διακτηρίαν εἶναι
5 χρήσιμον. ἐπεὶ δὲ εἰς τὴν πλάσιν μετέβησαν, καὶ ἡ τομὴ μετεκινήθη· τέμνουσι γὰρ δὴ νῦν τοῦ Σκιρροφοριῶνος καὶ Ἑκατομβαιῶνος ὥσπερ πρὸ τροπῶν μικρὸν ἢ ὑπὸ τροπάς. γίνεσθαι δέ φασι τρίενόν τε χρήσιμον καὶ καταυλήσεως βραχείας

---

[1] βομβυκίαν. In one kind of pipe the performer blew, not directly on to the 'reed,' but into a cap in which it was enclosed; this cap, from the resemblance in shape to a cocoon, was called βόμβυξ.

[2] εἶναι add. W.

[3] Plin. 16. 169–172. [4] September.

[5] *i.e.* between the free end of the vibrating 'tongue' and

suitable for making a 'cap.'[1] Such then, it is said, is[2] the reed's way of growth.

[3] Also it is said to differ from other reeds, to speak generally, in a certain luxuriance of growth, being of a fuller and more flesby character, and, one may say, 'female' in appearance. For it is said that even the leaf is broader and whiter, though the plume is smaller than that of other reeds, and some have no plume at all; these they call 'eunuch-reeds.' From these they say that the best mouthpieces are made, though many are spoiled in the making.

Till the time of Antigenidas, before which men played the pipe in the simple style, they say that the proper season for cutting the reeds was the month Boëdromion[4] about the rising of Arcturus; for, although the reed so cut did not become fit for use for many years after and needed a great deal of preliminary playing upon, yet the opening[5] of the reed-tongues is well closed, which is a good thing for the purpose of accompaniment.[6] But when a change was made to the more elaborate style of playing, the time of cutting the reeds was also altered; for in our own time they cut them in the months Skirrophorion[7] or Hekatombaion[8] about the solstice or a little earlier.[9] And they say that the reed becomes fit for use in three years and needs but little preliminary playing upon, and that the reed-tongues

the body or 'lay' of the reed mouthpiece: the instrument implied throughout is apparently one with a single vibrating 'tongue' (reed) like the modern clarinet.

[6] διακτηρίαν UMV; διακτορίαν Ald. ? πρὸς τὸ ἀκροατήριον, 'for the concert-room'; *quod erat illis theatrorum moribus utilius* Plin. *l.c.*

[7] June. [8] July.

[9] ὥσπερ conj. W.; ὡσπερεὶ UH.; ὡς περὶ MVAld.

δεῖσθαι καὶ κατασπάσματα τὰς γλώττας ἴσχειν· τοῦτο δὲ ἀναγκαῖον τοῖς μετὰ πλάσματος αὐλοῦσι. τοῦ μὲν οὖν ζευγίτου ταύτας εἶναι τὰς ὥρας τῆς τομῆς.

6 Ἡ δ' ἐργασία γίνεται τοῦτον τὸν τρόπον· ὅταν συλλέξωσι τιθέασιν ὑπαίθριον τοῦ χειμῶνος ἐν τῷ λέμματι· τοῦ δ' ἦρος περικαθάραντες καὶ ἐκτρίψαντες εἰς τὸν ἥλιον ἔθεσαν. τοῦ θέρους δὲ μετὰ ταῦτα συντεμόντες εἰς τὰ μεσογονάτια πάλιν ὑπαίθριον τιθέασι χρόνον τινά. προσλείπουσι δὲ τῷ μεσογονατίῳ τὸ πρὸς τοὺς βλαστοὺς γόνυ· τὰ δὲ μήκη τὰ τούτων οὐ γίνεται διπαλαίστων ἐλάττω. βέλτιστα μὲν οὖν εἶναι τῶν μεσογονατίων πρὸς τὴν ζευγοποιίαν ὅλου τοῦ καλάμου τὰ μέσα· μαλακώτατα δὲ ἴσχειν ζεύγη τὰ πρὸς τοὺς
7 βλαστούς, σκληρότατα δὲ τὰ πρὸς τῇ ῥίζῃ· συμφωνεῖν δὲ τὰς γλώττας τὰς ἐκ τοῦ αὐτοῦ μεσογονατίου, τὰς δὲ ἄλλας οὐ συμφωνεῖν· καὶ τὴν μὲν πρὸς τῇ ῥίζῃ ἀριστερὰν εἶναι, τὴν δὲ πρὸς τοὺς βλαστοὺς δεξιάν. τμηθέντος δὲ δίχα τοῦ μεσογονατίου τὸ στόμα τῆς γλώττης ἑκατέρας γίνεσθαι κατὰ τὴν τοῦ καλάμου τομήν· ἐὰν δὲ ἄλλον τρόπον ἐργασθῶσιν αἱ γλῶτται, ταύτας οὐ πάνυ συμφωνεῖν· ἡ μὲν οὖν ἐργασία τοιαύτη.

---

[1] *κατασπάσματα*: lit. 'convulsions'; *i.e.* the strong vibrations of a 'tongue,' the free end of which is kept away from the body or 'lay' of the mouthpiece. Such a 'reed' would have the effect of giving to the pipes a fuller and louder tone.

[2] *i.e.* so as to make a closed end.

have ample vibration,[1] which is essential for those who play in the elaborate style. Such, they tell us, are the proper seasons for cutting the reed used for the reed mouthpiece.

The manufacture is carried out in the following manner. Having collected the reed-stems they lay them in the open air during the winter, leaving on the rind; in the spring they strip this off, and, having rubbed the reeds thoroughly, put them in the sun. Later on, in the summer, they cut the sections from knot to knot into lengths and again put them for some time in the open air. They leave the upper knot on this internodal section[2]; and the lengths thus obtained are not less than two palmsbreadths long. Now they say that for making mouthpieces the best lengths are those of the middle of the reed, whereas the lengths towards the upper growths make very soft mouthpieces and those next to the root very hard ones. They say too that the reed-tongues made out of the same length are of the same quality, while those made from different lengths are not; also that the one from the length next to the root forms a left-hand[3] reed-tongue, and that from the length towards the upper growths a right-hand[3] reed-tongue. Moreover, when the length is slit, the opening of the reed-tongues in either case is made towards the point at which the reed was cut[4]; and, if the reed-tongues are made in any other manner, they are not quite of the same quality. Such then is the method of manufacture.

[3] *i.e.* the vibrating 'tongues' (reeds) for the left-hand and the right-hand pipe of the Double Pipe respectively.

[4] *i.e.* not at the closed end, but at the end which was 'lower' when the cane was growing: *cf.* § 6, προσλείπουσι δὲ κ.τ.λ.

8 Φύεται δὲ πλεῖστος μὲν μεταξὺ τοῦ Κηφισοῦ καὶ τοῦ Μέλανος· οὗτος δὲ ὁ τόπος προσαγορεύεται μὲν Πελεκανία· τούτου δ' ἔστιν ἄττα Χύτροι καλούμενοι βαθύσματα τῆς λίμνης, ἐν οἷς κάλλιστόν φασι γίνεσθαι· <γίνεσθαι> δὲ καὶ καθ' ὃ ἡ Προβατία καλουμένη καταφέρεται· τοῦτο δ' ἐστὶ ποταμὸς ῥέων ἐκ Λεβαδείας. κάλλιστος δὲ δοκεῖ πάντων γίνεσθαι περὶ τὴν Ὀξεῖαν καλουμένην Καμπήν· ὁ δὲ τόπος οὗτός ἐστιν ἐμβολὴ τοῦ Κηφισοῦ. γειτνιᾷ δ' αὐτῷ πεδίον εὔγειον, ὃ
9 προσαγορεύουσι Ἱππίαν. πρόσβορρος δὲ τόπος ἄλλος τῆς Ὀξείας Καμπῆς ἐστιν, ὃν καλοῦσι Βοηδρίαν· φύεσθαι δέ φασι καὶ κατὰ ταύτην εὐγενῆ τὸν κάλαμον. τὸ δὲ ὅλον, οὗ ἂν ᾖ βαθύγειον καὶ εὔγειον χωρίον καὶ ἰλυῶδες καὶ ὁ Κηφισὸς ἀναμίσγεται καὶ πρὸς τούτοις βάθυσμα τῆς λίμνης, κάλλιστον γίνεσθαι κάλαμον. περὶ γὰρ τὴν Ὀξεῖαν Καμπὴν καὶ τὴν Βοηδρίαν πάντα ταῦτα ὑπάρχειν. ὅτι δὲ ὁ Κηφισὸς μεγάλην ἔχει ῥοπὴν εἰς τὸ ποιεῖν καλὸν τὸν κάλαμον σημεῖον ἔχουσι· καθ' ὃν γὰρ τόπον ὁ Μέλας καλούμενος ἐμβάλλει βαθείας οὔσης τῆς λίμνης καὶ τοῦ ἐδάφους εὐγείου καὶ ἰλυώδους, ἢ ὅλως μὴ γίνεσθαι ἢ φαῦλον. ἡ μὲν οὖν γένεσις καὶ φύσις τοῦ αὐλητικοῦ καὶ ἡ κατεργασία καὶ τίνας ἔχει διαφορὰς πρὸς τοὺς ἄλλους ἱκανῶς εἰρήσθω.

10 Γένη δὲ οὐ ταῦτα μόνον ἀλλὰ πλείω τοῦ καλάμου τυγχάνει φανερὰς ἔχοντα τῇ αἰσθήσει διαφοράς· ὁ μὲν γὰρ πυκνὸς καὶ τῇ σαρκὶ καὶ τοῖς

---

[1] *cf.* Plut. *Sulla*, 20.
[2] *i.e.* the so-called 'Lake' Copaïs.
[3] καὶ add. W.

This reed grows in greatest abundance between the Kephisos and the Black River[1]; this district is called Pelekania, and in it are certain 'pots,' as they are called, which are deep holes in the marsh,[2] and in these holes they say that it grows fairest; it is also[3] said to be found[4] where the river called the 'Sheep River' comes down, which is a stream that flows from Lebadeia. But it appears to grow fairest of all near 'the Sharp Bend'; this place is the mouth of the Kephisos; near it is a rich plain called Hippias. There is another region north of the Sharp Bend called Boedrias; and here too they say that the reed grows fine, and in general that it is fairest wherever there is a piece of land with deep rich alluvial soil, where also Kephisos mingles[5] his waters with the soil, and where there is further a deep hole in the marsh; for that about the Sharp Bend and Boedrias all these conditions are found. As proof that the Kephisos has a great effect in producing the reed of good quality they have the fact that, where the river called the 'Black River' flows into the marsh, though the marsh is there deep and the bottom of good alluvial soil, it either does not grow at all or at best but of poor quality. Let this suffice for an account of the growth and character of the reed used for pipes, of the manufacture, and of its distinctive features as compared with other reeds.

But these are not the only kinds of reed; there are several others[6] with distinctive characters which are easily recognised; there is one that is of compact growth in flesh and has its joints close together;

[4] *γίνεσθαι* add. Sch.; *φασι· γίνεσθαι δὲ καθ' ὃ* UMVP; so Ald., but *καθ' ὃν*.

[5] *ἀναμίσγεται* : ? *ἀναμίσγηται*; *cf.* Plut. *Sull. l.c.*

[6] Plin. 16. 164–167; Diosc. 1. 85.

γόνασιν, ὁ δὲ μανὸς καὶ ὀλιγογόνατος· καὶ ὁ μὲν κοῖλος, ὃν καλοῦσί τινες συριγγίαν, οὐδὲν γὰρ ὡς εἰπεῖν ἔχει ξύλου καὶ σαρκός· ὁ δὲ στερεὸς καὶ συμπλήρης μικροῦ. καὶ ὁ μὲν βραχύς, ὁ δὲ εὐαυξὴς καὶ ὑψηλὸς καὶ παχύς. ὁ δὲ λεπτὸς καὶ πολύφυλλος, ὁ δὲ ὀλιγόφυλλος καὶ μονόφυλλος. ὅλως δὲ πολλαί τινές εἰσι διαφοραὶ κατὰ τὰς χρείας· ἕκαστος γὰρ πρὸς ἕκαστα χρήσιμος.

11 Ὀνόμασι δὲ ἄλλοι ἄλλοις προσαγορεύουσι· κοινότατον δέ πως ὁ δόναξ, ὃν καὶ λοχμωδέστατόν γέ φασιν εἶναι καὶ μάλιστα φύεσθαι παρὰ τοὺς ποταμοὺς καὶ τὰς λίμνας. διαφέρειν δ' ὅμως παντὸς καλάμου πολὺ τόν τε ἐν τῷ ξηρῷ καὶ τὸν ἐν τοῖς ὕδασι φυόμενον. ἴδιος δὲ καὶ ὁ τοξικός, ὃν δὴ Κρητικόν τινες καλοῦσιν· ὀλιγογόνατος μὲν σαρκωδέστερος δὲ πάντων καὶ μάλιστα κάμψιν δεχόμενος, και ὅλως ἄγεσθαι δυνάμενος ὡς ἂν θέλῃ τις θερμαινόμενος.

12 Ἔχουσι δέ, ὥσπερ ἐλέχθη, καὶ κατὰ τὰ φύλλα μεγάλας διαφορὰς οὐ πλήθει καὶ μεγέθει μόνον ἀλλὰ καὶ χροιᾷ. ποικίλος γὰρ ὁ Λακωνικὸς καλούμενος. ἔτι δὲ τῇ θέσει καὶ προσφύσει· κάτωθεν γὰρ ἔνιοι πλεῖστα φέρουσι τῶν φύλλων, αὐτὸς δὲ ὥσπερ ἐκ θάμνου πέφυκε. σχεδὸν δέ τινές φασι καὶ τῶν λιμναίων ταύτην εἶναι τὴν διαφοράν, τὸ πολύφυλλον καὶ παρόμοιον ἔχειν τρόπον τινὰ τὸ φύλλον τῷ τοῦ κυπείρου καὶ

another that is of open growth, with few joints; there is the hollow reed called by some the 'tube-reed,'[1] inasmuch as it has hardly any wood or flesh; there is another which is solid and almost entirely filled with substance; there is another which is short, and another which is of strong growth tall and stout; there is one which is slender and has many leaves, another which has few leaves or only one. And in general there are many differences in natural character and in usefulness, each kind being useful for some particular purpose.

Some distinguish the various kinds by different names; commonest perhaps is the pole-reed, which is said to be of very bushy habit, and to grow chiefly by rivers and lakes. And it is said that there is a wide difference in reeds in general between those that grow on dry land and those that grow in the water. Quite distinct again is the 'archer's' reed, which some call the 'Cretan': this has few joints and is fleshier than any of the others; it can also be most freely bent, and in general, when warmed, may be turned about as one pleases.

The various kinds have also, as was said, great differences in the leaves, not only in number and size, but also in colour. That called the 'Laconian' reed is parti-coloured. They also differ in the position and attachment of the leaves; some have most of their leaves low down, and the reed itself grows out of a sort of a bush. Indeed some say that this may be taken as the distinctive character of those which grow in lakes, namely, that these have many leaves, and that their foliage in a manner

[1] συριγγίαν conj. Sch. from Plin. *l.c.*, *syringiam*; *cf.* Diosc. *l.c.*, *Geop.* 2. 6. 23. συριγί U; σύριγγι MV; σύριγγα Ald.H.

φλεὼ καὶ θρύου καὶ βουτόμου· σκέψασθαι δὲ δεῖ τοῦτο.

13 Γένος δέ τι καλάμου φύεται καὶ ἐπίγειον, ὃ οὐκ εἰς ὀρθὸν ἀλλ' ἐπὶ γῆς ἀφίησι τὸν καυλόν, ὥσπερ ἡ ἄγρωστις, καὶ οὕτως ποιεῖται τὴν αὔξησιν. ἔστι δὲ ὁ μὲν ἄρρην στερεός, καλεῖται δὲ ὑπό τινων εἱλετίας. . . .

Ὁ δὲ Ἰνδικὸς ἐν μεγίστῃ διαφορᾷ καὶ ὥσπερ ἕτερον ὅλως τὸ γένος· ἔστι δὲ ὁ μὲν ἄρρην στερεός, ὁ δὲ θῆλυς κοῖλος· διαιροῦσι γὰρ καὶ τοῦτον τῷ ἄρρενι καὶ θήλει. φύονται δ' ἐξ ἑνὸς πυθμένος πολλοὶ καὶ οὐ λοχμώδεις· τὸ δὲ φύλλον οὐ μακρὸν ἀλλ' ὅμοιον τῇ ἰτέᾳ· τῷ δὲ μεγέθει μεγάλοι καὶ εὐπαγεῖς, ὥστε ἀκοντίοις χρῆσθαι. φύονται δὲ οὗτοι περὶ τὸν Ἀκεσίνην ποταμόν. ἅπας δὲ κάλαμος εὔζωος καὶ τεμνόμενος καὶ ἐπικαιόμενος καλλίων βλαστάνει· ἔτι δὲ παχύρριζος καὶ πολύρριζος, δι' ὃ καὶ δυσώλεθρος. ἡ δὲ ῥίζα γονατώδης, ὥσπερ ἡ τῆς ἀγρωστίδος, πλὴν οὐ παντὸς ὁμοίως. ἀλλὰ περὶ μὲν καλάμων ἱκανῶς εἰρήσθω.

XII. Κατάλοιπον δὲ εἰπεῖν ὡσὰν ἐκ τοῦ γένους τούτου περὶ σχοίνου· καὶ γὰρ καὶ τοῦτο τῶν ἐνύδρων θετέον. ἔστι δὲ αὐτοῦ τρία εἴδη, καθάπερ τινὲς διαιροῦσιν, ὅ τε ὀξὺς καὶ ἄκαρπος, ὃν δὴ καλοῦσιν ἄρρενα, καὶ ὁ κάρπιμος, ὃν μελαγκρανὶν

---

[1] θρύον, a kind of grass (see Index; *cf.* Hom. *Il.* 21. 351), conj. Sch.; βρύον MSS.; however Plut. *Nat. Quaest.* 2 gives βρύον along with τύφη and φλεώς in a list of marsh plants.
[2] δὲ δεῖ τοῦτο conj. W.; δὲ τοῦτο UMVAld.

resembles that of galingale *phleos thryon*[1] and sedge; but this needs[2] further enquiry.

There is also a kind of reed (bush-grass) which grows on land, and which is not erect, but sends out its stem over the ground, like the dog's-tooth grass, and so makes its growth. The 'male' reed is solid: some call it *eiletias*. . . . .[3]

The Indian reed (bamboo) is very distinct, and as it were a totally different kind; the 'male' is solid and the 'female' hollow (for in this kind too they distinguish a 'male' and a 'female' form); a number of reeds of this kind grow from one base and they do not form a bush; the leaf is not long, but resembles the willow leaf; these reeds are of great size and of good substance, so that they are used for javelins. They grow by the river Akesines.[4] All reeds are tenacious of life, and, if cut or burnt down, grow up again more vigorously; also their roots are stout and numerous, so that the plant is hard to destroy. The root is jointed, like that of the dog's-tooth grass, but this is not equally so in all kinds. However let this suffice for an account of reeds.

Of rushes.

XII. It remains to speak of the rush,[5] as though it belonged to this class of plants, inasmuch as we must reckon this also among water plants. Of this there are three kinds[6] as some distinguish, the 'sharp' rush, which is barren and is called the 'male'; the 'fruiting' kind which we call the 'black-

[3] Sch. marks a lacuna; there is nothing to correspond to ὁ μὲν ἄρρην. [4] Chenab.

[5] *cf.* 1. 5. 3; 1. 8. 1; Plin. 21. 112–115; Diosc. 4. 52.

[6] See Index.

καλοῦμεν διὰ τὸ μέλανα τὸν καρπὸν ἔχειν, παχύτερος δὲ οὗτος καὶ σαρκωδέστερος· καὶ τρίτος τῷ μεγέθει καὶ τῇ παχύτητι καὶ εὐσαρκίᾳ διαφέρων ὁ καλούμενος ὁλόσχοινος.

2 Ἡ μὲν οὖν μελαγκρανὶς αὐτός τις καθ' αὑτόν· ὁ δ' ὀξὺς καὶ ὁλόσχοινος ἐκ τοῦ αὐτοῦ φύονται· ὃ καὶ ἄτοπον φαίνεται, καὶ θαυμαστόν γ' ἦν ἰδεῖν ὅλης κομισθείσης τῆς σχοινιᾶς· οἱ πολλοὶ γὰρ ἦσαν ἄκαρποι πεφυκότες ἐκ τοῦ αὐτοῦ, κάρπιμοι δὲ ὀλίγοι. τοῦτο μὲν οὖν ἐπισκεπτέον. ἐλάττους δὲ ὅλως οἱ κάρπιμοι· πρὸς γὰρ τὰ πλέγματα χρησιμώτερος ὁ ὁλόσχοινος διὰ τὸ σαρκῶδες καὶ μαλακόν. κορυνᾷ δ' ὅλως ὁ κάρπιμος ἐξ αὐτοῦ τοῦ γραμμώδους ἐξοιδήσας, κἄπειτα ἐκτίκτει καθάπερ ᾠά. πρὸς μιᾷ γὰρ ἀρχῇ γραμμώδει ἔχει τοὺς περισταχυώδεις μίσχους, ἐφ' ὧν ἄκρων παραπλαγίους τὰς τῶν ἀγγείων ἔχει στρογγυλότητας ὑποχασκούσας· ἐν τούτοις δὲ τὸ σπερμάτιον ἀκιδῶδές ἐστι μέλαν ἑκάστῳ προσεμφερὲς

3 τῷ τοῦ ἀστερίσκου πλὴν ἀμενηνότερον. ῥίζαν δὲ ἔχει μακρὰν καὶ παχυτέραν πολὺ τοῦ σχοίνου· αὕτη δ' αὐαίνεται καθ' ἕκαστον ἐνιαυτόν, εἶθ' ἑτέρα πάλιν ἀπὸ τῆς κεφαλῆς τοῦ σχοίνου καθίεται· τοῦτο δὲ καὶ ἐν τῇ ὄψει φανερὸν ἰδεῖν τὰς μὲν αὔας τὰς δὲ χλωρὰς καθιεμένας· ἡ δὲ κεφαλὴ ὁμοία τῇ τῶν κρομύων καὶ τῇ τῶν γητείων, συμ-

---

[1] θ. γ' ἦν ἰδεῖν conj. W. from G; θ. ἐν γ' εἴδεῖν U; θ. ἔν γε ἰδεῖν MVP; θ. ἐνιδεῖν Ald.
[2] οἱ κάρπιμοι conj. R. Const.; οἱ καρποί Ald.H.
[3] γὰρ seems meaningless; G has *autem*.
[4] κορονᾷ; *cf.* 3. 5. 1.
[5] γραμμώδει conj. R. Const.; γοαμμώδεις Ald.H.

head' because it has black fruit; this is stouter and fleshier: and third the 'entire rush,' as it is called, which is distinguished by its size stoutness and fleshiness.

Now the 'black-head' grows by itself, but the 'sharp' rush and the 'entire' rush grow from the same stock, which seems extraordinary, and indeed it was strange to see it[1] when the whole clump of rushes was brought before me; for from the same stock there were growing 'barren' rushes, which were the most numerous, and also a few 'fruiting' ones. This then is a matter for further enquiry. The 'fruiting'[2] ones are in general scarcer, for[3] the 'entire rush' is more useful for wicker-work because of its fleshiness and pliancy. The 'fruiting' rush in general produces a club-like[4] head which swells straight from the wiry stem, and then bears egg-like bodies; for attached to a single wiry[5] base it has its very spike-like[6] branches all round it, and on the ends of these it has its round vessels borne laterally and gaping[7]; in each of these is the small seed, which is pointed and black, and like that of the Michaelmas daisy, except that it is less solid. It has a long root, which is stouter than that of the ordinary rush; this withers every year, and then another strikes down again from the 'head'[8] of the plant. And it is easy to observe that some of the roots as they are let down are withered, some green. The 'head' is like that of an onion or long onion,

[6] περισταχυώδεις seems an impossible word; ? περὶ αὐτὸν τοὺς σταχυώδεις.

[7] ὑποχασκούσας conj. Sch.; ἐπισχαζούσας Ald.H.

[8] *i.e.* the part above ground; *cf.* Plin. *l.c.* Sch. has disposed of the idea that κεφαλή is here a 'bulbous' root.

πεφυκυῖά πως ἐκ πλειόνων εἰς ταὐτὸ καὶ πλατεῖα κάτωθεν ἔχουσα κελύφη ὑπέρυθρα. συμβαίνει δ' οὖν ἴδιον ἐπὶ τῶν ῥιζῶν εἰ αὐαίνονται κατ' ἐνιαυτὸν καὶ ἐκ τοῦ ἄνωθεν πάλιν ἡ γένεσις. τῶν μὲν οὖν σχοίνων τοιαύτη τις φύσις.

Εἰ δὲ καὶ ὁ βάτος καὶ ὁ παλίουρος ἔνυδρά πώς ἐστιν ἢ πάρυδρα, καθάπερ ἐνιαχοῦ, φανεραὶ σχεδὸν καὶ αἱ τούτων διαφοραί· περὶ ἀμφοῖν γὰρ εἴρηται πρότερον.[1]

[Τῶν δὲ νήσων τῶν πλοάδων τῶν ἐν Ὀρχομενῷ τὰ μὲν μεγέθη παντοδαπὰ τυγχάνει, τὰ δὲ μέγιστα αὐτῶν ἐστιν ὅσον τριῶν σταδίων τὴν περίμετρον. ἐν Αἰγύπτῳ δὲ μάλιστα μεγάλα σφόδρα συνίσταται, ὥστε καὶ ὗς ἐν αὐταῖς ἐγγίνεσθαι πολλούς, οὓς καὶ κυνηγετοῦσι διαβαίνοντες.] καὶ περὶ μὲν ἐνύδρων ταῦτ' εἰρήσθω.

XIII. Περὶ δὲ βραχυβιότητος φυτῶν καὶ δένδρων τῶν ἐνύδρων ἐπὶ τοσοῦτον ἔχομεν ὡς ἂν καθ' ὅλου λέγοντες, ὅτι βραχυβιώτερα τῶν χερσαίων ἐστί, καθάπερ καὶ τὰ ζῷα. τοὺς δὲ καθ' ἕκαστον βίους ἱστορῆσαι δεῖ τῶν χερσαίων. τὰ μὲν οὖν ἄγριά φασιν οὐδεμίαν ἔχειν ὡς εἰπεῖν οἱ ὀρεοτύποι διαφοράν, ἀλλὰ πάντα εἶναι μακρόβια καὶ οὐθὲν βραχύβιον· αὐτὸ μὲν τοῦτο ἴσως ἀληθὲς λέγοντες· ἅπαντα γὰρ ὑπερτείνει πολὺ τὴν τῶν ἄλλων ζωήν. οὐ μὴν ἀλλ' ὅμως ἐστὶ τὰ μὲν μᾶλλον τὰ δ' ἧττον μακρόβια, καθάπερ ἐν τοῖς ἡμέροις· ποῖα

[1] 3. 18. 3 and 4; 4. 8. 1.

being, as it were, made up of several united together: it is broad, and underneath it has reddish scales. Now it is a peculiar fact about the roots of this plant that they wither every year and that the fresh growth of roots comes from the part of the plant which is above ground. Such is the character of rushes.

Bramble and Christ's thorn may be considered to some extent plants of the water or the waterside, as they are in some districts; but the distinctive characters of these plants are fairly clear, for we have spoken of both already.[1]

The floating islands of Orchomenos[2] are of various sizes, the largest being about three furlongs in circumference. But in Egypt very large ones form, so that even a number of boars are found in them, and men go across to the islands to hunt them. Let this account of water-plants suffice.

Of the length or shortness of the life of plants, and the causes.

XIII. As to the comparative shortness of life of plants and trees of the water we may say thus much as a general account, that, like the water-animals, they are shorter-lived than those of the dry land. But we must enquire into the lives of those of the dry land severally. Now the woodmen say that the wild kinds are almost[3] without exception long-lived, and none of them is short-lived: so far they may be speaking the truth; all such plants do live far longer than others. However, just as in the case of cultivated plants, some are longer-lived than others,

[2] *cf.* 4. 10. 2, to which § this note perhaps belongs.
[3] ὡς εἰπεῖν conj. Sch.; ὡς εἰπεῖ U; ὡς εἴποι MV; ὡς ἂν εἴποιεν Ald.H.

δὲ ταῦτα σκεπτέον. τὰ δὲ ἥμερα φανερῶς διαφέρει τῷ τὰ μὲν εἶναι μακρόβια τὰ δὲ βραχύβια· ὡς δ' ἁπλῶς εἰπεῖν τὰ ἄγρια τῶν ἡμέρων μακροβιώτερα καὶ ὅλως τῷ γένει καὶ τὰ ἀντιδιῃρημένα καθ' ἕκαστον, οἷον κότινος ἐλάας καὶ ἀχρὰς ἀπίου ἐρινεὸς συκῆς· ἰσχυρότερα γὰρ καὶ πυκνότερα καὶ ἀγονώτερα τοῖς περικαρπίοις.

2 Τὴν δὲ μακροβιότητα μαρτυροῦσιν ἐπί γέ τινων καὶ ἡμέρων καὶ ἀγρίων καὶ αἱ παραδεδομέναι φῆμαι παρὰ τῶν μυθολόγων· ἐλάαν μὲν γὰρ λέγουσι τὴν Ἀθήνῃσι, φοίνικα δὲ τὸν ἐν Δήλῳ, κότινον δὲ τὸν ἐν Ὀλυμπίᾳ, ἀφ' οὗ ὁ στέφανος· φηγοὺς δὲ τὰς ἐν Ἰλίῳ τὰς ἐπὶ τοῦ Ἴλου μνήματος· τινὲς δέ φασι καὶ τὴν ἐν Δελφοῖς πλάτανον Ἀγαμέμνονα φυτεῦσαι καὶ τὴν ἐν Καφύαις τῆς Ἀρκαδίας. ταῦτα μὲν οὖν ὅπως ἔχει τάχ' ἂν ἕτερος εἴη λόγος· ὅτι δέ ἐστι μεγάλη διαφορὰ τῶν δένδρων φανερόν· μακρόβια μὲν γὰρ τά τε προειρημένα καὶ ἕτερα πλείω· βραχύβια δὲ καὶ τὰ τοιαῦτα ὁμολογουμένως, οἷον ῥοιὰ συκῆ μηλέα, καὶ τούτων ἡ ἠρινὴ μᾶλλον καὶ ἡ γλυκεῖα τῆς ὀξείας, ὥσπερ τῶν ῥοῶν ἡ ἀπύρηνος. βραχύβια δὲ καὶ ἀμπέλων ἔνια γένη καὶ μάλιστα τὰ πολύκαρπα· δοκεῖ δὲ καὶ τὰ πάρυδρα βραχυβιώτερα

---

[1] καὶ τὰ ἀντ. conj. W.; κατὰ ἀντ. UMV; τὰ ἀντ. Ald.H.
[2] περικαρπίοις: *cf. C.P.* I. 17. 5.
[3] On the Acropolis: *cf.* Hdt. 8. 55; Soph. *O.C.* 694 foll.

and we must consider which these are. Cultivated plants plainly differ as to the length of their lives, but, to speak generally, wild plants are longer-lived than cultivated ones, both taken as classes, and also when one compares[1] the wild and cultivated forms of particular plants: thus the wild olive pear and fig are longer-lived than the corresponding cultivated trees; for the wild forms of these are stronger and of closer growth, and they do not produce such well-developed fruit-pulp.[2]

To the long-lived character of some plants, both cultivated and wild, witness is borne also by the tales handed down in mythology, as of the olive at Athens,[3] the palm in Delos,[4] and the wild olive at Olympia, from which the wreaths for the games are made; or again of the Valonia oaks at Ilium, planted on the tomb of Ilos. Again some say that Agamemnon planted the plane at Delphi, and the one at Kaphyai[5] in Arcadia. Now how this is may perhaps be another story, but anyhow it is plain that there is a great difference between trees in this respect; the kinds that have been mentioned, and many others besides, are long-lived, while the following are admittedly short-lived—pomegranate fig apple: and among apples the 'spring' sort and the 'sweet' apple are shorter-lived than the 'sour' apple, even as the 'stoneless' pomegranate is shorter-lived than the other kinds. Also some kinds of vine are short-lived, especially those which bear much fruit; and it appears that trees which grow by water are shorter-

[4] Under which Leto gave birth to Artemis and Apollo: *cf.* Paus. 8. 48. 3; Cic. *de Leg.* 1. 1.: Plin. 16. 238.

[5] Its planting is ascribed to Menelaus by Paus. 8. 23. 3.

τῶν ἐν τοῖς ξηροῖς εἶναι, οἷον ἰτέα λεύκη ἀκτὴ αἴγειρος.

3 Ἔνια δὲ γηράσκει μὲν καὶ σήπεται ταχέως, παραβλαστάνει δὲ πάλιν ἐκ τῶν αὐτῶν, ὥσπερ αἱ δάφναι καὶ αἱ μηλέαι τε καὶ αἱ ῥόαι καὶ τῶν φιλύδρων τὰ πολλά· περὶ ὧν καὶ σκέψαιτ' ἄν τις πότερα ταὐτὰ δεῖ λέγειν ἢ ἕτερα· καθάπερ εἴ τις τὸ στέλεχος ἀποκόψας, ὥσπερ ποιοῦσιν οἱ γεωργοί, πάλιν ἀναθεραπεύοι τοὺς βλαστούς, ἢ εἰ καὶ ὅλως ἐκκόψειεν ἄχρι τῶν ῥιζῶν καὶ ἐπικαύσειεν· καὶ γὰρ ταῦτα ποιοῦσιν, ὁτὲ δὲ καὶ ἀπὸ τοῦ αὐτομάτου συμβαίνει· πότερα δὴ τοῦτο ταὐτὸ δεῖ λέγειν ἢ ἕτερον; ᾗ μὲν γὰρ ἀεὶ τὰ μέρη τας αὐξήσεις καὶ φθίσεις φαίνεται παραλλάττοντα καὶ ἔτι τὰς διακαθάρσεις τὰς ὑπ' αὐτῶν, ταύτῃ μὲν ἂν δόξειε ταὐτὸν εἶναι· τί γὰρ ἂν ἐπὶ τούτων

4 ἢ ἐκείνων διαφέροι; ᾗ δ' ὥσπερ οὐσία καὶ φύσις τοῦ δένδρου μάλιστ' ἂν φαίνοιτο τὸ στέλεχος, ὅταν μεταλλάττῃ τοῦτο, κἂν τὸ ὅλον ἕτερον ὑπολάβοι τις, εἰ μὴ ἄρα διὰ τὸ ἀπὸ τῶν αὐτῶν ἀρχῶν εἶναι ταὐτὸ θείη· καίτοι πολλάκις συμβαίνει καὶ τὰς ῥίζας ἑτέρας εἶναι καὶ μεταβάλλειν τῶν μὲν σηπομένων τῶν δ' ἐξ ἀρχῆς βλαστανουσῶν. ἐπεί, ἐὰν ἀληθὲς ᾖ, ὥς γέ τινές φασι, τὰς ἀμπέλους μακρο-

---

1 *cf.* *C.P.* 2. 11. 5.

2 ἀναθεραπεύοι conj. W.; ἀναθεραπεύει Ald.

3 ἢ εἰ καὶ ὅλως conj. W.; ἃ εἰ καὶ καλῶς U; ἀεὶ καὶ καλῶς MV; καὶ εἰ καλῶς Ald.H.

4 Sc. and then encourage new growth.

lived than those which live in dry places: this is true of willow abele elder and black poplar.

Some trees, though they grow old and decay quickly, shoot up again from the same stock,[1] as bay apple pomegranate and most of the water-loving trees. About these one might enquire whether one should call the new growth the same tree or a new one; to take a similar case, if, after cutting down the trunk, one should, as the husband-men do, encourage[2] the new shoots to grow again, or if[3] one should cut the tree right down to the roots and burn the stump,[4] (for these things are commonly done, and they also sometimes occur naturally); are we then here too, to call the new growth the same tree, or another one? In so far as it is always the parts of the tree which appear to alternate their periods of growth and decay and also the prunings which they themselves thus make, so far the new and the old growth might seem to be the same tree; for what difference can there be in the one as compared with the other?[5] On the other hand, in so far as the trunk would seem to be above all the essential part of the tree, which gives it its special character, when this changes, one might suppose that the whole tree becomes something different—unless indeed one should lay down that to have the same starting-point constitutes identity; whereas it often[6] happens that the roots too are different and undergo a change, since some decay and others grow afresh.[7] For if it be true, as some assert, that the reason why the vine is the longest

[5] *i.e.* how can the substitution of one set of 'parts' for another destroy the identity of the tree as a whole?

[6] πολλάκις conj. Sch. from G; πολλὰ καὶ Ald. H.

[7] And so the 'starting-point' too is not constant.

βιωτάτας εἶναι τῷ μὴ φύειν ἑτέρας ἀλλ' ἐξ αὐτῶν
ἀεὶ συναναπληροῦσθαι, γελοῖον ἂν ἴσως δοκοίη τοιαύτη σύγκρισις ἐὰν <μὴ> μένῃ τὸ στέλεχος· αὕτη
γὰρ οἷον ὑπόθεσις καὶ φύσις δένδρων. τοῦτο μὲν
οὖν ὁποτέρως ποτὲ λεκτέον οὐθὲν ἂν διενέγκαι
5 πρὸς τὰ νῦν. τάχα δ' ἂν εἴη μακροβιώτατον τὸ
πάντως δυνάμενον ἀνταρκεῖν, ὥσπερ ἡ ἐλάα καὶ
τῷ στελέχει καὶ τῇ παραβλαστήσει καὶ τῷ
δυσωλέθρους ἔχειν τὰς ῥίζας. δοκεῖ δὲ ὁ βίος
τῆς γε μιᾶς εἶναι, καθ' ὃν τὸ στέλεχος δεῖ τὴν
ἀρχὴν τιθέντα μέτρον ἀναμετρεῖν τὸν χρόνον,
μάλιστα περὶ ἔτη διακόσια. εἰ δ' ὅπερ ἐπὶ τῶν
ἀμπέλων λέγουσί τινες, ὡς παραιρουμένων τῶν
ῥιζῶν κατὰ μέρος δύναται διαμένειν τὸ στέλεχος,
καὶ ἡ ὅλη φύσις ὁμοία καὶ ὁμοιοφόρος ὁποσονοῦν
χρόνον, μακροβιώτατον ἂν εἴη πάντων. φασὶ δὲ
δεῖν οὕτω ποιεῖν ὅταν ἤδη δοκῇ καταφέρεσθαι·
κλήματά τε ἐπιβάλλειν καὶ καρποῦσθαι τὸν
ἐνιαυτόν· μετὰ δὲ ταῦτα κατασκάψαντα ἐπὶ
θάτερα τῆς ἀμπέλου περικαθᾶραι πάσας τὰς
ῥίζας, εἶτ' ἐμπλῆσαι φρυγάνων καὶ ἐπαμήσασθαι
6 τὴν γῆν· τούτῳ μὲν οὖν τῷ ἔτει κακῶς φέρειν
σφόδρα, τῷ δ' ὑστέρῳ βέλτιον, τῷ δὲ τρίτῳ καὶ

---

1 ἐξ αὐτῶν Ald., sc. τῶν ῥιζῶν; ἐκ τῶν αὐτῶν conj. W.

2 *i.e.* such an argument practically assumes the permanence of the trunk, which in the case of the vine can hardly be considered apart from the root. δοκοίη τοιαύτη σύγκρισις I conj. from G; δικαιοτάτη σύγκρισις MVAld.; δικαιοτάτῃ συγκρίσεις U; δοκοίη εἶναι ἡ σύγκρισις conj. Sch.; so W. in his earlier edition: in his later editions he emends wildly.

lived of trees, is that, instead of producing new roots, it always renews itself from the existing ones,[1] such an illustration must surely lead to an absurd conclusion,[2] unless [3] we assume that the stock persists, as it must do, since it is, as it were, the fundamental and essential part of a tree. However it cannot matter much for our present purpose which account is the right one. Perhaps we may say that the longest-lived tree is that which in all ways is able to persist,[4] as does the olive by its trunk, by its power of developing sidegrowth, and by the fact that its roots are so hard to destroy. It appears that the life of the individual olive (in regard to which one should make the trunk the essential part and standard [5] in estimating the time), lasts for about two hundred years.[6] But if it is true of the vine, as some say, that, if the roots are partly removed, the trunk is able to survive, and the whole character of the tree remains the same and produces like fruits for any period, however long, then the vine will be the longest-lived of all trees. They say that, when the vine seems to be deteriorating, this is what one should do:—one should encourage the growth of branches and gather the fruit that year; and after that one should dig on one side of the vine and prune away all the roots on that side, and then fill the hole with brushwood and heap up the soil. In that year, they say, the vine bears very badly, but better in the next, while in the

[3] I have inserted μὴ, which G seems to have read.
[4] ἀνταρκεῖν U, *cf.* Ar. *Eq.* 540; αὐταρκεῖν Ald.
[5] καθ' ὃν τὸ στέλεχος δεῖ τὴν ἀρχὴν τιθέντα I conj.; so G; καθ' ὃν στέλεχος ἤδη τὴν ἀρχὴν τιθέντα μέτρον Ald.H.; εἰ δεῖ for ἤδη U; καθ' ὃ τοῦ στελέχους δεῖ τὸν ὄγκον τιθέντα μέτρον conj. W.; καθ' ὃν τὸ στ. ἤδη ἀρχὴν καὶ μέτρον χρὴ conj. Sch. *cf.* end of § 4. [6] Plin. 16. 241.

τετάρτῳ καθίστασθαι καὶ φέρειν πολλοὺς καὶ καλούς, ὥστε μηδὲν διαφέρειν ἢ ὅτε ἤκμαζεν· ἐπειδὰν δὲ πάλιν ἀποπληγῇ, θάτερον μέρος παρασκάπτειν καὶ θεραπεύειν ὁμοίως, καὶ οὕτως αἰεὶ διαμένειν· ποιεῖν δὲ τοῦτο μάλιστα δι' ἐτῶν δέκα· δι' ὃ καὶ κόπτειν οὐδέποτε τοὺς τοῦτο ποιοῦντας, ἀλλ' ἐπὶ γενεὰς πολλὰς ταὐτὰ τὰ στελέχη διαμένειν, ὥστε μηδὲ μεμνῆσθαι τοὺς φυτεύσαντας· τοῦτο μὲν οὖν ἴσως τῶν πεπειραμένων ἀκούοντα δεῖ πιστεύειν. τὰ δὲ μακρόβια καὶ βραχύβια διὰ τῶν εἰρημένων θεωρητέον.

XIV. Νοσήματα δὲ τοῖς μὲν ἀγρίοις οὔ φασι ξυμβαίνειν ὑφ' ὧν ἀναιροῦνται, φαύλως δὲ διατίθεσθαι καὶ μάλιστα ἐπιδήλως ὅταν χαλαζοκοπηθῇ ἢ βλαστάνειν μέλλοντα ἢ ἀρχόμενα ἢ ἀνθοῦντα, καὶ ὅταν ἢ πνεῦμα ψυχρὸν ἢ θερμὸν ἐπιγένηται κατὰ τούτους τοὺς καιρούς. ὑπὸ δὲ τῶν ὡραίων χειμώνων οὐδὲ ἂν ὑπερβάλλοντες ὦσιν οὐδὲν πάσχειν, ἀλλὰ καὶ ξυμφέρειν πᾶσι χειμασθῆναι· μὴ χειμασθέντα γὰρ κακοβλαστότερα γίνεσθαι.

2 τοῖς δὲ ἡμέροις ἐστὶ πλείω νοσήματα, καὶ τὰ μὲν ὥσπερ κοινὰ πᾶσιν ἢ τοῖς πλείστοις τὰ δ' ἴδια κατὰ γένη. κοινὰ δὴ τό τε σκωληκοῦσθαι καὶ ἀστροβολεῖσθαι καὶ ὁ σφακελισμός. ἅπαντα γὰρ ὡς εἰπεῖν καὶ σκώληκας

[1] ἀποπληγῇ: ἀπολήγῃ conj. Sch.
[2] Plin. 17. 216. [3] cf. C.P. 5. 8. 3.
[4] κατὰ γένη conj. W.; καὶ τὰ γένη UMV; καὶ κατὰ γένη Ald.

third and fourth it becomes normal again and bears many fair clusters, so that it is quite as good as when it was in its prime. And when it goes off again,[1] they say one should dig on the other side and apply the same treatment; and that so treated the tree lasts for ever; and this should be done at intervals of about ten years. And this is why those who adopt this treatment never cut down the vine, but the same stems remain for many generations, so that even those who planted the trees cannot remember doing so. However perhaps one should enquire of those who have had experience before accepting this statement. These examples may serve for considering which trees are long-lived and which short-lived.

Of diseases and injuries done by weather conditions.

XIV. [2] As to diseases—they say that wild trees are not liable to diseases which destroy them, but that they get into poor condition, and that most obviously when they are smitten with hail when either they are about to bud or are just budding or are in bloom; also when either a cold or a hot wind comes at such seasons: but that from seasonable storms, even if they be violent, they take no hurt,[3] but rather that it is good for them all to be exposed to weather: for, unless they are, they do not grow so well. Cultivated kinds however, they say, are subject to various diseases, some of which are, one may say, common to all or to most, while others are special to particular kinds.[4] General diseases are those[5] of being worm-eaten, of being sun-scorched, and rot.[6] All trees, it may be said,

[5] *κοινὰ δὴ τό τε* conj. W.; *κοινὰ καὶ τότε* UMV; *κοινά· οἷον τότε* Ald.H. [6] *cf.* 8. 10. 1.

ἴσχει πλὴν τὰ μὲν ἐλάττους τὰ δὲ πλείους, καθάπερ συκῆ μηλέα καὶ ἄπιος. ὡς δὲ ἁπλῶς εἰπεῖν ἥκιστα σκωληκοῦνται τὰ δριμέα καὶ ὀπώδη, καὶ ἀστροβολεῖται ὡσαύτως· μᾶλλον δὲ τοῖς νέοις ἢ τοῖς ἐν ἀκμῇ τοῦτο συμβαίνει, πάντων δὲ μάλιστα τῇ τε συκῇ καὶ τῇ ἀμπέλῳ.

3 Ἡ δ' ἐλάα πρὸς τῷ τοὺς σκώληκας ἴσχειν, οἳ δὴ καὶ τὴν συκῆν διαφθείρουσιν ἐντίκτοντες, φύει καὶ ἧλον· οἱ δὲ μύκητα καλοῦσιν, ἔνιοι δὲ λοπάδα· τοῦτο δ' ἐστὶν οἷον ἡλίου καῦσις. διαφθείρονται δ' ἐνίοτε καὶ αἱ νέαι ἐλάαι διὰ τὴν ὑπερβολὴν τῆς πολυκαρπίας. ἡ δὲ ψώρα καὶ οἱ προσφυόμενοι κοχλίαι συκῆς εἰσιν· οὐ πανταχοῦ δὲ τοῦτο συμβαίνει ταῖς συκαῖς, ἀλλ' ἔοικε καὶ τὰ νοσήματα γίνεσθαι κατὰ τοὺς τόπους, ὥσπερ τοῖς ζῴοις· ἐπεὶ παρ' ἐνίοις οὐ ψωριῶσι, καθάπερ οὐδὲ περὶ τὴν Αἰνείαν.

4 Ἁλίσκεται δὲ συκῆ μάλιστα καὶ σφακελισμῷ καὶ κράδῳ. καλεῖται δὲ σφακελισμὸς μὲν ὅταν αἱ ῥίζαι μελανθῶσι, κράδος δ' ὅταν οἱ κλάδοι· καὶ γὰρ καλοῦσί τινες κράδους, ὅθεν καὶ τοὔνομα τῇ νόσῳ· ὁ δ' ἐρινεὸς οὔτε κραδᾷ οὔτε σφακελίζει οὔτε ψωριᾷ οὔτε σκωληκοῦται ταῖς ῥιζαῖς ὁμοίως· οὐδὲ δὴ τὰ ἐρινά τινες ἀποβάλλουσιν οὐδ' ἐὰν ἐμφυτευθῶσιν εἰς συκῆν.

---

[1] ὀπώδη UMVAld.; εὐώδη H., evidently from Plin. 17. 221. *cf. C.P.* 5. 9. 4 and 5.

[2] λοπάδα: Plin. 17. 223, *patella*. The ἧλος is an abortive bud, called in Italian *novolo*.

[3] ἡλίου καῦσις conj. Scal. from Plin. *l.c. veluti solis exustio*: [illegible] G; ἡλοιαυτον U; ἧλοι αὐτὸν V; ἧλοι αὐτῶν M; ἧλοι αὐτῶν Ald. which W. prints provisionally.

have worms, but some less, as fig and apple, some more, as pear. Speaking generally, those least liable to be worm-eaten are those which have a bitter acrid[1] juice, and these are also less liable to sun-scorch. Moreover this occurs more commonly in young trees than in those which have come to their strength, and most of all it occurs in the fig and the vine.

The olive, in addition to having worms (which destroy the fig too by breeding in it), produces also a 'knot' (which some call a fungus, others a bark-blister[2]), and it resembles the effect of sun-scorch.[3] Also sometimes young olives are destroyed by excessive fruitfulness. The fig is also liable to scab, and to snails which cling to it. However this does not happen to figs everywhere, but it appears that, as with animals, diseases are dependent on local conditions; for in some parts, as about Aineia,[4] the figs do not get scab.

The fig is also often a victim to rot and to *krados*. It is called rot when the roots turn black, it is called *krados* when the branches do so; for some call the branches *kradoi*[5] (instead of *kladoi*), whence the name is transferred to the disease. The wild fig does not suffer from *krados* rot or scab, nor does it get so worm-eaten in its roots[6] as the cultivated tree; indeed some wild figs do not even shed their early fruit—not even if they are grafted[7] into a cultivated tree.

[4] *cf.* 5. 2. 1. [5] Evidently a dialectic form.

[6] ῥίζαις PAld.; συκαῖς W. after conj. of Sch.

[7] ἐμφυτευθῶσιν conj. Sch.; ἔνι φυτ. UMV; ἔνια φυτ. Ald. Apparently the object of such grafting was the 'caprification' of the cultivated tree (*cf.* 2. 8. 3); but grafting for this purpose does not seem to be mentioned elsewhere.

5 Ἡ δὲ ψώρα μάλιστα γίνεται ὅταν ὕδωρ ἐπὶ Πλειάδι γένηται μὴ πολύ· ἐὰν δὲ πολύ, ἀποκλύζεται· συμβαίνει δὲ τότε καὶ τὰ ἐρινὰ ἀπορρεῖν καὶ τοὺς ὀλύνθους. τῶν δὲ σκωλήκων τῶν ἐν ταῖς συκαῖς οἱ μὲν ἐξ αὐτῆς γίνονται οἱ δὲ ἐντίκτονται ὑπὸ τοῦ καλουμένου κεράστου· πάντες δὲ εἰς κεράστην ἀποκαθίστανται· φθέγγονται δὲ οἷον τριγμόν. νοσεῖ δὲ συκῆ καὶ ἐὰν ἐπομβρία γένηται· τά τε γὰρ πρὸς τὴν ῥίζαν καὶ αὐτὴ ἡ ῥίζα ὥσπερ μαδᾷ· τοῦτο δὲ καλοῦσι λοπᾶν.
6 ἡ δ' ἄμπελος τραγᾷ· τοῦτο δὲ μάλιστα αὐτῆς ἐστι πρὸς τῷ ἀστροβολεῖσθαι, ἢ ὅταν ὑπὸ πνευμάτων βλαστοκοπηθῇ ἢ ὅταν τῇ ἐργασίᾳ συμπάθῃ ἢ τρίτον ὑπτία τμηθῇ.

Ῥυὰς δὲ γίνεται, ὃ καλοῦσί τινες ψίνεσθαι, ὅταν ἐπινιφθῇ κατὰ τὴν ἀπάνθησιν ἢ ὅταν κρειττωθῇ· τὸ δὲ πάθος ἐστὶν ὥστε ἀπορρεῖν τὰς ῥᾶγας καὶ τὰς ἐπιμενούσας εἶναι μικράς. ἔνια δὲ καὶ ῥιγώσαντα νοσεῖ, καθάπερ ἡ ἄμπελος· ἀμβλοῦνται γὰρ οἱ ὀφθαλμοὶ τῆς πρωτοτόμου· καὶ πάλιν ὑπερθερμανθέντα· ζητεῖ γὰρ καὶ τούτων τὴν συμμετρίαν ὥσπερ καὶ τῆς τροφῆς. ὅλως δὲ πᾶν τὸ παρὰ φύσιν ἐπικίνδυνον.

---

[1] *cf. C.P.* 5. 9. 10; Col. 5. 9. 15.

[2] *cf.* 5. 4. 5; *C.P.* 5. 10. 5; Plin. 17. 221.

[3] αὐτὴ ἡ ῥίζα I conj.; αὐτὴν τὴν ῥίζαν U; om. Ald.

[4] *cf. C.P.* 5. 9. 12; Plin. 17. 225.

[5] *i.e.* shedding of the 'bark' of the roots λοπᾶν conj. Sch., *cf. C.P.* 5. 9. 9; λοπάδα Ald.H., *cf.* 4. 14. 3; but the word here points to a different disease.

[6] ὑπτία τομή seems to be a technical term for pruning in such a way that the growth of the new wood is encouraged

Scab[1] chiefly occurs when there is not much rain after the rising of the Pleiad; if rain is abundant, the scab is washed off, and at such times it comes to pass that both the spring and the winter figs drop off. Of the worms found in fig-trees some have their origin in the tree, some are produced in it by the creature called the 'horned worm'; but they all turn into the 'horned worm';[2] and they make a shrill noise. The fig also becomes diseased if there is heavy rain; for then the parts towards the root and the root itself[3] become, as it were, sodden,[4] and this they call 'bark-shedding.'[5] The vine suffers from over-luxuriance; this, as well as sun-scorch, specially happens to it either when the young shoots are cut by winds, or when it has suffered from bad cultivation, or, thirdly, when it has been pruned upwards.[6]

The vine becomes a 'shedder,'[7] a condition which some call 'casting of the fruit,' if the tree is snowed upon at the time when the blossom falls, or else when it becomes over lusty;[8] what happens is that the unripe grapes drop off, and those that remain on the tree are small. Some trees also contract disease from frost, for instance the vine; for then the eyes of the vine that was pruned early become abortive; and this also happens from excessive heat, for the vine seeks regularity in these conditions too, as in its nourishment. And in general anything is dangerous which is contrary to the normal course of things.

and so there is less fruit: exact sense obscure; ? 'from below' (*i.e.* with the blade of the knife pointing upwards). *cf. C.P. l.c.*; Col. 4. 24. 15; Plin. *l.c.*, *in supinum excisis.*

[7] *cf. C.P.* 5. 9. 13.

[8] κρειττωθῇ: *i.e.* the growth is over-luxuriant. The word occurs elsewhere only in the parallel passage *C.P. l.c.*, where occurs also the subst. κρείττωσις, evidently a technical term.

7 Μεγάλα δὲ ξυμβάλλεται καὶ τὰ τραύματα καὶ αἱ πληγαὶ τῶν περισκαπτόντων εἰς τὸ μὴ φέρειν τὰς μεταβολὰς ἢ καυμάτων ἢ χειμώνων· ἀσθενὲς γὰρ ὂν διὰ τὴν ἕλκωσιν καὶ τὸν πόνον εὐχειρωτότατόν ἐστι ταῖς ὑπερβολαῖς. σχεδὸν δέ, ὥς τινες οἴονται, τὰ πλεῖστα τῶν νοσημάτων ἀπὸ πληγῆς γίνεται· καὶ γὰρ τὰ ἀστρόβλητα καλούμενα καὶ τὰ σφακελίζοντα διὰ τὸ ἀπὸ ταύτης εἶναι τῶν ῥιζῶν τὸν πόνον. οἴονται δὲ καὶ δύο ταύτας εἶναι μόνας νόσους· οὐ μὴν ἀλλὰ τοῦτό γ' οὐκ ἄγαν ὁμολογούμενόν ἐστι.

[Πάντων δ' ἀσθενέστατον ἡ μηλέα ἡ ἠρινὴ καὶ τούτων ἡ γλυκεῖα.]

8 Ἔνιαι δὲ πηρώσεις οὐκ εἰς φθορὰν γίνονται ὅλων ἀλλ' εἰς ἀκαρπίαν· οἷον ἐάν τις τῆς πίτυος ἀφέλῃ τὸ ἄκρον ἢ τοῦ φοίνικος, ἄκαρπα γίνεσθαι ἄμφω δοκεῖ καὶ οὐχ ὅλως ἀναιρεῖσθαι.

Γίνονται δὲ νόσοι καὶ τῶν καρπῶν αὐτῶν, ἐὰν μὴ κατὰ καιρὸν τὰ πνεύματα καὶ τὰ οὐράνια γένηται· συμβαίνει γὰρ ὁτὲ μὲν ἀποβάλλειν γενομένων ἢ μὴ γενομένων ὑδάτων, οἷον τὰς συκᾶς, ὁτὲ δὲ χείρους γίνεσθαι σηπομένους καὶ καταπνιγομένους ἢ πάλιν ἀναξηραινομένους παρὰ τὸ δέον. χείριστον δὲ ἐὰν ἀπανθοῦσί τισιν ἐφύσῃ, καθάπερ ἐλάᾳ καὶ ἀμπέλῳ· συναπορρεῖ γὰρ ὁ καρπὸς δι' ἀσθένειαν.

---

[1] Plin. 17. 227.
[2] εὐχειρωτότατον conj. W. after Lobeck; εὐχειρότατον Ald.
[3] πόνον conj. H. from G; τόπον MVAld.
[4] This sentence is clearly out of place: the plural τούτων has nothing to refer to. *cf.* 4. 13. 2. It is represented however by Plin. *l.c.*

[1] Moreover the wounds and blows inflicted by men who dig about the vines render them less able to bear the alternations of heat and cold; for then the tree is weak owing to the wounding and to the strain put upon it, and falls an easy prey[2] to excess of heat and cold. Indeed, as some think, most diseases may be said to be due to a blow; for that even the diseases known as 'sun-scorch' and 'rot' occur because the roots have suffered in this way.[3] In fact they think that there are only these two diseases; but there is not general agreement on this point.

The 'spring apple' and especially the sweet form of it, has the weakest constitution.[4]

[5] Some mutilations however do not cause destruction of the whole[6] tree, but only produce barrenness; for instance, if one takes away the top of the Aleppo pine or the date-palm, the tree in both cases appears to become barren, but not to be altogether destroyed.

There are also diseases of the fruits themselves, which occur if the winds and rains do not come in due season. For it comes to pass[7] that sometimes trees, figs, for example, shed their fruit when rain does or does not come, and[8] sometimes the fruit is spoilt by being rotted and so choked off,[9] or again by being unduly dried up. It is worst of all for some trees, as olive and vine, if rain falls on them as they are dropping their blossom;[10] for then the fruit, having no strength, drops also.

[5] Plin. 17. 228 and 229.
[6] ὅλων conj. W.; τινων P_2Ald.H. *cf. C.P.* 5. 17. 3 and 6.
[7] *cf. C.P.* 5. 10. 5.
[8] δὲ add. Sch. [9] *cf. C.P. l.c.*
[10] ἀπανθοῦσι conj. Sch. from G and Plin. *l.c.*; ἐπανθοῦσι Ald.H.

9 Ἐν Μιλήτῳ δὲ τὰς ἐλάας, ὅταν ὦσι περὶ το ἀνθεῖν, κάμπαι κατεσθίουσιν, αἱ μὲν τὰ φύλλα αἱ δὲ τὰ ἄνθη, ἕτεραι τῷ γένει, καὶ ψιλοῦσι τὰ δένδρα· γίνονται δὲ ἐὰν ᾖ νότια καὶ εὐδιεινά· ἐὰν δὲ ἐπιλάβῃ καύματα ῥήγνυνται.

Περὶ δὲ Τάραντα προφαίνουσι μὲν ἀεὶ πολὺν καρπόν, ὑπὸ δὲ τὴν ἀπάνθησιν τὰ πολλ᾽ ἀπόλλυται. τὰ μὲν οὖν τοιαῦτα τῶν τόπων ἴδια.

10 Γίνεται δὲ καὶ ἄλλο νόσημα περὶ τὰς ἐλάας ἀράχνιον καλούμενον· φύεται γὰρ τοῦτο καὶ διαφθείρει τὸν καρπόν. ἐπικάει δὲ καὶ καύματά τινα καὶ ἐλάαν καὶ βότρυν καὶ ἄλλους καρπούς. οἱ δὲ καρποὶ σκωληκοῦνταί τινων, οἷον ἐλάας ἀπίου μηλέας μεσπίλης ῥόας. καὶ ὅ γε τῆς ἐλάας σκώληξ ἐὰν μὲν ὑπὸ τὸ δέρμα γένηται διαφθείρει τὸν καρπόν, ἐὰν δὲ τὸν πυρῆνα διαφάγῃ ὠφελεῖ. κωλύεται δὲ ὑπὸ τῷ δέρματι εἶναι ὕδατος ἐπ᾽ Ἀρκτούρῳ γενομένου. γίνονται δὲ καὶ ἐν ταῖς δρυπεπέσι σκώληκες, αἵπερ καὶ χείρους εἰς τὴν ῥύσιν· ὅλως δὲ καὶ δοκοῦσιν εἶναι σαπραί· δι᾽ ὃ καὶ γίνονται τοῖς νοτίοις καὶ μᾶλλον ἐν τοῖς ἐφύδροις. ἐγγίνονται δὲ καὶ κνῖπες ἔν τισι τῶν δένδρων, ὥσπερ ἐν τῇ δρυῒ καὶ τῇ συκῇ· καὶ δοκοῦσιν ἐκ τῆς ὑγρότητος συνίστασθαι τῆς ὑπὸ τὸν φλοιὸν συνισταμένης· αὕτη δέ ἐστι γλυκεῖα γευομένοις. γίνονται δὲ καὶ ἐν λαχάνοις τισίν,

[1] *cf. C.P.* 5. 10. 3.

[2] Tarentum: *cf. C.P. l.c.*

[3] ἀπάνθησιν conj. W.; ἄνθησιν Ald.

[4] Plin. 17. 229–231.

[5] ἀράχνιον conj. Sch. after Meurs.; ἀρίχνιον UP₂; ἀρχίχνιον MVP; ἀρχίνιον Ald. *cf. C.P.* 5. 10. 2.

[1] In Miletus the vines at the time of flowering are eaten by caterpillars, some of which devour the flowers, others, a different kind, the leaves; and they strip the tree; these appear if there is a south wind and sunny weather; if the heat overtakes them, the trees split.

About Taras[2] the olives always shew much fruit, but most of it perishes at the time when the blossom falls.[3] Such are the drawbacks special to particular regions.

[4] There is also another disease incident to the olive, which is called cobweb;[5] for this forms[6] on the tree and destroys the fruit. Certain hot[7] winds also scorch both olive vine-cluster and other fruits. And the fruits of some get worm-eaten,[8] as olive pear apple medlar pomegranate. Now the worm which infests the olive, if it appears below the skin, destroys the fruit; but if it devours the stone it is beneficial. And it is prevented from appearing under the skin if there is rain after[9] the rising of Arcturus. Worms also occur in the fruit which ripens on the tree, and these are more harmful as affecting the yield of oil. Indeed these worms seem to be altogether rotten; wherefore they appear when there is a south wind and particularly in damp places. The *knips*[10] also occurs in certain trees, as the oak and fig, and it appears that it forms from the moisture which collects under the bark, which is sweet to the taste. Worms also occur[11] in some

[6] φύεται Ald.; ἐμφύεται conj. Sch. from *C.P. l.c.*, but the text is perhaps defective.
[7] *cf. C.P.* 5. 10. 5.
[8] *cf. C.P.* 5. 10. 1.
[9] ἐπ' conj. Sch., *cf. C.P.* 5. 10. 1; ὑπ' U; ἀπ' Ald.H.
[10] *cf.* 2. 8. 3.
[11] The subject of γίνονται is probably σκώληκες, not κνῖπες.

ἔνθα δὲ κάμπαι διαφερούσης δῆλον ὅτι τῆς ἀρχῆς.

11 Καὶ τὰ μὲν νοσήματα σχεδὸν ταῦτα καὶ ἐν τούτοις ἐστίν. ἔνια δὲ πάθη τῶν κατὰ τὰς ὥρας καὶ τῶν κατὰ τοὺς τόπους γινομένων ἀναιρεῖν πέφυκεν, ἃ οὐκ ἄν τις εἴποι νόσους, οἷον λέγω τὴν ἔκπηξιν καὶ ὃ καλοῦσί τινες καυθμόν. ἄλλα δὲ παρ' ἑκάστοις πέφυκε πνεύματα ἀπολλύναι καὶ ἀποκάειν· οἷον ἐν Χαλκίδι τῆς Εὐβοίας Ὀλυμπίας ὅταν πνεύσῃ μικρὸν πρὸ τροπῶν ἢ μετὰ τροπὰς χειμερινὰς ψυχρός· ἀποκάει γὰρ τὰ δένδρα καὶ οὕτως αὖα ποιεῖ καὶ ξηρὰ ὡς οὐδ' ἂν ὑφ' ἡλίου καὶ χρόνου πολλοῦ γένοιτ' ἄν, δι' ὃ καὶ καλοῦσι καυθμόν· ἐγένετο δὲ πρότερον πολλάκις ἤδη καὶ ἐπ' Ἀρχίππου δι' ἐτῶν τετταράκοντα σφοδρός.

12 Πονοῦσι δὲ μάλιστα τῶν τόπων οἱ κοῖλοι καὶ οἱ αὐλῶνες καὶ ὅσοι περὶ τοὺς ποταμοὺς καὶ ἁπλῶς οἱ ἀπνευστότατοι· τῶν δένδρων δὲ μάλιστα συκῆ, δεύτερον δὲ ἐλάα. ἐλάας δὲ μᾶλλον ὁ κοτινος ἐπόνησεν ἰσχυρότερος ὤν, ὃ καὶ θαυμαστὸν ἦν· αἱ δὲ ἀμυγδαλαῖ τὸ πάμπαν ἀπαθεῖς· ἀπαθεῖς δὲ καὶ αἱ μηλέαι καὶ αἱ ἄπιοι καὶ αἱ ῥόαι ἐγένοντο· δι' ὃ καὶ τοῦτο ἦν θαυμαστόν. ἀποκάεται δὲ εὐθὺς ἐκ τοῦ στελέχους, καὶ ὅλως δὲ μᾶλλον καὶ πρότερον ὡς εἰπεῖν ἅπτεται <τὰ ἄνω> τῶν κάτω. φανερὰ δὲ γίνεται τὰ μὲν ἅμα περὶ τὴν βλάστησιν,

---

[1] Plin. 17. 232.

[2] τῶν κατὰ τοὺς τόπους conj. Sch. from Plin. *l.c.*; τῶν καθ' αὑτὰ Ald.

[3] ἔκπηξιν conj. Sch.; ἔκπληξιν UMP$_2$Ald. *cf. C.P.* 5. 12. 2, πῆξις.

[4] *cf. C.P.* 5. 12. 4.

pot-herbs, as also do caterpillars, though the origin of these is of course different.

Such are in general the diseases, and the plants in which they occur. Moreover[1] there are certain affections due to season or situation[2] which are likely to destroy the plant, but which one would not call diseases: I mean such affections as freezing[3] and what some call 'scorching.' Also[4] there are winds which blow in particular districts that are likely to destroy or scorch; for instance the 'Olympian' wind of Chalcis in Euboea, when it blows cold a little before or after the winter solstice; for this wind scorches up the trees and makes them more dry and withered than they would become from the sun's heat even in a long period; wherefore its effect is called 'scorching.' In old times it occurred very frequently, and it recurred with great violence in the time of Archippus, after an interval of forty years.

[5]The places which suffer most in this way are hollow places, valleys, the ground near rivers, and, in general, places which are least open to wind; the tree which suffers most is the fig, and next to that the olive. The wild olive, being stronger, suffered more than the cultivated tree, which was surprising. But the almonds were altogether unscathed, as also were the apples pears and pomegranates; wherefore this too was a surprising fact. The tree gets scorched by this wind right down to the trunk, and in general the upper are caught more and earlier than the lower parts.[6] The effects are seen partly at the actual

[5] *cf. C.P.* 5. 12. 7; Plin. 17. 232 and 233.

[6] κάτω UMVP; ἄνω W. after Sch.'s conj.: text probably defective; I have added τὰ ἄνω. *cf. C.P.* 5. 12. 5.

ἡ δὲ ἐλάα διὰ τὸ ἀείφυλλον ὕστερον· ὅσαι μὲν οὖν ἂν φυλλοβολήσωσιν ἀναβιώσκονται πάλιν, ὅσαι δ' ἂν μὴ τελέως ἀπόλλυνται. παρ' ἐνίοις δέ τινες ἀποκαυθεῖσαι καὶ τῶν φύλλων αὐανθέντων ἀνεβλάστησαν πάλιν ἄνευ τοῦ ἀποβαλεῖν καὶ τὰ φύλλα ἀνεβίωσεν. ἐνιαχοῦ δὲ καὶ πολλάκις τοῦτο συμβαίνει, καθάπερ καὶ ἐν Φιλίπποις.

13 Τὰ δ' ἐκπαγέντα, ὅταν μὴ τελέως ἀπόληται, τάχιστα ἀναβλαστάνει, ὥστε εὐθὺς τὴν ἄμπελον καρποφορεῖν, ὥσπερ ἐν Θετταλίᾳ. ἐν δὲ τῷ Πόντῳ περὶ Παντικάπαιον αἱ μὲν ἐκπήξεις γίνονται διχῶς, ὁτὲ μὲν ὑπὸ ψύχους ἐὰν χειμέριον ᾖ τὸ ἔτος, ὁτὲ δὲ ὑπὸ πάγων ἐάν γε πολὺν χρόνον διαμένωσι. ἀμφότερα δὲ μάλιστα γίγνονται μετὰ τροπὰς περὶ τὰς τετταράκοντα. γίνονται δὲ οἱ μὲν πάγοι ταῖς αἰθρίαις, τὰ δὲ ψύχη μάλιστα ὑφ' ὧν ἡ ἔκπηξις ὅταν αἰθρίας οὔσης αἱ λεπίδες καταφέρωνται. ταῦτα δ' ἐστὶν ὥσπερ τὰ ξύσματα πλὴν πλατύτερα, καὶ φερόμενα φανερὰ πεσόντα δὲ οὐ διαμένει· περὶ δὲ τὴν Θρᾴκην ἐκπήγνυνται.

14 Ἀλλὰ γὰρ αἱ μὲν νόσοι πόσαι τε καὶ ποῖαι καὶ τίνες γίνονται καὶ πάλιν αἱ δι' ὑπερβολὴν χειμῶνος ἢ καυμάτων φθοραὶ καὶ αἱ διὰ πνευμάτων ψυχρότητα ἢ θερμότητα διὰ τούτων θεωρείσθωσαν· ὧν ἐνίας οὐθὲν ἂν κωλύοι καὶ τοῖς ἀγρίοις εἶναι κοινὰς καὶ κατὰ τὴν ὅλην τῶν δένδρων φθορὰν καὶ ἔτι μᾶλλον κατὰ τὴν τῶν καρπῶν· ὃ καὶ συμβαῖνον ὁρῶμεν· οὐκ εὐκαρπεῖ

---

[1] Plin. 17. 233.

[2] ἐκπαγέντα conj. Sch.; ἐκπλαγέντα U; ἐκπληγέντα Ald.

[3] ἐάν γε conj. Sch.; ἐὰν δὲ U; ἐὰν π. χ. δ. γε Ald.

time of budding, but in the olive, because it is evergreen, they do not appear till later; those trees therefore which have shed their leaves come to life again, but those that have not done so are completely destroyed. In some places trees have been known, after being thus scorched and after their leaves have withered, to shoot again without shedding their leaves, and the leaves have come to life again. Indeed in some places, as at Philippi, this happens several times.

[1] Trees which have been frost-bitten,[2] when they are not completely destroyed, soon shoot again, so that the vine immediately bears fruit, for instance in Thessaly. In Pontus near Panticapaeum the frost-bite occurs in two ways, either just from cold, if the season is wintry, or from long[3] spells of frost; in either case this generally occurs in the[4] forty days after the winter solstice. The frosts occur in fine weather, but the cold spells, which cause the frost-bite, chiefly when in fine weather the 'flakes'[5] fall; these are like filings, but broader, and can be seen as they fall, but when they have fallen, they disappear—though in Thrace they freeze solid.

Let this suffice for consideration of the diseases, their number and nature, including the fatal effects of excessive cold and heat or of cold or hot winds. And it may well be that certain of these also affect wild trees, producing entire destruction of the tree and still more that of the fruit. Indeed we see this actually happen; for wild trees also often fail to

[4] περὶ conj. Sch., *cf.* *C.P.* 5. 12. 4; μετὰ UMVAld.
[5] λεπίδες conj. Scal. from G (*squammulae*); ῥεπίδες Ald. *cf.* Hdt. 4. 31.

γὰρ οὐδ' ἐκεῖνα πολλάκις, ἀλλ' οὐχ ὁμοίως οἶμαι παρατετήρηται.

XV. Λοιπὸν δ' εἰπεῖν ὅσα παραιρουμένων τινῶν μορίων ἀπόλλυται. κοινὴ μὲν δὴ πᾶσι φθορὰ τοῦ φλοιοῦ περιαιρεθέντος κύκλῳ· πᾶν γὰρ ὡς εἰπεῖν οὕτως ἀπόλλυσθαι δοκεῖ πλὴν ἀνδράχλη· καὶ αὕτη δὲ ἐάν τις τὴν σάρκα σφόδρα πιέσῃ καὶ τὸν μέλλοντα βλαστὸν διακόψῃ· πλὴν εἰ ἄρα φελλοῦ· τοῦτον γάρ φασι καὶ εὐσθενεῖν μᾶλλον περιαιρουμένου δῆλον ὅτι τοῦ ἔξω καὶ τοῦ κάτω πρὸς τῇ σαρκί, καθάπερ καὶ τῆς ἀνδράχλης. ἐπεὶ καὶ τοῦ κεράσου περιαιρεῖται καὶ τῆς ἀμπέλου καὶ τῆς φιλύρας, ἐξ οὗ τὰ σχοινία, καὶ μαλάχης τῶν ἐλαττόνων, ἀλλ' οὐχ ὁ κύριος οὐδ' ὁ πρῶτος, ἀλλ' ὁ ἐπιπολῆς, ὃς καὶ αὐτόματος ἐνίοτε ἀποπίπτει διὰ τὴν ὑπόφυσιν θατέρου.

2 Καὶ γὰρ φλοιορραγῆ ἔνια τῶν δένδρων ἐστίν, ὥσπερ καὶ ἡ ἀνδράχλη καὶ ἡ πλάτανος. ὡς δέ τινες οἴονται, πάλιν ὑποφύεται νέος, ὁ δὲ ἔξωθεν ἀποξηραίνεται καὶ ῥήγνυται καὶ αὐτόματος ἀποπίπτει πολλῶν, ἀλλ' οὐχ ὁμοίως ἐπίδηλος. φθείρονται μὲν οὖν' ὡς οἴονται, πάντα περιαιρουμένου, διαφέρει δὲ τῷ θᾶττον καὶ βραδύτερον καὶ

---

[1] Plin. 17. 234; *cf. C.P.* 5. 15. 1.
[2] *cf.* 1. 5. 2.
[3] βλαστὸν conj. Sch. from G; καρπὸν UAld.H.
[4] Plin. 17. 234–236.

produce a good crop of fruit; but, I imagine, they have not been so well observed.

Of the effects on trees of removing bark, head, heart-wood, roots, etc.; of various causes of death.

XV. 1 Next we must mention what trees perish when certain parts are removed. All perish alike, if the bark is stripped off all round; one may say that every tree, except the andrachne,[2] perishes under these circumstances; and this tree does so also, if one does violence to the flesh, and so breaks off the new growth[3] which is forming. However one should perhaps except the cork-oak; for this, they say, is all the stronger if its bark is stripped off, that is, the outer bark and also that which lies below it next the flesh—as with the andrachne. For the bark is also stripped from the bird-cherry the vine and the lime (and from this the ropes are made), and, among smaller plants, from the mallow; but in these cases it is not the real nor the first bark which is taken, but that which grows above that, which even of its own accord sometimes falls off because fresh bark is forming underneath.

4 In fact some trees, as andrachne and plane, have a bark which cracks.[5] As some think, in many cases a new bark forms[6] underneath, while the outer bark withers and cracks and in many cases falls off of its own accord; but the process is not so obvious as it is in the above mentioned cases. Wherefore, as they think, all trees are destroyed by stripping the bark, though the destruction is not in all cases equally

[5] *cf. C.P.* 3. 18. 3. φλοιορραγῆ ἔνια conj. Mold.; φλοιορραγία μία UMV; φυλλορογία μία Ald.
[6] ὑποφύεται conj. W.; ὑποφύει Ald.H.

μᾶλλον και ἧττον. ἔνια γὰρ πλείω χρόνον διαμένει, καθάπερ συκῆ καὶ φίλυρα καὶ δρῦς· οἱ δὲ
καὶ ζῆν φασι ταῦτα, ζῆν δὲ καὶ πτελέαν καὶ
φοίνικα· τῆς δὲ φιλύρας καὶ συμφύεσθαι τὸν
φλοιὸν πλὴν μικροῦ· τῶν δὲ ἄλλων οἷον πωροῦ-
σθαι καὶ ἰδίαν τινὰ φύσιν ἔχειν. βοηθεῖν δὲ
πειρῶνται διαπλάττοντες πηλῷ καὶ περιδοῦντες
φλοιοῖς καὶ καλάμοις καὶ τοῖς τοιούτοις, ὅπως μὴ
ψύχηται μηδ' ἀποξηραίνηται. καὶ ἤδη φασί που
ἀναφῦναι, καθάπερ καὶ ἐν Ἡρακλείᾳ τῇ Τραχινίᾳ,
3 τὰς συκᾶς. δεῖ δὲ ἅμα τῇ τῆς χώρας ἀρετῇ καὶ
τῇ τοῦ ἀέρος κράσει καὶ τὰ ἐπιγιγνόμενα τοιαῦτα
εἶναι· χειμώνων γὰρ ἢ καυμάτων ἐπιγινομένων
σφοδρῶν εὐθὺς ἀπόλλυνται· διαφέρουσι δὲ καὶ
αἱ ὧραι· περὶ γὰρ τὴν βλάστησιν ἐλάτης ἢ
πεύκης, ὅτε καὶ λοπῶσι, τοῦ Θαργηλιῶνος ἢ
Σκιρροφοριῶνος ἄν τις περιέλῃ, παραχρῆμα ἀπόλλυται. τοῦ δὲ χειμῶνος πλείω χρόνον ἀντ-
έχει καὶ ἔτι μᾶλλον τὰ ἰσχυρότατα, καθάπερ πρῖνος καὶ δρῦς· χρονιωτέρα γὰρ ἡ τούτων φθορά.
4 δεῖ δὲ καὶ τὴν περιαίρεσιν ἔχειν τι πλάτος,
πάντων μὲν μάλιστα δὲ τῶν ἰσχυροτάτων· ἐπεὶ
ἄν τις μικρὰν παντελῶς ποιήσῃ, οὐθὲν ἄτοπον τὸ
μὴ ἀπόλλυσθαι· καίτοι φασί γέ τινες, ἐὰν ὁποσονοῦν, συμφθείρεσθαι πάντως· ἀλλ' ἐπὶ τῶν
ἀσθενεστέρων τοῦτ' εἰκός. ἔνια γὰρ κἂν μὴ
κύκλῳ περιαιρεθῇ φθείρεσθαί φασιν, ἃ καὶ

[1] καὶ add. W. (text defective in MSS. except U).

rapid or complete. Some in fact, as fig lime and oak, survive for some time; indeed some say that these recover, and also the elm and date-palm, and that the bark even of the lime almost entirely closes up again, while in other trees it forms as it were a callus and[1] acquires a peculiar new character. Men try to help the tree by plastering it with mud and tying pieces of bark reeds or something of the kind about it, so that it may not take cold nor become dried up. And they say that the bark has been known to grow again;[2] for instance that that of the fig-trees at the Trachinian Heraclea did so. However this does not only depend on the quality of the soil and on the climate; the other circumstances which ensue must also be favourable; for, if great cold or heat ensues, the tree perishes at once. The season also makes a difference. For if one strips the bark of a silver-fir or fir at the time when the buds are shooting during Thargelion or Skirrophorion,[3] at which season it is separable, the tree dies at once. If it is done however in winter, the tree holds out longer; and this is especially true of the strongest trees, such as kermes-oak and oak; these it takes longer to kill. However the piece stripped off must be of a certain breadth to cause the death of the tree, especially in the case of the strongest trees; for, if one does it only a little, it is not surprising that the tree should not be killed; though some indeed say that, if it is done at all,[4] the tree certainly dies; this however is probably true only of the weaker kinds. For some, they say, if they are in bad barren

[2] ἀναφῦναι conj. Scal. from G; φῦναι Ald.H.
[3] May–June.
[4] ὁποσονοῦν conj. Sch. from G; ὁπωσοῦν Ald.

λυπρὰν ἔχει χώραν καὶ ἄτροφον. αὕτη μὲν δή, καθάπερ εἴρηται, κοινὴ φθορὰ πάντων.

XVI. Ἣν δὲ καλοῦσιν ἐπικοπὴν τῶν δένδρων, μόνον πεύκης ἐλάτης πίτυος φοίνικος, οἱ δὲ καὶ κέδρου καὶ κυπαρίττου φασί. ταῦτα γάρ, ἐὰν περιαιρεθῇ τὴν κόμην ἄνωθεν καὶ ἐπικοπῇ τὸ ἄκρον, φθείρεται πάντα καὶ οὐ βλαστάνει, καθάπερ οὐδ᾽ ἐπικαυθέντα ἢ πάντα ἢ ἔνια. τὰ δ᾽ ἄλλα πάντα καὶ περικοπέντα βλαστάνει, καὶ ἐνιά γε καλλίω γίνεται, καθάπερ ἐλάα. διαφθείρεται δὲ τὰ πολλὰ κἂν σχισθῇ τὸ στέλεχος· οὐδὲν γὰρ ὑπομένειν δοκεῖ πλὴν ἀμπέλου καὶ συκῆς καὶ ῥόας καὶ μηλέας· ἔνια δὲ κἂν ἑλκωθῇ καὶ μεῖζον καὶ βαθύτερον ἀπόλλυται. τὰ δ᾽ οὐδὲν πάσχει, καθάπερ ἡ πεύκη δᾳδουργουμένη, καὶ ἐξ ὧν δὴ τὰς ῥητίνας συλλέγουσιν, οἷον ἐλάτης τερμίνθου· καὶ γὰρ δὴ τούτων εἰς βάθος ἡ τρῶσις καὶ ἕλκωσις. καὶ γὰρ ἐξ ἀφόρων φοράδες γίνονται καὶ ἐξ ὀλιγοφόρων πολυφόροι.

2 Τὰ δὲ καὶ πελέκησιν ὑπομένει καὶ ὀρθὰ καὶ πεσόντα ὑπὸ πνεύματος, ὥστε πάλιν ἀνίστασθαι καὶ ζῆν καὶ βλαστάνειν, οἷον ἰτέα καὶ πλάτανος. ὅπερ συνέβη καὶ ἐν Ἀντάνδρῳ καὶ ἐν Φιλίπποις· ἐκπεσούσης γὰρ ὡς ἀπέκοψαν τοὺς ἀκρεμόνας καὶ ἐπελέκησαν, ἀνεφύη νύκτωρ ἡ πλάτανος κουφισθεῖσα τοῦ βάρους καὶ ἀνεβίω καὶ ὁ φλοιὸς περιέφυ πάλιν. παραπεπελεκημένη δ᾽ ἐτύγχανεν ἐκ τῶν δύο μερῶν· ἦν δὲ τὸ δένδρον μέγα μῆκος

---

[1] Plin. 17. 236; *cf.* 3. 7. 2; *C.P.* 5. 17. 3.
[2] *cf.* 3. 9. 5.
[3] ἄνωθεν καὶ conj. W.: καὶ ἄνωθεν Ald.
[4] *cf.* 1. 3. 3; 1. 14. 2.

soil, die even if the bark is not stripped all round. This then, as has been said, is a universal cause of death.

XVI. [1] The process which is called topping of trees is fatal only to fir silver-fir Aleppo pine[2] and date-palm, though some add prickly cedar and cypress. These, if they are stripped of their foliage at the top[3] and the crown is cut off, perish wholly and do not shoot again, as is the case with some, if not with all, if they are burnt. But all other trees shoot again after being lopped, and some, such as the olive,[4] become all the fairer. However most trees perish if the stem is split;[5] for no tree seems able to stand this, except vine fig pomegranate and apple; and some perish even if they are wounded severely and deeply. Some however take no harm[6] from this, as the fir when it is cut for tar, and those trees from which the resins are collected, as silver-fir and terebinth; though these trees are in fact then deeply wounded and mangled. Indeed they actually become fruitful[7] instead of barren, or are made to bear plentifully instead of scantily.

Some trees again submit to being hewn both when they are standing and when they have been blown down, so that they rise up again and live and shoot, for instance the willow and the plane. [8] This was known to happen in Antandros and at Philippi; a plane in Antandros having fallen and had its boughs lopped off and the axe applied to its trunk, grew again in the night when thus relieved of the weight, and the bark grew about it again. It happened that it had been hewn two thirds of the way round; it

[5] *cf. C.P.* 5. 16. 4; Plin. 17. 238.
[6] *cf. C.P.* 5. 16. 2.
[7] φοράδες conj. Sch.; φορίδες Ald.
[8] Plin. 16. 133.

μὲν μεῖζον ἢ δεκάπηχυ, πάχος δ' ὥστε μὴ ῥᾳδίως
3 ἂν περιλαβεῖν τέτταρας ἄνδρας. ἡ δὲ ἐν Φιλίπποις ἰτέα περιεκόπη μὲν τοὺς ἀκρεμόνας, οὐ μὴν
παρεπελεκήθη. μάντις δέ τις ἔπεισεν αὐτοὺς
θυσίαν τε ποιεῖσθαι καὶ τηρεῖν τὸ δένδρον ὡς
σημεῖον ἀγαθὸν γεγονός. ἀνέστη δὲ καὶ ἐν
Σταγείροις ἐν τῷ μουσείῳ λευκη τις ἐκπεσοῦσα.

4 Τῆς δὲ μήτρας ἐξαιρουμένης οὐθὲν ὡς εἰπεῖν φθείρεται δένδρον. σημεῖον δὲ ὅτι πολλὰ κοῖλα τῶν μέγεθος ἐχόντων δένδρων ἐστίν. οἱ δὲ περὶ Ἀρκαδίαν φασὶ μέχρι τινὸς μὲν ζῆν τὸ δένδρον, τελέως δὲ ἐξ ἅπαντος ἐξαιρεθείσης καὶ πεύκην φθείρεσθαι καὶ ἐλατην καὶ ἄλλο πᾶν.

5 Κοινὴ δὲ φθορὰ πάντων κἂν αἱ ῥίζαι περικοπῶσιν ἢ πᾶσαι ἢ αἱ πλεῖσται καὶ μέγισται καὶ κυριώταται τοῦ ζῆν. αὗται μὲν οὖν ἐξ ἀφαιρέσεως.

Ἡ δ' ὑπὸ τοῦ ἐλαίου προσθέσει τινὶ μᾶλλον ἢ ἀφαιρέσει· πολέμιον γὰρ δὴ καὶ τοῦτο πᾶσι· καὶ ἔλαιον ἐπιχέουσι τοῖς ὑπολείμμασι τῶν ῥιζῶν. ἰσχύει δὲ μᾶλλον τὸ ἔλαιον ἐν τοῖς νέοις καὶ ἄρτι φυομένοις· ἀσθενέστερα γάρ, δι' ὃ καὶ ἅπτεσθαι κωλύουσι.

Φθοραὶ δὲ καὶ ὑπ' ἀλλήλων εἰσὶ τῷ παραιρεῖσθαι τὰς τροφὰς καὶ ἐν τοῖς ἄλλοις ἐμποδίζειν. χαλεπὸς δὲ καὶ ὁ κιττὸς παραφυόμενος, χαλεπὸς δὲ καὶ ὁ κύτισος· ἀπόλλυσι γὰρ πάνθ' ὡς εἰπεῖν·

---

[1] τινὸς μὲν ζῆν τὸ δ. conj. W.; τινος ἐὰν (corrected) τοῦ δένδρου U; τινος ἐξηρέθη τοῦ δ. MVAld.

[2] *cf.* Plin. 17. 234; *C.P.* 5. 15. 6.

[3] πᾶσι· καὶ ἔλαιον ἐπιχέουσι conj. Sch.; πᾶσιν ἔλαιον ἐπιχεύουσιν UMP$_2$Ald.

was a large tree, more than ten cubits high, and of such girth that four men could not easily have encircled it. The willow at Philippi which grew again had had its branches lopped off, but the trunk had not been hewn. A certain seer persuaded the people to offer sacrifice and take care of the tree, since what had occurred was a good omen. Also at Stageira an abele in the school gardens which had fallen got up again.

Hardly any tree is destroyed by taking out the core; a proof of which is the fact that many large trees are hollow. The people of Arcadia say that the tree under these circumstances lives for a time,[1] but that, if the tree is entirely deprived of its core, fir or silver-fir or any other tree perishes.

All trees alike are destroyed when the roots are cut off, whether all or most of them, if those removed are the largest and the most essential to life. Such then are the causes of death which come from the removal of a part of the tree.

On the other hand the destruction which oil[2] causes is due rather to a kind of addition than to removal; for oil is hostile to all trees, and[3] so men pour it[4] over what remains of the roots. However oil is more potent with young trees which are just growing; for then they are weaker; wherefore men do not allow them to be touched at that time.

[5] Again trees may destroy one another, by robbing them of nourishment and hindering them in other ways. Again an overgrowth of ivy[6] is dangerous,[7] and so is tree-medick, for this destroys almost any-

[4] *i.e.* to complete the destruction of a tree. *cf.* Plut. *Quaest. Conv.* 2. 6. 2.

[5] Plin. 17. 239 and 240. [6] *cf. C.P.* 5. 15. 4.

[7] χαλεπὸς δὲ καὶ Ald.; χαλεπὸς δ' ἐστὶν conj. W.

ἰσχυρότερον δὲ τούτου τὸ ἅλιμον· ἀπόλλυσι γὰρ τὸν κύτισον.

6 Ἔνια δὲ οὐ φθείρει μὲν χείρω δὲ ποιεῖ ταῖς δυνάμεσι τῶν χυλῶν καὶ τῶν ὀσμῶν, οἷον ἡ ῥάφανος καὶ ἡ δάφνη τὴν ἄμπελον. ὀσφραίνεσθαι γάρ φασι καὶ ἕλκειν. δι᾽ ὃ καὶ ὅταν ὁ βλαστὸς πλησίον γένηται πάλιν ἀναστρέφειν καὶ ἀφορᾶν ὡς πολεμίας οὔσης τῆς ὀσμῆς. Ἀνδροκύδης δὲ καὶ παραδείγματι τούτῳ κατεχρήσατο πρὸς τὴν βοήθειαν τὴν ἀπὸ τῆς ῥαφάνου γινομένην πρὸς τὸν οἶνον, ὡς ἐξελαύνουσαν τὴν μέθην· φεύγειν γὰρ δὴ καὶ ζῶσαν τὴν ἄμπελον τὴν ὀσμήν. αἱ μὲν οὖν φθοραὶ πῶς τε γίνονται καὶ πόσαι καὶ ποσαχῶς φανερὸν ἐκ τῶν προειρημένων.

---

[1] ἕλκει: lit. 'draws it in'; *cf.* ἕλκειν ἀέρα, μέθυ, etc.

[2] *cf.* *C.P.* 2. 18. 4. ὁ βλαστὸς πλησίον conj. Dalec. from G; ὁ πλησίον βλαστός Ald.H.

thing. But *halimon* is more potent even than this, for it destroys tree-medick.

Again some things, though they do not cause death, enfeeble the tree as to the production of flavours and scents; thus cabbage and sweet bay have this effect on the vine. For they say that the vine scents the cabbage and is infected[1] by it. Wherefore the vine-shoot,[2] whenever it comes near this plant, turns back and looks away,[3] as though the smell were hostile to it. Indeed Androkydes[4] used this fact as an example to demonstrate the use of cabbage against wine, to expel the fumes of drunkenness for,[5] said he, even when it is alive, the vine avoids the smell. It is now clear from what has been said how the death of a tree may be caused, how many are the causes of death, and in what several ways they operate.

[3] ἀφορᾶν conj. Sch.; εὐφορεῖν U; ἀφορεῖν Ald.; *averti* G; *recedere* Plin. *l.c.*; ἐκχωρεῖν conj. W.

[4] A medical man who preached temperance to Alexander; *cf.* Plin. 14. 58; 17. 240.

[5] γὰρ δὴ καὶ conj. Dalec. from G; γὰρ δεῖ καὶ Ald.

BOOK V

E

I. Περὶ δὲ τῆς ὕλης, ποία τέ ἐστιν ἑκάστη, καὶ πόθ' ὡραία τέμνεσθαι, καὶ πρὸς ποῖα τῶν ἔργων χρησίμη, καὶ ποία δύσεργος ἢ εὔεργος, καὶ εἴ τι ἄλλο τῆς τοιαύτης ἱστορίας ἔχεται, πειρατέον ὁμοίως εἰπεῖν.

Ὡραῖα δὴ τέμνεσθαι τῶν ξύλων τὰ μὲν οὖν στρογγύλα καὶ ὅσα πρὸς φλοϊσμὸν ὅταν βλαστάνῃ· τότε γὰρ εὐπεριαίρετος ὁ φλοιός, ὃ δὴ καλοῦσι λοπᾶν, διὰ τὴν ὑγρότητα τὴν ὑπογινομένην αὐτῷ. μετὰ δὲ ταῦτα δυσπεριαίρετος καὶ τὸ ξύλον μέλαν γίνεται καὶ δυσειδές. τὰ δὲ τετράγωνα μετὰ τὸν λοπητόν· ἀφαιρεῖται γὰρ ἡ πελέκησις τὴν δυσείδειαν. ὅλως πᾶν πρὸς ἰσχὺν ὡραιότατον οὐ μόνον πεπαυμένον τῆς βλαστήσεως ἀλλ' ἔτι μᾶλλον ἐκπεπᾶναν τὸν καρπόν. ἀλλὰ διὰ τὸν φλοϊσμὸν ἀώροις οὖσιν ὡραίοις συμβαίνει γίνεσθαι τοῖς στρογγύλοις, ὥστε ἐναντίαι αἱ ὧραι κατὰ συμβεβηκός. εὐ-

[1] Plin. 16. 188. [2] *cf.* 3. 5. 1.
[3] δυσπεριαιρετός conj. Sch.; δυσπερικάθαρτος Ald.

BOOK V

Of the Timber of various Trees and its Uses.

I. In like manner we must endeavour to speak of timber, saying of what nature is that of each tree, what is the right season for cutting it, which kinds are hard or easy to work, and anything else that belongs to such an enquiry.

Of the seasons of cutting.

[1] Now these are the right seasons for cutting timber:—for 'round' timber and that whose bark is to be stripped the time is when the tree is coming into leaf. For then the bark is easily stripped (which process they call 'peeling' [2]) because of the moisture which forms beneath it. At a later time it is hard to strip,[3] and the timber obtained is black and uncomely. However square logs can be cut after the time of peeling, since trimming with the axe removes the uncomeliness. In general any wood is at the best season as to strength when it has not merely ceased coming into leaf, but has even ripened its fruit; however on account of the bark-stripping it comes to pass that 'round' timber is in season [4] when it is cut before it is ripe, so that, as it happens, the seasons are here reversed. Moreover the wood

[4] *i.e.* in practice the timber is cut before the ideally proper time.

χρούστερα δὲ τὰ ἐλάτινα γίνεται κατὰ τὸν πρῶτον λοπητόν.

2 Ἐπεὶ δὲ μάλιστ' ἢ μόνον περιαιροῦσι τὸν φλοιὸν ἐλάτης πεύκης πίτυος, ταῦτα μὲν τέμνεται τοῦ ἦρος· τότε γὰρ ἡ βλάστησις· τὰ δὲ ἄλλα ὁτὲ μὲν μετὰ πυροτομίαν, ὁτὲ δὲ μετὰ τρυγητὸν καὶ Ἀρκτοῦρον, οἷον ἀρία πτελέα σφένδαμνος μελία ζυγία ὀξύα φίλυρα φηγός τε καὶ ὅλως ὅσα κατορύττεται· δρῦς δὲ ὀψιαίτατα κατὰ χειμῶνα μετὰ τὸ μετόπωρον· ἐὰν δὲ ὑπὸ τὸν λοπητὸν τμηθῇ, σήπεται τάχιστα ὡς εἰπεῖν, ἐάν τε ἔμφλοιος ἐάν τε ἄφλοιος· καὶ μάλιστα μὲν τὰ ἐν τῷ πρώτῳ λοπητῷ, δεύτερα δὲ τὰ ἐν τῷ δευτέρῳ, τρίτα δὲ καὶ ἥκιστα τὰ ἐν τῷ τρίτῳ· τὰ δὲ μετὰ τὴν πέπανσιν τῶν καρπῶν ἄβρωτα διαμένει, κἂν ἀλόπιστα ᾖ· πλὴν ὑπὸ τὸν φλοιὸν ὑποδυόμενοι σκώληκες ἐπιπολῆς ἐγγράφουσι τὸ στέλεχος, οἷς καὶ σφραγῖσι χρῶνταί τινες· ὡραῖον δὲ τμηθὲν τὸ δρύϊνον ἀσαπές τε καὶ ἀθριπηδέστατον γίνεται καὶ σκληρὸν καὶ πυκνὸν ὥσπερ κέρας· πᾶν γὰρ ὅμοιόν ἐστιν ἐγκαρδίῳ· πλὴν τό γε τῆς ἁλιφλοίου καὶ τότε φαῦλον.

3 Συμβαίνει δὲ καὶ τοῦτο ὑπεναντίον, ὅταν τε κατὰ τὴν βλάστησιν τέμνωνται καὶ ὅταν μετα τοὺς καρπούς. τότε μὲν γὰρ ἀναξηραίνεται τὰ στελέχη καὶ οὐ βλαστάνει τὰ δένδρα· μετὰ δὲ τοὺς καρποὺς παραβλαστάνει. δυστομώτερα δὲ

---

[1] cf. 3. 5. 1. [2] ἢ add. Sch.

[3] φηγός τε conj. Scal.; πηγός τε U; φηγόσιν τε V; πηγόσιν τε MAld.

[4] κατορύττεται conj. Sch. from G; ὀρύττεται Ald. cf. 5. 4. 3; 5. 7. 5. [5] Plin. 16. 189.

of the silver-fir is of a better colour at the time[1] of the first peeling.

But since they strip the bark of[2] hardly any trees except silver-fir fir and pine, these trees are cut in the spring; for then is the time of coming into leaf. Other trees are cut sometimes after wheat-harvest, sometimes after the vintage and the rising of Arcturus, as *aria* (holm-oak) elm maple manna-ash *zygia* beech lime Valonia oak,[3] and in general all those whose timber is for underground use.[4] The oak is cut latest of all, in early winter at the end of autumn. [5] If it is cut at the time of peeling, it rots almost more quickly than at any other time, whether it has the bark on or not. This is especially so if it is cut during the first peeling, less so during the second, and least during the third. What is cut after the ripening of the fruit remains untouched by worms, even if it has not peeled: however worms get in under the bark and mark the surface of the stem, and such marked pieces of wood some use as seals.[6] Oak-wood if cut in the right season does not rot and is remarkably free from worms, and its texture is hard and close like horn; for it is like the heart of a tree throughout, except that that of the kind called sea-bark oak is even at that time of poor quality.[7]

Again, if the trees are cut at the time of coming into leaf, the result is the opposite of that which follows when they are cut after fruiting: for in the former case the trunks dry up and the trees do not sprout into leaf,[8] whereas after the time of fruiting they sprout at the sides. At this season however

[6] *cf.* Ar. *Thesm.* 427: θριπήδεστα σφραγίδια.
[7] *cf.* 3. 8. 5.
[8] βλαστάνει M; παραβλαστάνει W. with Ald.

διὰ τὴν σκληρότητα κατὰ ταύτην τὴν ὥραν. κελεύουσι δὲ καὶ δεδυκυίας τῆς σελήνης τέμνειν ὡς σκληροτέρων καὶ ἀσαπεστέρων γινομένων. ἐπεὶ δὲ αἱ πέψεις τῶν καρπῶν παραλλάττουσι, δῆλον ὅτι καὶ αἱ ἀκμαὶ πρὸς τὴν τομὴν παραλλάττουσιν· ἀεὶ γὰρ ὀψιαίτεραι αἱ τῶν ὀψικαρ-
4 ποτέρων. δι' ὃ καὶ πειρῶνταί τινες ὁρίζειν καθ' ἑκάστην· οἷον πεύκην μὲν καὶ ἐλάτην ὅταν ὑπολοπῶσιν· ἔτι δὲ ὀξύαν καὶ φίλυραν καὶ σφένδαμνον καὶ ζυγίαν τῆς ὀπώρας· δρῦν δέ, ὥσπερ εἴρηται, μετὰ τὸ φθινόπωρον. φασὶ δέ τινες πεύκην ὡραίαν εἶναι τοῦ ἦρος, ὅταν γε ἔχῃ τὴν καλουμένην κάχρυν, καὶ τὴν πίτυν ὅταν ὁ βότρυς αὐτῆς ἀνθῇ. ποῖα μὲν οὖν ὡραῖα καθ' ἕκαστον χρόνον οὕτω διαιροῦνται. πάντων δὲ δῆλον ὅτι βελτίω τὰ τῶν ἀκμαζόντων δένδρων ἢ τῶν νέων κομιδῇ καὶ γεγηρακότων· τὰ μὲν γὰρ ὑδατώδη, τὰ δὲ γεώδη.

5 Πλείστας δὲ χρείας καὶ μεγίστας ἡ ἐλάτη καὶ ἡ πεύκη παρέχονται, καὶ ταῦτα κάλλιστα καὶ μέγιστα τῶν ξύλων ἐστί. διαφέρουσι δὲ ἀλλήλων ἐν πολλοῖς· ἡ μὲν γὰρ πεύκη σαρκωδεστέρα τε καὶ ὀλιγόϊνος· ἡ δ' ἐλάτη καὶ πολύϊνος καὶ ἄσαρκος, ὥστε ἐναντίως ἑκάτερον ἔχειν τῶν μερῶν, τὰς μὲν ἶνας ἰσχυρὰς τὴν δὲ σάρκα

---

[1] αἱ add. Sch.

[2] ὑπολοπῶσιν conj. Sch.; εἰ πέλειν εἰσι U; ὑπελεινεισιν MV; ὑπελινῶσιν Ald.

[3] ταύτην conj. St.; καὶ τὴν Ald.H.

they are harder to cut because the wood is tougher It is also recommended to do the cutting when the moon has set, since then the wood is harder and less likely to rot. But, since the times when the fruit ripens are different for different trees, it is clear that the right moment for cutting also differs, being later for those[1] trees which fruit later. Wherefore some try to define the time for the cutting of each tree; for instance for fir and silver-fir the time is, they say, when they begin to peel[2]: for beech lime maple and *zygia* in autumn; for oak,[3] as has been said, when autumn is past. Some however say that the fir is ripe for cutting in spring, when it has on it the thing called 'catkin,'[4] and the pine when its 'cluster'[5] is in bloom. Thus they distinguish which trees are ripe for cutting at various times; however it is clear that in all cases the wood is better when the tree is in its prime than when it is quite young or has grown old, the wood of quite young trees being too succulent, and that of old ones too full of mineral matter.

Of the wood of silver-fir and fir.

Silver-fir and fir are the most useful trees and in the greatest variety of ways, and their[6] timber is the fairest and largest. Yet they differ from one another in many respects; the fir is fleshier and has few fibres, while the silver-fir has many fibres and is not fleshy, so that in respect of each component it is the reverse of the other, having stout fibres[7] but soft

[4] *cf.* 1. 1. 2 n.; 3. 5. 5.
[5] *i.e.* the male inflorescence.
[6] ταῦτα conj. Sch. from G; αὐτὰ Ald.H.
[7] *cf.* 3. 9. 7; Plin. 16. 184.

μαλακὴν καὶ μανήν· δι' ὃ τὸ μὲν βάρυ το δὲ
κοῦφον· τὸ μὲν γὰρ ἔνδαδον τὸ δὲ ἄδαδον, ἣ καὶ
6 λευκότερον. ἔχει δὲ καὶ ὄζους πλείους μὲν ἡ
πεύκη, σκληροτέρους δὲ ἡ ἐλάτη πολλῷ, μᾶλλον
δὲ καὶ σκληροτάτους πάντων· ἄμφω δὲ πυκνοὺς
καὶ κερατώδεις καὶ τῷ χρώματι ξανθοὺς καὶ
δᾳδώδεις. ὅταν δὲ τμηθῶσι, ῥεῖ καὶ ἐκ τῶν τῆς
ἐλάτης καὶ ἐκ τῶν τῆς πεύκης ἐπὶ πολὺν χρόνον
ὑγρότης καὶ μᾶλλον ἐκ τῶν τῆς ἐλάτης. ἔστι δὲ
καὶ πολύλοπον ἡ ἐλάτη, καθάπερ καὶ τὸ κρόμυον·
ἀεὶ γὰρ ἔχει τινὰ ὑποκάτω τοῦ φαινομένου, καὶ
7 ἐκ τοιούτων ἡ ὕλη. δι' ὃ καὶ τὰς κώπας ξύοντες
ἀφαιρεῖν πειρῶνται καθ' ἕνα καὶ ὁμαλῶς· ἐὰν γὰρ
οὕτως ἀφαιρῶσιν, ἰσχυρὸς ὁ κωπεών, ἐὰν δὲ
παραλλάξωσι καὶ μὴ κατασπῶσιν ὁμοίως, ἀσθε-
νής· πληγὴ γὰρ οὕτως, ἐκείνως δ' ἀφαίρεσις. ἔστι
δὲ καὶ μακρότατον ἡ ἐλάτη καὶ ὀρθοφυέστατον.
δι' ὃ καὶ τὰς κεραίας καὶ τοὺς ἱστοὺς ἐκ ταύτης
ποιοῦσιν. ἔχει δὲ καὶ τὰς φλέβας καὶ τὰς ἶνας
8 ἐμφανεστάτας πάντων. αὐξάνεται δὲ πρῶτον
εἰς μῆκος, ἄχρι οὗ δὴ ἐφίκηται τοῦ ἡλίου· καὶ
οὔτε ὄζος οὐδεὶς οὔτε παραβλάστησις οὔτε πάχος
γίνεται· μετὰ δὲ ταῦτα εἰς βάθος καὶ πάχος·
οὕτως αἱ τῶν ὄζων ἐκφύσεις καὶ παραβλαστήσεις.

---

1 τὸ μὲν γὰρ ἐνδ. conj. St. from G; ἐνδ. γὰρ Ald.
2 *cf.* 3. 9. 7.
3 *cf.* 3. 9. 7, μόνον οὐ διαφανεῖς, whence it appears that the epithet refers to colour.
4 Plin. 16. 195. 5 *i.e.* the annual rings. *cf.* 1. 5. 2; 5. 5. 3.
6 *cf.* Hom. *Od.* 12. 172.
7 κατασπῶσιν conj. W.; κατὰ πᾶσιν UMV; κατὰ πάντα Ald.
8 *cf.* Plin. *l.c.* 9 *cf.* 1. 2. 1.
10 ἐμφανέστατας conj. W.; εὐγενεστάτας Ald.
11 δὲ conj. Sch.; καὶ UAld.H.

flesh of open texture. Wherefore the timber of the one is heavy, of the other light, the one[1] being resinous, the other without resin; wherefore also it is whiter. Moreover the fir has more branches, but those of the silver-fir are much tougher, or rather they are tougher than those of any other tree; [2] the branches of both however are of close texture, horny,[3] and in colour brown and like resin-glutted wood. [4] When the branches of either tree are cut, sap streams from them for a considerable time, but especially from those of the silver-fir. Moreover the wood of the silver-fir has many layers, like an onion:[5] there is always another beneath that which is visible, and the wood is composed of such layers throughout. Wherefore, when men are shaving this wood to make oars,[6] they endeavour to take off the several coats one by one evenly: for, if they do this, they get a strong spar, while if they do the work irregularly and do not strip[7] off the coats evenly, they get a weak one; for the process in this case is hacking instead of stripping. The silver-fir also gives timber of the greatest lengths and of the straightest growth; wherefore yard-arms[8] and masts are made from it. Also the vessels[9] and fibre are more clearly[10] seen in it than in any other tree. At first[11] it grows in height only, until it has reached[12] the sunshine; and so far there is no branch nor sidegrowth nor density of habit; but after that the tree proceeds to increase in bulk[13] and density of habit, as[14] the outgrowing branches and sidegrowths develop.

[12] ἄχρι . . . ἐφίκηται conj. Sch.; ἄχρι οὗ δὴ κἀφίκηται U; ἄχρις οὐκ ἀφίκηται MV; ἄχρις οὗ ἀχίκηται Ald. H.
[13] *cf.* 4. 1. 4.
[14] Lit. 'this being the effect of the outgrowth.' πάχος· οὕτως Ald.; πάχος, ὅταν conj. W.

9 Ταῦτα μὲν οὖν ἴδια τῆς ἐλάτης, τὰ δὲ κοινὰ καὶ πεύκης καὶ ἐλάτης καὶ τῶν ἄλλων. ἔστι γὰρ ἡ μὲν τετράξοος ἡ δὲ δίξοος. καλοῦσι δὲ τετραξόους μὲν ὅσαις ἐφ' ἑκάτερα τῆς ἐντεριώνης δύο κτηδόνες εἰσὶν ἐναντίαν ἔχουσαι τὴν φύσιν· ἔπειτα καθ' ἑκατέραν τὴν κτηδόνα ποιοῦνται τὴν πελέκησιν ἐναντίας τὰς πληγὰς κατὰ κτηδόνα φέροντες, ὅταν ἐφ' ἑκάτερα τῆς ἐντεριώνης ἡ πελέκησις ἀναστρέφῃ. τοῦτο γὰρ ἐξ ἀνάγκης συμβαίνει διὰ τὴν φύσιν τῶν κτηδόνων. τὰς δὲ τοιαύτας ἐλάτας καὶ πεύκας τετραξόους καλοῦσι. εἰσὶ δὲ καὶ πρὸς τὰς ἐργασίας αὗται κάλλισται· πυκνότατα γὰρ ἔχουσι τὰ ξύλα καὶ τὰς αἰγίδας αὗται 10 φύουσιν. αἱ δίξοοι δὲ κτηδόνα μὲν ἔχουσι μίαν ἐφ' ἑκάτερα τῆς ἐντεριώνης, ταύτας δὲ ἐναντίας ἀλλήλαις, ὥστε καὶ τὴν πελέκησιν εἶναι διπλῆν, μίαν καθ' ἑκατέραν κτηδόνα ταῖς πληγαῖς ἐναντίαις· ἀπαλώτατα μὲν οὖν ταῦτά φασιν ἔχειν τὰ ξύλα, χείριστα δὲ πρὸς τὰς ἐργασίας· διαστρέφεται γὰρ μάλιστα. μονοξόους δὲ καλοῦσι τὰς ἐχούσας μίαν μόνον κτηδόνα· τὴν δὲ πελέκησιν αὐτῶν γίνεσθαι τὴν αὐτὴν ἐφ' ἑκάτερα τῆς ἐντεριώνης· φασὶ δὲ μανότατα μὲν ἔχειν τῇ φύσει τὰ ξύλα ταῦτα πρὸς δὲ τὰς διαστροφὰς ἀσφαλέστατα.

11 Διαφορὰς δὲ ἔχουσι τοῖς φλοιοῖς, καθ' ἃς γνωρίζουσιν ἰδόντες εὐθὺς τὸ δένδρον πεφυκὸς

---

[1] Plin. *l.c.*

[2] The meaning of 'four-cleft' etc. seems to be this:

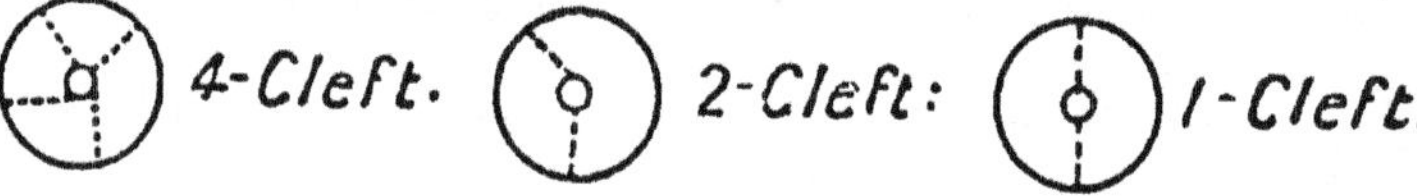

These are the characteristics peculiar to the silver-fir. Others it shares with the fir and the other trees of this class. [1] For instance, sometimes a tree is 'four-cleft,' sometimes 'two-cleft'; it is called 'four-cleft' when on either side of the heart-wood there are two distinct and diverse lines of fissure: in that case the blows of the axe follow these lines in cases where the hewing is stopped short on either side of the heart-wood.[2] For the nature of the lines of fissure compels the hewing to take this course. Silver-firs or firs thus formed are said to be 'four-cleft.' And these are also the fairest trees for carpentry, their wood being the closest and possessing the *aigis*.[3] Those which are 'two-cleft' have one single line of fissure on either side of the heart-wood, and the lines of fissure do not correspond to each other, so that the hewing also is performed by cuts which follow the two lines of fissure, so as to reach the two sides of the heart-wood at different angles. Now such wood, they say, is the softest, but the worst for carpentry, as it warps most easily. Those trees which have only a single[4] continuous line of fissure are said to be 'one-cleft,' though here too the cutting is done from either side of the heart-wood: and such wood has, they say, an open[5] texture, and yet[6] it is not at all apt to warp.

[7] There are also differences in the bark, by observation of which they can tell at once what the

[3] *cf.* 3. 9. 3. [4] μίαν conj. W.; μίαν δὲ P_2Ald.
[5] μανότατα conj. W.; μανότητα Ald.
[6] τὰ ξύλα . . . τὰς conj. Sch.; τὰ ξύλα· ταῦτα δὲ πρὸς τὰς Ald.H. [7] Plin. 16. 195 and 196.

ποῖόν τί ἐστι· τῶν μὲν γὰρ εὐκτηδόνων καὶ ἀστραβῶν καὶ ὁ φλοιὸς λεῖος καὶ ὀρθός, τῶν δ' ἐναντίων τραχύς τε καὶ διεστραμμένος· ὁμοίως δὲ καὶ ἐπὶ τῶν λοιπῶν. ἀλλ' ἔστι τετράξοα μὲν ὀλίγα μονόξοα δὲ πλείω τῶν ἄλλων. ἅπασα δὲ ἡ ὕλη μείζων καὶ ὀρθοτέρα καὶ ἀστραβεστέρα καὶ στιφροτέρα καὶ ὅλως καλλίων καὶ πλείων ἡ ἐν τοῖς προσβορείοις, ὥσπερ καὶ πρότερον ἐλέχθη· καὶ αὐτοῦ τοῦ δένδρου δὲ τὰ πρὸς βορρᾶν πυκνότερα καὶ νεανικώτερα. ὅσα δὲ ὑποπαράβορρα καὶ ἐν περίπνῳ, ταῦτα στρέφει καὶ παραλλάττει παρὰ μικρὸν ὁ βορέας, ὥστε εἶναι παρεστραμμένην αὐτῶν τὴν μήτραν καὶ
12 οὐ κατ' ὀρθόν. ἔστι δὲ ὅλα μὲν τὰ τοιαῦτα ἰσχυρὰ τμηθέντα δὲ ἀσθενῆ διὰ τὸ πολλὰς ἔχειν παραλλαγάς. καλοῦσι δὲ οἱ τέκτονες ἐπίτομα ταῦτα διὰ τὸ πρὸς τὴν χρείαν οὕτω τέμνειν. ὅλως δὲ χείρω τὰ ἐκ τῶν ἐφύγρων καὶ εὐδιεινῶν καὶ παλισκίων καὶ συνηρεφῶν καὶ πρὸς τὴν τεκτονικὴν χρείαν καὶ πρὸς τὴν πυρευτικήν. αἱ μὲν οὖν τοιαῦται διαφοραὶ πρὸς τοὺς τόπους εἰσὶν αὐτῶν τῶν ὁμογενῶν ὥς γε ἁπλῶς εἰπεῖν.

II. Διαιροῦσι γάρ τινες κατὰ τὰς χώρας, καί φασιν ἀρίστην μὲν εἶναι τῆς ὕλης πρὸς τὴν τεκτονικὴν χρείαν τῆς εἰς τὴν Ἑλλάδα παραγινομένης τὴν Μακεδονικήν· λεία τε γάρ ἐστι καὶ ἀστραβὴς καὶ ἔχουσα θυῖον. δευτέραν δὲ τὴν Ποντικήν, τρίτην δὲ τὴν ἀπὸ τοῦ Ῥυνδάκου,

---

[1] πεφυκὸς: *cf.* Xen. *Cyr.* 4. 3. 5.
[2] ὑποπαράβορρα conj. St.; ὑπὸ παράβορρα Ald.; ὑπόβορρα ἢ παράβορρα conj. Sch.

timber of the tree is like as it stands.[1] For if the timber has straight and not crooked lines of fissure, the bark also is smooth and regular, while if the timber has the opposite character, the bark is rough and twisted; and so too is it with other points. However few trees are 'four-cleft,' and most of those which are not are 'one-cleft.' All wood, as was said before, which grows in a position facing north, is bigger, more erect, of straighter grain, tougher, and in general fairer and more abundant. Moreover of an individual tree the wood on the northward side is closer and more vigorous. But if a tree stands sideways to the north[2] with a draught round it, the north wind by degrees twists and contorts[3] it, so that its core becomes twisted instead of running straight. The timber of such a tree while still in one piece is strong, but, when cut, it is weak, because the grain slants across the several pieces Carpenters call such wood 'short lengths,' because they thus cut it up for use. Again in general wood which comes from a moist, sheltered, shady or confined position is inferior both for carpentry and for fuel. Such are the differences, generally[4] speaking, between trees of the same kind as they are affected by situation.

Of the effects on timber of climate.

II. [5]Some indeed make a distinction between regions and say that the best of the timber which comes into Hellas for the carpenter's purposes is the Macedonian, for it is smooth and of straight grain, and it contains resin: second best is that from Pontus, third that

[3] παραλλάττει conj. Dalec.; παραλλάγει U; παραλήγει Ald.; παραλυγίζει conj. H. Steph.
[4] γε conj. Sch.; δὲ Ald. [5] Plin. 16. 197

τετάρτην δὲ τὴν Αἰνιανικήν· χειρίστην δὲ τήν τε Παρνασιακὴν καὶ τὴν Εὐβοϊκήν· καὶ γὰρ ὀζώδεις καὶ τραχείας καὶ ταχὺ σήπεσθαι. περὶ δὲ τῆς Ἀρκαδικῆς σκεπτέον.

2 Ἰσχυρότατα δὲ τῶν ξύλων ἐστὶ τὰ ἄοζα καὶ λεῖα· καὶ τῇ ὄψει δὲ ταῦτα κάλλιστα. ὀζώδη δὲ γίνεται τὰ κακοτροφηθέντα καὶ ἤτοι χειμῶνι πιεσθέντα ἢ καὶ ἄλλῳ τινὶ τοιούτῳ· τὸ γὰρ ὅλον τὴν πολυοζίαν εἶναι ἔνδειαν εὐτροφίας. ὅταν δὲ κακοτροφήσαντα ἀναλάβῃ πάλιν καὶ εὐσθενήσῃ, συμβαίνει καταπίνεσθαι τοὺς ὄζους ὑπὸ τῆς περιφύσεως· εὐτροφοῦν γὰρ καὶ αὐξανόμενον ἀναλαμβάνει καὶ πολλάκις ἔξωθεν μὲν λεῖον τὸ ξύλον διαιρούμενον δὲ ὀζῶδες ἐφάνη. δι' ὃ καὶ σκοποῦνται τῶν σχιστῶν τὰς μήτρας· ἐὰν γὰρ αὗται ἔχωσιν ὄζους, ὀζώδη καὶ τὰ ἐκτός· καὶ οὗτοι χαλεπώτεροι τῶν ἐκτὸς καὶ φανεροί.

3 Γίνονται δὲ καὶ αἱ σπεῖραι διὰ χειμῶνας τε καὶ κακοτροφίαν. σπείρας δὲ καλοῦσιν ὅταν ᾖ συστροφή τις ἐν αὐτῇ μείζων καὶ κύκλοις περιεχομένη πλείοσιν οὔθ' ὥσπερ ὁ ὄζος ἁπλῶς οὔθ' ὡς ἡ οὐλότης ἡ ἐν αὐτῷ τῷ ξύλῳ· δι' ὅλου γάρ πως αὕτη καὶ ὁμαλίζουσα· χαλεπώτερον δὲ τοῦτο πολὺ καὶ δυσεργότερον τῶν ὄζων. ἔοικε δὲ παραπλησίως καὶ ὡς ἐν τοῖς λίθοις ἐγγίνεσθαι

---

[1] A river which flows into the Propontis on the Asiatic side.

[2] Near Mount Oeta. Αἰνιανικήν conj. Palm. from Plin. *l.c.*; αἰανικὴν P_2Ald.H.

[3] ταῦτα κάλλιστα· ὀζώδη δὲ conj. Scal.; ταῦτα καὶ μάλιστα ὀζώδη γίν. Ald.H.; ταῦτα μάλιστα· ὀζώδη δὲ γίν. U.

from the Rhyndakos,[1] fourth that of the country of the Ainianes,[2] worst is that of Parnassus and that of Euboea, for it is full of knots and rough and quickly rots. As to Arcadian timber the case is doubtful.

Of knots and 'coiling' in timber.

The strongest wood is that which is without knots and smooth, and it is also the fairest in appearance.[3] Wood becomes knotty when it has been ill nourished and has suffered severely whether from winter or some such cause; for in general a knotty habit is supposed to indicate lack of nourishment. When however, after being ill nourished, the tree recovers and becomes vigorous, the result is that the knots are absorbed[4] by the growth which now covers them; for the tree, being now well fed and growing vigorously, recovers, and often the wood is smooth outside, though when split it is seen to have knots. And this is why they examine the core of wood that has been split; for, if this contains knots, the outward[5] parts will also be knotty, and these knots are harder to deal with than the outer ones, and are easily recognised.

[6]'Coiling' of the wood is also due to winter or ill nourishment. Wood is said to 'coil' when there is in it closer twisting[7] than usual, made up of an unusual number of rings: this is not quite like a knot, nor is it like the ordinary curling of the wood, which runs right through it and is uniform. 'Coiling' is much more troublesome and difficult to deal with than knots; it seems to correspond to the so-called

[4] καταπίνεσθαι: ? καταλαμβάνεσθαι. *cf.* below, § 3.
[5] *i.e.* outward in regard to the core. [6] Plin. 16. 198.
[7] ᾗ συστροφή conj. Scal.; ᾗ εὐστροφή U; ᾗ εὐτραφῆ Ald. etc.

τὰ καλούμενα κέντρα. ὅτι δ' ἡ περίφυσις καταλαμβάνει τοὺς ὄζους φανερώτατον ἐξ αὐτῆς τῆς αἰσθήσεως, οὐ μὴν ἀλλὰ καὶ ἐκ τῶν ἄλλων
4 τῶν ὁμοίων· πολλάκις γὰρ αὐτοῦ τοῦ δένδρου μερος τι συνελήφθη ὑπὸ θατέρου συμφυοὺς γενομένου· καὶ ἐάν τις ἐκγλύψας θῇ λίθον εἰς τὸ δένδρον ἢ καὶ ἄλλο τι τοιοῦτον, κατακρύπτεται περιληφθὲν ὑπὸ τῆς περιφύσεως· ὅπερ καὶ περὶ τὸν κότινον συνέβη τὸν ἐν Μεγάροις τὸν ἐν τῇ ἀγορᾷ· οὗ καὶ ἐκκοπέντος λόγιον ἦν ἁλῶναι καὶ διαρπασθῆναι τὴν πόλιν· ὅπερ ἐγένετο . . . . Δημήτριος. ἐν τούτῳ γὰρ διασχιζομένῳ κνημῖδες εὑρέθησαν καὶ ἄλλ' ἄττα τῆς Ἀττικῆς ἐργασίας κρεμαστά, τοῦ κοτίνου οὗ ἀνετέθη τὸ πρῶτον ἐγκοιλανθέντος. τούτου δ' ἔτι μικρὸν τὸ λοιπόν. πολλαχοῦ δὲ καὶ ἄλλοθι γίνεται πλείονα τοιαῦτα. καὶ ταῦτα μέν, ὥσπερ εἴρηται κοινὰ πλειόνων.

III. Κατὰ δὲ τὰς ἰδίας ἑκάστου φύσεις αι τοιαῦταί εἰσι διαφοραί, οἷον πυκνότης μανότης βαρύτης κουφότης σκληρότης μαλακότης, ὡσαύτως δὲ καὶ εἴ τις ἄλλη τοιαύτη· κοιναὶ δὲ ὁμοίως αὗται καὶ τῶν ἡμέρων καὶ τῶν ἀγρίων, ὥστε περὶ πάντων λεκτέον.

---

1 ὅτι δ' ἡ conj. W.; ὅτι δὴ UMV; ὅτι δὲ Ald.
2 *cf.* καταπίνεσθαι, above, § 2.
3 Plin. 16. 198 and 199.
4 ἐκγλύψας θῇ conj. W.; ἐκλύψας θῆι U; ἐκλιθασθῇ Ald.H.
5 Text defective.
6 *i.e.* the bark had grown over these. *cf.* Plin. *l.c.*

'centres' which occur in marbles. That[1] vigorous growth covers[2] up the knots is plain from simple observation of the fact and also from other similar instances. [3] For often some part of the tree itself is absorbed by the rest of the tree which has grown into it; and again, if one makes a hole in a tree and puts[4] a stone into it or some other such thing, it becomes buried, being completely enveloped by the wood which grows all round it: this happened with the wild olive in the market-place at Megara; there was an oracle that, if this were cut open, the city would be taken and plundered, which came to pass when Demetrius took it.[5] For, when this tree was split open, there were found greaves and certain other things[6] of Attic workmanship hanging there, the hole[7] in the tree having been made at the place where the things were originally hung on it as offerings. Of this tree a small part still exists, and in many other places further instances have occurred. Moreover, as has been said, such occurrences happen also with various other trees.

Of differences in the texture of different woods.

III. [8] Corresponding to the individual characters of the several trees we have the following kinds of differences in the wood:—it differs in closeness, heaviness, hardness or their opposites, and in other similar ways; and these differences are common to cultivated and wild trees. So that we may speak of all trees without distinction.

[7] ἐργασίας κρεμαστὰ τοῦ κοτίνου οὗ I conj. from G and Plin. *l.c.* (certain restoration perhaps impossible); κερμηστι δ' ἐστιν ἐν κοτίνῳ· οὗ U; Ald. has κερμηστὶ, M κρεμαστὶ, V κερμάστων; St. suggested κρεμαστῶν ὅπλων as words of the original text. [8] Plin. 16. 204–207.

Πυκνότατα μὲν οὖν δοκεῖ καὶ βαρύτατα πύξος εἶναι καὶ ἔβενος· οὐδὲ γὰρ οὐδ' ἐπὶ τοῦ ὕδατος ταῦτ' ἐπινεῖ. καὶ ἡ μὲν πύξος ὅλη, τῆς δὲ ἐβένου ἡ μήτρα, ἐν ᾗ καὶ ἡ τοῦ χρώματός ἐστι μελανία. τῶν δ' ἄλλων ὁ λωτός. πυκνὸν δὲ καὶ ἡ τῆς δρυὸς μήτρα, ἣν καλοῦσι μελάνδρυον· καὶ ἔτι μᾶλλον ἡ τοῦ κυτίσου· παρομοία γὰρ αὕτη δοκεῖ τῇ ἐβένῳ εἶναι.

2 Μέλαν δὲ σφόδρα καὶ πυκνὸν το τῆς τερμίνθου· περὶ γοῦν Συρίαν μελάντερόν φασιν εἶναι τῆς ἐβένου· καὶ ἐκ τούτου γὰρ καὶ τὰς λαβὰς τῶν ἐγχειριδίων ποιεῖσθαι, τορνεύεσθαι δὲ ἐξ αὐτῶν καὶ κύλικας Θηρικλείους, ὥστε μηδένα ἂν διαγνῶναι πρὸς τὰς κεραμέας· λαμβάνειν δὲ τὸ ἐγκάρδιον· δεῖν δὲ ἀλείφειν τὸ ξύλον· οὕτω γὰρ γίνεσθαι καὶ κάλλιον καὶ μελάντερον.

Εἶναι δὲ καὶ ἄλλο τι δένδρον, ὃ ἅμα τῇ μελανίᾳ καὶ ποικιλίαν τινὰ ἔχει ὑπέρυθρον, ὥστε εἶναι τὴν ὄψιν ὡσὰν ἐβένου ποικίλης· ποιεῖσθαι δ' ἐξ αὐτοῦ καὶ κλίνας καὶ δίφρους καὶ τὰ ἄλλα τὰ σπουδαζόμενα. τὸ <δὲ> δένδρον μέγα σφόδρα καὶ καλόφυλλον εἶναι ὅμοιον ταῖς ἀπίοις.

3 Ταῦτα μὲν οὖν ἅμα τῇ μελανίᾳ καὶ πυκνότητα ἔχει. πυκνὸν δὲ καὶ ἡ σφένδαμνος καὶ ἡ ζυγία καὶ ὅλως πάντα τὰ οὖλα· καὶ ἡ ἐλάα δὲ καὶ ὁ κότινος, ἀλλὰ κραῦρα. μανὰ δὲ τῶν μὲν ἀγρίων καὶ ἐρεψίμων τὰ ἐλάτινα μάλιστα,

---

[1] *cf.* Arist. *Meteor.* 4. 7 *ad fin.*
[2] *cf.* 1. 6. 1. [3] *cf.* 3. 15. 3.
[4] Probably so called from their resemblance in shape and

Box and ebony seem to have the closest and heaviest wood; for their wood does not even float on water. This applies to the box-tree as a whole, and to the core of the ebony, which contains the black pigment.[1] The nettle-tree also is very close and heavy, and so is the core of the oak, which is called 'heart of oak,' and to a still greater degree this is true of the core of laburnum[2]; for this seems to resemble the ebony.

The wood of the terebinth is also very black and close-grained; at least in Syria[3] they say that it is blacker than ebony, that in fact they use it for making their dagger handles; and by means of the lathe-chisel they also make of it 'Theriklean' cups,[4] so that no one could[5] distinguish these from cups made of pottery; for this purpose they use, it is said, the heart-wood, but the wood has to be oiled, for then it becomes comelier and blacker.

There is also, they say, another tree[6] which, as well as the black colour, has a sort of reddish variegation, so that it looks like variegated ebony, and of it are made beds and couches and other things of superior quality. This tree is very large and has handsome leaves and is like the pear.

These trees then, as well as the black colour, have close wood; so also have maple *zygia* and in general all those that are of compact growth; so also have the olive and the wild olive, but their wood is brittle.[7] Of wild trees which are used for roof-timbers the wood of the silver-fir is the least com-

colour to the cups made by Therikles, a famous Corinthian potter; see reff. to comedy in LS. *s.v.*

[5] μηδένα ἂν conj. W.; μηδ' ἂν ἕνα Ald.

[6] Sissoo wood. See Index App. (21).

[7] ἀλλὰ κραῦρα conj. Sch.; ἀλλὰ καὶ αὔρα MVAld.

τῶν δ' ἄλλων τὰ ἄκτινα καὶ τὰ σύκινα καὶ τὰ τῆς μηλέας καὶ τὰ τῆς δάφνης. σκληρότατα δὲ τὰ δρυϊνα καὶ τὰ ζύγινα καὶ τὰ τῆς ἀρίας· καὶ γὰρ ὑποβρέχουσι ταῦτα πρὸς τὴν τρύπησιν μαλάξεως χάριν. μαλακὰ δὲ καθ' ὅλου μὲν τὰ μανὰ καὶ χαῦνα· τῶν δὲ σαρκωδῶν μάλιστα φίλυρα. δοκεῖ δὲ καὶ θερμότατον εἶναι τοῦτο· σημεῖον δὲ ὅτι μάλιστα ἀμβλύνει τὰ σιδήρια· τὴν γὰρ βαφὴν ἀνίησι διὰ τὴν θερμότητα.

4 Θερμὸν δὲ καὶ κιττὸς καὶ δάφνη καὶ ὅλως ἐξ ὧν τὰ πυρεῖα γίνεται· Μενέστωρ δέ φησι καὶ συκάμινον. ψυχρότατα δὲ τὰ ἔνυδρα καὶ ὑδατώδη. καὶ γλίσχρα δὲ τὰ ἰτέϊνα καὶ ἀμπέλινα, δι' ὃ καὶ τὰς ἀσπίδας ἐκ τούτων ποιοῦσι· συμμύει γὰρ πληγέντα· κουφότερον δὲ τὸ τῆς ἰτέας, μανότερον γάρ, δι' ὃ καὶ τούτῳ μᾶλλον χρῶνται. τὸ δὲ τῆς πλατάνου γλισχρότητα μὲν ἔχει, φύσει δὲ ὑγρότερον τοῦτο καὶ τὸ τῆς πτελέας. σημεῖον δέ ἐστιν, μετὰ τὴν τομὴν ὀρθὸν ὅταν σταθῇ, πολὺ ὕδωρ ἀφίησι. τὸ δὲ τῆς συκαμίνου πυκνὸν ἅμα καὶ γλίσχρον.

5 Ἔστι δὲ καὶ ἀστραβέστατον τὸ τῆς πτελεας, δι' ὃ καὶ τοὺς στροφεῖς τῶν θυρῶν ποιοῦσι πτελεΐνους· ἐὰν γὰρ οὗτοι μένωσι, καὶ αἱ θύραι μένουσιν ἀστραβεῖς, εἰ δὲ μή, διαστρέφονται. ποιοῦσι δ' αὐτοὺς ἔμπαλιν τιθέντες τὰ ξύλα τό τε ἀπὸ τῆς ῥίζης καὶ τὸ ἀπὸ τοῦ φύλλου·

---

[1] ὑποβρέχουσι conj. Harduin from Plin. 16. 207; ἀποβρίθουσι Ald. H.; ἀποβρέχουσι mBas.

[2] *cf.* 5. 5. 1, which, referring to this passage, hardly agrees with it as now read.

pact, and among others that of the elder fig apple and bay. The hardest woods are those of the oak *zygia* and *aria* (holm-oak); in fact men wet[1] these to soften them for boring holes. In general, woods which are of open porous texture are soft, and of those of flesby texture the softest is the lime. The last-named seems also to be the hottest; the proof of which is that it blunts iron tools more than any other; for they lose their edge[2] by reason of its heat.

Ivy and bay are also hot woods, and so m general are those used for making fire-sticks; and Menestor[3] adds the wood of the mulberry. [4]The coldest woods are those which grow in water and are of succulent character. The wood again of willow and vine is tough; wherefore men make their shields of these woods; for they close up again after a blow; but that of the willow is lighter, since it is of less compact texture; wherefore they use this for choice. The wood of the plane is fairly tough, but it is moister in character, as also is that of the elm. A proof of this is that, if it is set upright[5] after being cut, it discharges much water.[6] The wood of the mulberry is at once of close grain and tough.

[7]The wood of the elm is the least likely to warp; wherefore they make the 'hinges'[8] of doors out of elm wood; for, if these hold, the doors also keep in place; otherwise they get wrenched out of place. They make the 'hinges' by putting wood from the root above[9] and wood 'from the foliage' below,[9] thus

[3] *cf.* 1. 2. 3 n. [4] Plin. 16. 209.
[5] ὀρθὸν ὅταν conj. W.: so G; ὀρθὸς ὅταν MV; ὅταν ὀρθὰ Ald.
[6] *cf.* 5. 1. 6. [7] Plin. 16. 210.
[8] Sc. an arrangement of cylindrical pivot and socket.
[9] *i.e.* as socket and pivot respectively; *cf.* 5. 5. 4.

καλοῦσι δὲ οἱ τέκτονες τὸ ἀπὸ τοῦ φύλλου τὸ ἄνω· ἐναρμοσθέντα γὰρ ἀλλήλοις ἑκάτερον κωλύει πρὸς τὴν ὁρμὴν ἐναντίως ἔχον. εἰ δὲ ἔκειτο κατὰ φύσιν, οὗπερ ἡ ῥοπὴ ἐνταῦθα πάντων ἂν ἦν ἡ φορά.

Τὰς δὲ θύρας οὐκ εὐθὺς συντελοῦσιν, ἀλλὰ πήξαντες ἐφιστᾶσι, κἄπειτα ὑστέρῳ οἱ δὲ τῷ τρίτῳ ἔτει συνετέλεσαν ἐὰν μᾶλλον σπουδάζωσι· τοῦ μὲν γὰρ θέρους ἀναξηραινομένων διΐστανται, τοῦ δὲ χειμῶνος συμμύουσιν. αἴτιον δ' ὅτι τῆς ἐλάτης τὰ μανὰ καὶ σαρκώδη ἕλκει τὸν ἀέρα ἔνικμον ὄντα.

6 Ὁ δὲ φοῖνιξ κοῦφος καὶ εὔεργος καὶ μαλακός, ὥσπερ ὁ φελλός, βελτίων δὲ τοῦ φελλοῦ ὅτι γλίσχρος· ἐκεῖνο δὲ θραυστόν. διὰ τοῦτο τὰ εἴδωλα νῦν ἐκ τοῦ τῶν φοινίκων ποιοῦσι, τὸν δὲ φελλὸν παρήκασι. τὰς ἶνας δὲ οὐ δι' ὅλου ἔχει οὐδ' ἐπὶ πολὺ καὶ μακρὰς οὐδ' ὡσαύτως τῇ θέσει ἐγκειμένας πάσας ἀλλὰ παντοδαπῶς. ἀναξηραίνεται δὲ καὶ λεαινόμενον καὶ πριόμενον τὸ ξύλον.

7 Τὸ δὲ θύον, οἱ δὲ θύαν καλοῦσι, παρ' Ἄμμωνι τε γίνεται καὶ ἐν τῇ Κυρηναίᾳ, τὴν μὲν μορφὴν ὅμοιον κυπαρίττῳ καὶ τοῖς κλάδοις καὶ τοῖς φύλλοις καὶ τῷ στελέχει καὶ τῷ καρπῷ, μᾶλλον δ' ὥσπερ κυπάριττος ἀγρία· πολὺ μὲν καὶ ὅπου

---

1 κωλύει: Sch. adds θάτερον from G.
2 ἔκειτο conj. W.; ἐκεῖνο Ald.
3 *i.e.* the 'upper' wood in the upper position.
4 πάντων MSS. (?); πάντως conj. W.
5 *i.e.* there would be no resistance. ἦν after ἂν add. Sch.

reversing the natural position: (by wood 'from the foliage' joiners mean the upper wood). For, when these are fitted the one into the other, each counteracts[1] the other, as they naturally tend in opposite directions: whereas, if the wood were set[2] as it grows,[3] all the parts[4] would give where the strain came.[5]

(They do not finish off the doors at once; but, when they have put them together, stand them up, and then finish them off the next year, or sometimes the next year but one,[6] if they are doing specially good work. For in summer, as the wood dries, the work comes apart, but it closes in winter. The reason is that the open fleshy texture of the wood of the silver-fir[7] drinks in the air, which is full of moisture.)

[8] Palm-wood is light easily worked and soft like cork-oak, but is superior to that wood, as it is tough, while the other is brittle. Wherefore men now make their images of palm-wood and have given up the wood of cork-oak. However the fibres do not run throughout the wood, nor do they run to a good length, nor are they all set symmetrically, but run in every direction. The wood dries while it is being planed and sawn.

[9] *Thyon* (thyine wood), which some call *thya*, grows near the temple of Zeus Ammon and in the district of Cyrene. In appearance the tree is like the cypress alike in its branches, its leaves, its stem, and its fruit; or rather it is like a wild cypress.[10] There

[6] *cf.* Plin. 16. 215.
[7] Of which the door itself is made.
[8] Plin. 16. 211. [9] Plin. 13. 100–102.
[10] κυπάριττος ἀγρία conj. Sch.; κυπάρισσον ἀγρίαν MAld.

νῦν ἡ πόλις ἐστί, καὶ ἔτι διαμνημονεύουσιν ὀροφάς τινας τῶν ἀρχαίων οὔσας. ἀσαπὲς γὰρ ὅλως τὸ ξύλον οὐλότατον δὲ τὴν ῥίζαν ἐστί· καὶ ἐκ ταύτης τὰ σπουδαιότατα ποιεῖται τῶν ἔργων. τὰ δὲ ἀγάλματα γλύφουσιν ἐκ τῶνδε, κέδρων κυπαρίττου λωτοῦ πύξου· τὰ δ' ἐλάττω καὶ ἐκ τῶν ἐλαΐνων ῥιζῶν· ἀρραγεῖς γὰρ αὗται καὶ ὁμαλῶς πως σαρκώδεις. ταῦτα μὲν οὖν ἰδιότητά τινα τόπων καὶ φύσεως καὶ χρείας ἀποδηλοῖ.

IV. Βαρέα δὲ καὶ κοῦφα δῆλον ὡς τῇ πυκνότητι καὶ μανότητι καὶ ὑγρότητι καὶ ξηρότητι καὶ τῷ γλοιώδει καὶ σκληρότητι καὶ μαλακότητι ληπτέον. ἔνια μὲν οὖν ἅμα σκληρὰ καὶ βαρέα, καθάπερ πύξος καὶ δρῦς· ὅσα δὲ κραῦρα καὶ τῇ ξηρότητι σκληρότατα, ταῦτ' οὐκ ἔχει βάρος. ἅπαντα δὲ τὰ ἄγρια τῶν ἡμέρων καὶ τὰ ἄρρενα τῶν θηλειῶν πυκνότερά τε καὶ σκληρότερα καὶ βαρύτερα καὶ τὸ ὅλον ἰσχυρότερα, καθάπερ καὶ πρότερον εἴπομεν. ὡς δ' ἐπὶ τὸ πᾶν καὶ τὰ ἀκαρπότερα τῶν καρπίμων καὶ τὰ χείρω τῶν καλλικαρποτέρων· εἰ μή που καρπιμώτερον τὸ ἄρρεν, ὥσπερ ἄλλα τέ φασι καὶ τὴν κυπάριττον καὶ τὴν κράνειαν. ἀλλὰ τῶν γε ἀμπέλων φανερῶς αἱ ὀλιγοκαρπότεραι καὶ πυκνοφθαλμότεραι καὶ στερεώτεραι· καὶ μηλεῶν δὲ καὶ τῶν ἄλλων ἡμέρων.

is abundance of it where now the city stands, and men can still recall that some of the roofs in ancient times were made of it. For the wood is absolutely proof against decay, and the root is of very compact texture, and they make of it the most valuable articles. Images are carved from these woods, prickly cedar cypress nettle-tree box, and the small ones also from the roots of the olive, which are unbreakable and of a more or less uniformly flesby character. The above facts illustrate certain special features of position, natural character and use.

Of differences in timber as to hardness and heaviness.

IV. Difference in weight is clearly to be determined by closeness or openness of texture, dampness or dryness, degree of glutinousness, hardness or softness. Now some woods are both hard and heavy, as box and oak, while those that are brittle and hardest owing to their dryness, are not heavy. [1] All wood of wild trees, as we have said before, is closer harder heavier, and in general stronger than that of the cultivated forms, and there is the same difference between the wood of 'male' and of 'female' trees, and in general between trees which bear no fruit and those which have fruit, and between those which bear inferior fruit and those whose fruit is better; on the other hand occasionally the 'male' tree is the more fruitful, for instance, it is said, the cypress the cornelian cherry and others. However of vines it is clear that those which bear less fruit have also more frequent knots and are more solid,[2] and so too with apples and other cultivated trees.

[1] Plin. 16. 211. [2] *cf. C.P.* 3. 11. 1.

2 Ἀσαπῆ δὲ φύσει κυπάριττος κέδρος ἔβενος λωτὸς πύξος ἐλάα κότινος πεύκη ἔνδᾳδος ἀρία δρῦς καρύα Εὐβοϊκή. τούτων δὲ χρονιώτατα δοκεῖ τὰ κυπαρίττινα εἶναι· τὰ γοῦν ἐν Ἐφέσῳ, ἐξ ὧν αἱ θύραι τοῦ νεωστὶ νεώ, τεθησαυρισμένα τέτταρας ἔκειτο γενεάς. μόνα δὲ καὶ στιλβηδόνα δέχεται, δι' ὃ καὶ τὰ σπουδαζόμενα τῶν ἔργων ἐκ τούτων ποιοῦσι. τῶν δὲ ἄλλων ἀσαπέστατον μετὰ τὰ κυπαρίττινα καὶ τὰ θυώδη τὴν συκάμινον εἶναί φασι, καὶ ἰσχυρὸν ἅμα καὶ εὔεργον τὸ ξύλον· γίνεται δὲ τὸ ξύλον [καὶ] παλαιούμενον μέλαν, ὥσπερ λωτός.

3 Ἔτι δὲ ἄλλο πρὸς ἄλλο καὶ ἐν ἄλλῳ ἀσαπές, οἷον πτελέα μὲν ἐν τῷ ἀέρι, δρῦς δὲ κατορυττομένη καὶ ἐν τῷ ὕδατι καταβρεχομένη· δοκεῖ γὰρ ὅλως ἀσαπὲς εἶναι· δι' ὃ καὶ εἰς τοὺς ποταμοὺς καὶ εἰς τὰς λίμνας ἐκ τούτων ναυπηγοῦσιν· ἐν δὲ τῇ θαλάττῃ σήπεται. τὰ δὲ ἄλλα διαμένει μᾶλλον, ὅπερ καὶ εὔλογον, ταριχευόμενα τῇ ἅλμῃ.

4 Δοκεῖ δὲ καὶ ἡ ὀξύη πρὸς τὸ ὕδωρ ἀσαπὴς εἶναι καὶ βελτίων γίνεσθαι βρεχομένη. καὶ ἡ καρύα δὲ ἡ Εὐβοϊκὴ ἀσαπής. φασὶ δὲ καὶ τὴν πεύκην ἐλάτης μᾶλλον ὑπὸ τερηδόνος ἐσθίεσθαι· τὴν μὲν γὰρ εἶναι ξηράν, τὴν δὲ πεύκην ἔχειν γλυκύτητα, καὶ ὅσῳ ἐνδᾳδωτέρα, μᾶλλον· πάντα

---

[1] Plin. 16. 213.

[2] τεθησαυρισμένα . . . ἔκειτο conj. Bentley; τεθησαυρισμένα . . . ἔκειντο Ald.H.; P has ἔκειτο.

Of differences in the keeping quality of timber.

[1] Naturally proof against decay are cypress prickly cedar ebony nettle-tree box olive wild olive resinous fir *aria* (holm-oak) oak sweet chestnut. Of these the wood of the cypress seems to last longest; at least the cypress-wood at Ephesus, of which the doors of the modern temple were made, lay stored up[2] for four generations. And this is the only wood which takes a fine polish, wherefore they make of it valuable articles. Of the others the least liable to decay after the wood of the cypress and thyine-wood is, they say, that of the mulberry, which is also strong and easily worked: when it becomes old, this wood turns black like that of the nettle-tree.

[3] Again whether a given wood is not liable to decay may depend on the purpose to which it is put and the conditions to which it is subjected: thus the elm does not decay if exposed to the air, nor the oak if it is buried or soaked in water; for it appears to be entirely proof against decay: wherefore they build vessels of it for use on rivers and on lakes, but in sea-water it rots, though other woods last all the better; which is natural, as they become seasoned with the brine.

[4] The beech also seems to be proof against decay in water and to be improved by being soaked. The sweet chestnut under like treatment is also proof against decay. They say that the wood of the fir is more liable to be eaten by the *teredon* than that of the silver-fir; for that the latter is dry, while the fir has a sweet taste, and that this is more so, the more the wood is soaked with resin[5]; they go on to

[3] Plin. 16. 218. [4] Plin. 16. 218 and 219.
[5] *cf.* 3. 9. 4.

δ' ἐσθίεσθαι τερηδόνι πλὴν κοτίνου καὶ ἐλάας· τὰ δὲ οὔ, διὰ τὴν πικρότητα. ἐσθίεται δὲ τὰ μὲν ἐν τῇ θαλάττῃ σηπόμενα ὑπὸ τερηδόνος, τὰ δ' ἐν τῇ γῇ ὑπὸ σκωλήκων καὶ ὑπὸ θριπῶν· οὐ γὰρ γίνεται τερηδὼν ἀλλ' ἢ ἐν τῇ θαλάττῃ. ἔστι δὲ ἡ τερηδὼν τῷ μὲν μεγέθει μικρόν, κεφαλὴν δ' ἔχει
5 μεγάλην καὶ ὀδόντας· οἱ δὲ θρίπες ὅμοιοι τοῖς σκώληξιν, ὑφ' ὧν τιτραίνεται κατὰ μικρὸν τὰ ξύλα. καὶ ἔστι ταῦτα εὐίατα· πιττοκοπηθέντα γὰρ ὅταν εἰς τὴν θάλατταν ἑλκυσθῇ στέγει· τὰ δὲ ὑπὸ τῶν τερηδόνων ἀνίατα. τῶν δὲ σκωλήκων τῶν ἐν τοῖς ξύλοις οἱ μέν εἰσιν ἐκ τῆς οἰκείας σήψεως, οἱ δ' ἐντικτόντων ἑτέρων· ἐντίκτει γάρ, ὥσπερ καὶ τοῖς δένδροις, ὁ κεράστης καλούμενος, ὅταν τιτράνῃ καὶ κοιλάνῃ περιστραφεὶς ὡσπερεὶ μυοδόχον. φεύγει δὲ τά τε ὀσμώδη καὶ πικρὰ καὶ σκληρὰ διὰ τὸ μὴ δύνασθαι τιτρᾶναι, καθάπερ
6 τὴν πύξον. φασὶ δὲ καὶ τὴν ἐλάτην φλοϊσθεῖσαν ὑπὸ τὴν βλάστησιν ἀσαπῆ διαμένειν ἐν τῷ ὕδατι· φανερὸν δὲ γενέσθαι ἐν Φενεῷ τῆς Ἀρκαδίας, ὅτε αὐτοῖς ἐλιμνώθη τὸ πεδίον φραχθέντος τοῦ βερέθρου· τότε γὰρ τὰς γεφύρας ποιοῦντες ἐλατίνας καί, ὅταν ἐπαναβαίνῃ τὸ ὕδωρ, ἄλλην καὶ ἄλλην ἐφιστάντες, ὡς ἐρράγη καὶ ἀπῆλθε, πάντα εὑρεθῆναι τὰ ξύλα ἀσαπῆ. τοῦτο μὲν οὖν ἐκ συμπτώματος.

---

[1] Plin. 16. 220 and 221.

[2] τιτραίνεται conj. Scal. from G; τιτρενεται UVo.; πεπαινεται MVAld.

[3] *cf* 4. 14. 5.

[4] ὡσπερεὶ μυοδόχον conj. W.; ὥσπερ οἱ μυόχοδοι MSS.; G omits. The word μυοδόχος does not occur elsewhere as a subst.

say that all woods are eaten by the *teredon* except the olive, wild or cultivated, and that these woods escape because of their bitter taste. [1] Now woods which decay in sea-water are eaten by the *teredon*, those which decay on land by the *skolex* and *thrips*; for the *teredon* does not occur except in the sea. It is a creature small in size, but has a large head and teeth; the *thrips* resembles the *skolex*, and these creatures gradually bore through [2] timber. The harm that these do is easy to remedy; for, if the wood is smeared with pitch, it does not let in water when it is dragged down into the sea; but the harm done by the *teredon* cannot be undone. Of the *skolekes* which occur in wood some come from the decay of the wood itself, some from other *skolekes* which engender therein. For these produce their young in timber, as the worm called the 'horned worm' [3] does in trees, having bored and scooped out a sort of mouse-hole [4] by turning round and round. But it avoids wood which has a strong smell or is bitter or hard, such as boxwood, since it is unable to bore through it. They say too that the wood of the silver-fir, if barked just before the time of budding, remains in water without decaying, and that this was clearly seen at Pheneos in Arcadia, when their plain was turned into a lake since the outlet was blocked up.[5] For at that time they made [6] their bridges of this wood, and, as the water rose, they placed more and more atop of them, and, when the water burst its way through and disappeared, all the wood was found to be undecayed. This fact then became known by means of an accident.

[5] *cf.* 3. 1. 2. φραχθέντος conj. Sch.; βραχέντος Ald. H.
[6] ποιοῦντες, ἐφιστάντες *nom. pendens.*

7 Ἐν Τύλῳ δὲ τῇ νήσῳ τῇ περὶ τὴν Ἀραβίαν εἶναί τί φασι ξύλον ἐξ οὗ τὰ πλοῖα ναυπηγοῦνται· τοῦτο δὲ ἐν μὲν τῇ θαλάττῃ σχεδὸν ἄσηπτον εἶναι· διαμένει γὰρ ἔτη πλείω ἢ διακόσια καταβυθιζόμενον· ἐὰν δὲ ἔξω, χρόνιον μὲν θᾶττον δὲ σήπεται. (θαυμαστὸν δὲ καὶ ἕτερον λέγουσι, οὐδὲν δὲ πρὸς τὴν σῆψιν. εἶναι γάρ τι δένδρον ἐξ οὗ τὰς βακτηρίας τέμνεσθαι, καὶ γίνεσθαι καλὰς σφόδρα ποικιλίαν τινὰ ἐχούσας ὁμοίαν τῷ τοῦ τίγριος δέρματι· βαρὺ δὲ σφόδρα τὸ ξύλον τοῦτο· ὅταν δέ τις ῥίψῃ πρὸς στερεώτερον τόπον, κατάγνυσθαι καθάπερ τὰ κεράμια.)

8 Καὶ τὸ τῆς μυρίκης δὲ ξύλον οὐχ ὥσπερ ἐνταῦθα ἀσθενές, ἀλλ' ἰσχυρὸν ὥσπερ πρίνινον ἢ καὶ ἄλλο τι τῶν ἰσχυρῶν. τοῦτο μὲν οὖν ἅμα μηνύει χώρας τε καὶ ἀέρος διαφορὰς καὶ δυνάμεις. τῶν δὲ ὁμογενῶν ξύλων, οἷον δρυΐνων πευκίνων, ὅταν ταριχεύωνται—ταριχεύουσι γὰρ οὐκ ἐν ἴσῳ βάθει πάντα δύοντες τῆς θαλάττης, ἀλλὰ τὰ μὲν πρὸς αὐτῇ τῇ γῇ, τὰ δὲ μικρὸν ἀνωτέρω, τὰ δ' ἐν πλείονι βάθει· πάντων δὲ τὰ πρὸς τὴν ῥίζαν θᾶττον δύεται καθ' ὕδατος, κἂν ἐπινῇ μᾶλλον ῥέπει κάτω.

V. Ἔστι δὲ τὰ μὲν εὔεργα τῶν ξύλων, τὰ δὲ δύσεργα· εὔεργα μὲν τὰ μαλακά, καὶ πάντων

---

[1] Plin. 16. 221; *cf.* 4. 7. 7.
[2] Teak. See Index App. (22).
[3] Calamander-wood. See Index App. (23).

[1] In the island of Tylos off the Arabian coast they say that there is a kind of wood[2] of which they build their ships, and that in sea-water this is almost proof against decay; for it lasts more than 200 years if it is kept under water, while, if it is kept out of water, it decays sooner, though not for some time. They also tell of another strange thing, though it has nothing to do with the question of decay: they say that there is a certain tree,[3] of which they cut their staves, and that these are very handsome, having a variegated appearance like the tiger's skin; and that this wood is exceedingly heavy, yet when one throws it down on hard ground[4] it breaks in pieces like pottery.

Moreover, the wood of the tamarisk[5] is not weak there, as it is in our country, but is as strong as kermes-oak or any other strong wood. Now this illustrates also the difference in properties caused by country and climate. Moreover when wood, such as that of oak or fir, is soaked in brine—not all being soaked at the same depth in the sea, but some of it close to shore, some rather further out, and some at a still greater depth—[6] in all cases the parts of the tree nearest the root (whichever tree it is) sink quicker under water, and even if they float, have a greater tendency to sink.

Which kinds of wood are easy and which hard to work. Of the core and its effects.

V. Some wood is easy to work, some difficult. Those woods which are soft are easy, and especially

[4] πρὸς στερ. τόπον can hardly be sound: ? 'on something harder than itself.'
[5] See Index, μυρίκη (2). [6] Plin. 16. 186.

μάλιστα φίλυρα· δύσεργα δὲ καὶ τὰ σκληρὰ καὶ τὰ ὀζώδη καὶ οὔλας ἔχοντα συστροφάς· δυσεργότατα δὲ ἀρία καὶ δρῦς, ὡς δὲ κατὰ μέρος ὁ τῆς πεύκης ὄζος καὶ τῆς ἐλάτης. ἀεὶ δὲ τῶν ὁμογενῶν τὸ μαλακώτερον τοῦ σκληροτέρου κρεῖττον· σαρκωδέστερον γάρ· καὶ εὐθὺ σκοποῦνται τὰς σανίδας οἱ τέκτονες οὕτως. τὰ δὲ μοχθηρὰ σιδήρια δύναται τέμνειν τὰ σκληρὰ μᾶλλον τῶν μαλακῶν· ἀνίησι γὰρ ἐν τοῖς μαλακοῖς, ὥσπερ ἐλέχθη περὶ τῆς φιλύρας, παρακονᾷ δὲ μάλιστα τὰ σκληρά· δι' ὃ καὶ οἱ σκυτοτόμοι ποιοῦνται τοὺς πίνακας ἀχράδος.

2 Μήτραν δὲ πάντα μὲν ἔχει φασὶν οἱ τέκτονες φανερὰν δ' εἶναι μάλιστα ἐν τῇ ἐλάτῃ· φαίνεσθαι γὰρ οἷον φλοιώδη τινὰ τὴν σύνθεσιν αὐτῆς τῶν κύκλων. ἐν ἐλάᾳ δὲ καὶ πύξω καὶ τοῖς τοιούτοις οὐχ ὁμοίως· δι' ὃ καὶ οὔ φασί τινες ἔχειν τῇ δυνάμει πύξον καὶ ἐλάαν· ἥκιστα γὰρ ἕλκεσθαι ταῦτα τῶν ξύλων. ἔστι δὲ τὸ ἕλκεσθαι τὸ συμπεριΐστασθαι κινουμένης τῆς μήτρας. ζῇ γὰρ ὡς ἔοικεν ἐπὶ χρόνον πολύν· δι' ὃ πανταχόθεν μὲν ἅμα μάλιστα δ' ἐκ τῶν θυρωμάτων ἐξαιροῦσιν, ὅπως ἀστραβῆ ᾖ· καὶ διὰ τοῦτο σχίζουσιν.

3 Ἄτοπον δ' ἂν δόξειεν ὅτι ἐν μὲν τοῖς ξύλοις τοῖς στρογγύλοις ἄλυπος ἡ μήτρα καὶ ἀκίνητος, ἐν δὲ τοῖς παρακινηθεῖσιν, ἐὰν μὴ ὅλως ἐξαιρεθῇ,

---

[1] 5. 3. 3.

[2] τὰ σκληρὰ conj. Sch. from G (?); ταῦτα P_2Ald.H.

[3] ἔχειν conj. Sch.; ἔχει ᾗ Ald.H.

[4] ἐλάαν conj. Scal. from G; ἐλάτην Ald.H.

[5] *i.e.* and this happens less in woods which have little core.

[6] ἅμα (? = ὁμοίως) MSS.; αὐτὴν conj. W.

that of the lime; those are difficult which are hard and have many knots and a compact and twisted grain. The most difficult woods are those of *aria* (holm-oak) and oak, and the knotty parts of the fir and silver-fir. The softer part of any given tree is always better than the harder, since it is fleshier: and carpenters can thus at once mark the parts suitable for planks. Inferior iron tools can cut hard wood better than soft: for on soft wood tools lose their edge, as was said[1] in speaking of the lime, while hard woods[2] actually sharpen it: wherefore cobblers make their strops of wild pear.

Carpenters say that all woods have[3] a core, but that it is most plainly seen in the silver-fir, in which one can detect a sort of bark-like character in the rings. In olive box and such woods this is not so obvious; wherefore they say that box and olive[4] lack this tendency; for that these woods are less apt to 'draw' than any others. 'Drawing' is the closing in of the wood as the core is disturbed.[5] For since the core remains alive, it appears, for a long time, it is always removed from any article whatever made of this wood,[6] but especially from doors,[7] so that they may not warp[8]: and that is why the wood is split.[9]

It might seem strange that in 'round'[10] timber the core does no harm and so is left undisturbed, while in wood whose texture has been interfered with,[11] unless it is taken out altogether, it causes

[7] θυρωμάτων conj. Sch.; γυρωμάτων Ald. *cf.* 4. 1. 2; Plin. 16. 225, *abietem valvarum paginis aptissimam.*

[8] ἀστραβῆ ᾖ conj. Dalec.; ἀστραβῆ UMVAld.

[9] *i.e.* to extract the core. [10] See below, § 5.

[11] παρακινηθεῖσι, *i.e.* by splitting or sawing. πελεκηθεῖσι conj. W.

κινεῖ καὶ παραστρέφει· μᾶλλον γὰρ εἰκὸς γυμνωθεῖσαν ἀποθνήσκειν. ὅμως δὲ οἵ γε ἱστοὶ καὶ
αἱ κεραῖαι ἐξαιρεθείσης ἀχρεῖοι. τοῦτο δὲ κατὰ
συμβεβηκός, ὅτι χιτῶνας ἔχει πλείους, ἰσχυρότατον δὲ καὶ λεπτότατον δὲ τὸν ἔσχατον, ξηρότα-
τον γάρ, καὶ τοὺς ἄλλους ἀνὰ λόγον. ὅταν οὖν
4 σχισθῇ, περιαιρεῖται τὰ ξηρότατα. εἰ δ' ἡ μήτρα
διὰ τὸ ξηρὸν σκεπτέον. διαστρέφει δὲ ἑλκομένη
τὰ ξύλα καὶ ἐν τοῖς σχιστοῖς καὶ πριστοῖς, ὅταν
μὴ ὡς δεῖ πρίωσι· δεῖ γὰρ ὀρθὴν τὴν πρίσιν εἶναι
καὶ μὴ πλαγίαν. οἷον οὔσης τῆς μήτρας ἐφ' ἣν
τὸ α, μὴ παρὰ τὴν βγ τέμνειν, ἀλλὰ παρὰ τὴν
βδ. φθείρεσθαι γὰρ οὕτω φασίν, ἐκείνως δὲ ζῆν.
ὅτι δὲ πᾶν ξύλον ἔχει μήτραν ἐκ τούτων οἴονται·
φανερὸν γάρ ἐστι καὶ τὰ μὴ δοκοῦντα πάντ' ἔχειν,
οἷον πύξον λωτὸν πρῖνον. σημεῖον δέ· τοὺς γὰρ
στρόφιγγας τῶν θυρῶν τῶν πολυτελῶν ποιοῦσι
μὲν ἐκ τούτων, συγγράφονται δὲ οἱ ἀρχιτέκτονες
οὕτως <μὴ> ἐκ μήτρας. ταὐτὸ δὲ τοῦτο σημεῖον
καὶ ὅτι πᾶσα μήτρα ἕλκεται, καὶ αἱ τῶν σκληρο-
5 τάτων, ἃς δή τινες καρδίας καλοῦσι. παντὸς δὲ

[1] And so cause no trouble.
[2] *cf.* 5. 1. 6. πλείους conj. Sch. from G; ἄλλους Ald.H.
[3] Text probably defective; ? insert ἐξηρέθη after ξηρὸν.
[4] The figure would seem to be

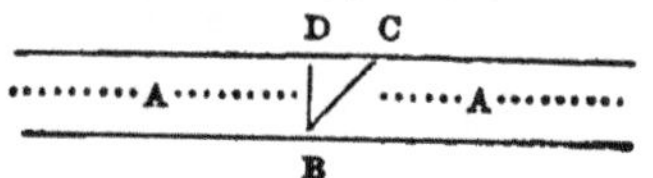

disturbance and warping: it were rather to be expected that it would die[1] when exposed. Yet it is a fact that masts and yard-arms are useless, if it has been removed from the wood of which they are made. This is however an accidental exception, because the wood in question has several coats,[2] of which the strongest and also thinnest is the outermost, since this is the driest, while the other coats are strong and thin in proportion to their nearness to the outermost. If therefore the wood be split, the driest parts are necessarily stripped off. Whether however in the other case the object of removing the core is to secure dryness is matter for enquiry.[3] However, when the core 'draws,' it twists the wood, whether it has been split or sawn, if the sawing is improperly performed: the saw-cut should be made straight and not slantwise. [4]Thus, if the core be represented by the line *A*, the cut must be made along the line *BD*, and not along the line *BC*: for in that case, they say, the core will be destroyed, while, if cut in the other way, it will live. For this reason men think that every wood has a core: for it is clear that those which do not seem to possess one nevertheless have it, as box nettle-tree kermes-oak: a proof of this is the fact that men make of these woods the pivots[5] of expensive doors, and accordingly[6] the headcraftsmen specify that wood with a core shall not[7] be used. This is also a proof that any core 'draws,' even those of the hardest woods, which some call the heart. In almost every wood, even

[5] *cf*, 5. 3. 5. *στρόφιγξ* here at least probably means 'pivot *and* socket.'
[6] *οὕτως* Ald.H.; *αὐτοὺς* conj. W. [7] *μὴ* add. W.

ὡς εἰπεῖν ξύλου σκληροτάτη καὶ μανοτάτη ἡ μήτρα, καὶ αὐτῆς τῆς ἐλάτης· μανοτάτη μὲν οὖν, ὅτι τὰς ἶνας ἔχει καὶ διὰ πολλοῦ καὶ τὸ σαρκῶδες τὸ ἀνὰ μέσον πολύ· σκληροτάτη δέ, ὅτι καὶ αἱ ἶνες σκληρόταται καὶ τὸ σαρκῶδες· δι' ὃ καὶ οἱ ἀρχιτέκτονες συγγράφονται παραιρεῖν τὰ πρὸς τὴν μήτραν, ὅπως λάβωσι τοῦ ξύλου τὸ πυκνότατον καὶ μαλακώτατον.

6 Τῶν δὲ ξύλων τὰ μὲν σχιστὰ τὰ δὲ πελεκητὰ τὰ δὲ στρογγύλα· σχιστὰ μέν, ὅσα διαιροῦντες κατὰ τὸ μέσον πρίζουσι· πελεκητὰ δέ, ὅσων ἀποπελεκῶσι τὰ ἔξω· στρογγύλα δὲ δῆλον ὅτι τὰ ὅλως ἄψαυστα. τούτων δὲ τὰ σχιστὰ μὲν ὅλως ἀρραγῆ διὰ τὸ γυμνωθεῖσαν τὴν μήτραν ξηραίνεσθαι καὶ ἀποθνήσκειν· τὰ δὲ πελεκητὰ καὶ τὰ στρογγύλα ῥήγνυται· μᾶλλον δὲ πολὺ τὰ στρογγύλα διὰ τὸ ἐναπειλῆφθαι τὴν μήτραν· οὐδὲν γὰρ ὅτι τῶν ἁπάντων οὐ ῥήγνυται. τοῖς δὲ λωτίνοις καὶ τοῖς ἄλλοις οἷς εἰς τοὺς στρόφιγγας χρῶνται πρὸς τὸ μὴ ῥήγνυσθαι βόλβιτον περιπλάττουσιν, ὅπως ἀναξηρανθῇ καὶ διαπνευσθῇ κατὰ μικρὸν ἡ ἐκ τῆς μήτρας ὑγρότης. ἡ μὲν οὖν μήτρα τοιαύτην ἔχει δύναμιν.

VI. Βάρος δὲ ἐνεγκεῖν ἰσχυρὰ καὶ ἡ ἐλάτη καὶ ἡ πεύκη πλάγιαι τιθέμεναι· οὐδὲν γὰρ ἐν-

---

[1] ξύλου σκληροτάτη conj. Sch. from G; ξύλον σκληρότατον UMV: so Ald. omitting καί.

[2] ἀποπελεκῶσι conj. Sch.; ἀποπλέκωσι UM; ἀποπλέκουσι Ald.; ἀποπελέκουσι mBas. [3] *cf. C.P.* 5. 17. 2.

in that of the silver-fir, the core is the hardest part,[1] and the part which has the least fibrous texture:—it is least fibrous because the fibres are far apart and there is a good deal of flesby matter between them, while it is the hardest part because the fibres and the fleshy substance are the hardest parts. Wherefore the headcraftsmen specify that the core and the parts next it are to be removed, that they may secure the closest and softest part of the wood.

Timber is either 'cleft,' 'hewn,' or 'round': it is called 'cleft,' when in making division they saw it down the middle, 'hewn' when they hew off[2] the outer parts, while 'round' clearly signifies wood which has not been touched at all. Of these, 'cleft' wood[3] is not at all liable to split, because the core when exposed dries and dies: but 'hewn' and 'round' wood are apt to split, and especially 'round' wood, because the core is included in it: no kind of timber indeed is altogether incapable of splitting. The wood of the nettle-tree and other kinds which are used for making pivots for doors are smeared[4] with cow-dung to prevent their splitting: the object being that the moisture due to the core may be gradually dried up[5] and evaporated. Such are the natural properties of the core.

Which woods can best support weight.

VI. [6] For bearing weight silver-fir and fir are strong woods, when set slantwise[7]: for they do not give like

[4] περιπλάττουσι conj. Sch. from G; περιπάττουσιν Ald.H. Plin. 16. 222. [5] ἀναξηρανθῇ conj. Sch.; ἀναξηραίνῃ Ald.H.
[6] Plin. 16. 222–224.
[7] *e.g.* as a strut. πλάγιαι conj. Sch. from Plin. *l.c.*; ἁπαλαὶ Ald.H.

διδόασιν, ὥσπερ ἡ δρῦς καὶ τὰ γεώδη, ἀλλ' ἀντωθοῦσι· σημεῖον δὲ ὅτι οὐδέποτε ῥήγνυνται, καθάπερ ἐλάα καὶ δρῦς, ἀλλὰ πρότερον σήπονται καὶ ἄλλως ἀπαυδῶσιν. ἰσχυρὸν δὲ καὶ ὁ φοῖνιξ· ἀνάπαλιν γὰρ ἡ κάμψις ἢ τοῖς ἄλλοις γίνεται· τὰ μὲν γὰρ εἰς τὰ κάτω κάμπτεται, ὁ δὲ φοῖνιξ εἰς τὰ ἄνω. φασὶ δὲ καὶ τὴν πεύκην καὶ τὴν ἐλάτην ἀντωθεῖν. τὸ δὲ τῆς Εὐβοϊκῆς καρύας, γίνεται γὰρ μέγα καὶ χρῶνται πρὸς τὴν ἔρεψιν, ὅταν μέλλῃ ῥήγνυσθαι ψοφεῖν ὥστε προαισθάνεσθαι πρότερον· ὅπερ καὶ ἐν Ἀντάνδρῳ συνέπεσεν ἐν τῷ βαλανείῳ καὶ πάντες ἐξεπήδησαν. ἰσχυρὸν δὲ καὶ τὸ τῆς συκῆς πλὴν εἰς ὀρθόν.

2 Ἡ δὲ ἐλάτη μάλιστα ὡς εἰπεῖν ἰσχυρόν. πρὸς δὲ τὰς τῶν τεκτόνων χρείας ἐχέκολλον μὲν μάλιστα ἡ πεύκη διά τε τὴν μανότητα καὶ τὴν εὐθυπορίαν· οὐδὲ γὰρ ὅλως οὐδὲ ῥήγνυσθαί φασιν ἐὰν κολληθῇ. εὐτορνότατον δὲ φιλύκη, καὶ ἡ λευκότης ὥσπερ ἡ τοῦ κηλάστρου. τῶν δὲ ἄλλων ἡ φίλυρα· τὸ γὰρ ὅλον εὔεργον, ὥσπερ ἐλέχθη, διὰ μαλακότητα. εὔκαμπτα δὲ ὡς μὲν ἁπλῶς εἰπεῖν ὅσα γλίσχρα. διαφέρειν δὲ δοκεῖ συκάμινος καὶ ἐρινεός, δι' ὃ καὶ τὰ ἴκρια καὶ τὰς στεφάνας καὶ ὅλως ὅσα περὶ τὸν κόσμον ἐκ τούτων ποιοῦσι.

3 Εὔπριστα δὲ καὶ εὔσχιστα τὰ ἐνικμότερα τῶν

---

[1] *i.e.* the strut becomes concave or convex respectively. *cf.* Xen. *Cyr.* 7. 5. 11.

[2] *i.e.* it cannot be used as a strut, or it would 'buckle,' though it will stand a vertical strain.

[3] Plin. 16. 225.

[4] *cf. C.P.* 5. 17. 3. εὐθυπορώτατα : εὐθυπορίαν.

oak and other woods which contain mineral matter, but make good resistance. A proof of this is that they never split like olive and oak, but decay first or fail in some other way. Palm-wood is also strong, for it bends the opposite way to other woods: they bend downwards, palm-wood upwards.[1] It is said that fir and silver-fir also have an upward thrust. As to the sweet chestnut, which grows tall and is used for roofing, it is said that when it is about to split, it makes a noise, so that men are forewarned: this occurred once at Antandros at the baths, and all those present rushed out. Fig-wood is also strong, but only when set upright.[2]

Of the woods best suited for the carpenter's various purposes.

3 The wood of the silver-fir may be called the strongest of all. But for the carpenter's purposes fir best takes glue because of its open texture and the straightness of its pores[4]; for they say that it never by any chance comes apart when it is glued. Alaternus[5] is the easiest wood for turning, and its whiteness is like that of the holly. Of the rest lime is the easiest, the whole tree, as was said, being easy to work because of the softness of the wood. In general those woods which are tough are easy to bend. The mulberry and the wild fig seem to be specially so; wherefore they make of these theatre-seats,[6] the hoops of garlands, and, in general, things for ornament.

[7] Woods which have a fair amount of moisture in them are easier to saw or split than those which

[5] *cf.* 5. 7. 7.

[6] Rendering doubtful. ἴκρια has probably here some unknown meaning, on which the sense of κόσμον depends.

[7] Plin. 16. 227.

πάμπαν ξηρῶν· τὰ μὲν γὰρ παύονται, τὰ δὲ ἵστανται· τὰ δὲ χλωρὰ λίαν συμμύει καὶ ἐνέχεται ἐν τοῖς ὀδοῦσι τὰ πρίσματα καὶ ἐμπλάττει, δι' ὃ καὶ παραλλάττουσιν ἀλλήλων τοὺς ὀδόντας ἵνα ἐξάγηται. ἔστι δὲ καὶ δυστρυπητότερα τὰ λίαν χλωρά· βραδέως γὰρ ἀναφερεται τὰ ἐκτρυπήματα διὰ τὸ βαρέα εἶναι· τῶν δὲ ξηρῶν ταχέως καὶ εὐθὺς ὁ ἀὴρ ἀναθερμαινόμενος ἀναδίδωσι· πάλιν δὲ τὰ λίαν ξηρὰ διὰ τὴν σκληρότητα δύσπριστα· καθάπερ γὰρ ὄστρακον συμβαίνει πρίειν, δι' ὃ καὶ τρυπῶντες ἐπιβρέχουσιν.

4 Εὐπελεκητότερα δὲ καὶ εὐτορνότερα καὶ εὐξοώτερα τὰ χλωρά· προσκάθηταί τε γὰρ τὸ τορνευτήριον μᾶλλον καὶ οὐκ ἀποπηδᾷ. καὶ ἡ πελέκησις τῶν μαλακωτέρων ῥᾴων, καὶ ἡ ξέσις δὲ ὁμοίως καὶ ἔτι λειοτέρα. ἰσχυρότατον δὲ καὶ ἡ κράνεια, τῶν δὲ ἄλλων οὐχ ἥκιστα ἡ πτελέα, δι' ὃ καὶ τοὺς στροφέας, ὥσπερ ἐλέχθη, ταῖς θύραις πτελείνους ποιοῦσιν. ὑγρότατον δὲ μελία καὶ ὀξύη· καὶ γὰρ τὰ κλινάρια τὰ ἐνδιδόντα ἐκ τούτων.

VII. Ὅλως δὲ πρὸς ποῖα τῆς ὕλης ἑκάστη χρησίμη καὶ ποία ναυπηγήσιμος καὶ οἰκοδομική, πλείστη γὰρ αὕτη ἡ χρεία καὶ ἐν μεγίστοις, πειρατέον εἰπεῖν, ἀφορίζοντα καθ' ἕκαστον τὸ χρήσιμον.

Ἐλάτη μὲν οὖν καὶ πεύκη καὶ κέδρος ὡς ἁπλῶς

---

[1] *παύονται* can hardly be right: Plin. *l.c.* seems to have had a fuller text.

[2] *ἐμπλάττει*: *cf. de Sens.* 66.

are altogether dry: for the latter give,[1] while the former resist. Wood which is too green closes up again when sawn, and the sawdust catches in the saw's teeth and clogs[2] them; wherefore the teeth of the saw are set alternate ways, to get rid of the sawdust. Wood which is too green is also harder to bore holes in; for the auger's dust is only brought up slowly, because it is heavy; while, if the wood is dry, the air gets warmed by the boring and brings it up readily and at once. On the other hand, wood which is over dry[3] is hard to saw because of its hardness: for it is like sawing through earthenware; wherefore they wet the auger when using it.

However green wood is easier to work with the axe the chisel or the plane; for the chisel gets a better hold and does not slip off. Again softer woods are easier for the axe and for smoothing,[4] and also a better polished surface is obtained. The cornelian cherry is also a very strong wood, and among the rest elm-wood is the strongest; wherefore, as was said,[5] they make the 'hinges' for doors of elm-wood. Manna-ash and beech have very moist wood, for of these they make elastic bedsteads.

Of the woods used in ship-building.

VII. Next we must endeavour to say in a general way, distinguishing the several uses, for which purposes each kind of timber is serviceable, which is of use for ship-building, which for house-building: for these uses extend far and are important.

Now silver-fir, fir and Syrian cedar[6] are, generally

[3] τὰ λίαν ξηρὰ conj. St.; λεῖα καὶ ξηρὰ Ald.H.
[4] Sc. with the carpenter's axe.
[5] 5. 3. 5. [6] See Index.

εἰπεῖν ναυπηγήσιμα· τὰς μὲν γὰρ τριήρεις καὶ τὰ μακρὰ πλοῖα ἐλάτινα ποιοῦσι διὰ κουφότητα, τὰ δὲ στρογγύλα πεύκινα διὰ τὸ ἀσαπές· ἔνιοι δὲ καὶ τὰς τριήρεις διὰ τὸ μὴ εὐπορεῖν ἐλάτης. οἱ δὲ κατὰ Συρίαν καὶ Φοινίκην ἐκ κέδρου· σπανίζουσι γὰρ καὶ πεύκης. οἱ δ' ἐν Κύπρῳ πίτυος· ταύτην γὰρ ἡ νῆσος ἔχει καὶ δοκεῖ κρείττων εἶναι τῆς
2 πεύκης. καὶ τὰ μὲν ἄλλα ἐκ τούτων· τὴν δὲ τρόπιν τριήρει μὲν δρυΐνην, ἵνα ἀντέχῃ πρὸς τὰς νεωλκίας, ταῖς δὲ ὁλκάσι πευκίνην· ὑποτιθέασι δ' ἔτι καὶ δρυΐνην ἐπὰν νεωλκῶσι, ταῖς δ' ἐλάττοσιν ὀξυΐνην· καὶ ὅλως ἐκ τούτου τὸ χέλυσμα.

Οὐχ ἅπτεται δὲ οὐδὲ κατὰ τὴν κόλλησιν ὁμοίως τὸ δρύϊνον τῶν πευκίνων καὶ ἐλατίνων· τὰ μὲν γὰρ πυκνὰ τὰ δὲ μανά, καὶ τὰ μὲν ὅμοια τὰ δ' οὔ. δεῖ δὲ ὁμοιοπαθῆ εἶναι τὰ μέλλοντα συμφύεσθαι καὶ μὴ ἐναντία, καθαπερανεὶ λίθον καὶ ξύλον.

3 Ἡ δὲ τορνεία τοῖς μὲν πλοίοις γίνεται συκαμίνου μελίας πτελέας πλατάνου· γλισχρότητα γὰρ ἔχειν δεῖ καὶ ἰσχύν. χειρίστη δὲ ἡ τῆς πλατάνου· ταχὺ γὰρ σήπεται. ταῖς δὲ τριήρεσιν ἔνιοι καὶ πιτυΐνας ποιοῦσι διὰ τὸ ἐλαφρόν. τὸ δὲ στερέωμα, πρὸς ᾧ τὸ χέλυσμα, καὶ τὰς ἐπωτίδας, μελίας καὶ συκαμίνου καὶ πτελέας· ἰσχυρὰ

---

[1] τριήρει conj. W.; τριήρη U; τριήρης MV; τριήρεσι Ald.

[2] ταῖς δ' ἐλάττοσιν ὀξυΐνην conj. W. (τοῖς Sch.); τοῖς μὲν ἐλάττοσιν ὀξύη Ald. *cf.* Plin. 16. 226.

[3] χέλυσμα, a temporary covering for the bottom: so Poll. and Hesych. explain.

speaking, useful for ship-building; for triremes and long ships are made of silver-fir, because of its lightness, and merchant ships of fir, because it does not decay; while some make triremes of it also because they are ill provided with silver-fir. The people of Syria and Phoenicia use Syrian cedar, since they cannot obtain much fir either; while the people of Cyprus use Aleppo pine, since their island provides this and it seems to be superior to their fir. Most parts are made of these woods; but the keel for a trireme[1] is made of oak, that it may stand the hauling; and for merchantmen it is made of fir. However they put an oaken keel under this when they are hauling, or for smaller vessels a keel of beech;[2] and the sheathing[3] is made entirely of this wood.

[4] (However oak-wood does not join well with glue on to fir or silver-fir; for the one is of close, the other of open grain, the one is uniform, the other not so; whereas things which are to be made into one piece should be of similar character, and not of opposite character, like wood and stone.)

The work of bentwood[5] for vessels is made of mulberry manna-ash elm or plane; for it must be tough and strong. That made of plane-wood is the worst, since it soon decays. For triremes some make such parts of Aleppo pine because of its lightness. The cutwater,[6] to which the sheathing is attached,[7] and the catheads are made of manna-ash mulberry

[4] This sentence is out of place; its right place is perhaps at the end of § 4.

[5] τορνεία; but the word is perhaps corrupt: one would expect the name of some part of the vessel.

[6] στερέωμα: apparently the fore part of the keel; =στεῖρα.

[7] πρὸς ᾧ τὸ χέλυσμα conj. W. after Scal,; πρόσω· τὸ σχέλυσμα Ald. (σχέλομα M, χέλυσμα U) πρόσω· τὸ δὲ χέλυσμα mBas.

γὰρ δεῖ ταῦτ' εἶναι. ναυπηγήσιμος μὲν οὖν ὕλη σχεδὸν αὕτη.

4 Οἰκοδομικὴ δὲ πολλῷ πλείων, ἐλάτη τε καὶ πεύκη καὶ κέδρος, ἔτι κυπάριττος δρῦς καὶ ἄρκευθος· ὡς δ' ἁπλῶς εἰπεῖν πᾶσα χρησίμη πλὴν εἴ τις ἀσθενὴς πάμπαν· οὐκ εἰς ταὐτὸ γὰρ πᾶσαι, καθάπερ οὐδ' ἐπὶ τῆς ναυπηγίας. αἱ δ' ἄλλαι πρὸς τὰ ἴδια τῶν τεχνῶν, οἷον σκεύη καὶ ὄργανα καὶ εἴ τι τοιοῦτον ἕτερον. πρὸς πλεῖστα δὲ σχεδὸν ἡ ἐλάτη παρέχεται χρείαν· καὶ γὰρ πρὸς τοὺς πίνακας τοὺς γραφομένους. τεκτονικῇ μὲν οὖν ἡ παλαιοτάτη κρατίστη, ἐὰν ᾖ ἀσαπής· εὐθετεῖ γὰρ ὡς εἰπεῖν πᾶσι χρῆσθαι· ναυπηγικῇ δὲ διὰ τὴν κάμψιν ἐνικμοτέρᾳ ἀναγκαῖον· ἐπεὶ πρός γε τὴν κόλλησιν ἡ ξηροτέρα συμφέρει. ἵσταται γὰρ καινὰ τὰ ναυπηγούμενα καὶ ὅταν συμπαγῇ καθελκυσθέντα συμμύει καὶ στέγει, πλὴν ἐὰν μὴ παντάπασιν ἐξικμασθῇ· τότε δὲ οὐ δέχεται κόλλησιν ἢ οὐχ ὁμοίως.

5 Δεῖ δὲ καὶ καθ' ἕκαστον λαμβάνειν εἰς ποῖα χρήσιμός ἐστιν. ἐλάτη μὲν οὖν καὶ πεύκη, καθάπερ εἴρηται, καὶ πρὸς ναυπηγίαν καὶ πρὸς

---

[1] ἐλάτη . . . ἄρκευθος conj. W.; ἐλάτη τε καὶ πεύκη καὶ κέδρος ἔτι κυπάριττος δρῦς πεύκη καὶ κέδρος ἄρκευθος U; ἐλάτη τε καὶ πεύκη καὶ κέδρος καὶ ἄρκευθος Ald.H.: so also MV, omitting καὶ before ἀρκ.

[2] ὡς δ' ἁπλῶς conj. Sch.; ἁπλῶς δ' ὡς Ald.

[3] καινὰ conj. Sch.; καὶ νῦν Ald.

[4] συμπαγῇ conj. W., which he renders 'when it has been glued together'; συμπίῃ Ald. G's reading was evidently different.

and elm; for these parts must be strong. Such then is the timber used in ship-building.

Of the woods used in house-building.

For house-building a much greater variety is used, silver-fir fir and prickly cedar; also cypress oak and Phoenician cedar.[1] In fact, to speak generally,[2] any wood is here of service, unless it is altogether weak: for there are various purposes for which different woods are serviceable, just as there are in ship-building. While other woods are serviceable for special articles belonging to various crafts, such as furniture tools and the like, the wood of silver-fir is of use for almost more purposes than any other wood; for it is even used for painters' tablets. For carpentry the oldest wood is the best, provided that it has not decayed; for it is convenient for almost anyone to use. But for ship-building, where bending is necessary, one must use wood which contains more moisture (though, where glue is to be used, drier wood is convenient). For timber-work for ships is set to stand when it is newly[3] made: then, when it has become firmly united,[4] it is dragged down to the water, and then it closes up and becomes watertight,—unless[5] all the moisture has been dried out of it, in which case it will not take the glue, or will not take it so well.

Of the uses of the wood of particular trees.

But we must consider for what purposes[6] each several wood is serviceable. Silver-fir and fir, as has been said, are suitable both for ship-building house-

[5] πλὴν ἐὰν μὴ conj. W.; π. ἐάν τε M; π. ἐάν γε Ald.
[6] *i.e.* apart from ship-building and house-building, in which *several* woods are used.

οἰκοδομίαν καὶ ἔτι πρὸς ἄλλα τῶν ἔργων, εἰς πλείω δὲ ἡ ἐλάτη. πίτυϊ δὲ χρῶνται μὲν εἰς ἄμφω καὶ οὐχ ἧττον εἰς ναυπηγίαν, οὐ μὴν ἀλλὰ ταχὺ διασήπεται. δρῦς δὲ πρὸς οἰκοδομίαν καὶ πρὸς ναυπηγίαν ἔτι τε πρὸς τὰ κατὰ γῆς κατορυττόμενα. φίλυρα δὲ πρὸς τὰ σανιδώματα τῶν μακρῶν πλοίων καὶ πρὸς κιβώτια καὶ πρὸς τὴν τῶν μέτρων κατασκευήν. ἔχει δὲ καὶ τὸν φλοιὸν χρήσιμον πρός τε τὰ σχοινία καὶ πρὸς τὰς κίστας· ποιοῦσι γὰρ ἐξ αὐτῆς.

6 Σφένδαμνός τε καὶ ζυγία πρὸς κλινοπηγίαν καὶ πρὸς τὰ ζυγὰ τῶν λοφούρων. μίλος δὲ εἰς παρακολλήματα κιβώτοις καὶ ὑποβάθροις καὶ ὅλως τοῖς τοιούτοις. πρῖνος δὲ πρὸς ἄξονας ταῖς μονοστρόφοις ἁμάξαις καὶ εἰς ζυγὰ λύραις καὶ ψαλτηρίοις. ὀξύη δὲ πρὸς ἁμαξοπηγίαν καὶ διφροπηγίαν τὴν εὐτελῆ. πτελέα δὲ πρὸς θυροπηγίαν καὶ γαλεάγρας· χρῶνται δὲ καὶ εἰς τὰ ἁμαξικὰ μετρίως. πηδὸς δὲ εἰς ἄξονάς τε ταῖς ἁμάξαις καὶ εἰς ἔλκηθρα τοῖς ἀρότροις. ἀνδράχλη δὲ ταῖς γυναιξὶν εἰς τὰ περὶ τοὺς ἱστούς. ἄρκευθος δὲ εἰς τεκτονίας καὶ εἰς τὰ ὑπαίθρια καὶ εἰς τὰ κατορυττόμενα κατὰ γῆς διὰ τὸ ἀσαπές.

7 ὡσαύτως δὲ καὶ ἡ Εὐβοϊκὴ καρύα, καὶ πρός γε τὴν κατόρυξιν ἔτι μᾶλλον ἀσαπής. πύξῳ δὲ χρῶνται μὲν πρὸς ἔνια, οὐ μὴν ἀλλ' ἥ γε ἐν τῷ Ὀλύμπῳ γινομένη διὰ τὸ βραχεῖά τε εἶναι καὶ ὀζώδης ἀχρεῖος. τερμίνθῳ δὲ οὐδὲν χρῶνται

[1] κίστας: *cf.* 3. 13. 1; perhaps 'hampers,' *cf.* 5. 7. 7.
[2] παρακολλήματα: lit. 'things glued on.'
[3] Plin. 16. 229.
[4] ταῖς μονοστρόφοις ἁμάξαις: or, perhaps, 'the wheels of

building and also for other kinds of work, but silver-fir is of use for more purposes than fir. Aleppo pine is used for both kinds of building, but especially for ship-building, yet it soon rots. Oak is used for house-building, for ship-building, and also for underground work; lime for the deck-planks of long ships, for boxes, and for the manufacture of measures; its bark is also useful for ropes and writing-cases,[1] for these are sometimes made of it.

Maple and *zygia* are used for making beds and the yokes of beasts of burden: yew for the ornamental work attached[2] to chests and footstools and the like: kermes-oak[3] for the axles of wheel-barrows[4] and the cross-bars of lyres and psalteries: beech for making waggons and cheap carts: elm for making doors and weasel-traps, and to some extent it is also used for waggon work; *pedos*[5] for waggon-axles and the stocks of ploughs: andrachne is used for women for parts of the loom: Phoenician cedar for carpenters' work[6] and for work which is either to be exposed to the air or buried underground, because it does not decay. Similarly the sweet chestnut is used, and it is even less likely to decay if it is used for underground work. Box is used for some purposes; however that which grows on Mount Olympus[7] is useless, because only short pieces can be obtained and the wood[8] is full of knots. Terebinth is not used,[9] except the fruit and the resin.

carts with solid wheels.' ταῖς conj. Sch.; τε καὶ UMV; τε καὶ μονοστρόφους ἁμάξας Ald.

[5] πηδος (with varying accent) MSS.: probably = πάδος, 4. 1. 3; πύξος Ald., but see § 7.

[6] τεκτονίας can hardly be right.

[7] *cf.* 3. 15. 5.

[8] *cf.* 1. 8. 2, of box in general; Plin. 16. 71.

[9] Inconsistent with 5. 3. 2.

πλὴν τῷ καρπῷ καὶ τῇ ῥητίνῃ. οὐδὲ φιλυκῃ πλὴν τοῖς προβάτοις· ἀεὶ γάρ ἐστι δασεῖα. τῇ δὲ ἀφάρκῃ εἰς χάρακάς τε καὶ τὸ καίειν. κηλάστρῳ δὲ καὶ σημύδᾳ πρὸς βακτηρίας. ἔνιοι δὲ καὶ δάφνῃ· τὰς γὰρ γεροντικὰς καὶ κούφας ταύτης ποιοῦσιν. ἰτέᾳ δὲ πρός τε τὰς ἀσπίδας καὶ τὰς κίστας καὶ τὰ κανᾶ καὶ τἆλλα. προσαναλαβεῖν δέ ἐστι καὶ τῶν ἄλλων ἕκαστον ὁμοίως.

8 Διῄρηται δὲ καὶ πρὸς τὰ τεκτονικὰ τῶν ὀργάνων ἕκαστα κατὰ τὴν χρείαν· οἷον σφυρίον μὲν καὶ τερέτριον ἄριστα μὲν γίνεται κοτίνου· χρῶνται δὲ καὶ πυξίνοις καὶ πτελεΐνοις καὶ μελεΐνοις· τὰς δὲ μεγάλας σφύρας πιτυΐνας ποιοῦσιν. ὁμοίως δὲ καὶ τῶν ἄλλων ἕκαστον ἔχει τινὰ τάξιν. καὶ ταῦτα μὲν αἱ χρεῖαι διαιροῦσιν.

VIII. Ἑκάστη δὲ τῆς ὕλης, ὥσπερ καὶ πρότερον ἐλέχθη, διαφέρει κατὰ τοὺς τόπους· ἔνθα μὲν γὰρ λωτὸς ἔνθα δὲ κέδρος γίνεται θαυμαστή, καθάπερ καὶ περὶ Συρίαν· ἐν Συρίᾳ γὰρ ἔν τε τοῖς ὄρεσι διαφέροντα γίνεται τὰ δένδρα τῆς κέδρου καὶ τῷ ὕψει καὶ τῷ πάχει· τηλικαῦτα γάρ ἐστιν ὥστ' ἔνια μὲν μὴ δύνασθαι τρεῖς ἄνδρας περιλαμβάνειν· ἔν τε τοῖς παραδείσοις ἔτι μείζω καὶ καλλίω. φαίνεται δὲ καὶ ἐάν τις ἐᾷ καὶ μὴ τέμνῃ τόπον οἰκεῖον ἕκαστον ἔχον γίνεσθαι θαυμαστὸν τῷ μήκει καὶ πάχει. ἐν Κύπρῳ γοῦν οὐκ ἔτεμνον οἱ βασιλεῖς, ἅμα μὲν τηροῦντες καὶ ταμιευόμενοι, ἅμα

---

1 Inconsistent with 5. 6. 2. φιλυρέα conj. Sch.

2 καὶ σημύδα conj. Sch.; καὶ μυῖα U; καὶ μύα Ald. *cf.* 3. 14. 4.

[1] Alaternus is only useful for feeding sheep; for it is always leafy. Hybrid arbutus is used for making stakes and for burning: holly and Judas-tree[2] for walking-sticks: some also use bay for these; for of this[3] they make light sticks and sticks for old men. Willow is used for shields hampers baskets and the like. We might in like manner add the several uses of the other woods.

[4] Distinction is also made between woods according as they are serviceable for one or other of the carpenter's tools: thus hammers and gimlets are best made of wild olive, but box elm and manna-ash are also used, while large mallets are made of Aleppo pine. In like manner there is a regular practice about each of the other tools. Such are the differences as to the uses of various woods.

Of the localities in which the best timber grows.

VIII. Each kind of timber, as was said before, differs according to the place[5] where it grows; in one place nettle-tree, in another the cedar is remarkably fine, for instance in Syria; for in Syria and on its mountains the cedars grow to a surpassing height and thickness: they are sometimes so large that three men cannot embrace the tree. And in the parks they are even larger and finer. It appears that any tree, if it is left alone in its natural position and not cut down, grows to a remarkable height and thickness. For instance in Cyprus the kings used not to cut the trees, both because they took great care of them and hus-

[3] ταύτης conj. H.; ταύτας UMVAld.
[4] Plin. 16. 230.
[5] τόπους conj. Scal. from G; πόδας Ald.

δὲ καὶ διὰ τὸ δυσκόμιστον εἶναι. μῆκος μὲν ἦν τῶν εἰς τὴν ἑνδεκήρη τὴν Δημητρίου τμηθέντων τρισκαιδεκαόργυιον, αὐτὰ δὲ τὰ ξύλα τῷ μήκει θαυμαστὰ καὶ ἄοζα καὶ λεῖα. μέγιστα δὲ καὶ παρὰ πολὺ τὰ ἐν τῇ Κύρνῳ φασὶν εἶναι· τῶν γὰρ ἐν τῇ Λατίνῃ καλῶν γινομένων ὑπερβολῇ καὶ τῶν ἐλατίνων καὶ τῶν πευκίνων—μείζω γὰρ ταῦτα καὶ καλλίω τῶν Ἰταλικῶν—οὐδὲν εἶναι
2 πρὸς τὰ ἐν τῇ Κύρνῳ. πλεῦσαι γάρ ποτε τοὺς Ῥωμαίους βουλομένους κατασκευάσασθαι πόλιν ἐν τῇ νήσῳ πέντε καὶ εἴκοσι ναυσί, καὶ τηλικοῦτον εἶναι τὸ μέγεθος τῶν δένδρων ὥστε εἰσπλέοντας εἰς κόλπους τινὰς καὶ λιμένας διασχισθεῖσι τοῖς ἱστοῖς ἐπικινδυνεῦσαι. καὶ ὅλως δὲ πᾶσαν τὴν νῆσον δασεῖαν καὶ ὥσπερ ἠγριωμένην τῇ ὕλῃ· δι' ὃ καὶ ἀποστῆναι τὴν πόλιν οἰκίζειν· διαβάντας δέ τινας ἀποτεμέσθαι πάμπολυ πλῆθος ἐκ τόπου βραχέος, ὥστε τηλικαύτην ποιῆσαι σχεδίαν ἣ ἐχρήσατο πεντήκοντα ἱστίοις· οὐ μὴν ἀλλὰ διαπεσεῖν αὐτὴν ἐν τῷ πελάγει. Κύρνος μὲν οὖν εἴτε διὰ τὴν ἄνεσιν εἴτε καὶ τὸ ἔδαφος καὶ τὸν ἀέρα πολὺ διαφέρει τῶν ἄλλων.

3 Ἡ δὲ τῶν Λατίνων ἔφυδρος πᾶσα· καὶ ἡ μὲν πεδεινὴ δάφνην ἔχει καὶ μυρρίνους καὶ ὀξύην θαυμαστήν· τηλικαῦτα γὰρ τὰ μήκη τέμνουσι ὥστ' εἶναι διανεκῶς τῶν Τυρρηνίδων ὑπὸ τὴν τρόπιν· ἡ δὲ ὀρεινὴ πεύκην καὶ ἐλάτην. τὸ δὲ

---

[1] Demetrius Poliorcetes. *cf.* Plut. *Demetr.* 43; Plin. 16. 203.

[2] ἐπικινδυνεῦσαι conj. W.; ἐπὶ τὸν πύκνον Ald.; so U, but πυκνον.

[3] *i.e.* against the overhanging trees. ? ἱστίοις, to which διασχ. is more appropriate.

banded them, and also because the transport of the timber was difficult. The timbers cut for Demetrius'[1] ship of eleven banks of oars were thirteen fathoms long, and the timbers themselves were without knots and smooth, as well as of marvellous length. But largest of all, they say, are the trees of Corsica; for whereas silver-fir and fir grow in Latium to a very great size, and are taller and finer than the silver-firs and firs of South Italy, these are said to be nothing to the trees of Corsica. For it is told how the Romans once made an expedition to that island with twenty-five ships, wishing to found a city there; and so great was the size of the trees that, as they sailed into certain bays and creeks, they got into difficulties[2] through breaking their masts.[3] And in general it is said that the whole island is thickly wooded and, as it were, one wild forest; wherefore the Romans gave up the idea of founding their city: however some of them made an excursion[4] into the island and cleared away a large quantity of trees from a small area, enough to make a raft with fifty sails;[5] but this broke up in the open sea. Corsica then, whether because of its uncultivated condition or because of its soil and climate, is very superior in trees to other countries.

The country of the Latins is all well watered; the lowland part contains bay, myrtle, and wonderful beech: they cut timbers of it of such a size that they will run the whole length[6] of the keel of a Tyrrhenian vessel. The hill country produces fir and silver-fir. The district called by Circe's name is, it

[4] *διαβάντας δέ τινας* conj. St. from G; *διαβάντα δέ τινα* Ald.H.
[5] *ἣ ἐχρήσατο πεντ. ἱστ.* conj. Sch.; *ᾗ ἐχρήσαντο οἱ* Ald.H.
[6] *διανεκῶς* conj. Sch.; *διὰ νεὼς* Ald.

Κιρκαῖον καλούμενον εἶναι μὲν ἄκραν ὑψηλήν, δασεῖαν δὲ σφόδρα καὶ ἔχειν δρῦν καὶ δάφνην πολλὴν καὶ μυρρίνους. λέγειν δὲ τοὺς ἐγχωρίους ὡς ἐνταῦθα ἡ Κίρκη κατῴκει καὶ δεικνύναι τὸν τοῦ Ἐλπήνορος τάφον, ἐξ οὗ φύονται μυρρίναι καθάπερ αἱ στεφανώτιδες τῶν ἄλλων ὄντων μεγάλων μυρρίνων. τὸν δὲ τόπον εἶναι καὶ τοῦτον νέαν πρόσθεσιν, καὶ πρότερον μὲν οὖν νῆσον εἶναι τὸ Κιρκαῖον, νῦν δὲ ὑπὸ ποταμῶν τινων προσκεχῶσθαι καὶ εἶναι ἠϊόνα. τῆς δὲ νήσου τὸ μέγεθος περὶ ὀγδοήκοντα σταδίους. καὶ τὰ μὲν τῶν τόπων ἴδια πολλὴν ἔχει διαφοράν, ὥσπερ εἴρηται πολλάκις.

IX. Τὸ δὲ καὶ πρὸς τὴν πύρωσιν πῶς ἑκάστη τῆς ὕλης ἔχει λεκτέον ὁμοίως καὶ πειρατέον λαβεῖν. ἄνθρακες μὲν οὖν ἄριστοι γίνονται τῶν πυκνοτάτων, οἷον ἀρίας δρυὸς κομάρου· στερεώτατοι γάρ, ὥστε πλεῖστον χρόνον ἀντέχουσι καὶ μάλιστα ἰσχύουσι· δι' ὃ καὶ ἐν τοῖς ἀργυρείοις τούτοις χρῶνται πρὸς τὴν πρώτην τούτων ἕψησιν. χείριστοι δὲ τούτων οἱ δρύϊνοι· γεωδέστατοι γάρ· χείρους δὲ καὶ οἱ τῶν πρεσβυτέρων τῶν νέων, καὶ μάλιστα οἱ τῶν γερανδρύων διὰ ταὐτό· ξηρότατοι γάρ, δι' ὃ καὶ πηδῶσι καιόμενοι· δεῖ δὲ ἔνικμον εἶναι.

2 Βέλτιστοι δὲ οἱ τῶν ἐν ἀκμῇ καὶ μάλιστα οἱ

---

[1] cf. Hom. *Od.* 10. 552 foll., 11. 51–80, 12. 8–15; Plin. 15. 119.

[2] νέαν πρόσθεσιν conj. Sch.; εἰς ἀνδρὸς θέσιν Ald.

is said, a lofty promontory, but very thickly wooded, producing oak, bay in abundance, and myrtle. There, according to the natives, dwelt Circe, and they shew Elpenor's tomb,[1] on which grow myrtles like those used for garlands, though other kinds of myrtle are large trees. Further it is said that the district is a recent addition[2] to the land, and that once this piece of land was an island, but now the sea has been silted up by certain streams and it has become united to the coast, and the size of the 'island'[3] is about eighty furlongs in circumference. There is[4] then much difference in trees, as has been said repeatedly, which is due to the individual character of particular districts.

Of the uses of various woods in making fire: charcoal, fuel, fire-sticks.

IX. Next we must state in like manner and endeavour to determine the properties of each kind of timber in relation to making fire. The best charcoal is made from the closest wood, such as *aria* (holm-oak) oak arbutus; for these are the most solid, so that they last longest and are the strongest; wherefore these are used in silver-mines for the first smelting of the ore. Worst of the woods mentioned is oak, since it contains most mineral matter,[5] and the wood of older trees is inferior to that of the younger, and for the same reason that of really old trees[6] is specially bad. For it is very dry, wherefore it sputters as it burns; whereas wood for charcoal should contain sap.

The best charcoal comes from trees in their prime

[3] *cf.* Plin. 3. 57. [4] ἔχει conj. Sch.; εἶναι Ald.
[5] *i.e.* and so makes much ash. [6] *cf.* 2. 7. 2.

τῶν κολοβῶν· συμμέτρως γὰρ ἔχουσι τῷ πυκνῷ καὶ γεώδει καὶ τῷ ὑγρῷ· βελτίους δὲ καὶ ἐκ τῶν εὐείλων καὶ ξηρῶν καὶ προσβόρρων ἢ ἐκ τῶν παλισκίων καὶ ὑγρῶν καὶ πρὸς νότον· καὶ εἰ ἐνικμοτέρας ὕλης, πυκνῆς· ὑγροτέρα γὰρ ἡ πυκνή. καὶ ὅλως, ὅσα ἢ φύσει ἢ διὰ [τὸν] τόπον ξηρότερον πυκνότερα, ἐξ ἁπάντων βελτίω διὰ τὴν αὐτὴν αἰτίαν. χρεία δὲ ἄλλων ἄλλη· πρὸς ἔνια γὰρ ζητοῦσι τοὺς μαλακούς, οἷον ἐν τοῖς σιδηρείοις τοὺς τῆς καρύας τῆς Εὐβοϊκῆς, ὅταν ἤδη κεκαυμένος ᾖ, καὶ ἐν τοῖς ἀργυρείοις τοὺς πιτυΐνους.

3 χρῶνται δὲ καὶ αἱ τέχναι τούτοις. ζητοῦσι δὲ καὶ οἱ χαλκεῖς τοὺς πευκίνους μᾶλλον ἢ δρυΐνους· καίτοι ἀσθενέστεροι ἀλλ᾽ εἰς τὴν φύσησιν ἀμείνους ὡς ἧσσον καταμαραινόμενοι· ἔστι δὲ ἡ φλὸξ ὀξυτέρα τούτων. τὸ δὲ ὅλον ὀξυτέρα φλὸξ καὶ ἡ τούτων καὶ ἡ τῶν ξύλων τῶν μανῶν καὶ κούφων καὶ ἡ τῶν αὔων· ἡ δ᾽ ἐκ τῶν πυκνῶν καὶ χλωρῶν νωθεστέρα καὶ παχυτέρα· πασῶν δὲ ὀξυτάτη ἡ ἐκ τῶν ὑλημάτων· ἄνθρακες δὲ ὅλως οὐ γίνονται διὰ τὸ μὴ ἔχειν τὸ σωματῶδες.

4 Τέμνουσι δὲ καὶ ζητοῦσι εἰς τὰς ἀνθρακιὰς τὰ

---

[1] κολοβῶν conj. Palm.; κολλάβων U; κολάβων Ald.

[2] δὲ καὶ ἐκ τῶν conj. W.; δὲ καὶ οἱ τῶν UMVP; δὲ οἱ τῶν Ald.H.

[3] καὶ εἰ ἐνικμοτέρας conj. W.; καὶ οἱ ἐνακμοτέρας U; καὶ ἡ ἐν ἀκμητέρας MV; καὶ οἱ ἐν ἀκμητέρας Ald.Bas.Cam. The sense seems to require ὑγροτέρας for ἐνικμοτέρας and ἐνικμοτέρα for ὑγροτέρα. G seems to have had a fuller text.

[4] *i.e.* from growing in a damper place. *cf.* 5. 9. 4.

and especially from trees which have been topped[1]: for these contain in the right proportion the qualities of closeness admixture of mineral matter and moisture. Again better charcoal comes from trees[2] in a sunny dry position with a north aspect than from those grown in a shady damp position facing south. Or, if the wood[3] used contains a good deal of moisture,[4] it should be of close texture; for such wood contains more sap.[5] And, for the same reason, that which is of closer texture either from its own natural character or because it was grown in a drier spot,[6] is, whatever the kind of tree, better.[7] But different kinds of charcoal are used for different purposes: for some uses men require it to be soft; thus in iron-mines they use that which is made of sweet chestnut when the iron has been already smelted, and in silver-mines they use charcoal of pine-wood: and these kinds are also used by the crafts. Smiths[8] require charcoal of fir rather than of oak: it is indeed not so strong, but it blows up better into a flame, as it is less apt to smoulder: and the flame from these woods is fiercer. In general the flame is fiercer not only from these but from any wood which is of open texture and light, or which is dry: while that from wood which is of close texture or green is more sluggish and dull. The fiercest flame of all is given by brushwood; but charcoal cannot be made from it at all, since it has not the necessary substance.

They cut and require for the charcoal-heap straight

[5] *cf.* § 1 *ad fin.*

[6] ξηρότερον conj. W.; ξηρότερα UMV; πυκνότερα ξηρότερα Ald. I have bracketed τὸν.

[7] βελτίω conj. Sch.; βελτίων UM; βέλτιον Ald. H.

[8] *cf.* Plin. 16. 23.

εὐθέα καὶ τὰ λεῖα· δεῖ γὰρ ὡς πυκνότατα συνθεῖναι πρὸς τὴν κατάπνιξιν. ὅταν δὲ περιαλείψωσι τὴν κάμινον, ἐξάπτουσι παρὰ μέρος παρακεντοῦντες ὀβελίσκοις. εἰς μὲν τὴν ἀνθρακιὰν τὰ τοιαῦτα ζητοῦσι.

Δύσκαπνα δὲ τῷ γένει μὲν ὅλως τὰ ὑγρά· καὶ τὰ χλωρὰ διὰ τοῦτο δύσκαπνα. λέγω δὲ τὰ ὑγρὰ τὰ ἕλεια, οἷον πλάτανον ἰτέαν λεύκην αἴγειρον· ἐπεὶ καὶ ἡ ἄμπελος ὅτε ὑγρὰ δύσκαπνος. ἐκ δὲ τῆς ἰδίας φύσεως ὁ φοῖνιξ, ὃν δὴ καὶ μάλιστά τινες ὑπειλήφασι δύσκαπνον· ὅθεν καὶ Χαιρήμων ἐποίησε "τοῦ τε δυσκαπνοτάτου φοίνικος ἐκ γῆς
5 ῥιζοφοιτήτους φλέβας." δριμύτατος δὲ ὁ καπνὸς συκῆς καὶ ἐρινεοῦ καὶ εἴ τι ἄλλο ὀπῶδες· αἰτία δὲ ἡ ὑγρότης· φλοϊσθέντα δὲ καὶ ἀποβρεχθέντα ἐν ὕδατι ἐπιρρύτῳ καὶ μετὰ ταῦτα ξηρανθέντα πάντων ἀκαπνότατα καὶ φλόγα μαλακωτάτην ἀνίησιν, ἅτε καὶ τῆς οἰκείας ὑγρότητος ἐξῃρημένης. δριμεῖα δὲ καὶ ἡ τέφρα καὶ ἡ κονία ἡ ἀπ' αὐτῶν. μάλιστα δέ φασι τὴν ἀπὸ τῆς ἀμυγδαλῆς.

6 Πρὸς δὴ τὰς καμινίας καὶ τὰς ἄλλας τέχνας ἄλλη ἄλλοις χρησίμη. ἐμπυρεύεσθαι δὲ ἄριστα συκῆ καὶ ἐλάα· συκῆ μέν, ὅτι γλίσχρον τε καὶ μανόν, ὥστε ἕλκει τε καὶ οὐ δίεισιν· ἐλάα δέ, ὅτι πυκνὸν καὶ λιπαρόν.

---

[1] λεῖα conj. Scal. from G; νέα Ald.
[2] With sods. *cf.* Plin., *l.c.*, who seems to have had a fuller text.
[3] An Athenian tragic poet. Scal. restores the quotation

smooth[1] billets: for they must be laid as close as possible for the smouldering process. When they have covered[2] the kiln, they kindle the heap by degrees, stirring it with poles. Such is the wood required for the charcoal-heap.

In general damp wood makes an evil smoke, and for this reason green wood does so: I mean the damp woods which grow in marshy ground, such as plane willow abele black poplar: for even vine-wood, when it is damp, gives an evil smoke. So does palm-wood of its own nature, and some have supposed it to give the most evil smoke of all: whence Chaeremon[3] speaks of "Veins issuing underground from roots of palm with its malodorous smoke." Most pungent is the smoke of fig-wood, whether wild or cultivated, and of any tree which has a curdling juice; the reason lies in the sap; when such wood has been barked and soaked in running water and then dried, it gives as little smoke as any other, and sends up a very soft[4] flame, since its natural moisture also has been removed. The cinders and ashes of such wood are also pungent, and especially, they say, those of almond-wood.

For the crafts requiring a furnace and for other crafts various woods are serviceable according to circumstances.[5] For kindling fig and olive are best: fig, because it is tough and of open texture, so that it easily catches fire and does not let it through,[6] olive, because it is of close texture and oily.

thus: *τοῦ τε δυσκαπνωτάτου | φοίνικος ἐκ γῆς ῥιζοφοιτήτους φλέβας* (*ῥιζοφιτύτους* conj. Schneidewin).

[4] *i.e.* not sputtering.

[5] *καὶ . . . χρησίμη* conj. W.; *τέχναις ἀλλήλοις χρησίμη* U; *τ. ἀλλήλας χρ.* MV; *τέχνη ἄλλη ἐστι χρ.* P; *τ. ἀλλήλοις ἐστὶ χρησίμη* Ald.

[6] *i.e.* burn out quickly.

Πυρεῖα δὲ γίνεται μὲν ἐκ πολλῶν, ἄριστα δέ,
ὥς φησι Μενέστωρ, ἐκ κιττοῦ· τάχιστα γὰρ καὶ
πλεῖστον ἀναπνεῖ. πυρεῖον δέ φασιν ἄριστον
μὲν ἐκ τῆς ἀθραγένης καλουμένης ὑπό τινων·
τοῦτο δ' ἐστὶ δένδρον ὅμοιον τῇ ἀμπέλῳ καὶ τῇ
οἰνάνθῃ τῇ ἀγρίᾳ· ὥσπερ ἐκεῖνα καὶ τοῦτο ἀνα-
7 βαίνει πρὸς τὰ δένδρα. δεῖ δὲ τὴν ἐσχάραν ἐκ
τούτων ποιεῖν τὸ δὲ τρύπανον ἐκ δάφνης· οὐ γὰρ
ἐκ ταὐτοῦ τὸ ποιοῦν καὶ πάσχον, ἀλλ' ἕτερον
εὐθὺ δεῖ κατὰ φύσιν, καὶ τὸ μὲν δεῖ παθητικὸν
εἶναι τὸ δὲ ποιητικόν. οὐ μὴν ἀλλὰ καὶ ἐκ τοῦ
αὐτοῦ γίνεται καί, ὥς γέ τινες ὑπολαμβάνουσιν,
οὐδὲν διαφέρει. γίνεται γὰρ ἐκ ῥάμνου καὶ
πρίνου καὶ φιλύρας καὶ σχεδὸν ἐκ τῶν πλείστων
πλὴν ἐλάας· ὃ καὶ δοκεῖ ἄτοπον εἶναι· καὶ γὰρ
σκληρότερον καὶ λιπαρὸν ἡ ἐλάα· τοῦτο μὲν οὖν
ἀσύμμετρον ἔχει δῆλον ὅτι τὴν ὑγρότητα πρὸς
τὴν πύρωσιν. ἀγαθὰ δὲ τὰ ἐκ ῥάμνου· ποιεῖ δὲ
τοῦτο καὶ τὴν ἐσχάραν χρηστήν· πρὸς γὰρ τῷ
ξηρὰν καὶ ἄχυμον εἶναι δεῖ καὶ μανοτέραν, ἵν' ἡ
τρίψις ἰσχύῃ, τὸ δὲ τρύπανον ἀπαθέστερον· δι'
ὃ τὸ τῆς δάφνης ἄριστον· ἀπαθὲς γὰρ ὂν ἐργά-
ζεται τῇ δριμύτητι. πάντα δὲ τὰ πυρεῖα βορείοις
μὲν θᾶττον καὶ μᾶλλον ἐξάπτεται, νοτίοις δὲ
ἧττον· καὶ ἐν μὲν τοῖς μετεώροις μᾶλλον, ἐν δὲ
τοῖς κοίλοις ἧττον.
8 Ἀνίει δὲ τῶν ξύλων τὰ κέδρινα καὶ ἁπλῶς ὧν

---

1 π. δὲ γίνεται μὲν conj. Sch.; π. μὲν γίνεται δὲ UMVAld.
2 *cf.* 1. 2. 3 n.
3 κιττοῦ conj. Bod. from *de igne* 64, Plin. 16. 208; καρύου Ald.
4 πυρεῖον conj. Salm.; πυροὶ UMVAld.

Fire-sticks are made[1] from many kinds of wood, but best, according to Menestor,[2] from ivy[3]: for that flares up most quickly and freely. They say also that a very good fire-stick[4] is made of the wood which some call traveller's joy; this is a tree like the vine or the 'wild vine,' which, like these, climbs up trees. The stationary piece[5] should be made of one of these, the drill of bay; for the active and passive parts of the apparatus should not be of the same wood, but different in their natural properties to start with, one being of active, the other of passive character. Nevertheless they are sometimes made of the same wood, and some suppose that it makes no difference. They are made in fact of buckthorn kermes-oak lime and almost any wood except olive; which seems surprising, as olive-wood is rather hard and oily; however it is plainly its moisture which makes it less suitable for kindling. The wood of the buckthorn is also good, and it makes a satisfactory stationary piece; for, besides being dry and free from sap it is necessary that this should also be of rather open texture, that the friction may be effectual; while the drill should be one which gets little worn by use. And that is why one made of bay is best; for, as it is not worn by use, it is effective through its biting quality. All fire-sticks take fire quicker and better in a north than in a south wind, and better in an exposed spot than in one which is shut in.

Some woods, such as prickly cedar, exude[6] moisture, and, generally speaking, so do those

[5] *i.e.* the piece of wood to be bored. *cf. de igne, l.c.*
[6] ἀνίει. ? ἀνιδίει.

ἐλαιώδης ἡ ὑγρότης· δι' ὃ καὶ τὰ ἀγάλματά φασιν ἰδίειν ἐνίοτε· ποιοῦσι γὰρ ἐκ τούτων. ὃ δὲ καλοῦσιν οἱ μάντεις Εἰλειθυίας ἄφεδρον, ὑπὲρ οὗ καὶ ἐκθύονται, πρὸς τοῖς ἐλατίνοις γίνεται συνισταμένης τινὸς ὑγρότητος, τῷ σχήματι μὲν στρογγύλον μέγεθος δὲ ἡλίκον ἄπιον ἢ καὶ μικρῷ μεῖζον ἢ ἔλαττον. ἐκβλαστάνει δὲ μάλιστα τὰ ἐλάϊνα καὶ ἀργὰ κείμενα καὶ εἰργασμένα πολλάκις, ἐὰν ἰκμάδα λαμβάνῃ καὶ ἔχῃ τόπον νοτερόν· ὥσπερ ἤδη τις στροφεὺς τῆς θύρας ἐβλάστησε, καὶ εἰς κυλίκιον πλίνθινον τεθεῖσα κώπη ἐν πήλῳ.

---

[1] *cf.* *C.P.* 5. 4. 4. οἱ μάντεις . . . ἐλατίνοις conj. Lobeck.; οἱ λεῖαν . . . τοῖς ἑκατίνοις U; οἰλείαν . . . τοὺς ἐκματίνοις V; οἱ λεῖαν τῆς εἰληθήας . . . τοῖς ἐκματίνοις M; οἱ λεῖαν τῆς ἀληθυίας ἔφαιδρον . . . τοὺς ἑκατίνους P_2; ἰλεῖαν τῆς εἰληθυίας ἔφυδρον . . τοὺς ἑκατίνους Ald.

whose sap is of an oily character; and this is why statues are sometimes said to 'sweat'; for they are made of such woods. That which seers call the *menses* of Eileithuia,'[1] and for the appearance of which they make atonement,[2] forms on the wood of the silver-fir when some moisture gathers on it: the formation is round[3] in shape, and in size about as large as a pear, or a little larger or smaller. Olive-wood is more apt than other woods to produce shoots even when lying idle or made into manufactured articles; this it often does, if it obtains moisture and lies in a damp place; thus the socket of a door-'hinge'[4] has been known to shoot, and also an oar which was standing in damp earth in an earthenware vessel.[5]

[2] *i.e.* as a portent. *cf. Char.* 16. 2.
[3] *στρόγγυλον* conj. Sch.; *στρογγύλης* UMVP_2Ald.
[4] *cf.* 5. 6. 4; Plin. 16. 230.
[5] *πλινθ. τεθ. κώπη ἐν πήλῳ* conj. Spr.; *πλίνθινον τεθεὶς τῇ κώπῃ πηλός* P_2Ald.H.

CPSIA information can be obtained
at www.ICGtesting.com
Printed in the USA
BVOW11s1218291217
504008BV00018B/562/P